POLYMERS: CHEMISTRY AND PHYSICS OF MODERN MATERIALS

THIRD EDITION

POLYMERS: CHEMISTRY AND PHYSICS OF MODERN MATERIALS

THIRD EDITION

J.M.G. COWIE

Heriot-Watt University
Scotland, UK

VALERIA ARRIGHI

Heriot-Watt University
Scotland, UK

CRC Press
Taylor & Francis Group
6000 Broken Sound Parkway NW, Suite 300
Boca Raton, FL 33487-2742

© 2008 by Taylor & Francis Group, LLC
CRC Press is an imprint of Taylor & Francis Group, an Informa business

Library of Congress Cataloging-in-Publication Data

Cowie, J. M. G. (John McKenzie Grant)
    Polymers : chemistry and physics of modern materials. -- 3rd ed. / authors, J.M.G. Cowie and Valeria Arrighi.
        p. cm.
    Includes bibliographical references and index.
    ISBN-13: 978-0-8493-9813-1 (hardcover : acid-free paper)
    ISBN-10: 0-8493-9813-4 (hardcover : acid-free paper)
    1. Polymers--Textbooks. 2. Polymerization--Textbooks. I. Arrighi, V. (Valeria)
II. Title.

QD381.C68 2007
547'.7--dc22   2007005187

Visit the Taylor & Francis Web site at
http://www.taylorandfrancis.com

and the CRC Press Web site at
http://www.crcpress.com

Contents

Preface

When the first edition of this book appeared in 1973, it was meant to serve two major functions; the first was to provide a broad-based text on polymer science at an introductory level that would illustrate the interdisciplinary nature of the subject, and the second was to create a high-information, inexpensive text that students would be able to afford. The response to the book over the intervening years has been both surprising and gratifying, and seems to indicate that the stated aims have been achieved.

The same principles are still applied, in this, the third edition, and I have simply attempted to keep abreast of the recent advances in various areas of the subject. Although the basics of polymer science remain unchanged, significant discoveries have been made in the area of control over molecular weight, macromolecular structure, and architecture, and the consequent ability to prepare materials with specific properties. To account for this, the relevant chapters have been expanded to include controlled radical polymerizations, metallocene chemistry, and the preparation of block and graft copolymers, and multiarmed and dendritic structures. Work has also moved steadily into extending the areas where polymeric materials can be used in electronic, biological, and medical applications, and examples of this are included. The chapter on characterization has also been modified to reflect the current approaches where newer techniques have tended to replace or improve on the older methods. Perhaps the most noticeable change is the addition of a series of problems at the end of each chapter. These are designed to test whether the reader has understood the various points raised in each chapter, and in some cases to expand on that knowledge. This feature has been introduced in response to comments from several lecturers who use the book as a text for their courses in polymer science, and I hope this will prove useful to them and their students.

In preparing this third edition I have been ably assisted by Dr. Valeria Arrighi, who, among other contributions, has been largely responsible for compiling all the problems and exercises, and restructuring one or two of the chapters.

The addition of new material has inevitably meant that some sections, present in previous editions, have been omitted, if no longer deemed to be in common use, or have been superseded by modern techniques. However, I hope that the revisions will be acceptable to those who use the text and that it will continue to satisfy the educational needs it was originally designed to meet.

Finally, I would like to dedicate this third edition to my family, Ann, Graeme, and Christian, and grandchildren Emma and Lauren.

The Authors

J.M.G. Cowie, B.Sc., Ph.D., D.Sc., C.Chem., FRSC, FRSE, was educated at Edinburgh University in Scotland, where he graduated in 1955 with a first class honors degree in chemistry. He was appointed to the post of assistant lecturer there in 1956 and was awarded a Ph.D. in 1958 for studies in polysaccharides, particularly starches.

In 1958, he went as a postdoctoral fellow to the National Research Council in Ottawa, Canada, where he worked with Dr. Stan Bywater, and in 1960 he joined the permanent staff of the National Research Council as an associate research officer.

He held this post until 1967 when he returned to the UK as a lecturer at the University of Essex at Colchester. Here he assisted Professor Manfred Gordon to establish a Polymer Research Consortium, involving physicists, mathematicians, and chemists in polymer science problems of common interest. After two years in Essex, he was appointed to a senior lectureship in the new University of Stirling in Scotland, in 1969 and subsequently succeeded Professor R. P. Bell to the chair of chemistry in 1973. From 1974 to 1988 he was Head of Department at Stirling, but moved to Heriot-Watt University as foundation professor of chemistry of materials, the post he held until retirement in 1998. From then to the present, he has been Professor Emeritus (Research) and has continued working with his research group.

In 1976, he was elected a Fellow of the Royal Society of Chemistry and in the following year he was elected Fellow of the Royal Society of Edinburgh. He was visiting professor at the University of Victoria, Canada, in 1979 and the University of Toronto in 1980.

In 1983 he joined the editorial board of *Polymer*, of which he then became main editor, and is also on the editorial boards of *Journal of Applied Polymer Science*, *Polymer Contents*, *Polymers for Advanced Technologies*, *Korean Polymer Journal*, and *Trends in Polymer Science*. He was awarded a D.Sc. from the University of Edinburgh for Research Work on the hydrodynamics of polymer solutions, studies of relaxation processes in the glassy state of polymers, and on physical characterization of polymers. His current interests are centered on phase equilibria in polymer blends, polymer liquid crystals, ion conduction in polymer electrolytes, physical aging, and liquid crystalline cellulose/polymer blends and composites.

He has obtained research grants from industry and Research Councils and has been a consultant for ICI, Courtaulds, International Paint, Hoechst Celanese, DSM, Akzo, API Foils, and a number of smaller companies.

He is author of the textbook, *Polymers: Chemistry and Physics of Modern Materials* and of *Alternating Copolymers* and the editor of several other texts. He has published over 260 papers, together with numerous book chapters, articles for encyclopedias, and general articles.

He has served as chairman of the MacroGroup, British High Polymer Forum, Royal Society of Edinburgh fellowship selection committee (and externally, Council of Disability, Spinal Injuries Scotland and vice chairman of Disability Scotland). He was awarded the MacroGroup Medal for the advancement of polymer science and technology in 2001, and an honorary D.Sc. from Heriot-Watt University in 2005.

Valeria Arrighi, Laurea, Ph.D., D.I.C., C.Chem., FRSC, graduated in Italy from the University of Padova, having completed a five-year degree course in chemistry. After a year spent teaching chemistry at a high school in Trento, Italy, and a short period of employment in industry, she joined Imperial College, UK, in 1988 as a Ph.D. student, under the supervision of Professor Julia Higgins and supported by funding from Enichem (Italy). She was awarded a Ph.D. and a D.I.C. in 1991 for studies on liquid crystalline polymers using neutron and x-ray scattering, and solid state NMR.

From 1992 to 1995, she worked with Professor Higgins as a postdoctoral research associate, and in 1996 she joined the Department of Chemistry at Heriot-Watt University in Edinburgh, as a lecturer in polymer chemistry — the post she is currently holding. She was elected Fellow of the Royal Society of Chemistry in 2001.

Her current research interests include studies of miscibility and physical aging in blends, nanophase separation in polymers with long side-chains, polymer dynamics, liquid crystalline polymers, composites, and systems containing nanoparticles. A common feature of these studies is the use of scattering techniques, especially neutron scattering, to study the local structure, conformation, and dynamics in polymers. She has written various reviews and book chapters in this area and has served on selection panels to allocate beam time at neutron facilities.

Dr. Arrighi has obtained research grants from Research Councils and the European Commission and has over 80 publications, including book chapters and articles for encyclopedias.

1 Introduction

1.1 BIRTH OF A CONCEPT

What is a polymer? If that question had been asked during the latter half of the 19th century and the first quarter of the 20th, it would have been met with either a blank uncomprehending stare or, worse, by derision from sections of the scientific community. This question, which is very much pertinent today, concerns substances that are so pervasive in our everyday lives that we would have difficulty in avoiding them but that may be handled, used, ignored, commented on, and normally taken for granted. Some of these substances are new and recent products resulting from the ingenuity of the chemist; some are naturally occurring and have been used by humans for several thousand years, and some form part of our bodies. All the substances, referred to as polymers or macromolecules, are giant molecules with molar masses ranging from several thousands to several millions.

Today, the concept of a giant molecule is universally accepted by scientists, but this was not always so, and the initial antagonism toward the idea that very large covalently bonded molecules could exist was deep-seated and difficult to dispel. It appears to have stemmed from the different approaches to the interpretation of colloidal behavior. In 1861, the Scotsman Thomas Graham distinguished between *crystalloid* substances, which could diffuse easily when in solution, and *colloids* or gluelike substances, which refused to crystallize, exhibited high viscosities in solution, and diffused slowly when dissolved in liquids. He explained this difference in behavior by assuming that crystalloids were small particles whereas colloids were composed of large particles. This was acceptable to most scientists, but disagreement became apparent when there were attempts at further analysis on the molecular level. This divergence of opinion is embodied in the physical approach as opposed to the chemical approach.

The chemical approach assumed that colloidal substances were, in fact, large molecules and that their behavior could be explained in terms of the size of the individual molecules. The physical approach favored the concept that the molecular sizes were no different in magnitude from those of the crystalloid materials, but that colloidal behavior was a consequence of the formation of aggregates of these smaller molecules in solutions that were held together by physical forces rather than chemical bonds.

The physical approach prevailed because it suited the chemical methodology of the period. Classical organic chemistry demanded the careful preparation and investigation of pure substances with well-defined melting points and molar masses. Even when experimental measurements pointed to the existence of large molecules, the data were rationalized to fit the physical approach. Thus, whereas rubber latex, which showed colloidal behavior, was assigned the correct structural formula I for the individual units, it was postulated to have the ring formation II.

$$CH_3$$
$$-CH_2-C=CH-CH_2-$$

$$\begin{bmatrix} CH_3 \\ CH_2-C=CH-CH_2 \\ CH_2-C=CH-CH_2 \\ CH_3 \end{bmatrix}_x$$

I   II

These rings were thought to form large aggregates in the latex particle. This idea was essential if particle masses of 6500 and 10^5, which had been calculated from ebullioscopic and cryoscopic measurements of rubber particles in solution, were to be explained in accord with the physical approach to the problem.

The majority of scientists were so locked into the mind-warp of believing that only small molecules could exist as chemical entities, that the possibility of the structure I forming long chains, rather than rings, as an alternative way of explaining the high molar masses did not seem to have been seriously considered. Similar work on starch, cellulose, and in the protein field showed the existence of high molar mass species, but here, too, interpretation favored the aggregate hypothesis.

One should not, however, be overly critical of this failure to accept a concept that to us may be obvious. Received wisdom is a deceptively comfortable framework to work within, and it takes a strong-minded, and perhaps equally dogmatic, person to break out of its strictures; the German organic chemist Hermann Staudinger proved to be that person. Building on observations by the English chemist Pickles (who was a fellow skeptic), which cast doubts on the presence of physical forces of aggregation in colloidal systems, and on his own work on the viscosity of materials exhibiting colloidal behavior, he began a long battle of conversion. From 1927 onward, he started to convince other chemists, albeit slowly, that colloidal substances like rubber, starch, and cellulose were, in fact, long, linear, threadlike molecules of variable length, composed of small definable molecular units, covalently bonded to one another to form macromolecules or polymers.

This was no easy task. He was asked by colleagues why he wished to abandon the "beautiful area of the low molecular chemistry" and turn to work in Schmier-enchemie (greasy chemistry). Even at the end of the 1920s, he was given the following advice: "Dear colleague, let me advise you to dismiss the idea of large molecules; there are no organic molecules with a molecular mass over 5000. Purify your products, as for instance rubber, and they will crystallize and reveal themselves as low-molecular-weight substances."

Whereas this "greasy" chemistry image of polymer science was a difficult one to erase from some areas of chemical academia, the grease has turned out to be a rich vein of scientific gold. How rich can only be judged by digging deeply into one of the most exciting and diverse areas of science in which the possibilities for innovation seem endless.

1.2 SOME BASIC DEFINITIONS

To place polymer science in the proper perspective, we must examine the subject on as broad a basis as possible. It is useful to consider polymers first on the molecular level, than as materials. These considerations can be interrelated by examining the various aspects in the sequence: synthesis, characterization, mechanical behavior, and application; but before discussing the detailed chemistry and physics, some of the fundamental concepts must be introduced to provide essential background to such a development. We need to know what a polymer is and how it is named and prepared. It is also useful to identify which physical properties are important, and so it is necessary to define the molar mass and the molar mass distribution, obtain an appreciation of the molecular size and shape, and recognize the important transition temperatures.

A *polymer* is a large molecule constructed from many smaller structural units called *monomers*, covalently bonded together in any conceivable pattern. In certain cases, it is more accurate to call the structural or repeat unit a *monomer residue* because atoms are eliminated from the simple monomeric unit during some polymerization processes.

The essential requirement for a small molecule to qualify as a monomer or "building block" is the possession of two or more bonding sites, through which each can be linked to other monomers to form the polymer chain. The number of bonding sites is referred to as the *functionality*. Monomers such as hydroxyacids (HO—R—COOH) or vinyl chloride (CH_2=CHCl) are bifunctional. The hydroxyacid will condense with the other hydroxyacid molecules through the —OH and —COOH groups to form a linear polymer, and the polymerization reaction in this case consists of a series of simple organic reactions similar to

$$ROH + R'COOH \rightleftharpoons R'COOR + H_2O$$

The double bond of the vinyl compound is also bifunctional as activation by a free radical or an ion leads to polymer formation

$$H_2C=CHCl + R^\bullet \longrightarrow RCH_2-CHCl-CH_2-CHCl\rightsquigarrow$$

Bifunctional monomers form linear macromolecules, but if the monomers are polyfunctional, i.e., they have three or more bonding sites as in glycerol (CH_2OH. CHOH. CH_2OH), branched macromolecules can be produced. These may even develop into large three-dimensional networks containing both branches and cross-links.

When only one species of monomer is used to build a macromolecule, the product is called a *homopolymer*, normally referred to simply as a polymer. If the chains are composed of two types of monomer unit, the material is known as a *copolymer*, and if three different monomers are incorporated in one chain, a *terpolymer* results.

Copolymers prepared from bifunctional monomers can be subdivided further into four main categories (see Figure 1.1):

1. *Statistical copolymers* in which the distribution of the two monomers in the chain is essentially random, but influenced by the individual monomer reactivities.
2. *Alternating copolymers* with a regular placement along the chain.
3. *Block copolymers* comprised of substantial sequences or blocks of each.
4. *Graft copolymers* in which blocks of one monomer are grafted on to a backbone of the other as branches.

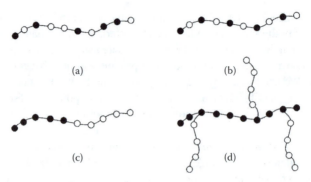

FIGURE 1.1 Schematic diagram of (a) statistical, (b) alternating, (c) block, and (d) graft copolymers.

1.3 SYNTHESIS OF POLYMERS

A process used to convert monomer molecules into a polymer is called *polymerization*, and the two most important groups are step-growth and addition. A step-growth polymerization is used for monomers with functional groups such as —OH, —COOH, —COCl, etc., and is normally, but not always, a succession of condensation reactions. Consequently, the majority of polymers formed in this way differ slightly from the original monomers because a small molecule is eliminated in the reaction, e.g., the reaction between ethylene glycol and terephthalic acid produces a polyester better known as *terylene*.

$$n\text{HO(CH}_2)_2\text{OH} + n\text{HOOC}-\bigcirc-\text{COOH} \longrightarrow \left(\text{O(CH}_2)_2\text{O}-\underset{\text{O}}{\overset{\text{O}}{\text{C}}}-\bigcirc-\underset{\text{O}}{\overset{\text{O}}{\text{C}}} \right)_n$$
$$+(2n-1)\text{H}_2\text{O}$$

The addition polymerizations, for olefinic monomers, are chain reactions that convert the monomers into polymers by stimulating the opening of the double bond with a free radical or ionic initiator. The product then has the same chemical composition as the starting material, e.g., acrylonitrile produces polyacrylonitrile without the elimination of a small molecule.

$$n\text{CH}_2=\text{CHCN} \longrightarrow \sim(\text{CH}_2\text{CHCN})_n\sim$$

The length of the molecular chains, which will depend on the reaction conditions, can be obtained from measurements of molar masses.

1.4 NOMENCLATURE

The least ambiguous method of naming a polymer is based on its source. However, a wide variety of trade names are commonly used. The prefix "poly" is attached to

the name of the monomer in addition polymers, and so polyethylene, polyacrylonitrile, and polystyrene denote polymers prepared from these single monomers. When
the monomer has a multiworded name or has a substituted parent name, then this
is enclosed in parentheses and prefixed with poly, e.g., poly(methylmethacrylate),
poly(vinyl chloride), poly(ethylene oxide), etc.

Polymers prepared by self-condensation of a single monomer such as ω-amino
lauric acid are named in a similar manner, but this polymer, poly(ω-amino lauric
acid) (sometimes known as *nylon-12*), can also be prepared by a ring-opening
reaction using lauryl lactam and could then be called poly(lauryl lactam). Both names
are correct.

International Union of Pure and Applied Chemistry (IUPAC) has attempted to
formalize the nomenclature of regular, single-stranded organic polymers and has
proposed a set of procedures, some of which are described briefly as follows.

The first step is to select a constitutional repeat unit (CRU), which may contain
one or more subunits. The name of the polymer is then the name of the CRU, in
parentheses prefixed by poly. Before naming the CRU, it must be orientated correctly.
This involves placing the constituent parts in order of seniority with the highest to
the left. In descending order, this would be heterocyclic rings, chains with hetero
atoms, carbocyclic rings, and chains with only carbon atoms, if such an order is
possible chemically.

Thus, $\left(\text{O}-\text{CH}_2-\text{CH}_2\right)$ would be poly(oxy ethylene) rather than $\left(\text{CH}_2\text{CH}_2-\text{O}\right)$
poly(ethylene oxy). If there is a substituent on part of the CRU, then orientation
will place the substituent closest to the left of the substituted portion, thus, poly(oxy
1-methyl ethylene):

$$\left(\text{O}-\underset{\underset{\text{CH}_3}{|}}{\text{CH}}-\text{CH}_2\right)$$

is preferred, rather than

$$\left(\text{O}-\text{CH}_2-\underset{\underset{\text{CH}_3}{|}}{\text{CH}}\right)$$

Similarly, a more complex CRU might be orientated as

and named poly(3,5 pyridine diyl-1,3-cyclohexylene oxydimethylene). Other examples are shown in Table 1.1 and the reader is referred to the references listed at the
end of the chapter for a more comprehensive coverage.

TABLE 1.1
Nomenclature of Some Common Polymers

Name	Structure	Trivial Name
Poly(methylene)	$\left[CH_2CH_2\right]_n$	Polyethylene
Poly(propylene)	$\left[CH\!-\!CH_2\right]_n$ $\quad CH_3$	Polypropylene
Poly(1,1-dimethylethylene)	$\begin{array}{c} CH_3 \\ \left[C\!-\!CH_2\right]_n \\ CH_3 \end{array}$	Polyisobutylene
Poly(1-methyl-1-butenylene)	$\left[C\!=\!CHCH_2CH_2\right]_n$ $\quad CH_3$	Polyisoprene
Poly(1-butenylene)	$\left[CH\!=\!CHCH_2CH_2\right]_n$	Polybutadiene
Poly(1-phenylethylene)	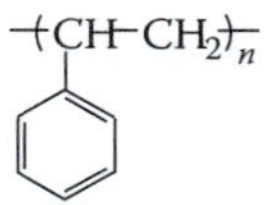 $\left[CH\!-\!CH_2\right]_n$ ⬡	Polystyrene
Poly(1-cyanoethylene)	$\left[CH\!-\!CH_2\right]_n$ $\quad CN$	Polyacrylonitrile
Poly(1-hydroxyethylene)	$\left[CH\!-\!CH_2\right]_n$ $\quad OH$	Poly(vinylalcohol)
Poly(1-chloroethylene)	$\left[CH\!-\!CH_2\right]_n$ $\quad Cl$	Poly(vinylchloride)

Poly(1-acetoxyethylene)	$+\text{CH--CH}_2\frac{}{}_n$ OOCCH_3	Poly(vinylacetate)
Poly(1,1-difluoroethylene)	$+\overset{F}{\underset{F}{C}}\text{--CH}_2\frac{}{}_n$	Poly(vinylidenefluoride)
Poly(1-(methoxycarbonyl) ethylene)	$+\text{CH--CH}_2\frac{}{}_n$ COOCH_3	Poly(methylacrylate)
Poly(1-(methoxycarbonyl)-1-methyl-ethylene)	$+\overset{CH_3}{\underset{COOCH_3}{C}}\text{--CH}_2\frac{}{}_n$	Poly(methylmethacrylate)
Poly(oxymethylene)	$+\text{OCH}_2\frac{}{}_n$	Polyformaldehyde
Poly(oxyethylene)	$+\text{OCH}_2\text{CH}_2\frac{}{}_n$	Poly(ethylene oxide) (sometimes called polyethylene glycol)
Poly(oxyphenylene)	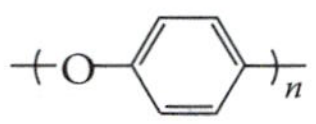	Poly(phenyleneoxide)
Poly(oxyethylene-oxyterephthaloyl)	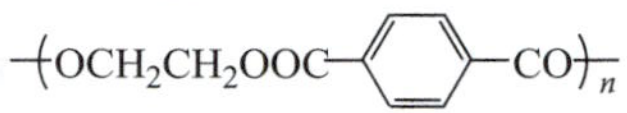 $+\text{OCH}_2\text{CH}_2\text{OOC}\,\text{---}\,\text{CO}\frac{}{}_n$	Poly(ethylene terephthalate)
Poly(iminohexamethyl-eneiminoadipoyl)	$+\text{NH(CH}_2)_6\text{NHCO(CH}_2)_4\text{CO}\frac{}{}_n$	Poly(hexamethylene adipamide)
Poly(difluoromethylene)	$+\overset{F}{\underset{F}{C}}\text{--}\overset{F}{\underset{F}{C}}\frac{}{}_n$	Poly(tetrafluoroethylene)

1.5 AVERAGE MOLAR MASSES AND DISTRIBUTIONS[1]

One of the most important features that distinguishes a synthetic high polymer from a simple molecule is the inability to assign an exact molar mass to a polymer. This is a consequence of the fact that in a polymerization reaction, the length of the chain formed is determined entirely by random events. In a condensation reaction, it depends on the availability of a suitable reactive group and, in an addition reaction, on the lifetime of the chain carrier. Inevitably, because of the random nature of the growth process, the product is a mixture of chains of differing length — a *distribution* of chain lengths — which in many cases can be calculated statistically.

The polymer is characterized best by a molar mass distribution and the associated molar mass averages, rather than by a single molar mass. The typical distributions, shown in Figure 1.2, can be described by a variety of averages. As the methods used for estimating the molar mass of polymers employ different averaging procedures, it is safer to use more than one technique to obtain two or more averages, thereby characterizing the sample more fully.

A colligative method, such as osmotic pressure, effectively counts the number of molecules present and provides a *number-average* molar mass $<M>_n$ defined by

$$\langle M \rangle_n = \frac{\Sigma N_i M_i}{\Sigma N_i} = \frac{\Sigma w_i}{\Sigma (w_i / M_i)} \tag{1.1}$$

where N_i is the number of molecules of species i of molar mass M_i. The brackets $\langle \ \rangle$ indicate that it is an average value, but by convention these are normally omitted.

The alternative expression is in terms of the mass $w_i = N_i M_i / N_A$ if required, where N_A is Avogadro's constant.

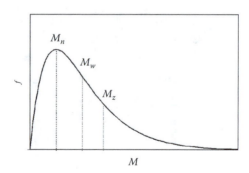

FIGURE 1.2 Typical distribution of molar masses for a synthetic polymer sample, where f is the fraction of polymer in each interval of M considered.

---

[1] The quantity molar mass is used throughout this text instead of the dimensionless quantity molecular weight, which is usual in polymer chemistry. All the equations in later sections evaluate molar mass rather than the dimensionless quantity molecular weight.

From light-scattering measurements, a method depending on the size rather than the number of molecules, a weight-average molar mass $\langle M \rangle_w$ is obtained. This is defined as

$$\langle M \rangle_w = \frac{\sum N_i M_i^2}{\sum N_i M_i} = \frac{\sum w_i M_i}{\sum w_i} \tag{1.2}$$

Statistically $\langle M \rangle_n$ is simply the first moment, and $\langle M \rangle_w$ is the ratio of the second to the first moment, of the number distribution.

A higher average, the z-average given by

$$\langle M \rangle_z = \frac{\sum N_i M_i^3}{\sum N_i M_i^2} = \frac{\sum w_i M_i^2}{\sum w_i M_i}, \tag{1.3}$$

can be measured in the ultracentrifuge, which also yields another useful average, the (z + 1)-average,

$$\langle M \rangle_{z+1} = \frac{\sum N_i M_i^4}{\sum N_i M_i^3}, \tag{1.4}$$

often required when describing mechanical properties.

A numerical example serves to highlight the differences in the various averages. Consider a hypothetical polymer sample composed of chains of four distinct molar masses, 100,000, 200,000, 500,000, and 1,000,000 g mol^{-1} in the ratio 1:5:3:1, than

$$M_n/\text{g mol}^{-1} = \frac{(1 \times 10^5) + (5 \times 2 \times 10^5) + (3 \times 5 \times 10^5) + (1 \times 10^6)}{1 + 5 + 3 + 1} = 3.6 \times 10^5$$

$$M_w/\text{g mol}^{-1} = \frac{\{1 \times (10^5)^2\} + \{5 \times (2 \times 10^5)^2\} + \{3 \times (5 \times 10^5)^2\} + \{1 \times (10^6)^2\}}{(1 \times 10^5) + (5 \times 2 \times 10^5) + (3 \times 5 \times 10^5) + (1 \times 10^6)}$$

$$= 5.45 \times 10^5$$

and $M_z = 7.22 \times 10^5$ g mol^{-1}.

The breadth of the distribution can often be gauged by establishing the *heterogeneity index* (M_w/M_n). For many polymerizations, the most probable value is about 2.0, but both larger and smaller values can be obtained, and it is at best only a rough guide.

An alternative method of describing the chain length of a polymer is to measure the *average degree of polymerization x*. This represents the number of monomer units or residues in the chain and is given by

$$x = M/M_0, \tag{1.5}$$

where M_0 is the molar mass of monomer or residue and M is the appropriate average molar mass. Hence, the x average depends on which average is used for M. (To avoid confusion between the mole fraction x and the average degree of polymerization x, the latter will always be subscripted as x_n or x_w to indicate the particular M used in Equation 1.5.)

1.6 SIZE AND SHAPE

Some measure of the polymer size is obtained from the molar mass, but what is the actual length of a chain and what shape does it adopt? We can begin to answer these questions by first considering a simple molecule such as butane and examining the behavior when the molecule is rotated about the bond joining carbon 2 to carbon 3.

The Newman and "saw horse" projections show the trans position in Figure 1.3a with the "dihedral angle" $\phi = 180°$. This is the most stable conformation with the greatest separation between the two methyl groups. Rotation about the C_2—C_3 bond alters ϕ and moves the methyl groups past the opposing hydrogen atoms so that an extra repulsive force is experienced when an eclipsed position (Figure 1.3b) is reached.

The progress of rotation can be followed by plotting the change in potential energy $V(\phi)$ as a function of the dihedral angle, as shown in Figure 1.4. The resulting diagram for butane exhibits three minima at $\phi = \pi$, $\pi/3$, and $5\pi/3$ called the *trans* and $\pm$ *gauche* states, respectively, and the greater depth of the trans position indicates that this is the position of maximum stability. Although the gauche states are slightly less stable, all three minima can be regarded as discrete rotational states. The maxima correspond to the eclipsed positions and —CH_3 are angles of maximum instability. These diagrams will vary with the type of molecule and need not be symmetrical, but the butane diagram is very similar to that for the simple polymer polyethylene $-(\,CH_2$—$CH_2\,)_{\overline{n}}$, if the groups are replaced by the two sections of the chain adjoining the bond of rotation. The backbone of this polymer is composed of a chain of tetrahedral carbon atoms covalently bonded to each other so that the molecule can be represented as an extended all trans zigzag chain. For a typical value of $M = 1.6 \times 10^5$ g mol^{-1}, the chain

contains 10,000 carbon atoms; thus in the extended zigzag state, assuming a tetrahedral angle of 109° and a bond length of 0.154 nm, the chain would be about 1260 nm long and 0.3 nm diameter. Magnified one million times, the chain could be represented by a piece of wire 126×0.03 cm. This means that polyethylene is a long threadlike molecule, but how realistic is the extended all trans conformation? As every group of four atoms in the chain has a choice of three possible stable rotational states, a total of $3^{10,000}$ shapes are available to this particular chain, only one of which is the all trans state. So, in spite of the fact that the all trans extended conformation has the lowest energy, the most probable conformation will be some kind of randomly coiled state, assuming that no external ordering forces are present and that the rotation about the carbon bonds is in no way impeded. The many possible coiled forms are generated

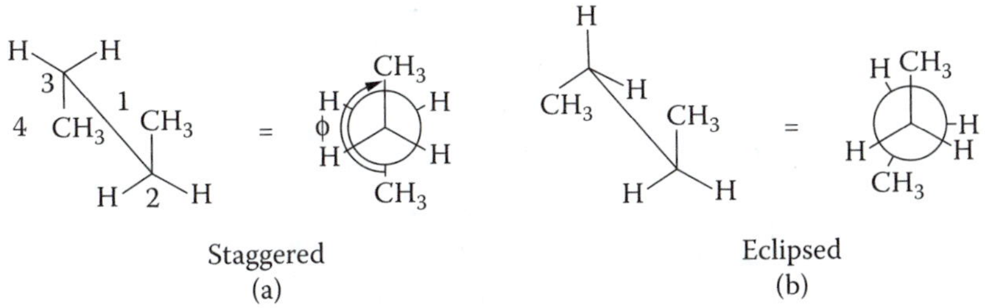

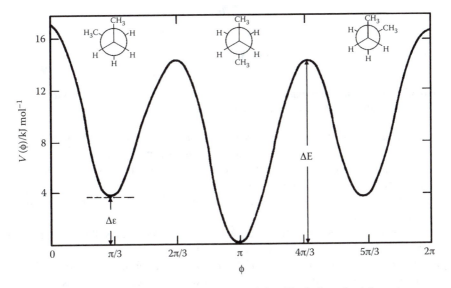

FIGURE 1.3 Newman and "saw horse" projection for n-butane, (a) a staggered state with $\phi = \pi$ and (b) an eclipsed position.

FIGURE 1.4 Potential energy $V(\phi)$ as a function of the dihedral angle ϕ for n-butane.

simply by allowing the chain to rotate into a gauche position which moves the atom out of the plane of the adjacent bonds. This is shown more clearly (see Figure 1.5) by considering the various cones of revolution available to a chain over only two bonds. The distribution of trans (t) and gauche (g) states along a chain will be a function of the temperature and the relative stability of these states. Consequently, there is an unequal distribution of each. The ratio of the number of trans n_t to gauche n_g states is then governed by a Boltzmann factor and

$$n_g/n_t = 2 \exp(-\Delta\varepsilon/kT), \tag{1.6}$$

where k is the Boltzmann constant, $\Delta\varepsilon$ is the energy difference between the two minima, and the 2 arises because of the $\pm$ gauche states available. For polyethylene, $\Delta\varepsilon$ is about 3.34 kJ mol^{-1}, and values of (n_g/n_t) for 100, 200, and 300 K are 0.036, 0.264, and 0.524, respectively, showing that the chain becomes less extended and

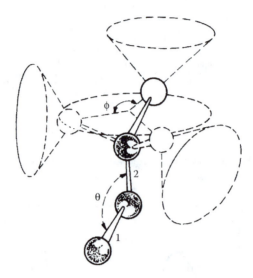

FIGURE 1.5 Diagrammatic representation of the cones or revolution available to the third and fourth bonds of a simple carbon chain with a fixed bond angle θ.

more coiled as the temperature increases. Because of the possibility of rotation about the carbon bonds, the chain is in a state of perpetual motion, constantly changing shape from one coiled conformation to another form, equally probable at the given temperature. The speed of this wriggling varies with temperature (and from one polymer to another) and dictates many of the physical characteristics of the polymer, as we shall see later.

The height of the potential energy barrier ΔE determines the rate of bond interchange between the t and the g states, and for polyethylene it is about 16.7 kJ mol^{-1}. When ΔE is very high (about 80 kJ mol^{-1}), rotation becomes very difficult, but as the temperature is raised, the fraction of molecules that possess energy in excess of ΔE increases and rotation from one state to another becomes easier.

Realistically, then, a polymer chain is better represented by a loosely coiled ball (Figure 1.6) than an extended rod. For the magnified-polyethylene chain considered earlier, a ball of about 4 cm diameter is a likely size.

The term *conformation* has been used here when referring to a three-dimensional geometric arrangement of the polymer, which changes easily when the bonds are rotated.

There is a tendency to use the term *configuration* in a synonymous sense, but as far as possible, this will be reserved for the description of chains in which the geometric variations can only be interchanged by breaking a bond. Configurational isomers of importance for polymers are discussed in the following section

1.7 CONFIGURATION

For polymers, two types of configurational isomers are of importance: (1) geometric isomerism, e.g., cis and trans and (2) stereoisomers.

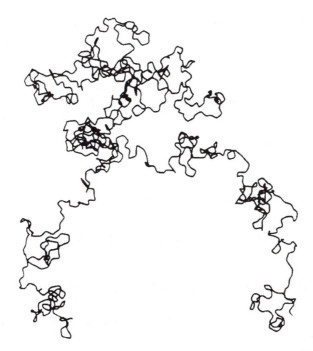

FIGURE 1.6 Random arrangement of a polyethylene chain containing 1000 freely rotating C-C bonds, in which each successive bond has been given a random choice of six equally spaced angular positions. (From Treloar, L.R.G., *Introduction to Polymer Science*, Wykeham Publications, 1970. With permission.)

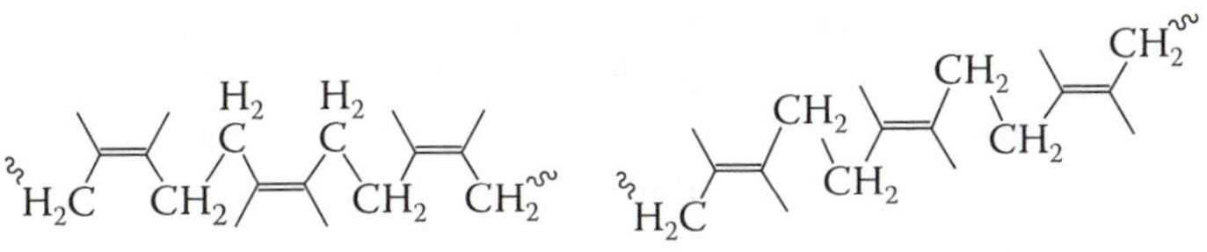

(a)   (b)

FIGURE 1.7 Geometric isomers of 1,4-polybutadiene: (a) cis and (b) trans configuration.

As shown in Figure 1.7, the polymerization of monomers with two double bonds (e.g., butadiene and isoprene) leads to polymer chains with a residual double bond per monomer unit: cis–trans isomerism is possible. Two important polymers that show this type of isomerism are 1,4-polybutadiene and 1,4-polyisoprene. The regularity of the trans configuration makes this type of isomer more crystalline, with a higher melting point compared to the cis configuration.

If the polymer chain contains carbon atoms with two different substituents, then the C atom is asymmetrical. This situation is encountered for vinyl monomers such as polypropylene. For a chain containing asymmetric centers, three different stereoisomers are possible (Figure 1.8):

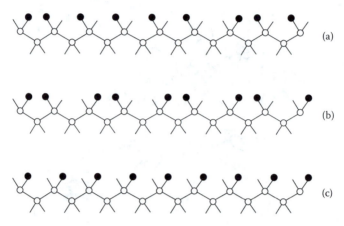

FIGURE 1.8 Sections of polymer chains having different tacticity: (a) atactic, (b) syndiotactic, and (c) isotactic.

1. The *isotactic* configuration in which each substituent is placed on the same side of the chain.
2. The *syndiotactic* structure in which substituents lie on alternate sides of the backbone.
3. A disordered, termed *atactic*, configuration in which substituent groups are placed randomly on either side of the chain.

Of the three stereoisomers, atactic polypropylene was the first to be synthesized. This is because it is the atactic structure that is obtained by conventional polymerization of the monomers when no optically active catalyst is used. It was only when the Ziegler–Natta catalysts were introduced in the 1950s that the production of stereoregular polymers became possible. It is interesting to note that polypropylene did not find any commercial use until the Ziegler–Natta catalysts became available. In fact, the atactic structure that was originally produced leads to a viscous liquid at room temperature of limited use. Isotactic polypropylene instead is one of the most important commercial polymers.

1.8 THE GLASS TRANSITION TEMPERATURE T_g AND THE MELTING TEMPERATURE T_m

At sufficiently low temperatures, all polymers are hard rigid solids. As the temperature rises, each polymer eventually obtains sufficient thermal energy to enable its chains to move freely enough for it to behave like a viscous liquid (assuming no degradation has occurred).

There are two ways in which a polymer can pass from the solid to the liquid phase, depending on the internal organization of the chains in the sample. The different types of thermal response, illustrated by following the change in specific volume, are shown schematically in Figure 1.9.

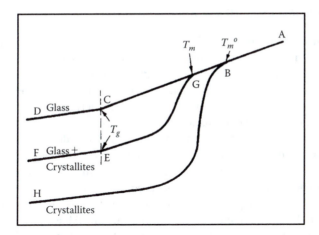

FIGURE 1.9 Schematic representation of the change of specific volume v of a polymer with temperature T for (i) a completely amorphous sample (A–C–D), (ii) a semicrystalline sample (A–G–F), and (iii) a perfectly crystalline material (A–B–H).

A polymer may be completely amorphous in the solid state, which means that the chains in the specimen are arranged in a totally random fashion. The volume change in amorphous polymers follows the curve A–D. In the region C–D the polymer is a glass, but as the sample is heated, it passes through a temperature T_g, called the *glass transition temperature*, beyond which it softens and becomes rubberlike. This is an important temperature because it represents the point where important property changes take place, i.e., the material may be more easily deformed or become ductile above T_g. A continuing increase in temperature along C–B–A leads to a change of the rubbery polymer to a viscous liquid.

In a perfectly crystalline polymer, all the chains would be incorporated in regions of three-dimensional order, called *crystallites*, and no glass transition would be observed because of the absence of disordered chains in the sample. The crystalline polymer, on heating, would follow curve H–B–A; at T_m°, melting would be observed, and the polymer would become a viscous liquid.

Perfectly crystalline polymers are not encountered in practice, and instead polymers may contain varying proportions of ordered and disordered regions in the sample. These semicrystalline polymers usually exhibit both T_g and T_m, corresponding to the ordered and disordered portions and follow curves similar to F–E–G–A. As T_m° is the melting temperature of a perfectly crystalline polymer of high molar mass, T_m is lower and more often represents a melting range, because the semicrystalline polymer contains a spectrum of chain lengths and crystallites of various sizes with many defects. These imperfections act to depress the melting temperature, and experimental values of T_m can depend on the previous thermal history of the sample.

Nevertheless, both T_g and T_m are important parameters, which serve to characterize a given polymer.

1.9 ELASTOMERS, FIBERS, AND PLASTICS

A large number of synthetic polymers now exist covering a wide range of properties. These can be grouped into three major classes: plastics, fibers, and elastomers; but there is no firm dividing line between the groups. However, some classification is useful from a technological viewpoint, and one method of defining a member of these categories is to examine a typical stress–strain plot (Figure 1.10). Rigid plastics and fibers are resistant to deformation and are characterized by a high modulus and low percentage elongations. Elastomers readily undergo deformation and exhibit large reversible elongations under small applied stresses, i.e., they exhibit elasticity. The flexible plastics are intermediate in behavior. An outline of the structure–property relations will be presented later, but before proceeding further with the more detailed science, we can profitably familiarize ourselves with some of the more common polymers and their uses. Some of these are presented in Table 1.2, where an attempt is made to show that the lines of demarcation used to divide polymers into the three major groups are not clear-cut.

A polymer normally used as a fiber may make a perfectly good plastic if no attempt is made to draw it into a filament. Similarly, a plastic, if used at a temperature above its glass transition and suitably cross-linked, may make a perfectly acceptable elastomer. In the following text, a brief account of some of the more common plastics, fibers, and elastomers is given. The classification is based essentially on their major technological application under standard working conditions.

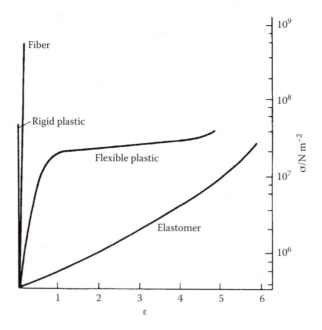

FIGURE 1.10 Typical stress–strain ($\sigma - \varepsilon$) plots for a rigid plastic, a fiber, a flexible plastic, and an elastomer.

TABLE 1.2
Some Common Plastics, Elastomers, and Fibers

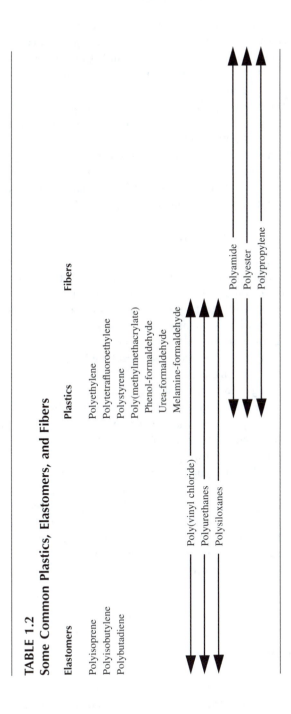

Elastomers	Plastics	Fibers
Polyisoprene	Polyethylene	Polyamide
Polyisobutylene	Polytetrafluoroethylene	Polyester
Polybutadiene	Polystyrene	Polypropylene
Poly(vinyl chloride)	Poly(methylmethacrylate)	
Polyurethanes	Phenol-formaldehyde	
Polysiloxanes	Urea-formaldehyde	
	Melamine-formaldehyde	

1.10 FIBER-FORMING POLYMERS

Whereas there are many fiber-forming polymers, only a few have achieved great technological and commercial success. It is significant that these are polymers of long standing, and it has been suggested that further fiber research may involve the somewhat prosaic task of attempting to improve, modify, or reduce the cost of existing fibers, rather than to look for new and better alternatives. The commercially important fibers are listed in Table 1.3; all are thermoplastic polymers.

The polyamides are an important group of polymers, which include the naturally occurring proteins in addition to the synthetic nylons. The term *nylon*, originally a trade name, has now become a generic term for the synthetic polyamides, and the numerals which follow, e.g., nylon-6,6, distinguish each polymer by designating the number of carbon atoms lying between successive amide groups in the chain. Thus, nylon-6,10 is prepared from two monomers and has the structure

$$-\!\left[NH(CH_2)_6NHCO(CH_2)_8CO\right]_n\!-$$

with alternative sequences of six and ten carbon atoms between the nitrogen atoms, whereas nylon-6 is prepared from one monomer and has the repeat formula $-\!\left[NH(CH_2)_5CO\right]_n\!-$ with regular sequences of six carbon atoms between the nitrogen atoms. A nylon with two numbers is termed *dyadic* indicating that it contains both dibasic acid (or acid chloride) and diamine moieties, in which the first number represents the diamine and the second the diacid used in the synthesis. The *monadic* nylons have one number, indicating that synthesis involved only one type of monomer. This terminology means that a poly (α-amino acid) would be nylon-2.

Terylene is an important polyester. It exhibits high resilience, durability, and low moisture absorption, properties that contribute to its desirable "wash and wear" characteristics. The harsh feel of the fiber, caused by the stiffness of the chain, is overcome by blending it with wool and cotton.

The acrylics and modacrylics are among the most important of the amorphous fibers. They are based on the acrylonitrile unit $-CH_2CH(CN)-$ and are usually manufactured as copolymers. When the acrylonitrile content is 85% or higher, the polymer is an *acrylic* fiber; but if this drops to between 35 and 85%, it is known as a *modacrylic* fiber. Vinyl chloride and vinylidene chloride are the most important comonomers, and the copolymers produce high-bulk yarns, which can be subjected to a controlled shrinking process after fabrication. Once shrunk, the fibers are dimensionally stable.

1.11 PLASTICS

A plastic is rather inadequately defined as an organic high polymer capable of changing its shape on the application of a force and retaining this shape on removal of this force, i.e., a material in which a stress produces a nonreversible strain.

The main criterion is that plastic materials can be formed into complex shapes, often by the application of heat or pressure, and a further subdivision into those that

TABLE 1.3
Chemical Structure of Synthetic Fibers

Polymer	Repeat Unit	Trade Names
Polyamides (nylons) (Uses: drip-dry fabrics, cordage, braiding, bristles, and surgical sutures)		
Polycaprolactam	$\left[NH(CH_2)_5CO\right]_n$	Nylon-6, Perlon
Poly(decamethylene carboxamide)	$\left[NH(CH_2)_{10}CO\right]_n$	Nylon-11, Rilsan
Poly(hexamethylene adipamide)	$\left[NH(CH_2)_6NHCO(CH_2)_4CO\right]_n$	Nylon-6,6, Bri-nylon
Poly(m-phenylene isophthalamide)	$\left[NH\text{—}\bigcirc\text{—}NHCO\text{—}\bigcirc\text{—}CO\right]_n$	Nomex
Polyesters (Uses: fabrics, tire-cord yarns, and yacht sails)		
Poly(ethylene terephthalate)	$\left[OC\text{—}\bigcirc\text{—}COO(CH_2)_2O\right]_n$	Terylene, Dacron
Polyureas		
Poly(nonamethylene urea)	$\left[NHCONH(CH_2)_9\right]_n$	Urylon
Acrylics (Uses: fabrics and carpeting)		
Polyacrylonitrile	$\left[CH_2\underset{\underset{CN}{\mid}}{CH}\right]_n$	Orlon, Courtelle, Acrilan, Creslan
	(often as copolymer with >85% acrylonitrile	
Acrylonitrile copolymers	35% < acrylonitrile < 85% + vinyl chloride + vinylidene chloride	Dynel Verel
Hydrocarbons (Uses: carpets and upholstery)		
Polyethylene	$\left[CH_2CH_2\right]_n$	Courlene, Vestolen
Polypropylene (isotactic)	$\left(CH_2\underset{\underset{CH_3}{\mid}}{CH}\right)_n$	Ulstron, Herculon, Meraklon
Halogen-Substituted Olefins (Uses: knitwear and protective clothing)		
Poly(vinyl chloride)	$\left[CH_2CHCl\right]_n$	Rhovyl, Valren
Poly(vinylidene chloride)	$\left[CH_2\underset{\underset{Cl}{\mid}}{\overset{\overset{Cl}{\mid}}{C}}\right]_n$	Saran, Tygan
Poly(tetrafluoroethylene)	$\left[CF_2CF_2\right]_n$	Teflon, Polifen
Vinyl (Uses: fibers, adhesives, paint, sponges, films, and plasma extender)		
Poly(vinyl alcohol)	$\left[CH_2\underset{\underset{OH}{\mid}}{CH}\right]_n$ (normally cross-linked)	Vinylon, Kuralon, Mewlon

are *thermosetting* and those that are *thermoplastic* is useful. The thermosetting materials become permanently hard when heated above a critical temperature and will not soften again on reheating. They are usually cross-linked in this state. A thermoplastic polymer will soften when heated above T_g. It can then be shaped and, on cooling, will harden in this form. However, on reheating, it will soften again and can be reshaped if required before hardening when the temperature drops. This cycle can be carried out repeatedly.

A number of the important thermoplastics are shown in Table 1.4, together with a few examples of their more important uses, determined by the outstanding properties of each. Thus, polypropylene, poly(phenylene oxide), and TPX have good thermal stability and can be used for items requiring sterilization. The optical qualities of polystyrene and poly(methyl methacrylate) are used in situations where

TABLE 1.4
Thermoplastics

Polymer	Repeat Unit	Density (g cm^{-3})	Uses
Polyethylene	$+CH_2CH_2+_n$		Household products,
High density		0.94–0.96	insulators, pipes, toys,
Low density		0.92	bottles
Polypropylene	$+CH_2-CH+(CH_3)_n$	0.90	Water pipes, integral hinges, sterilizable hospital equipment
Poly(4-methylpentene-1) (TPX)	$+CH_2CH+_n$ CH_2 CH H_3C CH_3	0.83	Hospital and laboratory ware
Poly(tetrafluoroethylene) (PTFE)	$+CF_2CF_2+_n$	2.20	Nonstick surfaces, insulation, gaskets
Poly(vinyl chloride) (PVC)	$+CH_2CHCl+_n$	1.35–1.45	Records, bottles, house siding, and eaves
Polystyrene	$-(CH_2CH(C_6H_5))-_n$	1.04–1.06	Lighting panels, lenses, wall tiles, flower pots
Poly(methylmethacrylate) (PMMA)	CH_3 $-(CH_2-C-)-_n$ $COOCH_3$	1.17–1.20	Bathroom fixtures, knobs, combs, illuminated signs
Polycarbonates	$-(R.O.COO)-$	1.20	Cooling fans, marine propellers, safety helmets
Poly(2,6-dimethylphenylene oxide)	CH_3 ... O ... CH_3 $_n$	1.06	Hot water fittings, sterilizable, medical, and surgical equipment

transparency is a premium, whereas the low-frictional coefficient and superb chemical resistance of poly(tetrafluoro-ethylene) make it useful in nonstick cookware and protective clothing. Low-density polyethylene, although mechanically inferior to the high-density polymer, has better impact resistance and can be used when greater flexibility is required, whereas the popularity of poly(vinyl chloride) lies in its unmatched ability to form a stable, dry, flexible material when plasticized. The polyamides and terylene are also important thermoplastics.

1.12 THERMOSETTING POLYMERS

The thermoset plastics generally have superior abrasion and dimensional stability characteristics compared with the thermoplastics, which have better flexural and impact properties. In contrast to the thermoplastics, thermosetting polymers, as the name implies, are changed irreversibly from fusible, soluble products into highly intractable cross-linked resins that cannot be molded by flow and so must be fabricated during the cross-linking process. Typical examples are:

Phenolic resins, prepared by reacting phenols with aldehydes. They are used for electrical fitments, radio and television cabinets, heat resistant knobs for cooking utensils, game parts, buckles, handles, and a wide variety of similar items.

Amino resins are related polymers formed from formaldehyde and either urea or melamine. In addition to many of the uses listed earlier, they can be used to manufacture lightweight tableware, and counter and table surfaces. Being transparent they can be filled and colored using light pastel shades, whereas the phenolics are already rather dark and, consequently, have a more restricted color range.

Thermosetting *polyester resins* are used in paints and surface coatings. In these, oxidation during drying forms a cross-linked film, which provides a tough, resistant finish.

Epoxy resins are polyethers prepared from glycols and dihalides and are extensively used as surface coatings, adhesives, and flexible enamel-like finishes because of their combined properties of toughness, chemical resistance, and flexibility.

1.13 ELASTOMERS

The modern elastomer industry was founded on the naturally occurring product isolated from the latex of the tree *Hevea brasiliensis*. It was first used by indigenous South Americans and was called *caoutchouc*, but, later, simply *rubber*, when it was discovered by Priestley that the material rubbed out pencil marks.

From the early 20th century, chemists have been attempting to synthesize materials whose properties duplicate or at least simulate those of natural rubber, and this has led to the production of a wide variety of synthetic elastomers. Some of these have become technologically important and are listed in Table 1.5, together with their general uses.

TABLE 1.5
Some Common Elastomers and Their Uses

Polymer	Formula	Uses
Natural rubber (polyisoprene-*cis*)	$-(CH_2-\overset{\displaystyle CH_3}{\underset{\displaystyle }{C}}=CH-CH_2)_n-$	General purposes
Polybutadiene	$-(CH_2-CH=CH-CH_2)_n-$	Tire treads
Butyl	$-\left(CH_2-\overset{\displaystyle CH_3}{\underset{\displaystyle CH_3}{C}}\right)_n-$	Inner tubes, cable sheathing, roofing, tank liners
SBR	$-\left(CH_2-CH=CH-CH_2-CH_2-CH-\right)_n-$	Tires, general purposes
ABS	$-(CH_2-\underset{\displaystyle CN}{CH}-CH_2-CH-C_6H_5)_n$ $-(CH-CH=CH-CH_2)_m-$	Oil hoses, gaskets, flexible fuel tanks
Polychloroprene	$-\left(CH_2-\overset{\displaystyle }{\underset{\displaystyle Cl}{C}}=CH-CH_2\right)_n-$	Used when oil resistance, good weathering, and inflammability characteristics are needed
Silicones	$-\left(O-\overset{\displaystyle R}{\underset{\displaystyle R}{Si}}\right)_n-$	Gaskets, door seals, medical application flexible molds
Polyurethanes	$-(R_1-NHCOOR_2\,OOCHN)_n-$	Printing rollers, sealing and jointing
EPR	$-\left((CH_2-CH_2)_m-(CH_2-\underset{\displaystyle CH_3}{CH})_p\right)_n-$	Window strips and channeling

Although a large number of synthetic elastomers are now available, natural rubber must still be regarded as the standard elastomer because of the excellently balanced combination of desirable qualities. Presently, it accounts for almost 36% of the total world consumption of elastomers, and its gradual replacement by synthetic varieties is partly a result of demand outstripping natural supply.

The most important synthetic elastomer is styrene-butadiene (SBR) which accounts for 41% of the world market in elastomers. It is used predominantly for vehicle tires when reinforced with carbon black. Nitrile rubber (NBR) is a random copolymer of acrylonitrile (mass fraction 0.2 to 0.4) and butadiene, and it is used when an elastomer that is resistant to swelling in organic solvents is required. The range of properties can be extended when styrene is also incorporated in the chain, forming ABS rubber. Butyl rubber (IIR) is prepared by copolymerizing small quantities of isoprene (3 parts) with isobutylene (97 parts). The elastic properties are poor, but it is resistant to corrosive fluids and has a low permeability to gases. Polychloroprene possesses the desirable qualities of being a fire retardant and resistant to weathering, chemicals, and oils. More recently, ABA triblock copolymers of styrene-(ethene-*stat*-butene)-styrene have become commercially available. These are thermoplastic elastomers, which are unaffected by polar solvents and nonoxidizing acids (alkalis).

Elastomers that fail to crystallize on stretching must be strengthened by the addition of filters such as carbon black. SBR, poly(ethylene-*stat*-propylene), and the silicone elastomers fall into this category. Whereas polyethylene is normally highly crystalline, copolymerization with propylene destroys the ordered structure; and if carried out in the presence of a small quantity of nonconjugated diene (e.g., dicyclopentadiene), a cross-linking site is introduced. The material is an amorphous random

$$\text{www}\left(\text{CH}_2\text{–CH}_2\right)\text{www}\left(\text{H}_2\text{C–CH}\atop\text{CH}_3\right)\text{CH–CH–www}$$

terpolymer, which when cross-linked forms an elastomer with a high resistance to oxidation. Unfortunately, it is incompatible with other elastomers and is unsuitable for blending.

The silicone elastomers have a low cohesive energy between the chains, which results in poor thermoplastic properties and an unimpressive mechanical response. Consequently, they are used predominantly in situations requiring temperature stability over a range of 190 to 570 K when conditions are unsuitable for other elastomers.

Extensive use has been made of room temperature vulcanizing silicone rubbers. These are based on linear poly(dimethyl siloxane) chains, with M ranging from 10^4 to 10^5 g mol^{-1}, and hydroxyl terminal groups. Curing can be achieved in a number of ways, either by adding a cross-linking agent and a metallic salt catalyst, such as tri- or tetra-alkoxysilane with stannous octoate, by exposure to light, or by incorporating in the mixture a cross-linking agent sensitive to atmospheric water, which initiates vulcanization. The products are good sealing, encapsulating, and caulking materials; they make good flexible molds and are excellent insulators. They have found a wide application in the building, aviation, and electronics industries. Other room temperature curing adhesives and sealants are listed in Table 1.6.

Having briefly introduced the diversity of structure and property encountered in the synthetic polymers, we can now examine more closely the fundamental chemistry and physics of these materials.

TABLE 1.6
Room Temperature Curing Adhesives and Sealants

Type	General Description	Particular Advantages	Limitations (General Operating Temperature Range)
Moisture curing polyurethanes (PUs)	Sealants (rather than adhesives), which cure by an isocyanate reaction to atmospheric H_2O	Adhesive/sealing effect to a wide range of substrates	Slow curing; usually low modulus (−80°C to +120°C)
RTV (room temperature vulcanizing) silicones	Sealants (rather than adhesives), which cure on exposure to atmospheric moisture by a condensation mechanism that results in release of side products such as acetic acid, alcohols, or amines	Excellent thermal, oxidative, and hydrolytic stability	Unpleasant side products; limited adhesion (−80°C to +200°C)
Anaerobic adhesives and sealants	Fluids, which cure in the absence of air and the presence of metals, heat of UV light by the free-radical mechanism	Very good adhesion to metals and ceramics; resistant to organic solvents	Relatively brittle when cured; curing is sensitive to substrate and to joint geometry (−50°C to +150°C)
Cyanoacrylate adhesives	Relatively low viscosity, which cure anionically in response to substrate-borne atmospheric moisture	Excellent adhesion to a wide range of substrates; very effective on rubber and on most plastics	Brittle when cured; limited thermal and hydrolytic stability (−50°C to +80°C)
Acrylics	Methacrylic adhesives, which cure free radically in response to substrates treated with a primer/hardener; the adhesives usually contain rubber toughener	Forms durable adhesive joints to metallic substrates; high peel strength	Inhibited by atmospheric oxygen; limit cure-through-gap (−50°C to +100°C)

PROBLEMS

1. Consider the following monomers or monomer combinations, and in each case,
   a. Indicate the functionality
   b. Draw the structure of the corresponding polymer or polymers
   c. Indicate whether you would expect polymerization to proceed via a step-growth or chain-growth mechanism.

(a)

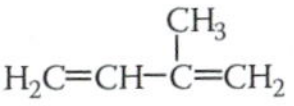

$$\underset{\text{isoprene}}{H_2C{=}CH{-}\overset{\overset{\displaystyle CH_3}{|}}{C}{=}CH_2}$$

(b)

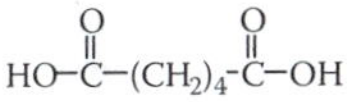

$$HO{-}\overset{\overset{\displaystyle O}{\|}}{C}{-}(CH_2)_4{-}\overset{\overset{\displaystyle O}{\|}}{C}{-}OH \qquad HO{-}(CH_2)_{10}{-}OH$$

adipic acid (hexanedioic acid) and   1,10-decamethylene glycol

(c)

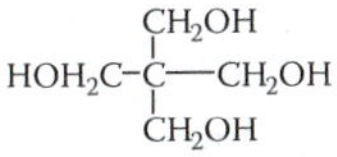

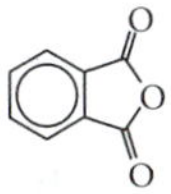

$$HOH_2C{-}\overset{\overset{\displaystyle CH_2OH}{|}}{\underset{\underset{\displaystyle CH_2OH}{|}}{C}}{-}CH_2OH$$

Pentaerythritol   and   phthalic anhydride

(d) $H_2N(CH_2)_5NH_2$ $ClCO(CH_2)_5COCl$

Pentamethylenediamine   and   heptanedioyl chloride

(e)

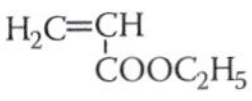

$$\underset{\overset{\displaystyle |}{\displaystyle COOC_2H_5}}{H_2C{=}CH}$$ ethyl acrylate

(f)

$$\overset{\overset{\displaystyle O}{\|}}{C}{-}O{-}CH_2{-}CH{=}CH_2$$
$$\underset{\underset{\displaystyle O}{\|}}{C}{-}O{-}CH_2{-}CH{=}CH_2$$

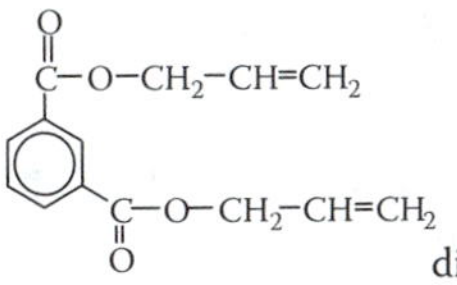

diallyl isophthalate

2. Consider the reaction of adipic acid with one of the following compounds:

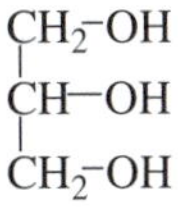

$$\underset{\text{glycerol}}{\overset{\overset{\displaystyle CH_2{-}OH}{|}}{\underset{\underset{\displaystyle CH_2{-}OH}{|}}{CH{-}OH}}} \qquad \underset{\text{1,4-butanediol}}{HO{-}CH_2CH_2CH_2CH_2{-}OH} \qquad \underset{\text{butanol}}{HO{-}CH_2CH_2CH_2CH_3}$$

In each case, indicate whether a linear polymer or a network is formed.

3. Draw the repeat units of the two polymers obtained by the polymerization of the following two monomers:

$$HO-(CH_2)_5-\overset{\overset{\displaystyle O}{||}}{C}-OH \qquad\qquad H_2N-(CH_2)_5-\overset{\overset{\displaystyle O}{||}}{C}-OH$$

Which of the two would you expect to exhibit the higher melting point and why?

4. Calculate the number-average and weight-average molar mass of the following ternary mixture of monodisperse polymer samples, each characterized by molar mass, M_i, and number of moles, N_i, as reported in the following table:

i	N_i	M_i
1	0.053	10,000
2	0.035	50,000
3	0.012	100,000

Determine the polydispersity of the sample.

5. Name each of the following polymers by the IUPAC system:

(a) $\left(\!\!\begin{array}{c} CH-CH_2 \\ | \\ (CH_2)_9CH_3 \end{array}\!\!\right)_{\!n}$

(b) $\left(\!\!\begin{array}{c} O-CH-CH_2 \\ | \\ Br \end{array}\!\!\right)_{\!n}$

(c) $\left(O-CH_2-CH_2\right)_n$

(d) $\left(HNCO(CH_2)_5CONH(CH_2)_5\right)_n$

6. The following polymers from DuPont are known by the registered trademarks listed below the structure. Rename them using the IUPAC nomenclature.

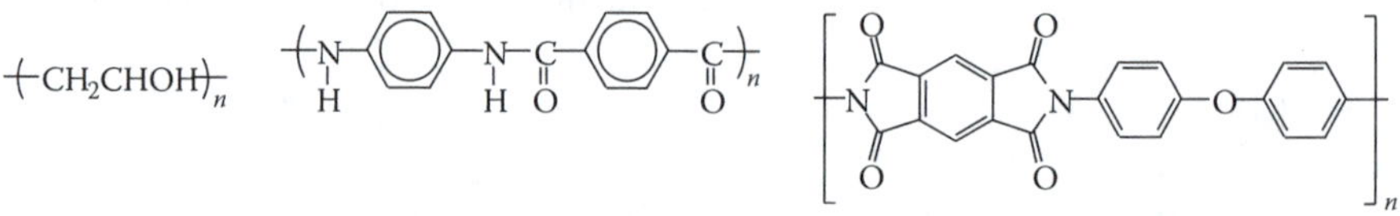

$\left(CH_2CHOH\right)_n$ Elvanol®

Kevlar®

Kapton®

7. A polymer sample is composed of a series of fractions:

M × 10⁻³	Mass (mg)
20–60	40
60–100	160
100–140	360
140–180	720
180–220	440
220–260	228
260–300	140

Calculate the number average, weight average from these data. Draw the molecular weight distribution curve and indicate in the plot the position of the weight- and number-average molecular weights. Determine the polydispersity of the sample.

8. Identify all the possible isomers (structural and stereoisomers) resulting from the polymerization of 2-methyl butadiene.

9. Polybutadiene samples prepared by anionic polymerization contain a random distribution of *cis*-1,4, *trans*-1,4 and vinyl-1,2 units. Draw structures of a portion of polymer chains consisting of approximately 20 units for PB samples of the following composition:
   a. 38% *cis*-1,4, 51% *trans*-1,4 and 11% vinyl-1,2 units
   b. 11% *cis*-1,4, 13% *trans*-1,4 and 76% vinyl-1,2 units

10. Hydrogenation of the polybutadiene samples in Problem 9 yields a random copolymer. Determine the extended length of the polymers containing 100 repeat units.

REFERENCES

Generic source-based nomenclature for polymers, *Pure Appl. Chem.*, 73(9), 1511, 2001.

Hounshell, D.A. and Smith, J.K., The Nylon Drama, *American Heritage of Invention and Technology*, 4, p. 40, 1988.

Morawetz, H., *Polymers: The Origins and Growth of a Science*, John Wiley and Sons, 1985.

Nomenclature of regular single-strand organic polymers, *Pure Appl. Chem.*, 74(10), 1921, 2002.

Seymour, R.B., *History of Polymer Science and Technology*, Marcel Dekker, 1982.

Treloar, L.R.G., *Introduction to Polymer Science*, Wykeham Publications, 1970.

Use of abbreviations for names of polymeric substances (1986), *Pure Appl. Chem.*, 59, 691, 1987.

Zandvoort, H., *Studies in the History and Philosophy of Science*, 19(4), 489, 1988.

BIBLIOGRAPHY

Allcock, H.R., Mark, J.E., and Lampe, F.W., *Contemporary Polymer Chemistry*, 3rd ed., Prentice Hall, 2003.

Allen, G. and Bevington, J.C., Eds., *Comprehensive Polymer Science: The Synthesis, Characterization, Reactions, and Applications of Polymers*, Pergamon Press, 1989.

Billmeyer, F.W., *Textbook of Polymer Science*, 3rd ed., Interscience Publishers, 1984.

Bovey, F.A. and Winslow, F.H., *Macromolecules: An Introduction to Polymer Science*, Academic Press, 1979.

Brandrup, J., Immergut, E.H., Grulke, E.A., Abe, A., and Bloch, D.R., Eds., *Polymer Handbook*, 4th ed., John Wiley and Sons, 2005.

Carraher, C.E., Jr., *Seymour/Carraher's Polymer Chemistry*, 6th ed., CRC Press, 2003.

Elias, H.G., *An Introduction to Polymer Science*, John Wiley and Sons, 1996.

Elias, H.G., *Macromolecules*, Vol. 1, 2, John Wiley and Sons, 2006.

Flory, P.J., *Principles of Polymer Chemistry*, Cornell University Press, 1953.

Hiemenz, P.C. and Lodge, T., *Polymer Chemistry*, 2nd ed., CRC Press, 2007.

Mark, H.F., *Encyclopedia of Polymer Science and Technology*, 12 volume set, 3rd ed., John Wiley and Sons, 2004.

Munk, P. and Aminabhavi, T.M., *Introduction to Macromolecular Science*, 2nd ed., John Wiley and Sons, 2002.

Odian, G., *Principles of Polymerization*, 4th ed., John Wiley and Sons, 2004.

Painter, P.C. and Coleman, M.M., *Fundamentals of Polymer Science: An Introductory Text*, Technomic Publishing, 1997.

Ravve, A., *Principles of Polymer Chemistry*, Plenum Press, 2000.

Rodriguez, F., Cohen, C., Ober, C.K., and Archer, L., *Principles of Polymer Systems*, 5th ed., Taylor and Francis, 2003.

Rosen, S.L., *Fundamental Principles of Polymeric Materials*, 2nd ed., John Wiley and Sons, 1993.

Sperling, L.H., *Introduction to Physical Polymer Science*, 4th ed., Wiley-Interscience, 2005.

Stevens, M., *Polymer Chemistry: An Introduction*, 3rd ed., Oxford University Press, 1998.

Young, R.J. and Lovell, P.A., *Introduction to Polymers*, 2nd ed., Chapman and Hall, 1991.

2 Step-Growth Polymerization

The classical subdivision of polymers into two main groups was made around 1929 by W.H. Carothers, who proposed that a distinction be made between polymers prepared by the stepwise reaction of monomers and those formed by chain reactions. These he called:

1. *Condensation polymers*, characteristically formed by reactions involving the elimination of a small molecule such as water at each step.
2. *Addition polymers*, in which no such loss occurred.

Although these definitions were perfectly adequate at the time, it soon became obvious that notable exceptions existed and that a fundamentally sounder classification should be based on a description of the chain-growth mechanism. It is preferable to replace the term *condensation* with *step-growth* or *step-reaction*. Reclassification as step-growth polymerization now logically includes polymers such as polyurethanes, which grow by a step-reaction mechanism without elimination of a small molecule.

In this chapter, we shall examine the main features of step-growth polymerization, beginning with the simpler reactions that produce linear chains exclusively. This type of polymerization is used to produce some of the industrially important fibers such as nylon and terylene. A brief discussion of the more complex branching reactions follows to illustrate how the thermosetting plastics are formed.

2.1 GENERAL REACTIONS

In any reaction resulting in the formation of a chain or network of high molar mass, the functionality (see Section 1.2) of the monomer is of prime importance. In step-growth polymerization, a linear chain of monomer residues is obtained by the stepwise intermolecular condensation or addition of the reactive groups in bifunctional monomers. These reactions are analogous to simple reactions involving monofunctional units as typified by a polyesterification reaction involving a diol and a diacid.

$$HO-R-OH + HOOC-R'-COOH \rightleftharpoons HO-R-OCO-R'-COOH + H_2O$$

If the water is removed as it is formed, no equilibrium is established and the first stage in the reaction is the formation of a dimer, which is also bifunctional. As the reaction proceeds, longer chains, trimers, tetramers, and so on will form through other esterification reactions, all essentially identical in rate and mechanism until, ultimately, the reaction contains a mixture of polymer chains of large molar masses M. However, the formation of samples with significantly large values of M is subject to a number of rather stringent conditions, which will be examined in greater detail later in this chapter.

Two major groups, both distinguished by the type of monomer involved, can be identified in step-growth polymerization. In the first group, two polyfunctional monomers take part in the reaction, and each possesses only one distinct type of functional group as in the previous esterification reaction, or more generally:

$$A\text{--}A + B\text{--}B \longrightarrow \left(A\text{--}AB\text{--}B\right)$$

The second group is encountered when the monomer contains more than one type of functional group such as a hydroxyacid (HO—R—COOH), represented generally as A–B, where the reaction is

$$nA\text{--}B \longrightarrow \left(AB\right)_n$$

or

$$n\left(HO\text{--}R\text{--}COOH\right) \longrightarrow H\left(ORCO\right)_n OH$$

A large number of step-growth polymers have the basic structure

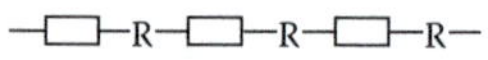

$$-\square\text{--}R\text{--}\square\text{--}R\text{--}\square\text{--}R\text{--}$$

where R can be $\left(CH_2\right)_x$ or 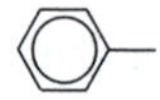 ⬡— and the —$\square$— link is one of three important groups:

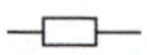

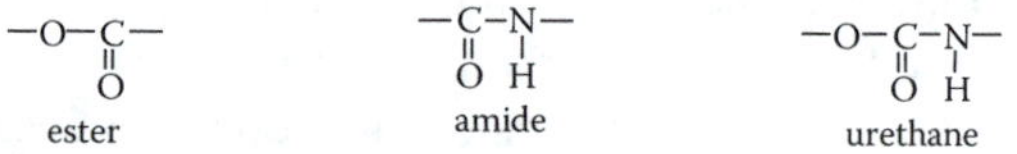

—O—C— ‖ O ester	—C–N— ‖ ⎥ O H amide	—O–C–N— ‖ ⎥ O H urethane

Other links and groups are involved in these reactions, and some typical step-reaction polymers are shown in Table 2.1.

2.2 REACTIVITY OF FUNCTIONAL GROUPS

One basic simplifying assumption proposed by Flory, when analyzing the kinetics of step-growth systems, was that all functional groups can be considered as being equally reactive. This implies that a monomer will react with both monomer or polymer species with equal ease.

The progress of the reaction can be illustrated in Figure 2.1 where, after 25% reaction, the number-average chain length x_n is still less than two because monomers,

TABLE 2.1
Typical Step-Growth Polymerization Reactions

Polymer	Reaction
Polyester	$n\text{HO}(CH_2)_x\text{COOH} \longrightarrow \text{HO}\{(CH_2)_x\text{-}\overset{\overset{O}{\parallel}}{C}\text{-}O\}_n H + (n-1)H_2O$
Polyamide	$n\text{NH}_2\text{-R-COOH} \longrightarrow H\{\text{NH-R-CO}\}_n OH + (n-1)H_2O$
	$n\text{NH}_2\text{-R-NH}_2 + n\text{HOOC-R'-COOH} \longrightarrow H\{\text{NH-R-NHCO-R'-CO}\}OH + (2n-1)H_2O$
Polyurethanes	$n\text{HO-R-OH} + \text{NCO-R'-NCO} \longrightarrow \{\text{OROCONH-R'-NHCO}\}_n$
Polyanhydride	$n\text{HOOC-R-COOH} \longrightarrow \text{HO}\{OC\text{-}R\text{-}CO\cdot O\}_n H + (n-1)H_2O$

Polysiloxane

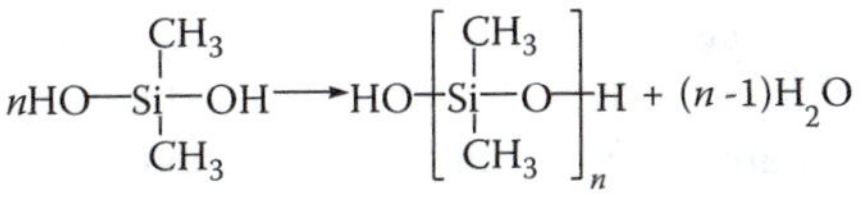

$$n\text{HO}\overset{\overset{CH_3}{|}}{\underset{\underset{CH_3}{|}}{Si}}\text{-OH} \longrightarrow \text{HO}\left[\overset{\overset{CH_3}{|}}{\underset{\underset{CH_3}{|}}{Si}}\text{-}O\right]_n H + (n-1)H_2O$$

Phenol-
formaldehyde

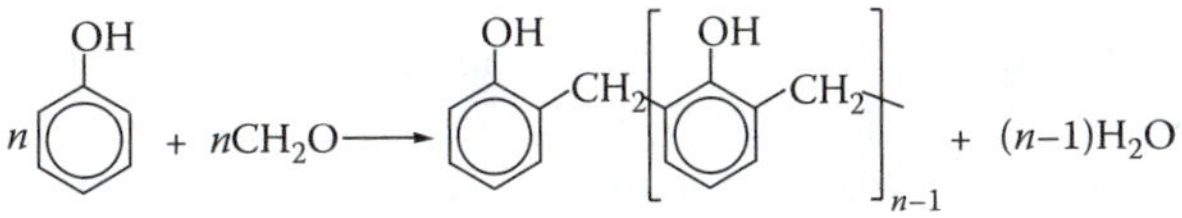

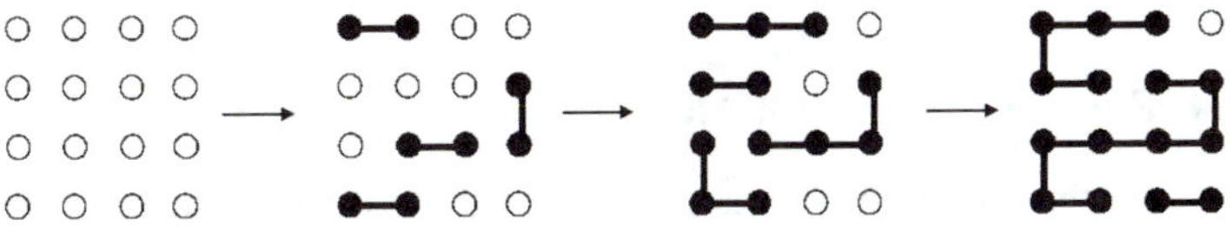

FIGURE 2.1 Diagrammatic representation of a step-growth polymerization.

being the most predominant species, will tend to react most often, and the reaction is mainly the formation of dimers and trimers. Even after 87.5% of the reaction, x_n will only be about 8, and it becomes increasingly obvious that if long chains are required, the reaction must be pushed toward completion.

2.3 CAROTHERS EQUATION

W.H. Carothers, the pioneer of step-growth reactions, proposed a simple equation relating x_n to a quantity p describing the extent of the reaction for linear polycondensations or polyadditions.

If N_0 is the original number of molecules present in an A–B monomer system and N is the number of all molecules remaining after time t, then the total number of functional groups of either A or B that have reacted is $(N_0 - N)$. At that time t, the extent of reaction p is given by

$$P = (N_0 - N)/N_0 \quad \text{or} \quad N = N_0(1 - p). \tag{2.1}$$

If we remember that $x_n = N_0/N$, a combination of expressions gives the *Carothers equation*,

$$x_n = 1/(1 - p). \tag{2.2}$$

This equation is also valid for an A–B + B–B reaction when one considers that, in this case, there are initially $2N_0$ molecules.

The Carothers equation is particularly enlightening when we examine the numerical relation between x_n and p; thus, for $p = 0.95$ (i.e., 95% conversion), $x_n = 50$ and when $p = 0.99$, then $x_n = 100$. In practical terms, it has been found that for a fiber-forming polymer such as nylon-6.6 $+ NH(CH_2)_6NHCO(CH_2)_4CO +_{\overline{n}}$ the value of M_n has to be about 12,000 to 13,000 g mol^{-1} if a high-tenacity fiber is to be spun, and as this corresponds to $x_n = 106$ to 116, the polymerization has to proceed beyond 99% completion. Similarly, for polyesters derived from ω-hydroxydecanoic acid, $H + O(CH_2)_9CO +_{\overline{n}} OH$, x_n of about 150 is optimum for good fibers, and so p must exceed 0.99.

Note that when using the Carothers equation for an A–A, B–B system, half the average molar mass of the [A–AB–B] repeat unit is used to calculate the degree of polymerization. This is because Equation 2.2 gives a measure of the average number of monomers in a polymer and, as two different types are being used in systems such as the nylon-6.6 example, the average-molar mass of these is $(114 + 112)/2 = 113$.

2.4 CONTROL OF THE MOLAR MASS

Quite obviously, the control of the molar mass of the product of these reactions is very important. Very-high-molar-mass material may be too difficult to process, whereas low-molar-mass polymer may not exhibit the properties desired in the end product, and one must be able to stop the reaction at the required value of p. Consequently, the reactions are particularly demanding with respect to the purity of the reagents, and accurate control of the amount of each species in the mixture is cardinal. It is symptomatic of these critical requirements that only the following four types of reaction usefully produce linear polymers with $M_n > 25,000$ g mol^{-1}:

1. *Schotten–Baumann reaction*: This involves the use of an acid chloride in an esterification or amidation; for example, the so-called nylon rope trick reaction is an interfacial condensation between sebacoyl chloride and hexamethylenediamine, producing a polyamide known as nylon-6,10.

$$nClCO(CH_2)_8COCl + nH_2N(CH_2)_6NH_2 \longrightarrow$$
$$+CO(CH_2)_8CONH(CH_2)_6NH +_{\overline{n}} + (2n-1)HCl$$

The bifunctional acyl chloride is dissolved in a suitable solvent such as dichloromethane and placed in a beaker. An aqueous alkaline solution of the bifunctional amine is layered on top, and the nylon-6,10, which forms immediately at the interface, can be drawn off as a continuous filament until the reagents are exhausted. The reaction has an S_N2 mechanism with the release of a proton to the base in the aqueous phase; the condensation is a successive stepwise reaction by this S_N2 route.

2. *Salt dehydration*: Direct esterification requires high purity materials in equimolar amounts because esterifications rarely go beyond 98% completion in practice. To overcome this, hexamethylene diamine and a dibasic acid such as adipic acid can be reacted to produce a nylon salt, hexamethylene diammonium adipate. A solution of 0.5-mol diamine in a mixture of 95% ethanol (160 cm³) and distilled water (60 cm³) is added to 0.5-mol diacid dissolved in 600 cm³ of 95% ethanol over a period of 15 min. The mixture is stirred for 30 min during which time the nylon salt precipitates as a white crystalline solid. This can be recrystallized and should melt at 456 K. The pure salt can be converted into a polyamide by heating it under vacuum in a sealed tube, protected by wire gauze, at about 540 K in the presence of a small quantity of the diacid, e.g., 10 g salt to 0.55-g adipic acid is a suitable mixture. If a lower molar mass is desired, a monofunctional acid can replace the adipic acid and act as a chain terminator.

3. *Ester interchange*: An alternative reaction is a trans-esterification in the presence of a proton donating or weak base catalyst such as sodium methoxide, e.g.,

In this way, ethylene glycol and dimethyl terephthalate produce terylene. Ester interchange is the most practical approach to polyester formation because of the faster reaction rate and use of more easily purified products. The formation of poly(ethylene terephthalate) is essentially a two-stage process. The first stage, at 380 to 470 K, is the formation of dimers and trimers, each with two hydroxyl end-groups, and during this formation the methanol is being distilled off. To complete the reaction, the temperature is raised to 530 K, and condensation of these oligomers produces a polymer with large M_n. The major advantage is that the stoichiometry is self-adjusting in the second stage.

4. *Step polyaddition (urethane formation)*: Polyurethanes with large M_n can be prepared using a method based on the Wurtz alcohol test. In the presence of a basic catalyst such as a diamine, ionic addition takes place, for example, between 1,4-butanediol and 1,6-hexanediisocyanate:

This reaction produces a highly crystalline polymer, but a vast number of urethanes can be formed by varying the reactants, and a series of polymers covering a wide spectrum of properties can be prepared.

2.5 STOICHIOMETRIC CONTROL OF M_n

Chain growth, to produce high-molar-mass material, becomes increasingly difficult as the reaction proceeds. This is due to a number of reasons: (1) the difficulty in ensuring the precise equivalence of the reactive groups in the starting materials when two or more types of monomer are used; (2) the decreasing frequency of functional groups meeting and reacting as their concentration diminishes; and (3) the increasing likelihood of interference from side reactions.

Often, it is preferable to avoid production of high-molar-mass polymer, and control can be effected by rapidly cooling the reaction at the appropriate stage or by adding calculated quantities of monofunctional materials as in the preparation of nylon-6,6 from the salt.

More usefully, a precisely controlled stoichiometric imbalance of the reactants in the mixture can provide the desired result. For example, an excess of diamine over an acid chloride would eventually produce a polyamide with two amine end-groups incapable of further growth when the acid chloride was totally consumed. This can be expressed in an extension of the Carothers equation as,

$$x_n = (1 + r)/(1 + r - 2rp), \tag{2.3}$$

where r is the ratio of the number of molecules of the reactants. Thus, for quantitative reaction ($p = 0.999$) between N molecules of phenolphthalein and $1.05N$ molecules of terephthaloyl chloride to form poly(terephthaloyl phenolphthalein) (see scheme below), the value of $r = N_{AA}/N_{BB} = 1/1.05 = 0.952$ and $x_n = (1 + 0.952)/(1 + 0.952 - 2 \times 0.999 \times 0.952) \approx 39$, rather than 1000 for $r = 1$. This reaction is an interfacial polycondensation whose progress may be followed by noting the color change from the red phenolphthalein solution to the colorless polyester. The interface can remain stationary in the experiment, but the uniformity of the polymer is improved by increasing the reaction surface using high-speed stirring.

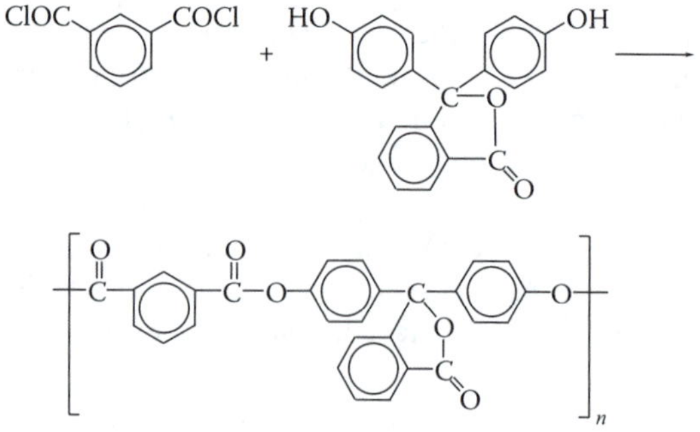

However, the purity of the starting monomers is also crucial, and if one of them contains only 95% of the expected difunctional material, then $r = 0.95$ and the attainable x_n will again be ≈ 40.

In practice, $p = 1$ is rarely achieved, nor is it a perfect stoichiometric balance. The consequences of this are shown in Figure 2.2 where x_n is plotted as (a) a function of p, calculated from Equation 2.2, and (b) as a function of the stoichiometric ratio for

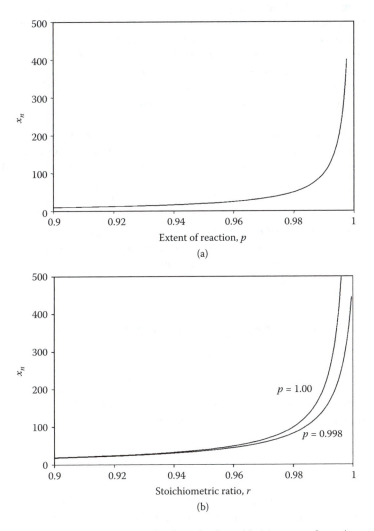

FIGURE 2.2 Variation of the degree of polymerization with (a) extent of reaction p, calculated from Equation 2.2, and (b) the stoichiometric ratio r for

$$p = 1.00, \quad x_n = \frac{1+r}{1-r}, \quad \text{and } 0.998, \quad x_n = \frac{1+r}{1+r-2rp},$$

calculated from Equation 2.3.

$p = 1.00$ and 0.998, calculated from Equation 2.3. The corresponding equation for a monofunctional additive is similar to Equation 2.3, only now r is defined as the ratio $N_{AA}/(N_{BB} + 2N_B)$ where N_B is the number of monofunctional molecules added.

2.6 KINETICS

The assumption that functional group reactivity is independent of chain length can be verified kinetically by following a polyesterification. The simple esterification is an acid-catalyzed process in which protonation of the acid is followed by interaction with the alcohol to produce an ester and water. If significant polymer formation is to be achieved, the water must be removed continuously from the reaction to displace the equilibrium, and the water eliminated can be used to estimate the extent of the reaction. Alternatively, the rate of disappearance of carboxylic groups can be measured by titrating aliquots of the mixture.

Self-catalyzed reaction. If no acid catalyst is added, the reaction will still proceed because the acid can act as its own catalyst. The rate of condensation at any time t can then be derived from the rate of disappearance of $-$COOH groups and

$$- d[COOH]/dt = k[COOH]^2[OH]. \tag{2.4}$$

The second-order [COOH] term arises from its use as a catalyst, and k is the rate constant. For a system with equivalent quantities of acid and glycol, the functional group concentration can be written simply as c and

$$- dc/dt = kc^3 \tag{2.5}$$

This expression can be integrated under the conditions that $c = c_0$ at time $t = 0$ and

$$2kt = 1/c^2 - 1/c_0^2. \tag{2.6}$$

The water formed is removed and can be neglected. From the Carothers equation it follows that $c = c_0(1 - p)$, leading to the final form

$$2c_0^2 kt = \left[1/(1-p)^2\right] - 1. \tag{2.7}$$

Acid-catalyzed reaction. The uncatalyzed reaction is rather slow, and a high x_n is not readily attained. In the presence of a catalyst, there is an acceleration of the rate, and the kinetic expression is altered to

$$- d[COOH]/dt = k'[COOH][OH], \tag{2.8}$$

which is kinetically first order in each functional group. The new rate constant k' is then a composite of the rate constant k and the catalyst concentration, which also remains constant. Hence,

$$-dc/dt = k'c^2, \tag{2.9}$$

and integration gives finally

$$c_0k't = [1/(1 - p)] - 1. \tag{2.10}$$

Both equations have been verified experimentally by Flory as shown in Figure 2.3.

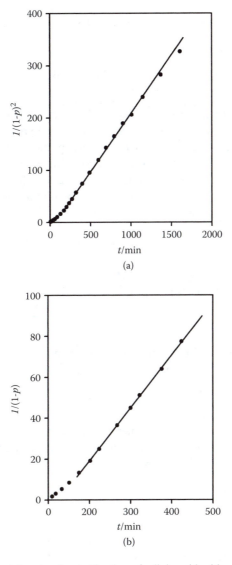

(a)

(b)

FIGURE 2.3(a) Self-catalyzed polyesterification of adipic acid with ethylene glycol at 439 K; (b) polyesterification of adipic acid with ethylene glycol at 382 K, catalyzed by 0.4 mol percent of p-toluene sulfonic acid. (Data from Flory.)

2.7 MOLAR MASS DISTRIBUTION IN LINEAR SYSTEMS

The creation of long-chain polymers by the covalent linking of small molecules is a random process, leading to chains of widely varying lengths. Because of the random nature of the process, the distribution of chain lengths in a sample can be arrived at by simple statistical arguments.

The problem is to calculate the probability of finding a chain composed of x basic structural units in the reaction mixture at time t, for either of the two reactions

$$xA - A + xB - B \rightarrow A - A[B - BA - A]_{x-1} B - B,$$

or

$$xA - B \qquad \rightarrow A \left[BA \right]_{x-1} B,$$

that is, to calculate the probability that a functional group A or B has reacted. For the sake of clarity, we can consider one of the functional groups to be carboxyl and determine the probability that $(x - 1)$ carboxyl groups have reacted to form a chain. This is p^{x-1}, where p is the extent of the reaction, defined in Equation 2.1.

It follows that if a carboxyl group remains unreacted, the probability of finding this uncondensed group is $(1 - p)$, and so the probability P_x of finding one chain x units long (i.e., an x-mer) is

$$P_x = (1 - p)p^{x-1}. \tag{2.11}$$

As the fraction of x-mers in any system equals the probability of finding one, the total number N_x present is given by

$$N_x = N(1 - p)p^{x-1}, \tag{2.12}$$

where N is the total number of polymer molecules present in the reaction. Substitution of the Carothers equation (Equation 2.2) gives

$$N_x = N_0 (1 - p)^2 p^{x-1}, \tag{2.13}$$

where N_0 is the total number of monomer units present initially. The variation of N_x for various values of p and x is shown in Figure 2.4. A slightly different set of curves is obtained if the composition is expressed in terms of mass fraction w; in this case, $w_x = xN_x/N_0$ to give

$$w_x = x(1 - p)^2 p^{x-1} \tag{2.14}$$

Both reveal that very high conversions are necessary if chains of significant size are to be obtained and that although monomer is normally the most numerous species, the proportion of low-molar-mass material decreases as p exceeds 0.95.

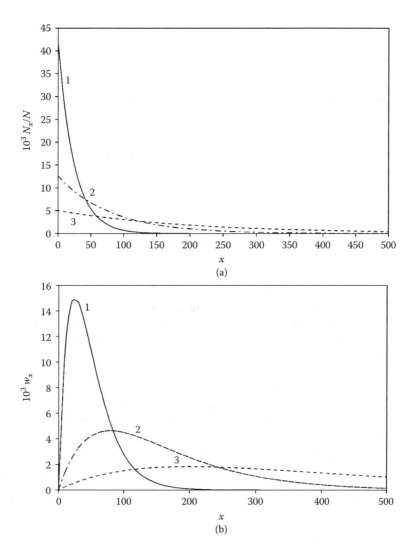

FIGURE 2.4 (a) Number fraction distribution curves for linear step-growth polymerizations: curve 1, $p = 0.9600$; curve 2, $p = 0.9875$; curve 3, $p = 0.9950$. (b) Corresponding weight fraction distribution for the same system.

2.8 AVERAGE MOLAR MASSES

Number- and weight-average molar masses can be calculated from the equations if M_0 is taken as the molar mass of the repeat unit. Thus,

$$M_n = N_x \Sigma(M_0 N_x)/N = M_0/(1-p) \tag{2.15}$$

and

$$M_w = M_0(1+p)/(1-p). \tag{2.16}$$

It can be seen that a heterogeneity index (M_w/M_n) for the most probable distribution when $p = 1$ is

$$(M_w / M_n) = \{M_0(1+p)/(1-p)\}\{(1-p)/M_0\} = 2. \tag{2.17}$$

2.9 CHARACTERISTICS OF STEP-GROWTH POLYMERIZATION

It might be appropriate at this stage to summarize the main features of the step reactions.

1. Any two molecular species in the mixture can react.
2. The monomer is almost all incorporated in a chain molecule in the early stages of the reaction, i.e., about 1% of monomer remains unreacted when $x_n = 10$. Hence, polymer yield is independent of the reaction time in the later stages.
3. Initiation, propagation, and termination reactions are essentially identical in rate and mechanism.
4. The chain length increases steadily as the reaction proceeds.
5. Long reaction times and high conversions are necessary for the production of a polymer with large x_n.
6. Reaction rates are slow at ambient temperatures but increase with a rise in temperature, although this has little effect on the chain length of the final product.
7. Activation energies are moderately high, and reactions are not excessively exothermic.

2.10 TYPICAL STEP-GROWTH REACTIONS

Reactions are normally carried out in bulk in the temperature range 420 to 520 K to encourage fast reactions and promote the removal of the low-molar-mass condensation product. Activation energies are about 80 kJ mol^{-1}.

Low-temperature polycondensations. The advantage of high-temperature reactions are partly counteracted by the increasing danger of side-reactions, and room-temperature reactions using highly energetic reactants provide routes to a variety of polymers. The use of the Schotten–Baumann reaction for polyamides has already been outlined. This is an example of an unstirred interfacial reaction in which the diamine is soluble in both phases and diffuses across the interface into the organic layer in which polymerization takes place. Continuous polymer production is achieved by withdrawing the film formed at the interface to allow continued diffusion of the reactants. Alternatively, the continuity of the polymerization reaction can be maintained by stirring the system vigorously; this ensures a constantly changing interface and increases the surface area available for the reaction. As both methods are diffusion controlled, the need for stringent stoichiometric control is obviated.

When the diamine used is aromatic, only low-molar-mass polymer is formed because of the lower-reaction rates. To produce longer chains, conditions must be readjusted so that both phases are polar and miscible, and vigorous high-speed stirring of the system is necessary. This is used in the reaction between isophthaloyl chloride and *m*-phenylene diamine. These aromatic polyamides are particularly versatile materials and of considerable interest.

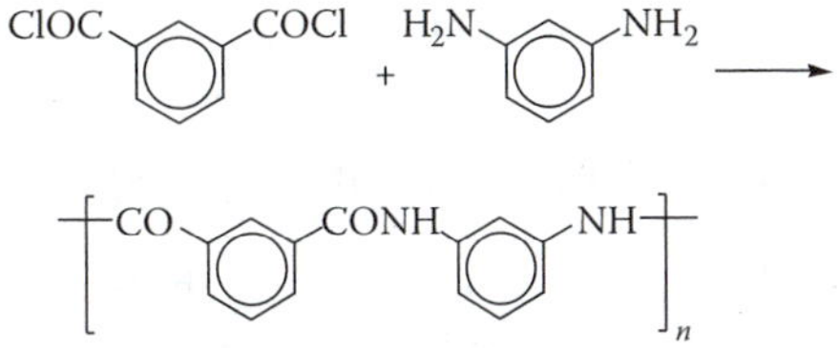

The ultimate extension to a homogeneous system with inert polar solvents is used in the synthesis of polyimides. Poly(methylene 4,4'-diphenylene pyromellitamide acid) is prepared by mixing equal amounts of pyromellitic dianhydride and *bis*(4-aminophenyl methane) in N,N'-dimethylformamide. The reaction is maintained at 288 K, with stirring, for an hour, and the polymer is isolated by precipitation in vigorously stirred water.

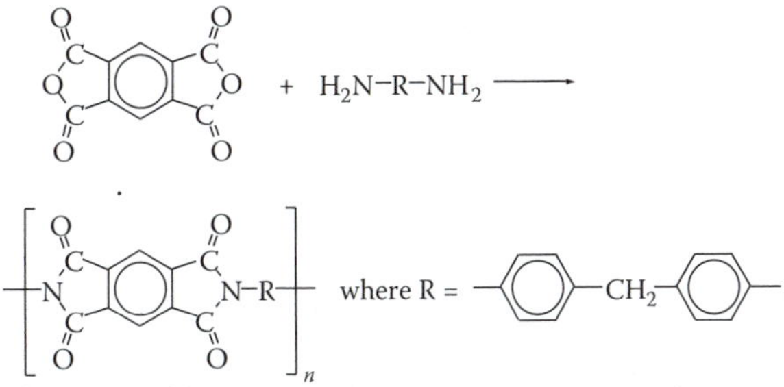

Polysulfonamides, polyanhydrides, and polyurethanes can also be formed in homogeneous low-temperature reactions.

2.11 RING FORMATION

The assumption so far has been that all bifunctional monomers in step-growth reactions form linear polymers. This is not always true; competitive side reactions such as cyclization may occur, as with certain hydroxyacids that may form lactones or lactams if an amino acid is used.

$$HO-R-COOH \longrightarrow R{\overset{\displaystyle CO}{\underset{\displaystyle O}{\big<}}} + H_2O$$

lactone

To gauge the importance of such reactions, consideration must be given to the thermodynamic and kinetic aspects of ring formation. A study of ring strain in cycloalkanes has shown that 3- and 4-membered rings are severely strained but that this decreases dramatically for 5-, 6-, or 7-membered rings, then increases again up to 11 before decreasing for very large rings. In addition to the thermodynamic stability, the kinetic feasibility of two suitable functional groups being in juxtaposition to react must also be considered. This probability decreases with increasing ring size, and again 5-, 6-, or 7-membered rings are favored and will form in preference to a linear chain when possible.

2.12 NONLINEAR STEP-GROWTH REACTIONS

In systems containing bifunctional monomers, a high degree of polymerization is attained only when the reaction is forced almost to completion. The introduction of a trifunctional monomer into the reaction produces a rather startling change, which is best illustrated using a modified form of the Carothers equation. A more general *functionality factor* f_{av} is introduced, defined as the average number of functional groups present per monomer unit. For a system containing N_0 molecules initially and equivalent numbers of two function groups A and B, the total number of functional groups is $N_0 f_{av}$. The number of groups that have reacted in time t to produce N molecules is then $2(N_0-N)$ and

$$p = 2(N_0 - N)/N_0 f_{av}, \tag{2.18}$$

The expression for x_n then becomes

$$x_n = 2/(2 - p\, f_{av}), \tag{2.19}$$

but this is only valid when equal numbers of both functional groups are present in the system.

For a completely bifunctional system such as an equimolar mixture of phthalic acid and ethylene glycol, $f_{av} = 2$, and $x_n = 20$ for $p = 0.95$. If, however, a trifunctional alcohol, glycerol, is added so that the mixture is composed of 2 mol diacid, 1.4 mol diol, and 0.4 mol glycerol, f_{av} increases to

$$f_{av} = (2 \times 2 + 1.4 \times 2 + 0.4 \times 3)/3.8 = 2.1.$$

The value of x_n is now 200 after 95% conversion, but only a small increase to 95.23% is required for x_n to approach infinity — a most dramatic increase. This is a direct result of incorporating a trifunctional unit in a linear chain in which the unreacted hydroxyl provides an additional site for chain propagation. This leads to the formation of a highly branched structure, and the greater the number of multi-functional units, the faster the growth into an insoluble three-dimensional network. When this happens, the system is said to have reached its *gel point*.

If the stoichiometry of the system is unbalanced, the definition of the average functionality must be modified and becomes

$$f' = 2rf_A f_B f_C / \{f_A f_C + r\rho f_A f_B + r(1-\rho)f_B f_C\}, \tag{2.20}$$

when monomers A and C have the same functional group but different functionalities f_A and f_C, f_B is the functionality of monomer B. Also,

$$r = (n_A f_A + n_C f_C)/n_B f_B \leq 1, \tag{2.21}$$

$$\rho = n_C f_C / (n_A f_A + n_C f_C), \tag{2.22}$$

and n_A, n_B, and n_C are the amounts of each component. A commonly encountered system has monomers with $f_A = f_B = 2$ and $f_C > 2$. In this more general system, the onset of gelation can be predicted by establishing the critical conversion limit beyond which a gel is sure to form. This is derived on the basis of the observation that x_n approaches infinity at or beyond the gel point. In the stoichiometric case, the equation

$$p = (2/f_{av}) - (2/x_n f_{av}),$$

becomes at the gel point, when x_n tends to infinity,

$$p_G = 2/f_{av}, \tag{2.23}$$

whereas the critical extent of reaction p_G for the general case is

$$p_G = (1-\rho)/2 + 1/2r + \rho/f_C. \tag{2.24}$$

2.13 STATISTICAL DERIVATION

An expression for p_G can be derived from statistical arguments. First, a branching coefficient ζ is introduced, defined as the probability that a multifunctional monomer ($f > 2$) is connected either by a linear chain segment or directly to a second multifunctional monomer (or branch point) rather than to a chain segment terminating in a single functional group.

The critical value ζ_G necessary for incipient gelation can be calculated from the probability that at least one of the $(f-1)$ chain segments extending from a branch unit will be connected to another branch unit, and as this is simply $(f-1)^{-1}$ we have

$$\zeta_G = 1/(f-1), \tag{2.25}$$

where f is now the functionality of the branching unit and not an average as defined previously. If more than one monomer in the system is multifunctional, then an average of all monomers with $f \geq 3$ is used.

It follows that if a chain ends in a branch point, $\zeta(f-1)$ chains on the average will emanate from that point, and as ζ of these will also end in a branching unit, a further $[\zeta(f-1)]^2$ are generated and so on. This occurs when $\zeta(f-1)$ is greater than 1 and the growth of the network is limited only by the boundaries of the system. For $\zeta(f-1) < 1$, no gelation is observed.

The arguments can be extended and for a system

$$A{-}A + B{-}B + \overset{A}{\underset{A \quad A}{\diagdown\!\diagup}}$$

the critical extent of reaction of the A group at the gel point is

$$p_G = \left[r + r\rho(f-2) \right]^{-1/2}. \tag{2.26}$$

2.14 COMPARISON WITH EXPERIMENT

The gel point in a branching system is usually detected by a rapid increase in viscosity η, indicated by the inability of bubbles to rise in the medium. It is also characterized by a rapid increase in x. The values of these quantities for the reaction of diethylene glycol + succinic acid + 1,2,3-propanetricarboxylic acid are shown in Figure 2.5. The increase in x_n is not as dramatic as x_w, which can be identified with the η curve. The divergence of x_w and x_n is illustrated in Table 2.2 for the preceding system when the reaction mixture has been adjusted hypothetically to provide a ratio of carboxyl to amine groups of 1. This situation can be achieved by assuming that the mixture contains 98.5 mol of diacid A, 100 mol of diol B, and 1 mol of triacid C. From Equation 2.21, it follows that $r = 98.5 \times 2 + (1 \times 3)/(2 \times 100) = 1$. These reaction conditions lead to the appearance of a gel point at $p_G = 0.9925$, and the ratio (x_w/x_n) increases sharply as the reaction approaches the critical point. The distribution is readily compared with that for a totally bifunctional system for which $(x_w/x_n) = 1 + p$, and the broadening of the molar-mass distribution, characteristic of these branching polymerizations, is well illustrated.

A comparison of experimental and theoretical values of p_G has been reported by Flory for a mixture of a tricarboxylic acid with a diol and a diacid; the results are given in Table 2.3.

The statistical equation underestimates p_G whereas Equation 2.24 overestimates the experimental value. The Carothers equation leads to a high p_G because molecules larger than the observed x_n exist in the mixture, and these undergo gelation before the predicted value is attained. This difficulty is overcome in the statistical treatment,

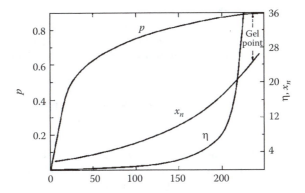

FIGURE 2.5 Variation of the viscosity η and the degree of polymerization x_n with the extent of network polymer formation in the system diethylene glycol + succinic acid + 1,2,3-propanetricarboxylic acid. (Data from Flory.)

TABLE 2.2
Branching System with $p_G = 0.9925$

P	x_w	x_w/x_n
0.100	1.2	1.1
0.500	3.0	1.5
0.700	5.8	1.7
0.900	20.4	2.0
0.950	45.6	2.2
0.980	153.8	2.7
0.990	747.2	5.6
0.992	3306.8	18.2

TABLE 2.3
1,2,3-Propanetricarboxylic Acid, Diethylene Glycol, and Adipic Acid or Succinic Acid System

			p_G	
R	ρ	Exponent	Equation 2.24	Equation 2.26
0.800	0.375	0.991	1.063	0.955
1.000	0.293	0.911	0.951	0.879
1.002	0.404	0.894	0.933	0.843

---

but now differences are attributable to intramolecular cyclization in the system, and the loops that are formed are nonproductive in a branching sense. This means that the reaction must proceed further to overcome waste.

2.15 POLYURETHANES

An important and versatile family of polymers, whose diverse uses include foams, fibers, elastomers, adhesives, and coatings, is formed by the interaction of diisocyanates with diols.

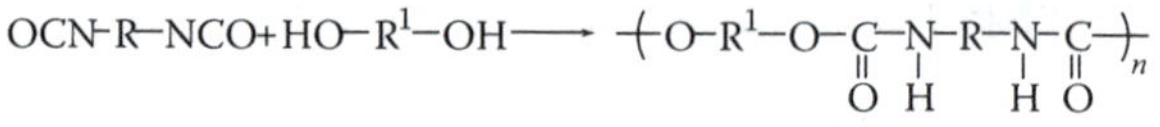

$$\text{OCN-R-NCO} + \text{HO-R}^1\text{-OH} \longrightarrow \left[\text{O-R}^1\text{-O-}\underset{\underset{\text{O}}{\|}}{\text{C}}\text{-}\underset{\underset{\text{H}}{|}}{\text{N}}\text{-R-}\underset{\underset{\text{H}}{|}}{\text{N}}\text{-}\underset{\underset{\text{O}}{\|}}{\text{C}}\right]_n$$

This leads to the production of linear polymers, although branched or cross-linked structures can also be prepared by using multifunctional starting materials or, inadvertently, through side reactions. The nature and stiffness of the groups R and R^1 will control the type of material formed and its rigidity or flexibility. This, in turn, will control the properties, and ultimately, the use of the polyurethanes formed.

The basic chain-growth reaction is addition without elimination of a small molecule and is a consequence of the highly electrophilic nature of the carbon atom in the isocyanate group.

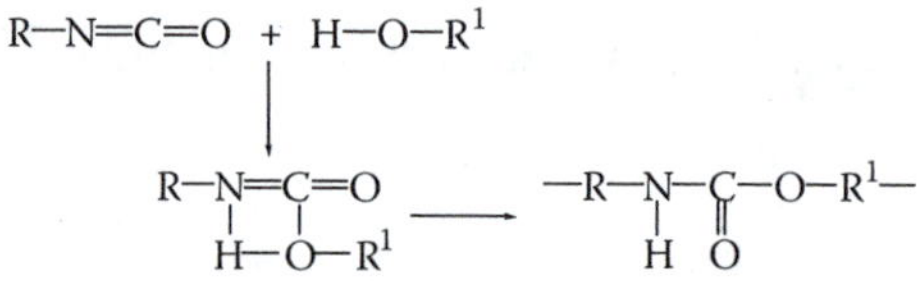

$$\text{R-N=C=O} + \text{H-O-R}^1$$

$$\underset{\text{H-O-R}^1}{\overset{\text{R-N=C=O}}{|}} \longrightarrow -\text{R-N-}\underset{\overset{\|}{\text{O}}}{\text{C}}\text{-O-R}^1-$$

This makes it susceptible to attack by nucleophilic reagents, such as alcohols, acids, water, amines, and mercaptans. The electrophilicity of the carbon atom can be increased if R is an aromatic ring which can conjugate with the isocyanate group. Thus, aromatic diisocyanates are more reactive than aliphatic ones.

Isocyanate groups may not all have equal reactivity, for example, in 1-methyl-2,4-diisocyanatobenzene

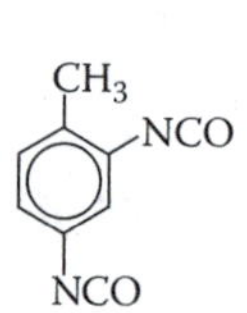

the isocyanate group nearest the methyl group has a different reactivity from the other; also when one isocyanate group is reacted, the susceptibility of the second to a reaction may be altered.

Although it is expected that difunctional monomers will give a linear polyurethane, the polymerization reaction is subject to possible side reactions. The formation of allophanate groups can occur, particularly if reaction temperatures exceed 400 K. Here, an isocyanate group adds onto the secondary amine in the urethane link, and a branched or cross-linked structure is formed.

$$wwR-N-C-O-Rww \quad + \quad OCN-Rww$$
$$\quad\quad |\;\;|| $$
$$\quad\quad H\;\;O$$

$$\downarrow$$

$$wwR-N-C-O-Rww$$
$$\quad\quad |\;\;||$$
$$\quad\quad O{=}C\;\;O \qquad\qquad \text{Allophanate link}$$
$$\quad\quad |$$
$$\quad\quad H-N-Rww$$

Polyurethanes are often prepared in two stages: first, the production of a prepolymer and, second, a chain extension reaction. During the first step, diisocyanate is reacted with a dihydroxy-terminated short-chain polyether or polyester, such as poly(ethylene adipate), poly(ε-caprolactone), or poly(tetramethylene glycol) of approximate molar mass 1000–3000. Throughout this stage, the diisocyanate is used in excess to give isocyanate end-capped blocks. These are then chain-extended by subsequent reaction with a short-chain diol (e.g., ethylene glycol or 1,4-butane diol) or a diamine (see Table 2.4). The reaction with the diamine gives a urea linkage and a poly(urethane-*co*-urea) structure, which can also react with other (NCO) groups leading to biuret formation.

$$wwN-C-N\,ww \quad + \quad OCN-Rww$$
$$\quad |\;\;||\;\;|$$
$$\quad H\;\;O\;\;H$$

$$\downarrow$$

$$wwN-C-N\,ww$$
$$\quad |\;\;||\;\;|$$
$$\quad H\;\;O\;\;C{=}O \qquad \text{Biuret link}$$
$$\quad\quad\quad\;\; |$$
$$\quad\quad\quad H-N-Rww$$

Cross-linked systems can also be obtained by the use of multifunctional mono-mers. In the polyurethanes such as those illustrated in Table 2.4 and Figure 2.6, the isocyanate monomer unit is regarded as the "hard" segment, and the polyol forms the "soft" segment. The structures shown are particularly useful in biomedical research and have been demonstrated to be suitable for cardiovascular applications.

A major use of polyurethanes is in the manufacture of foams, both rigid and flexible. The chain-extension reactions described previously are normally used when elastomeric-type products are required, but for foams, the chain-extender molecule can be omitted and polyols with an average functionality in excess of three are used. The reaction is base-catalyzed by tertiary amines or organo tin compounds (e.g., stannous octanoate) and must also include a "blowing" agent in the reaction mixture. This can be achieved by adding controlled quantities of water to the system and making use of the reaction:

$$wwRNCO + H_2O \longrightarrow wwR-N-C-OH \longrightarrow wwRNH_2 + CO_2$$
$$\qquad\qquad\qquad\qquad\quad |\;\;||$$
$$\qquad\qquad\qquad\qquad\quad H\;\;O$$

where water reacts with an isocyanate group to form the unstable carbamic acid, which decomposes to produce an amine with evolution of CO_2. The released gas forms spherical bubbles, which increase in size, and eventually impinge on one

TABLE 2.4
Formation of (a) Aromatic Polyurethane, Pellethane, (b) Aromatic Polyurethane Biomer, and (c) Aliphatic Polyurethane Tecoflex

(a)

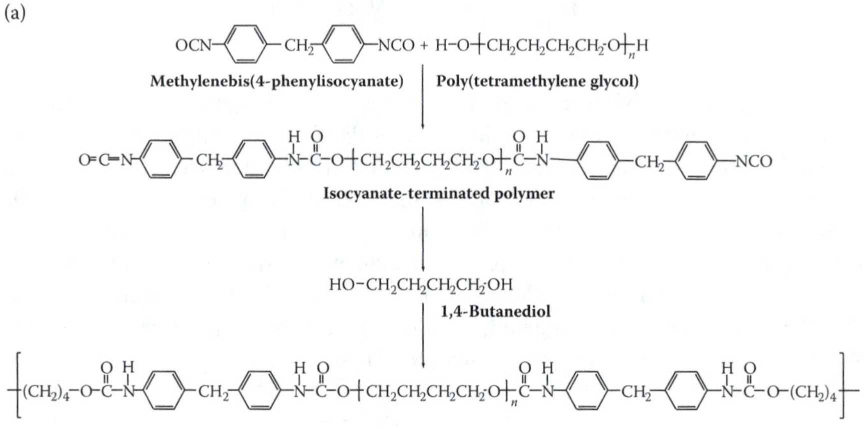

Methylenebis(4-phenylisocyanate) | Poly(tetramethylene glycol)

Isocyanate-terminated polymer

HO–CH$_2$CH$_2$CH$_2$CH$_2$OH
1,4-Butanediol

Pellethane (Upjohn product)

(b)

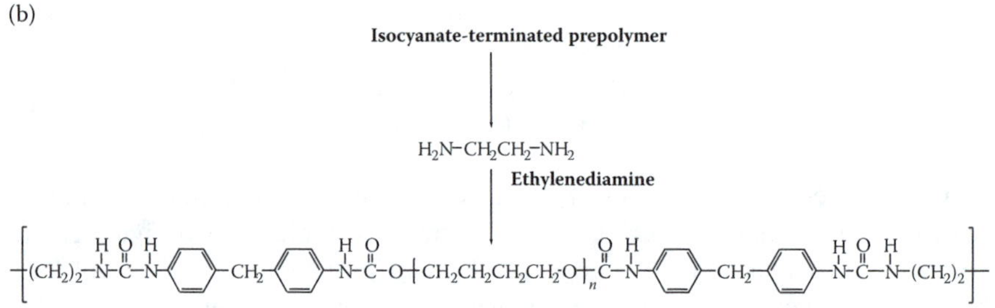

Isocyanate-terminated prepolymer

H$_2$N–CH$_2$CH$_2$–NH$_2$
Ethylenediamine

Biomer (Ethicon product)

(c)

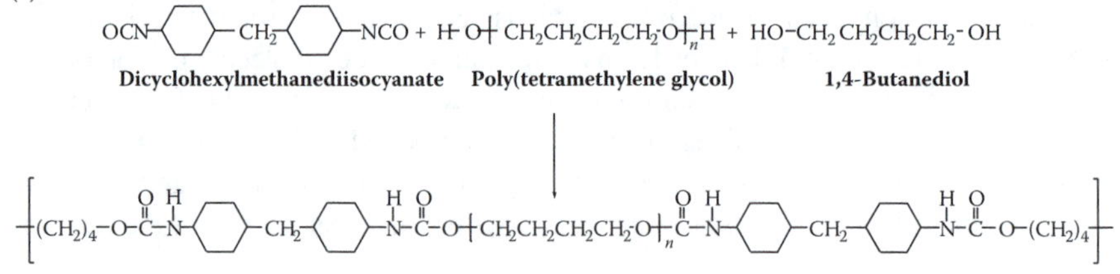

Dicyclohexylmethanediisocyanate   Poly(tetramethylene glycol)   1,4-Butanediol

Tecoflex (Thermedics product)

another to form a polyhedral cell structure in the polymer matrix. A volatile liquid can be used as an alternative "blowing" agent, and Freon (CFCl$_3$) with a boiling point of 294 K can be included in the reaction mixture. As the polymerization is exothermic, liberating about 80 kJ mol^{-1}, the heat is sufficient to vaporize the CFCl$_3$ and create the foam. Use of these chlorofluorocarbon compounds is now discouraged on environmental grounds as they are believed to attack and destroy the ozone layer.

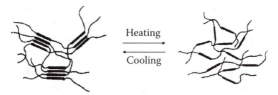

FIGURE 2.6 Schematic representation of a polyurethane, with hard blocks shown as the bold lines and soft blocks as the thin lines, phase separating in the solid state and undergoing disordering on heating. (Adapted from Pearson, R.G., in *Speciality Polymers*, Dyson, R.W. (Ed.), Blackie, 1987.)

If a flexible foam is required, then longer, more flexible polyols and trifunctional cross-linking monomers are used, whereas higher-cross-link densities and short-chain polyols tend to form more rigid foams.

In the absence of a blowing agent and with conditions favoring the formation of linear chains, thermoplastic polyurethane elastomers can be formed. Materials with different properties can be obtained by altering the ratio of hard to soft blocks, and it is also observed that, in many cases, the hard segments can crystallize. This can give rise to phase-separated structures, shown schematically in Figure 2.6, which behave as thermoplastic elastomers (see also Section 15.6).

2.16 THERMOSETTING POLYMERS

The production of highly branched network polymers is commercially important, but as cross-linking results in a tough and highly intractable material, the fabrication process is usually carried out in two stages.

The first stage is the production of an incompletely reacted prepolymer; this is either solid or liquid and of moderately low-molar mass. The second stage involves conversion of this into the final cross-linked product *in situ* (i.e., a mold or form of some description). The prepolymers are either random or structoset and are discussed separately:

Phenol-formaldehyde: Random prepolymers are prepared by reacting phenol ($f = 3$ for the ortho and para positions in the ring) with bifunctional formaldehyde. The base-catalyzed reaction produces a mixture of methylol phenols.

The composition of the mixture can be varied by altering the phenol to formaldehyde ratio. At this stage, the methylol intermediates are dried and ground, and in some cases, a filler such as mica, glass fiber, or sawdust may be added. A cross-linking agent such as hexamine is mixed with the prepolymer together with CaO as a catalyst. When further heating takes place during molding, the hexamine decomposes to form HCHO and ammonia, which acts as a catalyst in the final cross-linking process by the HCHO.

Amino resins: A related family of polymers is made from random prepolymers prepared by reacting either urea or melamine (shown below) with HCHO. The products are known as *aminoplasts*.

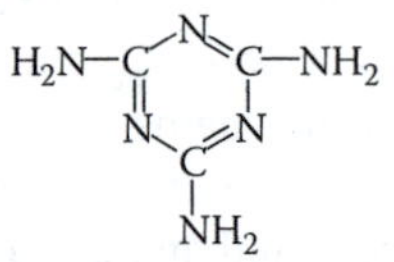

Epoxides: Structoset prepolymers are designed to have controlled and defined structures with functional groups located either at the chain-ends (*structo-terminal*) or located along the chain (*structopendant*). Epoxy resins make use of the epoxy group for the former, and hydroxyl groups for the latter. One of the major resin prepolymers formed by the reaction of bisphenol A and epichlorohydrin can be treated as either.

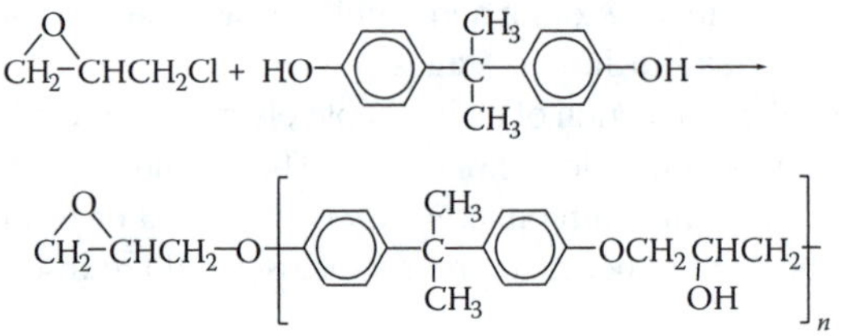

These prepolymers can subsequently react with a cross-linking (or curing) agent to turn them into strong intractable network structures. The most commonly used group of low-cost, room-temperature, cross-linking agents is the aliphatic amines, e.g., diethylene triamine ($H_2NCH_2CH_2NHCH_2CH_2NH_2$) and triethylene tetramine ($H_2N[CH_2CH_2NH]_2CH_2CH_2NH_2$). These react with the epoxy groups.

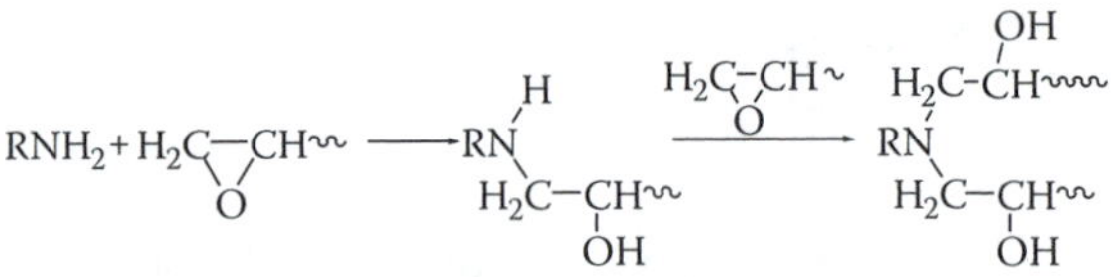

Anhydrides are also used as curing agents and have the advantage that they do not irritate the skin as do some of the amines. The reaction usually proceeds best in the presence of a catalyst such as a Lewis base (or acid).

One proposed mechanism is that of a two-step reaction with ring opening at stage one and addition of the epoxy at stage two. After this, the anion reacts with another molecule of anhydride.

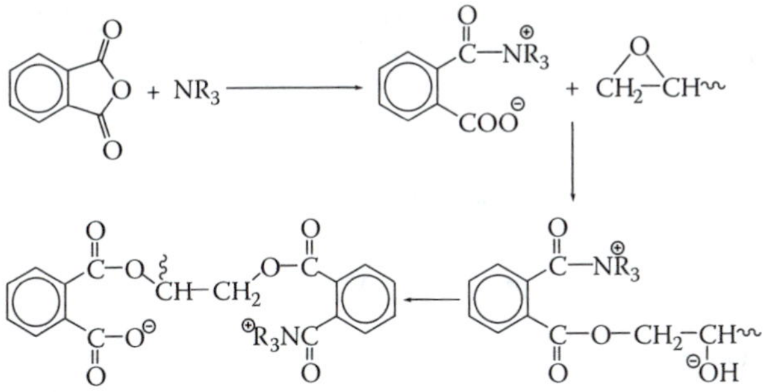

In the uncatalyzed reaction, the anhydride ring is opened by reaction with a hydroxyl group, but the reaction is much slower. The hydroxyl group also reacts with the epoxy group.

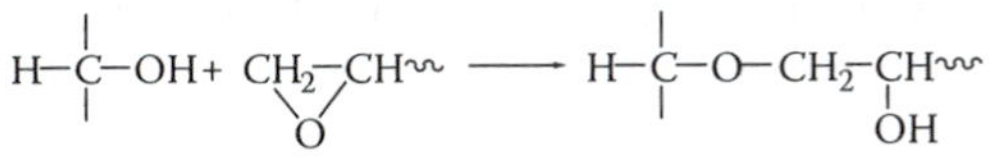

Other structures can be generated, and an important group consists of the epoxy novolac resins, which find use as molding compounds.

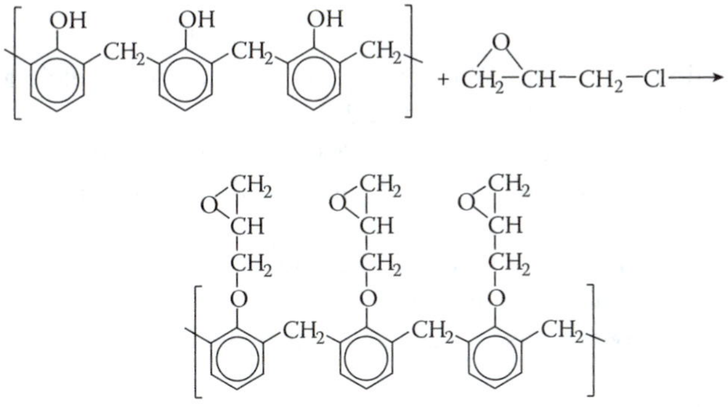

Epoxide novolac

The high-cross-link densities that can be achieved in these systems make the resulting materials mechanically stable and resistant to high temperatures.

Flame-retardant resins can be made using tetrabromobisphenol A as part of the prepolymer. When degraded at high temperatures, this compound releases halogen compounds, which trap the free radicals formed, thereby helping to quench the flame.

Epoxy resins are used in tool making, as adhesives, insulators, tough surface coatings and, importantly, in composites in combination with reinforcing fibers. They have also found extensive use in the encapsulation of electronic devices.

PROBLEMS

1. A list of monomers capable of undergoing condensation reactions is given in the following text. For each, indicate the functionality. Using different combinations, write as many polymeric structures as possible. Indicate whether a linear chain or a network is formed.

$$HO-(CH_2)_2-OH \qquad HO-(CH_2)_2-\overset{\overset{\displaystyle O}{\|}}{C}-OH \qquad \begin{array}{l} CH_2-OH \\ | \\ CH-OH \\ | \\ CH_2-OH \end{array}$$

(a)   (b)   (c)

$$HO-\overset{\overset{\displaystyle O}{\|}}{C}-(CH_2)_4-\overset{\overset{\displaystyle O}{\|}}{C}-OH \qquad CH_3COOH \qquad CH(COOH)_3$$

(d)   (e)   (f)

2. The condensation reaction between maleic anhydride and propylene glycol leads to a polyester. Draw the structure of the polymer that is formed.

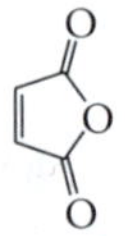

maleic anhydride

What is the architecture of the polymer produced after addition and polymerization of styrene? What change do you expect if maleic anhydride were to be replaced by sebacic acid?

3. Draw the structure of the polymers obtained from the monomer or monomers listed in the following text, and classify the nature of the linkage between monomer units.
   a. Ethylenediamine and terephthalic acid
   b. Sebacic acid and 1,6-hexanediol
   c. Hexamethylene diisocyanate and 2-methyl-1,3-propanediol
   d. Glycine

4. Draw the repeat units of the two polymers obtained by the polymerization of the following two monomers:

$$HO-(CH_2)_2-\overset{\overset{\displaystyle O}{\|}}{C}-OH \qquad H_2N-(CH_2)_5-\overset{\overset{\displaystyle O}{\|}}{C}-OH$$

Which of the two would you expect to exhibit the higher melting point and why?

5. A nylon-6 sample prepared by ring-opening polymerization of caprolactam was analyzed to determine its molar mass. The acid- and amine-group concentrations were obtained by end-group analysis (i.e., titration of the acid

groups with KOH/ benzyl alcohol to neutral point and potentiometric titration of the amine groups) (Oshinski et al., 1996). Draw the structure of the polymer, and calculate the number average molar mass, given that [COOH] $= 60.8\ \mu$eq g^{-1} and [NH$_2$] $= 59.0\ \mu$eq g^{-1}. Do the same for a different nylon 6 sample with [COOH] $= 50.7\ \mu$eq g^{-1} and [NH$_2$] $= 35.8\ \mu$eq g^{-1}. Comment on the difference between these two samples and suggest reasons for this.

6. Titration of polymers containing acidic groups with alcoholic (water or methanol) KOH to a phenolphthalein end point gives the acid value AV.

$$AV = \text{mg of } KOH\ /\ \text{g of polymer}$$

The equivalent weight is then

$$EW = \frac{56100\ \text{mg } KOH\ /\ \text{g}}{AV} = \frac{M_n}{functionality}$$

Determine the acid value and M_n of the polyester prepared from 124 g (I), 104 g of (II), and 460 g of (III).

HO—(CH$_2$)$_2$—OH   HO–CH$_2$C(CH$_3$)$_2$CH$_2$–OH   HO–$\overset{\overset{\text{O}}{\|}}{\text{C}}$–(CH$_2$)$_4$-$\overset{\overset{\text{O}}{\|}}{\text{C}}$–OH

(I)   (II)   (III)

7. The following data were obtained for the uncatalyzed equimolar esterification (Hamann et al.,1968) of lauryl alcohol ($b = 0.200$ mol) with lauric acid ($a = 0.200$ mol) in lauryl laurate at 163°C.

Time (Min)	Percentage Reaction
0	0
30	5.48
60	9.82
120	18.1
180	23.8
240	27.1
300	32.4
360	35.2
420	38.0
480	40.7
540	42.3
600	44.6
720	48.0
840	50.1
960	53.2
1080	54.8
1200	57.5
1380	59.8
1560	61.7

Determine the order of the reaction, given that for equimolar esterifications and $n > 1$ the general rate formula is

$$(n - 1)kt = \frac{1}{(a - x)^{n-1}} - \frac{1}{a^{n-1}}$$

where a is the initial concentration of acid groups, equal to the initial concentration of hydroxyl groups, x is the concentration of ester groups at time t, and k is the rate constant.

8. For the polyesterification reaction between diethylene glycol and adipic acid, calculate the number-average molecular weight of the polyester that is formed when the extent of reaction p is equal to 0.90. Note that, for condensation polymers that are synthesized from two reactants, half the average-molar-mass of the repeat unit is used to calculate the degree of polymerization.

9. Kinetic data for the uncatalyzed equimolar polyesterification (Hamann et al., 1968) of 1,10-decamethylene glycol and adipic acid at 161°C and 190°C are tabulated below. In this experiment, time and percentage of reaction were measured using, as a starting point, an equimolar mixture of the acid and alcohol dissolved in their ester. Thus, the extent of reaction at $t = 0$ was equal to 82%.

T = 161°C		T = 190°C	
Time (min)	**Percentage Reaction**	**Time (min)**	**Percentage Reaction**
0	0	0	0
20	9.1	30	20.6
40	16.0	60	39.0
100	31.6	90	50.2
150	41.1	150	61.2
210	47.9	225	66.8
270	52.5	300	71.5
330	57.0	370	74.4
390	60.0	465	77.2
450	62.6	510	78.2
510	64.6	550	78.8
550	65.5	600	79.6
700	69.2	660	80.6
840	71.9	730	81.7
880	72.4	800	82.5
1060	74.8		
1200	76.2		
1320	77.2		

a. By assuming that $[COOH]_0$ is equal to 1.25 mol per kg of mixture, determine the rate constants and activation energy of the uncatalyzed reaction.

    b. Compare results for the polyesterification reaction in the preceding text with those obtained in Problem 7 for the uncatalyzed equimolar esterification of lauryl alcohol with lauric acid. Which conclusion can you draw on the dependence of reactivity on molecular size to which groups are attached?

    c. By comparing the kinetic data reported in the preceding text to those of Figure 2.4(a) for the polyesterification of adipic acid with ethylene glycol, discuss reasons why the preceding reaction was followed starting from $p = 0.82$.

10. The differential rate law for the catalyzed esterification is given by:

$$\frac{-d[A]}{dt} = k_c[A][B] = \frac{-dx}{dt}$$

Find the integrated rate law when the initial concentrations of A and B groups, $[A]_o$ and $[B]_o$, are different, i.e., $[A]_o \neq [B]_o$.

11. The second-order rate constant for the catalyzed polyesterification of 12-hydroxystearic acid at 133.5°C is 9.23×10^{-5} mol^{-1} L sec^{-1}. Using an initial concentration of ester groups equal to 3.5 mol L^{-1}, determine the time required for p to reach values of 0.5, 0.7, and 0.9. Determine the number- and weight-average molecular weights corresponding to these p values. The activation energy for this reaction is reported to be 47.9 kJ mol^{-1}. Find the reaction constant at 160°C. How would this value change if the concentration of catalyst employed in the reaction (0.0354 mol l^{-1}, in this case) was to double?

12. The graph of Figure 2.5 shows that N_x/N is larger for $x = 100$ at $p = 0.9875$ than for a degree of conversion of either 0.9600 or 0.9950. At any given extent of reaction p, there is a distribution of chain sizes. The maximum probability of a given chain-length occurring depends on the extent of reaction. Show that the relationship between the extent of reaction at which the probability of occurrence of x-mers is at a maximum, p_{max} and x is

$$p_{max} = \frac{(x-1)}{(x+1)}$$

13. Consider a mixture of 1 mol of 1,6-hexane diol reacting with 1.0 mol of adipic acid.

    a. By neglecting end-group contributions, determine the extent of reaction p that is necessary to achieve $M_n = 5700$.

    b. If a 1% excess of diol is used, what would the M_n value be if the reaction was to be carried out to the same extent of reaction p as in (a)? What is the p value now required to achieve the same M_n as in (a)?

14. Nylon-6,6 is prepared by reacting 1,6-diaminohexane and adipic acid. Determine the amount of acetic acid that should be added to the reaction

mixture to produce a polymer of number average molar mass equal to 10,000 at 99.5% conversion.

15. Consider the condensation reaction taking place between 1.5 mol of phthalic acid, 0.2 mol of glycerol, and 1.2 mol of ethylene glycol.

    a. Use the modified Carothers equation to determine the number-average degree of polymerization at $p = 0.950$ and compare this value with x_n obtained if glycerol were to be replaced by an equivalent amount of ethylene glycol to provide the same number of functional groups.

    b. Calculate the extent of reaction at which gelation is expected to occur on the basis of either the modified Carothers equation or the Flory statistical theory.

REFERENCES

Flory, P.J., *J. Am. Chem. Soc.,* 61, 3334, 1939; 62, 2261, 1940; 63, 3083, 1941.

Hamann, S.D., Solomon, D.H., and Swift, J.D., *J. Macromol. Sci. — Chem.,* A2, 153, 1968.

Oshinski, A.J., Keskkula, H., and Paul, D.R., *Polymer,* 37, 4891, 1996.

Pearson, R.G., in *Speciality Polymers,* Dyson, R.W. (Ed.), Blackie, 1987.

BIBLIOGRAPHY

Allen, G. and Bevington, J.C., Eds., *Comprehensive Polymer Science*, Vol. 5, Pergamon Press, 1989.

Flory, P.J., *Principles of Polymer Chemistry*, Cornell University Press, 1953, chap. 3.

Frazer, A.H., *High Temperature Resistant Polymers*, Interscience Publishers, 1968.

Lenz, R.W., *Organic Chemistry of Synthetic High Polymers*, Interscience Publishers, 1967, chap. 3.

Mark, H.F. and Whitby, G.S., *The Collected Papers of Wallace Hume Carothers*, Interscience Publishers, 1940.

Nielson, P.O., Properties of Epoxy Resins, Hardeners and Modifiers, *Adhesives Age*, Vol. 42, 1982.

Odian, G., *Principles of Polymerization*, 4th ed., John Wiley and Sons, 2004, chap. 2.

Oertel, G., Ed., *Polyurethane Handbook*, 2nd ed., Hanser Gardner Publications, Munich, 1994.

Potter, W.G., *Epoxide Resins*, Butterworth, 1970.

Rempp, P. and Merrill, E.W., *Polymer Synthesis*, 2nd ed., Hüthig and Wepf, 1991.

Rogers, M.E. and Long, T.E., *Synthetic Methods in Step Growth Polymers*, John Wiley and Sons, 2003.

Solomon, D.H., *The Chemistry of Organic Film Formers*, John Wiley and Sons, 1967.

3 Free-Radical Addition Polymerization

3.1 ADDITION POLYMERIZATION

In step-growth polymerization reactions it is often necessary to use multifunctional monomers if polymers with high molar masses are to be formed; this is not the case when addition reactions are employed. Long chains are readily obtained from monomers such as vinylidene compounds with the general structure $CH_2=CR_1R_2$. These are bifunctional units, where the special reactivity of π-bonds in the carbon to carbon double bond makes them susceptible to rearrangement if activated by free-radical or ionic initiators. The active center created by this reaction then propagates a kinetic chain, which leads to the formation of a single macromolecule whose growth is stopped when the active center is neutralized by a termination reaction. The complete polymerization proceeds in three distinct stages: (1) *initiation*, when the active center which acts as a chain carrier is created; (2) *propagation*, involving growth of the macromolecular chain by a kinetic chain mechanism and characterized by a long sequence of identical events, namely the repeated addition of a monomer to the growing chain; and (3) *termination*, whereby the kinetic chain is brought to a halt by the neutralization or transfer of the active center. Typically the polymer formed has the same chemical composition as the monomer, i.e., each unit in the chain is a complete monomer, and not a residue as in most step-growth reactions.

3.2 CHOICE OF INITIATORS

A variety of chain initiators is available to the polymer chemist. These fall into three general categories: *free radical*, *cationic*, and *anionic*. The choice of the most appropriate one depends largely on the groups R_1 and R_2 in the monomer and their effects on the double bond. This arises from the ability of the alkene π-bond to react in a different way with each initiator species to produce either heterolytic (I) or

$$\overset{+}{\underset{|}{\overset{|}{C}}}-\overset{-}{\underset{|}{\overset{|}{C}}} \rightleftharpoons \overset{|}{\underset{|}{C}}=\overset{|}{\underset{|}{C}} \rightleftharpoons \overset{|}{\underset{|}{\overset{|}{C}}}-\overset{|}{\underset{|}{\overset{|}{C}}}\cdot$$

$$\text{I} \qquad\qquad\qquad \text{II}$$

homolytic (II) fission. In most olefinic monomers of interest the group R_1 is either H or CH_3, and for simplicity we can consider it to be H. The group R_2 is then classifiable as an electron-withdrawing group

$$\overset{\delta+}{H_2C=CH}\overset{\delta-}{\longrightarrow R_2}$$

or an electron-donating group

$$\overset{\delta-}{H_2C=CH}\overset{\delta+}{\longleftarrow R_2}$$

Both alter the negativity of the π-bond electron cloud and thereby determine whether a radical, an anion, or a cation will be stabilized preferentially.

In general, electron withdrawing substituents, $-CN, -COOR, -CONH_2$, reduce the electron density at the double bond and favor propagation by an anionic species. Groups that tend to increase the double-bond nucleophilicity by donating electrons, such as alkenyl, alkoxyl, and phenyl, encourage attack by cationic initiators and, in addition, the active centers formed are resonance stabilized. Alkyl groups do not stimulate cationic initiation unless in the form of 1,1-dialkyl monomers or alkyl dienes, and then heterogeneous catalysts are necessary. As resonance stabilization of the active center is an important factor, monomers like styrene and 1,3-butadiene can undergo polymerization by both ionic methods because the anionic species can also be stabilized.

Because of its electrical neutrality, the free radical is a less selective and more generally useful initiator. Most substituents can provide some resonance stabilization for this propagating species. Some examples are shown in Table 3.1.

3.3 FREE-RADICAL POLYMERIZATION

A free radical is an atomic or molecular species whose normal bonding system has been modified such that an unpaired electron remains associated with the new structure. The radical is capable of reacting with an olefinic monomer to generate a

TABLE 3.1
Effect of Substituent on Choice of Initiator

Monomer	Initiator		
	Free Radical	Anionic	Cationic
Ethylene, $H_2C=CH_2$	+	−	+
1,1'-Dialkylolefin, $H_2C=CR_1R_2$	−	−	+
Vinyl ethers, $H_2C=CHOR$	−	−	+
Vinyl halides, $H_2C=CH(Hal)$	+	−	−
Vinyl esters, $H_2C=CHOCOR$	+	−	−
Methacrylic esters, $H_2C=C(CH_3)COOR$	+	+	−
Acrylonitrile, $H_2C=CHCN$	+	+	−
Styrene, $H_2C=CHPh$	+	+	+
1,3-Butadiene, $H_2C=CH–CH=CH_2$	+	+	+

chain carrier that can retain its activity long enough to propagate a macromolecular chain under the appropriate conditions.

$$R^{\bullet} + H_2C{=}CHR_1 \longrightarrow RCH_2CHR_1^{\bullet}$$

3.4 INITIATORS

An effective initiator is a molecule that, when subjected to heat, electromagnetic radiation, or chemical reaction, will readily undergo homolytic fission into radicals of greater reactivity than the monomer radical. These radicals must also be stable long enough to react with a monomer and create an active center. Particularly useful for kinetic studies are compounds containing an azonitrile group, as the decomposition kinetics are normally first order, and the rates are unaffected by the solvent environment.

Typical radical-producing reactions are the following:

1. *Thermal decomposition* can be usefully applied to organic peroxides or azo compounds, e.g., benzoyl peroxide when heated eventually forms two phenyl radicals with loss of CO_2:

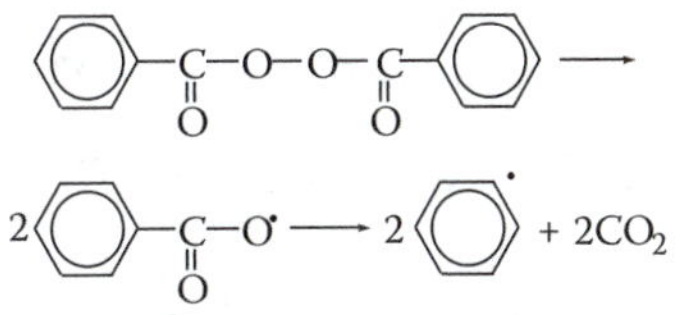

   A simpler one-stage decomposition is obtained when dicumyl peroxide is used:

$$C_6H_5{-}C(CH_3)_2{-}O{-}O{-}(CH_3)_2C{-}C_6H_5 \longrightarrow 2C_6H_5{-}C(CH_3)_2O^{\bullet}$$

2. *Photolysis* is applicable to metal iodides, metal alkyls, and azo compounds, e.g., α,α'-azobisisobutyronitrile (AIBN) is decomposed by radiation with a wavelength of 360 nm.

   Note that in each reaction there are two radicals $R^{\bullet}$ produced from one initiator molecule I; in general:

$$I \longrightarrow 2R^{\bullet} \tag{3.1}$$

3. *Redox reactions,* e.g., the reaction between the ferrous ion and hydrogen peroxide in solution produces hydroxyl radicals,

$$H_2O_2 + Fe^{2+} \longrightarrow Fe^{3+} + OH^- + OH^\bullet$$

Alkyl hydroperoxides may be used in place of H_2O_2. A similar reaction is observed when cerium(IV) sulfate oxidizes an alcohol:

$$RCH_2OH + Ce^{4+} \longrightarrow Ce^{3+} + H^+ + RC(OH)H^\bullet$$

4. *Persulfates* are useful in emulsion polymerizations where decomposition occurs in the aqueous phase, and the radical diffuses into a hydrophobic, monomer containing, droplet.

$$S_2O_8^{2-} \longrightarrow 2SO_4^{\bullet-}$$

5. *Ionizing radiation*, such as α, -β-, γ-, or x-rays, may be used to initiate a polymerization by causing the ejection of an electron, followed by dissociation and electron capture to produce a radical.

$$\text{Ejection: } C \rightsquigarrow C^\bullet + e^-$$

$$\text{Dissociation: } C^+ \longrightarrow A^\bullet + Q^\bullet$$

$$e^- \text{-Capture: } Q^+ + e^- \longrightarrow Q^\bullet$$

Initiators that undergo thermal decomposition must be selected to ensure that at the polymerization temperature they provide a suitable source of free radicals that will sustain the reaction. For an initiator concentration of 0.1 M, a useful guide is that the rate of radical production should be 10^{-6} to 10^{-7} mol dm^{-3} s^{-1}, which is equivalent to a $k_d \approx 10^{-5}$ to 10^{-6} s^{-1} in the temperature range 320–420 K, normally regarded as the most suitable for radical polymerizations. The rate equations for several commonly used initiators are shown in Table 3.2, and the operating temperatures are calculated on this basis. The half-lives are also shown in Figure 3.1, plotted as a function of temperature.

Reference to these data provides the chemist with the information needed to make the correct selection for the chosen polymerization conditions. It should be noted, however, that the rate equations for the decomposition of the initiator molecules can depend on the solvent, and the values shown in Table 3.2 should be regarded as good approximations.

3.4.1 INITIATOR EFFICIENCY

Although the decomposition of an initiator molecule can be quantitative, chain initiation may be less than 100% efficient. In a kinetic analysis the effective radical concentration is represented by an *efficiency factor f*, which is less than unity when only a proportion of the radicals generated are effective in the creation of a kinetic chain. Inefficient chain propagation may be due to several side reactions.

TABLE 3.2
Some Radical Initiator Decomposition Rate Equations and the
Corresponding Suggested Temperature Range for Use

Initiator	Rate Equation (s⁻¹)	Temperature Range (K)
RC(O)OO(O)CR		
R = Et	$k = 10^{14} \exp(-146 \text{ kJ}/RT)$	382–402
R = Bu^t	$k = 6.3 \times 10^{15} \exp(-157 \text{ kJ}/RT)$	377–395
RN=NR		
R = Me$_2$C(CN)	$k = 1.8 \times 10^{15} \exp(-128.7 \text{ kJ}/RT)$	310–340
R = PhCHMe	$k = 1.3 \times 10^{15} \exp(-152.6 \text{ kJ}/RT)$	378–398
R = Me$_2$CH	$k = 5 \times 10^{13} \exp(-170.5 \text{ kJ}/RT)$	453–473

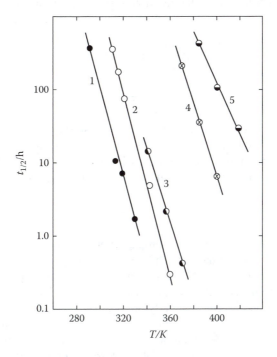

FIGURE 3.1 Half-lives $t_{1/2}$ of selected peroxide initiators. Curve **1,** isopropyl percarbonate; curve **2,** 2,2′-azo-bis-isobutyronitrile; curve **3,** benzoyl peroxide; curve **4,** di-tertiary butyl peroxide; curve **5,** cumene hydroperoxide.

Primary recombination can occur if the diffusion of radical fragments is impeded in the solution and a cage effect leads to

$$2(CH_3)_2\overset{\cdot}{\underset{CN}{C}} \longrightarrow (CH_3)_2\underset{CN}{C}-\underset{CN}{C}(CH_3)_2$$

The solvent usually plays an important part, and the extent of the decomposition of benzoyl peroxide is limited to 35% in tetrachloroethylene, 50% in benzene, and 85% in ethyl acetate, after refluxing for 4 h.

Further wastage occurs when induced decomposition is effected by the attack of an active center

$$R^\bullet + R'-O-O-R' \longrightarrow ROR' + R'O^\bullet$$

This effectively produces one radical instead of the three potential radical species.

For a 100% efficient initiator $f = 1$, but most initiators have efficiencies in the range 0.3 to 0.8.

3.5 CHAIN GROWTH

A chain carrier is formed from the reaction of the free radical and a monomer unit; chain propagation then proceeds rapidly by addition to produce a linear polymer.

$$RM_1^\bullet + M_1 \rightarrow RM_2^\bullet \tag{3.2}$$

$$RM_n^\bullet + M_1 \rightarrow RM_{n+1}^\bullet \tag{3.3}$$

The average life time of the growing chain is short, but a chain of over 1000 units can be produced in 10^{-2} to 10^{-3} s. Bamford and Dewar have estimated that the thermal polymerization of styrene at 373 K leads to chains of $x = 1650$ in approximately 1.24 s, i.e., a monomer adds on once in every 0.75 ms.

3.6 TERMINATION

In theory, the chain could continue to propagate until all the monomer in the system has been consumed, but for the fact that free radicals are particularly reactive species and interact as quickly as possible to form inactive covalent bonds. This means that short chains are produced if the radical concentration is high because the probability of radical interaction is correspondingly high, and the radical concentration should be kept small if long chains are required. Termination of chains can take place in several ways by: (1) the interaction of two active chain ends, (2) the reaction of an active chain end with an initiator radical, (3) termination by transfer of the active center to another molecule which may be solvent, initiator, or monomer, or (4) interaction with impurities (e.g., oxygen) or inhibitors.

The most important termination reaction is the first, a biomolecular interaction between two chain ends. Two routes are possible.

1. *Combination,* where two chain ends couple together to form one long chain.

$$\sim\!\!\sim\!CH_2\!-\!\overset{\bullet}{C}H + H\overset{\bullet}{C}\!-\!CH_2\!\!\sim\!\!\sim \longrightarrow \sim\!\!\sim\!CH_2\!-\!CH\!-\!CH\!-\!CH_2\!\!\sim$$
$$\qquad\quad \underset{Cl}{|} \qquad \underset{Cl}{|} \qquad\qquad\qquad \underset{Cl}{|} \quad \underset{Cl}{|}$$

2. *Disproportionation*, with hydrogen abstraction from one end to give an unsaturated group and two dead polymer chains.

$$
\underset{\substack{|\\ \text{COOCH}_3}}{\text{~CH}_2\text{-}\overset{\substack{\text{CH}_3\\|}}{\text{C}}^{\bullet}} \;+\; \underset{\substack{|\\ \text{COOCH}_3}}{\overset{\substack{\text{CH}_3\\|}}{\text{C}}\text{-CH}_2\text{~}} \longrightarrow \underset{\substack{|\\ \text{COOCH}_3}}{\text{~CH}\!=\!\overset{\substack{\text{CH}_3\\|}}{\text{C}}} \;+\; \underset{\substack{|\\ \text{COOCH}_3}}{\overset{\substack{\text{CH}_3\\|}}{\text{CH}}\text{-CH}_2\text{~}}
$$

One or both processes may be active in any system depending on the monomer and polymerizing conditions. Experimental evidence suggests that polystyrene terminates predominantly by combination, whereas poly(methyl methacrylate) terminates mainly by disproportionation when the reaction is above 333 K but by both mechanisms below this temperature (see Table 3.3). The mechanism can be determined by measuring the number of initiator fragments per chain using a radioactive initiator. One fragment per chain is counted when disproportionation is operative and two when combination occurs. Alternatively, the number-average molar mass of the product can be measured.

3.7 STEADY-STATE KINETICS

The three basic steps in the polymerization process can be expressed in general terms as follows:

Initiation is a two-stage reaction. Initiator decomposition

$$
I \xrightarrow{k_d} 2R^{\bullet}, \tag{3.4}
$$

followed by radical attack on a monomer unit to form a chain carrier,

$$
R^{\bullet} + M \xrightarrow{k_i} RM^{\bullet} \tag{3.5}
$$

As the initial decomposition is slow compared with both the rate of addition of a primary radical to a monomer and the termination reaction, it is the rate

---

TABLE 3.3
Termination Mechanisms for Polymer Radicals

Monomer	Temperature	Mechanism
Styrene	330–370	Combination
Acrylonitrile	330	Combination
Methyl methacrylate	273	Mainly combination
Methyl methacrylate	>330	Mainly disproportionation
Methyl acrylate	360	Mainly disproportionation
Vinyl acetate	360	Mainly disproportionation

determining step. The rate of initiation v_i is then the rate of production of chain radicals

$$v_i = d[RM^\bullet]/dt = 2k_d f [I], \tag{3.6}$$

where the factor 2 is introduced because two potentially effective radicals are produced in the decomposition; f also measures the ability of these to propagate chains. This expression is valid for a thermo-initiation, but many reactions can be photo-initiated when the monomer absorbs radiation and acts as its own initiator. The rate v_{ip} is then dependent on the intensity of light absorbed

$$v_{ip} = 2\phi I_a. \tag{3.7}$$

The quantum yield ϕ replaces f and defines the initiator efficiency; I_a is related to the incident light intensity I_o, the monomer concentration, and the extinction coefficient ε, so that

$$v_{ip} = 2\phi\varepsilon I_o[M]. \tag{3.8}$$

When the monomer is a poor absorber of radiation, small quantities of a photosensitizer may be added to absorb the energy and then transfer this to the monomer to create an active center. In this case, [M] is replaced by the concentration of photosensitizer.

Propagation is the addition of a monomer to the growing radical

$$RM_n^\bullet + M_1 \xrightarrow{k_p} RM_{n+1}^\bullet \tag{3.9}$$

The rate of bimolecular propagation is assumed to be the same for each step so that

$$v_p = k_p[M][M^\bullet], \tag{3.10}$$

where [M$^\bullet$] represents the concentration of growing ends and [M$^\bullet$] is usually low at any particular time. The reaction is essentially the conversion of a monomer to a polymer and can be followed from the rate of disappearance of monomer.

Termination is also a bimolecular process depending only on [M$^\bullet$]; v_t for both mechanisms is

$$v_t = 2k_t[M^\bullet][M^\bullet]. \tag{3.11}$$

The rate constant k_t is actually $(k_{tc} + k_{td})$ where the two mechanisms, combination and disproportionation, are possible but will be written as k_t for convenience. If the chain reaction does not lead to an explosion, a steady state is reached where the rate of radical formation is exactly counterbalanced by the rate of destruction, i.e., $v_i = v_t$, and for a thermal reaction

$$2k_t[M^\bullet]^2 = 2k_d f [I] \tag{3.12}$$

From this an expression for [M•] is obtained

$$[M^\bullet] = \{f\,k_d\,[I]\,/k_t\}^{1/2}, \tag{3.13}$$

and because the concentration of radicals is usually too low to be determined accurately, it is replaced in the kinetic expression. The overall rate of polymerization is then

$$v_p = k_p\,\{fk_d[I]/k_t\}^{1/2}[M], \tag{3.14}$$

which shows that the rate is proportional to the monomer concentration and to $[I]^{1/2}$ if f is high, but for a low efficiency initiator f becomes a function of [M], and the rate is then proportional to $[M]^{3/2}$.

By analogy the rate of a photo-polymerization becomes

$$v_{pp} = k_{pp}\,\{\phi\,\varepsilon\,I_o/k_t\}^{1/2}[M]^{3/2}. \tag{3.15}$$

Two parameters of interest can be derived from this analysis, the kinetic chain length $\bar{v}$ and the average degree of polymerization x.

The kinetic chain length $\bar{v}$ is a measure of the average number of monomer units reacting with an active center during its lifetime and is related to x_n through the mechanism of the termination. Thus, combination means $x_n = 2\bar{v}$, but $x_n = \bar{v}$ if disproportionation is the only termination reaction. Under steady-state conditions

$$\bar{v} = v_p/v_i = v_p/v_t = k_p^2[M]^2/2k_tv_p. \tag{3.16}$$

This means that as $\bar{v}$ is inversely proportional to the rate of polymerization, an increase in temperature produces an increase in v_p and a corresponding decrease in chain length. The equation shows that $\bar{v}$ is inversely proportional to the radical concentration, hence the chain length is short for high radical concentrations and vice versa. This means that a certain degree of control over the polymer chain length can be exercised by altering the initiator concentration.

3.8 HIGH-CONVERSION BULK POLYMERIZATIONS

In many polymerizations, a marked increase in rate is observed toward the end of the reaction instead of the expected gradual decrease caused by the depletion of the monomer and initiator. This *auto-acceleration* is a direct result of the increased viscosity of the medium, and the effect is most dramatic when polymerizations are carried out in the bulk phase or in concentrated solutions. The phenomenon, sometimes known as the *Trommsdorff-Norrish* or *gel effect*, is caused by the loss of the steady state in the polymerization kinetics.

When the viscosity of the reaction medium increases, the different steps in the polymerization reaction become diffusion controlled at stages that depend on the activation energy for that step. The initiation and propagation reactions have higher activation energies relative to that of the termination step, and so only become diffusion controlled in the latter stages of the reaction. The termination step has the lowest activation energy and becomes diffusion controlled at lower viscosities (earlier in the reaction). It is also a bimolecular process requiring fruitful collision between two radical ends attached to long, highly entangled polymer chains, and with increase in the viscosity of the medium, this chain movement becomes progressively more hindered, with the results that the active chain ends have greater difficulty finding one another. This leads to a significant reduction in the rate of termination and, as this is the main process responsible for removing radicals in the polymerization, the overall radical concentration gradually increases. A consequence of this is an increase in the number of propagation steps with the release of more heat. This speeds up the rate of decomposition of the initiator, thereby producing even more radicals and resulting in auto-acceleration. A major hazard of this sequence of events is that if sufficient energy is released that is not dissipated rapidly and efficiently, then an explosion may occur.

The simple classical kinetic analysis outlined in Section 3.7 is an oversimplified model based on the assumptions that (1) steady-state conditions prevail throughout the reaction at all conversions, (2) radical reactivity is independent of chain length, (3) no chain transfer occurs, and (4) there is no auto-acceleration or gel effect. A schematic diagram of the rate of a typical bulk polymerization is presented in Figure 3.2(a) as a function of percentage conversion. The rate rises initially to the steady-state condition at about 0.1% conversion and obeys the steady-state approximation until about 10% conversion. It then dips through a minimum before the gel effect begins to manifest itself by a rapid acceleration in rate. If the polymer has a glass transition temperature that is higher than the polymerization temperature, then as the remaining monomer no longer depresses the polymer glass transition below the reaction temperature, the mixture becomes glassy and the rate falls. This is illustrated

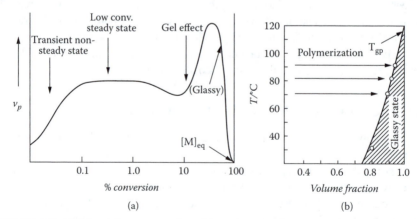

FIGURE 3.2 (a) Schematic representation of the polymerization rate vs. conversion. (b) Schematic representation of the vitrification points for a bulk polymerization of poly(methyl methacrylate) at various polymerization temperatures. The solid line represents the volume fraction of polymer at which polymerization stops, as a function of polymerization temperature.

in Figure 3.2(b) where the polymerization of poly(methyl methacrylate) in the bulk ceases when the glass transition temperature is no longer depressed by the plasticizing effect of the monomer below the polymerization reaction temperature, and vitrification occurs.

Careful experimental work has shown that the rate constant for termination is dependent on the chain length when $x_n \leq 100$, but is essentially independent of chain length for $x_n > 1000$. Close analysis of this effect suggests that k_t is then a function of coil size and will be large when the polymer coil is small or tightly bound. The consequence of this is that polymerization will be slower in a thermodynamically poor solvent than in a good one, and as the reaction progresses it ought to slow down not only because of monomer depletion but also because the reaction medium will become progressively less favorable for the polymer. However, this effect may not be noticed.

O'Driscoll has proposed that the auto-acceleration can be modeled by recognizing that the termination reaction is diffusion controlled but will also depend on the size of the chain involved. The critical chain length for entanglement n_c then becomes an important parameter, and two termination rate constants can also be defined, one for chains smaller than n_c and one for large entangled chains. These are, respectively, k_t and k_{te}. If $\bar{v}$ is the kinetic chain length and v_p is the conventional steady-state polymerization, then the observed rate v_p^* is given by

$$\frac{v_p^*}{v_p} - 1 = \left[\left(\frac{k_t}{k_{te}} \right)^{1/2} - 1 \right] \exp(-n_c \bar{v}) \qquad (3.17)$$

Good agreement between the predicted rate from Equation 3.17 and the experimental values obtained for the bulk polymerization of methyl methacrylate has been reported.

The auto-acceleration can be avoided by performing the polymerization in more dilute solutions or by stopping the reaction before the diffusion effect grows to noticeable proportions. However, polymerizations in solvents can be influenced by the choice of medium, and this feature is treated in the chain transfer reactions.

3.9 CHAIN TRANSFER

Termination in a free-radical polymerization normally occurs by collision between two active centers attached to polymer chains, but the chain length of the product in many systems is lower than one would expect if this was the mechanism solely responsible for limiting the kinetic chain length $\bar{v}$. Usually x_n will lie within the expected limits of $\bar{v}$ (disproportionation) and $2\bar{v}$ (combination), but not always, and Flory (1953) found that attenuation of chain growth takes place if there is premature termination of the propagating chain by a transfer of activity to another species through a collision. This is a competitive process involving the abstraction of an atom by a chain carrier from an inactive molecule XY with replaceable atoms and is dependent on the strength of the X–Y bond.

$$\sim\!M_m^{\bullet} + XY \longrightarrow \sim\!M_mX + Y^{\bullet}$$

It is important to note that the free radical is not destroyed in the reaction; it is merely transferred, and if the new species is sufficiently active, another chain will emanate from the new center. This is known as *chain transfer* and is a reaction resulting in the exchange of an active center between molecules during a bimolecular collision. Several types of chain transfer have been identified.

Transfer to monomer. The two important reactions in this group both involve hydrogen abstraction. Two competitive alternatives exist in the first group

$$R^{\bullet} + H_2C{=}CHX \overset{\displaystyle RH + H_2C{=}C^{\bullet}X \quad (I)}{\underset{\displaystyle RH_2CHX^{\bullet} \quad (II)}{\bigg\langle}}$$

If the radical formed in reaction (II) is virtually unstabilized by resonance, then the reaction with the parent unreactive monomer may produce little chain propagation due to the tendency for stabilization to occur by removal of hydrogen from the monomer. This leads to rapid chain termination and is known as *degradative transfer*. Allylic monomers are particularly prone to this type of reaction

$$\sim\!\!\sim\!R^{\bullet} + H_2C{=}CHCH_2OCOCH_3 \longrightarrow \sim\!\!\sim\!RH + H_2\overset{\bullet}{C}{=\!\!=}CH{=\!\!=}\overset{\bullet}{C}H{-}OCOCH_3$$

where abstraction of the α-hydrogen leads to a resonance-stabilized allylic radical capable only of bimolecular combination with another allyl radical. This is effectively an auto-inhibition by the monomer. Propylene also reacts in this manner and both monomers are reluctant to polymerize by a free-radical mechanism.

A second group of transfer reactions can occur by hydrogen abstraction from the pendant group. The relevant kinetic expression is

$$v_{tr} = k_{tr}^{M}[M][M^{\bullet}] \tag{3.18}$$

Transfer to initiator. Organic peroxides, when used as initiators, are particularly susceptible to chain transfer. Azo initiators are not vulnerable in this respect and are more useful when a kinetic analysis is required. For peroxides

$$v_{tr} = k_{tr}^{I}[I][M^{\bullet}] \tag{3.19}$$

Transfer to polymer. The transfer reaction with a polymer chain leads to branching rather than initiation of a new chain so that the average molar mass is relatively unaffected. The long- and short-chain branching detected in polyethylene is believed to arise from this mode of transfer.

Transfer to modifier. Molar masses can be controlled by addition of a known and efficient chain transfer agent such as an alkyl mercaptan.

$$\sim\!CH_2CHX^\bullet + RSH \longrightarrow \sim\!CH_2CH_2X + RS^\bullet$$

$$RS^\bullet + H_2C\!=\!CHX \longrightarrow RSCH_2CHX^\bullet$$

Mercaptans are commonly used because the S—H bond is weaker and more susceptible to chain transfer than a C—H bond.

Transfer to solvent. A significant decrease in polymer chain length is often found when polymerizations are carried out in solution rather than in the undiluted state, and this variation is a function of both the extent of dilution and the type of solvent used. The effectiveness of a solvent in a transfer reaction depends largely on the amount present, the strength of the bond involved in the abstraction step, and the stability of the solvent radical formed. With the exception of fluorine, halogen atoms are easily transferred, and the reaction of styrene in CCl_4 is a good example of this chain transfer.

$$\sim\!CH_2\!-\!\underset{\underset{C_6H_5}{|}}{CH^\bullet} + CCl_4 \longrightarrow \sim\!CH_2\underset{\underset{C_6H_5}{|}}{CHCl} + CCl_3^\bullet \qquad (I)$$

$$CCl_3^\bullet + H_2C\!=\!CHC_6H_5 \longrightarrow Cl_3CCH_2\!-\!\underset{\underset{C_6H_5}{|}}{CH^\bullet} \qquad (II)$$

$$Cl_3C\!\sim\!\underset{\underset{C_6H_5}{|}}{CH^\bullet} + CCl_4 \longrightarrow Cl_3C\!\sim\!\underset{\underset{C_6H_5}{|}}{CHCl} + CCl_3^\bullet \qquad (III)$$

When the solvent is present in significant quantity, step (I) is of minor importance and the resulting polymer contains four chlorine atoms which can be detected by analysis.

Hydrogen is normally the atom abstracted and, as radical stability enhances the transfer reaction, we find that toluene, which forms a primary radical, is less efficient than ethyl benzene, which forms a secondary radical. Both are inferior to isopropyl benzene, which forms a tertiary radical. All are much better than *t*-butyl benzene, whose radical is unstable, so that virtually no chain transfer takes place in this solvent. It is interesting to note that even benzene acts as a chain transfer agent on a modest scale.

The kinetic expression is

$$v_{tr} = k_{tr}^S[S][M^\bullet] \qquad (3.20)$$

TABLE 3.4
Chain Transfer Constant of Various Agents
to Styrene at 333 K

Agent	$10^4 C_s$
Benzene	0.023
n-Heptane	0.42
sec-Butyl benzene	6.22
m-Cresol	11.0
CCl$_4$	90
CBr$_4$	22000
n-Butylmercaptan	210000

3.9.1 CONSEQUENCES OF CHAIN TRANSFER

The primary effect is a decrease in the polymer chain length, but other less obvious occurrences can be detected. If k_{tr} is much larger than k_p, then a very small polymer is formed with x_n between 2 and 5. This is known as *telomerization*. The chain reinitiation process can also be slower than the propagation reaction, and a decrease in v_p is observed. However, the influence on x_n is most important, and it can be estimated by considering all the transfer processes in a form known as the Mayo equation:

$$1/x_n = (1/x_n)_0 + C_s[S]/[M] \qquad (3.21)$$

This is a simplified form in which the main assumption is that solvent transfer predominates and all other terms are included in $(1/x_n)_0$. The chain transfer constant C_s is then (k_{tr}^s/k_p).

A plot of $1/x_n$ against $\{[S]/[M]\}$ for a variety of agents is shown in Figure 3.3. The slope is a measure of C_s and the intercept is $(1/x_n)_0$. If the activation energy of the process is required, log C_s can be plotted against $1/T$.

3.10 INHIBITORS AND RETARDERS

Chain transfer agents can lower the average chain length and in extreme cases, when used in large proportions, may lead to the formation of telomers.

Some chain transfer agents yield radicals with low activity and if the reinitiation reaction is slow, the polymerization rate decreases because there is a buildup of radicals leading to increased termination by coupling. When this happens the substance responsible is said to be a *retarder*, e.g., nitrobenzene acts in this way with styrene.

In extreme cases an added reagent may suppress polymerization completely by reacting with the initiating radical species and converting them all efficiently into unreactive substances. This is known as *inhibition*, but the difference between an inhibitor and a retarder is merely in the degree of efficiency.

The phenomena are typified by the reaction of styrene with benzoquinone, nitrobenzene, and nitrosobenzene, studied by Schulz (1947), and shown in Figure 3.4.

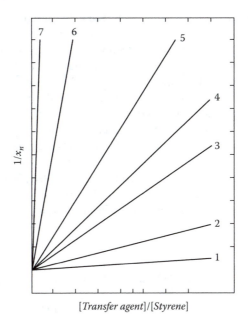

FIGURE 3.3 Effect of chain transfer to various solvents on the degree of polymerization of polystyrene at 333 K. 1, Benzene; 2, *n*-heptane; 3, *sec*-butyl benzene; 4, *m*-cresol; 5, CCl$_4$; 6, CBr$_4$; 7, *n*-butyl mercaptan.

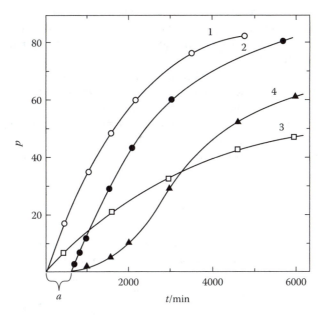

FIGURE 3.4 Polymerization of styrene at 373 K in the presence of: curve 1, no inhibitor; curve 2, 0.1% benzoquinone; curve 3, 0.5% nitrobenzene; curve 4, 0.2% nitrosobenzene. The time *t* and percentage conversion *p* are plotted.

Curve 1 represents the polymerization of styrene in the absence of any agents. When benzoquinone is added, the polymerization is completely inhibited until all the benzoquinone has been consumed and then the reaction proceeds normally (curve 2). The time interval a is the induction period and reflects the time taken for the benzoquinone to react with all the radicals formed until no more benzoquinone is left unreacted. In the presence of nitrobenzene, curve 3, the polymerization continues but at a much reduced rate. The action of nitrosobenzene is much more complex (curve 4). It acts first as an inhibitor but probably produces a substance during this period, which then acts as a retarder, and both effects are observed.

Monomers are usually transported and stored in the presence of inhibitors to prevent premature polymerization, and so must be redistilled and purified prior to use.

An excellent inhibitor is the resonance stabilized radical diphenyl picryl hydrazyl (DPPH), used extensively as a radical scavenger because the stoichiometry of the reaction is 1:1.

3.11 ACTIVATION ENERGIES AND THE EFFECT OF TEMPERATURE

The influence of temperature on the course of a polymerization reaction depends on initiator efficiency and decomposition rate, chain transfer, and chain propagation, but it is important to have some knowledge of its effect in order to formulate optimum conditions for a reaction.

The energy of activation of a polymerization reaction is easily determined using an Arrhenius plot when the rate constants have been determined at several temperatures, but even for a simple reaction, the overall rate is still a three-stage process, and the total activation energy is the sum of the three appropriate contributions for initiation, propagation, and termination.

Remembering that v_p is proportional to $k_p(k_d/k_t)^{1/2}$, the overall activation energy E_a is

$$E_a = \tfrac{1}{2} E_d + (E_p - \tfrac{1}{2} E_t). \tag{3.22}$$

The term $(E_p - \tfrac{1}{2} E_t)$ provides a measure of the energy required to polymerize a particular monomer and has been estimated for styrene to be 27.2 kJ mol^{-1} and for vinyl acetate, 19.7 kJ mol^{-1}. Initiators have values of E_d in the range 125 to 170 kJ mol^{-1}, and this highlights the controlling role of the initiation step in free-radical polymerization. Consequently, values of E_a are generally in the range 85 to 150 kJ mol^{-1}.

As the temperature term in the rate equation is $\exp\{(\tfrac{1}{2} E_t - \tfrac{1}{2} E_d - E_p)/RT\}$, the exponent will normally be negative so that the polymerization rate will increase as the temperature is raised. The change in molar mass can also be examined in this way, and now the quantity $\{k_p/(k_d k_t)^{1/2}\}$ is the one of interest. The required energy term is $\exp\{(E_p - \frac{1}{2}E_d - \frac{1}{2}E_t)/RT\}$, and in thermal polymerizations this is negative and usually about -60 kJ mol^{-1}. As the temperature increases, the chain length decreases rapidly, and only in pure photochemical reactions where E_d is zero is the activation energy slightly positive, leading to a modest increase in x_n as the temperature goes up.

3.12 THERMODYNAMICS OF RADICAL POLYMERIZATION

The conversion of an alkene to a long chain polymer has a negative enthalpy (ΔH_p is negative) because the formation of a σ-bond from a π-bond is an exothermic process. Although the enthalpy change favors the polymerization, the change in entropy is unfavorable and negative because the monomer becomes incorporated into a covalently bonded chain structure. However, examination of the relative magnitudes of the two effects shows that whereas $-\Delta S_p$ is in the range 100 to 130 J K^{-1} mol^{-1}, $-\Delta H_p$ is normally in the range 30 to 150 kJ mol^{-1}. The overall Gibbs free energy change $\Delta G_p = \Delta H_p - T\Delta S_p$ is then negative, and the polymerization is thermodynamically feasible.

These conditions favor the formation of polymer, but it is obvious from the general treatment of the energetics of the reaction that the chain length decreases as the temperature rises. This can be understood if we postulate the existence of a depolymerization reaction.

When the temperature increases, the depolymerization reaction becomes more important, and ΔG_p becomes less negative. Eventually, a temperature is reached at which $\Delta G_p = 0$, and the overall rate of polymerization is zero. This temperature is known as the *ceiling temperature* T_c.

If both the forward and reverse reactions are treated as chain reactions, then

$$\text{M}_n^\bullet + \text{M} \underset{k_{dp}}{\overset{k_p}{\rightleftharpoons}} \sim \text{M}_{n+1}^\bullet$$

where k_{dp} is the rate constant for the depropagation. The overall rate expression is then obtained by modifying Equation 3.10,

$$v_p = k_p[\text{M}^\bullet][\text{M}] - k_{dp}[\text{M}^\bullet] \tag{3.23}$$

whereas the degree of polymerization x is

$$x = (k_p[\text{M}] - k_{dp}[\text{M}^\bullet])/v_t. \tag{3.24}$$

At the ceiling temperature $v_p = 0$ and so

$$K = (k_p/k_{dp}) = 1/[\text{M}_e] \tag{3.25}$$

where $[\text{M}_e]$ is the equilibrium monomer concentration.

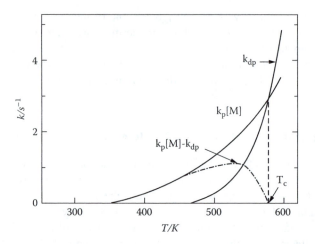

FIGURE 3.5 The temperature dependence of $k_p[M]$ and k_{dp} for styrene (From Dainton, F.S. and Ivin, K.J., *Q. Rev.*, 12, 61, 1958. With permission.)

The ceiling temperature, T_c, can be described schematically as the intersection of the propagation and depropagation rate curves (see Figure 3.5). Hence, above T_c it is impossible for any polymeric material to form.

The thermodynamic significance of T_c can be related to this kinetic analysis by using the Arrhenius expressions for the rate constants. At equilibrium

$$A_p \exp(-E_p/RT_c)[M_e] = A_{dp} \exp(-E_{dp}/RT_c) \qquad (3.26)$$

By rearranging and considering equilibrium conditions it follows that

$$T_c = \Delta H_p/(\Delta S_p^\circ + R \ln[M_e]) \qquad (3.27)$$

Here, ΔH_p is the enthalpy change for the polymerization, which equals the difference in activation energies for the forward and reverse reaction, and ΔS_p° is the standard entropy change.

Equation 3.27 shows that the ceiling temperature is a function of the free monomer concentration and that for a given monomer concentration there will be a specific ceiling temperature at which the particular $[M_e]$ will be in equilibrium with polymer chains. Thus, for any polymerization temperature an equilibrium monomer concentration can be estimated for a selected temperature of 298 K (Table 3.5).

As ΔH_p is negative, a rise in temperature will cause $[M_e]$ to increase, thus at 405 K, methyl methacrylate has a value of $[M_e] \approx 0.5$ mol dm^{-3}, whereas α-methylstyrene will not polymerize at all. Ceiling temperatures then refer to a given monomer concentration, and it is more convenient to refer it to a standard state. This can either be referred to pure liquid monomer or a concentration of 1 mol dm^{-3}; typical examples for pure liquid monomers are given in Table 3.6. Whereas the ceiling temperature alters with monomer concentration, it is also sensitive to pressure. As ΔH and ΔS are both negative, an increase in T_c is obtained if $-\Delta S$ can be decreased.

TABLE 3.5
Equilibrium Monomer Concentrations for 298 K,
Estimated from Equation 3.27

Monomer	$[M_e]$ (mol dm^{-3})
Vinyl acetate	10^{-9}
Styrene	10^{-6}
Methyl methacrylate	10^{-3}
α-Methylstyrene	2.6

TABLE 3.6
Ceiling Temperatures Based on Pure Liquid Monomer as the Standard
State and Heats of Polymerization for Selected Monomers

Monomer	T_c (K)	$-\Delta H_p{}^a$ (kJ mol^{-1})	$-\Delta H_p{}^b$ (kJ mol^{-1})
Tetrafluoroethylene	853	155.5	
Styrene	583	68.5	
Methyl methacrylate	493	58.1	56.0
Thioacetone	368		
Tetrahydrofuran	353		
α-Methylstyrene	334	35.2	34.1
Acetaldehyde	242		

ᵃ Calorimetric.
ᵇ Equation 3.27.

This can be achieved either by raising the monomer concentration (in a solution polymerization) or by decreasing the volume change normally observed during polymerization. Experimental data show that there is a linear dependence of log T_c on pressure and the Clapeyron–Clausius equation

$$\frac{dT_c}{dP} = \frac{T_c \Delta V}{\Delta H} \tag{3.28}$$

is applicable. Typical values for the rate of increase of T_c with pressure are 0.17 K MPa^{-1} for α-methylstyrene and 0.2 K MPa^{-1} for tetrahydrofuran.

It should be noted that, whereas the preceding discussion has been cast in terms of free-radical polymerizations, the thermodynamic argument is independent of the nature of the active species. Consequently, the analysis is equally valid for ionic polymerizations. A further point to note is that for the concept to apply, an active species capable of propagation and depropagation must be present. Thus, inactive polymer can be stable above the ceiling temperature for that monomer, but the polymer will degrade rapidly by a depolymerization reaction if main chain scission is stimulated above T_c.

3.13 HEATS OF POLYMERIZATION

Addition polymerization is an exothermic process, and the change in enthalpy is typically in the range 34 to 160 kJ mol^{-1}. The particular values differ for each monomer and are influenced by several factors, i.e., (1) the energy difference between monomer and polymer resulting from resonance stabilization of the double bond by the substituent or by conjugation, (2) steric strains in the polymer imposed on the new single bonds by substituent interactions, and (3) polar or secondary bonding effects.

The most important factors are (1) and (2) and while the general observation is that the higher the resonance stabilization in the monomer, the less exothermic the reaction, it is believed that steric factors have the greatest effect on ΔH_p. Thus, the unusually high steric strains generated on forming poly(α-methylstyrene) are caused by interactions between the phenyl rings and α-methyl units, and result in low values for ΔH_p and T_c. These also seem to be responsible for the facile unzipping degradation reaction. Polymerization of tetrafluoroethylene, on the other hand, is a highly exothermic reaction, and produces a polymer with little steric strain. Some values for ΔH_p that have been measured calorimetrically and can be compared favorably with data calculated from Equation 3.27 are shown in Table 3.6.

The large heat of polymerization can have serious practical consequences, especially when polymerizations are rapid, and can even lead to thermal explosions. To avoid these defects, the rate of the process must be controlled or other practical expediencies adopted. Heat removal is particularly problematic in bulk polymerizations taken to high conversions as the reaction mixtures become very viscous, and efficient stirring becomes difficult.

The generation of dangerous hot spots in the reaction may be avoided by keeping path lengths for heat loss low, by performing polymerizations in solution, or providing a heat sink by carrying out polymerizations in emulsions and dispersions where a large volume of inert liquid phase is present.

3.14 POLYMERIZATION PROCESSES

Industrial radical initiated polymerizations can be carried out in one of four different ways:

1. With monomer only — *bulk*
2. In a solvent — *solution*
3. With monomer dispersed in an aqueous phase — *suspension*
4. As an *emulsion*

Bulk polymerization is used in the production of polystyrene, poly(methylmethacrylate), and poly(vinyl chloride). The reaction mixture contains only monomer and initiator, but because the reaction is exothermic, hot spots tend to develop when heat removal is inefficient. Auto-acceleration occurs in the highly viscous medium, making control difficult and impeding efficient monomer conversion. To overcome some of the disadvantages, low conversions are used, after which the unreacted monomer is stripped off and recycled. The main advantages of the technique lie in the optical clarity of the product and its freedom from contaminations.

Quiescent mass polymerization is an unstirred reaction used for casting sheets of poly(methyl methacrylate). A low-molar-mass prepolymer is prepared, and then the main polymerization is carried out *in situ* making use of the Trommsdorff effect to obtain high-molar-mass material and tougher sheets. The two-stage approach helps to control the heat evolved.

In *solution polymerization*, the presence of the solvent facilitates heat transfer and reduces the viscosity of the medium. Unfortunately, the additional complication of chain transfer arises and solvents must be selected with care.

Ethylene, vinyl acetate, and acrylonitrile are polymerized in this way. The redox initiated polymerization of acrylonitrile is an example of precipitation polymerization where the polyacrylonitrile formed is insoluble in water and separates as a powder. This can lead to undesirable side reactions known as *popcorn polymerizations* when tough cross-linked nodules of polymer grow rapidly and foul the feed lines in industrial plants.

Suspension polymerization counteracts the heat problem by suspending droplets of water-insoluble monomer in an aqueous phase. The droplets are obtained by vigorous agitation of the system and are in the size range 0.01 to 0.5 cm diameter. The method is, in effect, a bulk polymerization, which avoids the complications of heat and viscosity build up.

Emulsion polymerization is an important technological process widely used to prepare acrylic polymers, poly(vinyl chloride), poly(vinyl acetate), and a large number of copolymers. The technique differs from the suspension method in that the particles in the system are much smaller, 0.05 to 5 μm diameter, and the initiator is soluble in the aqueous phase rather than in the monomer droplets. The process offers the unique opportunity of being able to increase the polymer chain length without altering the reaction rate. This can be achieved by changing either the temperature or the initiator concentration, and the reasons for this will become more obvious when we examine the technique more closely.

The essential ingredients are monomer, emulsifying agent, water, and a water-soluble initiator. The surfactant is normally an amphipathic long chain fatty acid salt with a hydrophilic "head" and a hydrophobic "tail." In aqueous solutions these form aggregates or micelles (0.1 to 0.3 μm long), consisting of 50 to 100 molecules oriented with the tails inward, thereby creating an interior hydrocarbon environment and a hydrophilic surface of heads in contact with the water. The micelles exist in equilibrium with free molecules in the aqueous phase, and the concentration must exceed the critical micelle concentration of the emulsifier.

When monomer is added to the dispersion, the bulk of it remains in the aqueous phase as droplets, but some dissolves in the micelles, swelling them. Free radicals are generated from a water-soluble redox system such as persulfate + ferrous ions

$$S_2O_8^{2-} + Fe^{2+} \longrightarrow Fe^{3+} + SO_4^{2-} + SO_4^{-\bullet}$$

at a rate of 10^{16} dm^{-3} s^{-1}. The radicals diffuse through the aqueous phase and penetrate both the micelles and droplets, but as the concentration of micelles (about 10^{21} dm^{-3})

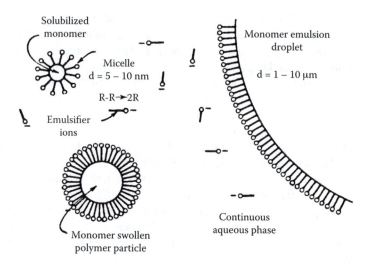

FIGURE 3.6 Schematic representation of an emulsion polymerization system.

far exceeds that of the droplets (10^{13} to 10^{14} dm^{-3}), polymerization is centered almost exclusively in the micelle interior. After only 2 to 10% conversion, the character of the system has changed markedly. Constant replenishment of the polymer swollen micelles takes place by diffusion from the droplets, which decrease steadily in size until at about 50 to 80% conversion they have been totally consumed. Polymerization then continues at a steadily decreasing rate until all the remaining monomer in the micelle is converted into polymer.

A schematic diagram of a typical emulsion system is shown in Figure 3.6. The polymerization process can be described using the model developed by Smith and Ewart. This assumes that a micelle-containing monomer is penetrated by a radical diffusing in through the aqueous phase, and this initiates a chain propagation reaction with rate v_p. The chain will continue to grow until another radical reaches the micelle and enters, where it will encounter the growing chain end and terminate it. The micelle will then remain quiescent until another chain is initiated by the entry of another radical, and the process of chain growth continues until the next radical arrives and once more terminates the chain, i.e., the model assumes that only one active radical can be tolerated in a micelle, and so this will contain either one or zero radicals at any one time. The result is that the polymerization process within any micelle in the system comprises a series of start–stop reactions, and the rate of the on–off switching is controlled by both the rate of radical production and the number of micelles in the reaction medium. As the entry of a radical into a micellar particle is random, the chances of a chain growing in a micelle at any particular time are 50:50. This means that if there are N^* micellar particles containing monomer and polymer in the system, then on average only $N^*/2$ will be active at any period during the course of the complete polymerization reaction, so the rate v_p will be proportional to the concentration of the monomer in the micelle [M*] and to the number of active micelles, ($N^*/2$), i.e.,

$$v_p = k_p[M^*][N^*/2] \tag{3.29}$$

where k_p [M*] is the rate of polymerization within a single micelle. If the rate of radical production is v_i, then the rate at which they enter a micelle is $(v_i/N*)$, which is the rate of initiation (or termination) in the micelle. The kinetic chain length in a micellar particle is then

$$\bar{v} = \frac{k_p[M^*]}{(v_i/N^*)} = \frac{k_p[M^*][N^*]}{2 f k_d[I]} \tag{3.30}$$

The consequences of this analysis are (1) an increase in the initiator concentration decreases the polymer chain length while leaving the rate of polymerization unaffected and, more surprisingly, (2) for a fixed initiator concentration both v_p and the chain length are a function of the number of micelles in the system. Thus, an increase in the surfactant concentration alone is sufficient to increase the polymerization rate and the molar mass of the product. This can be understood on the basis of the model, as increasing the number of micelles while holding the rate of radical production constant means that the time between the penetration of the micelle by successive radicals will be increased, thereby allowing the chain propagation to continue for longer periods before termination.

A combination of high rates and large x_n can be obtained without temperature variation, and this provides the system with its particular appeal. Control of the chain length can be achieved, when desired, by adding a chain transfer agent such as dodecyl mercaptan.

3.15 FEATURES OF FREE-RADICAL POLYMERIZATION

The main features of a radical polymerization can now be summarized and contrasted with the corresponding step-growth reactions (Section 2.9).

1. A high-molar-mass polymer is formed immediately when the reaction begins, and the average chain length shows little variation throughout the course of the polymerization.
2. The monomer concentration decreases steadily throughout the reaction.
3. Only the active center can react with the monomer and add units onto the chain one after the other.
4. Long reaction times increase the polymer yield, but not the molar mass of the polymer.
5. An increase in temperature increases the rate of the reaction but decreases the molar mass.

3.16 CONTROLLED RADICAL POLYMERIZATION

Conventional free-radical polymerization reactions are used widely on an industrial scale to produce important commodity polymers and copolymers from vinyl-based monomers. As processes are relatively insensitive to impurities and can be used to polymerize a wide range of monomers, they are difficult to control because of the numerous, fast, irreversible, termination and chain transfer reactions that occur. In theory, the steady-state kinetic analysis outlined in Section 3.1 indicates that the molecular weight of the product can be calculated from Equation 3.16, but in practice the various side reactions tend to lead to a product with a broad molecular weight distribution, and molecular weights lower than predicted.

As modern polymer science is interested in producing materials with function-alized, controlled structures, interest in exercising greater control over free-radical polymerization has increased in recent years. This has resulted in the development of a group of free-radical reactions that produce structurally well-defined polymers with low polydispersities. These are grouped under the general heading "controlled radical polymerizations" (CRP), often called *living radical polymerizations* (LRP) even though they still undergo bimolecular termination reactions, unlike the living anionic systems described in Chapter 4, which have no termination reaction other than those caused by impurities in the system.

The significant difference between conventional radical polymerizations and all of the LRP techniques developed so far is the establishment of a rapid dynamic equilibrium between a very small amount of chain-growing free radicals and a large excess of the dormant species. The general scheme for this reversible activation process is represented schematically below:

$$\left(P_n\!\!-\!\!X\right) \underset{k_{da}}{\overset{k_a}{\rightleftharpoons}} \underset{\text{active}}{P_n^{\bullet}} + X^{\bullet}$$

$$\underset{\text{dormant}}{} \qquad \downarrow M \qquad\qquad \text{(scheme I)}$$

$$\left(P_m\!\!-\!\!X\right) \rightleftharpoons P_m^{\bullet} + X^{\bullet}$$

or more compactly as:

$$\left(P_n\!\!-\!\!X\right) \underset{k_{da}}{\overset{k_a}{\rightleftharpoons}} \underset{(M)}{P_n^{\bullet}} + X^{\bullet}$$

$$\underset{\text{dormant}}{} \qquad\qquad \underset{\text{active}}{}$$

where k_a and k_{da} are the activation and deactivation rate constants, respectively.

These reactions involve a dormant species (P_n—X) end capped with a group X that is capable of undergoing a reversible termination reaction with a propagating polymer chain end $P_n^{\bullet}$, without itself acting as a chain initiator. The inability of $X^{\bullet}$ to initiate chain growth is of primary importance as it means that chain growth is confined to the $P_n^{\bullet}$ radical. Thus, the structure of this mediating radical $X^{\bullet}$ is critical in ensuring the success of a CRP process.

Three principle processes have been developed that provide the activa-tion/growth/deactivation cycle, which can be repeated many times throughout the polymerization reactions.

Dissociation–combination (DC)

$$P_n\!\!-\!\!X \underset{k_c}{\overset{k_d}{\rightleftharpoons}} P_n^{\bullet} + X^{\bullet} \qquad\qquad (3.31)$$

Atom transfer (AT)

$$P_n\!\!-\!\!X + A \underset{k_{da}}{\overset{k_a}{\rightleftharpoons}} P_n^{\bullet} + AX^{\bullet} \qquad (3.32)$$

Degenerative chain transfer (DT)

$$P_n\!\!-\!\!X + P_m^{\bullet} \underset{k'_{ex}}{\overset{k_{ex}}{\rightleftharpoons}} P_m\!\!-\!\!X + P_n^{\bullet} \qquad (3.33)$$

In each of these mechanisms, the reverse reaction dominates the equilibrium and keeps the overall concentration of the propagating radical $(P_n^{\bullet})$ low, typically $[P_n^{\bullet}]/[P_n\!\!-\!\!X] < 10^{-5}$. If this reversible radical trapping process occurs frequently, it minimizes the irreversible termination reactions but also means that the polymer chains all have an equal chance to grow, resulting in polymers with a narrow molecular weight distribution. It also follows that, unlike conventional free-radical polymerizations, the polymer chain length will increase steadily with the reaction time, similar to living anionic polymerizations.

In any CRP, the polymerization reaction is initiated by a species $(P_o\!\!-\!\!X)$, which in several cases may be a low-molecular-weight homologue of the polymer itself, produced in the early stages of the reaction.

3.17   NITROXIDE-MEDIATED POLYMERIZATIONS

Dissociation–combination living polymerizations are typified by nitroxide-mediated polymerizations, the first example of which used 2,2,6,6 tetramethyl piperidinyl-1-oxy (TEMPO) as the mediating stable free radical. The reaction steps for the polymerization of styrene, using benzoyl peroxide as the initiator, are given below.

Initiation

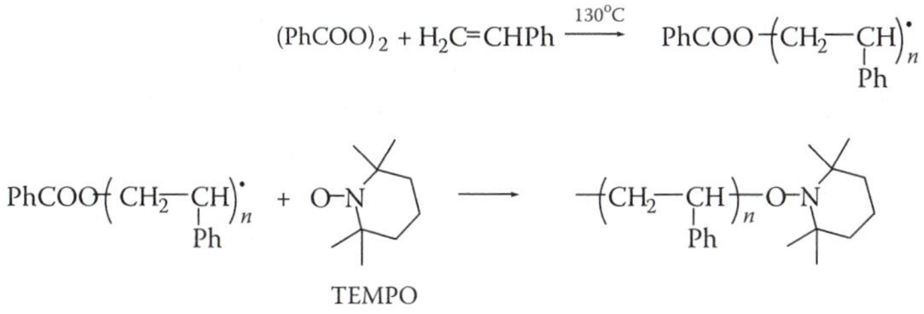

Propagation

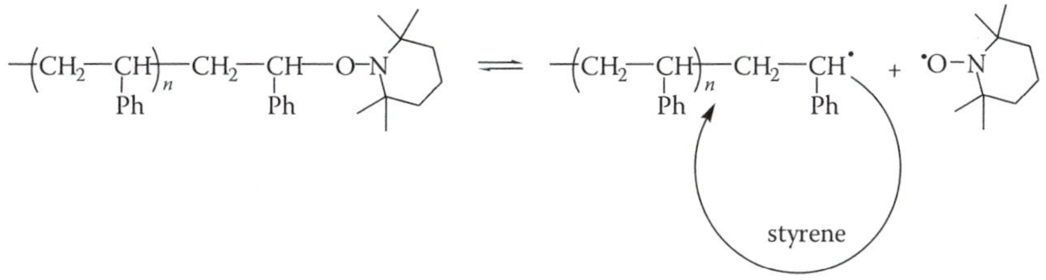

The polystyrene produced had a polydispersity index (M_w/M_n) of 1.2 to 1.3, which is much lower than that obtained using conventional free-radical processes.

Unfortunately, TEMPO is not the ideal nitroxide, and reactions give poor yields of polymer with several by-products that contaminate the product and are difficult to remove. Improvements can be achieved by using TEMPO-like structures, such as di-*tert*-butyl nitroxides, and additives, such as acetic acid, that improve the rate of polymerization. More important has been the development of new alkoxyamine structures such as phosphonate (1) and arene (2) nitroxides

(R = H, NH$_2$COOH)

(1)   (2)

which have greater versatility and can be used to polymerize acrylates, acrylamides, and acrylonitrile type monomers. These second-generation alkoxyamines give much better control of the molecular weight of the polymer and can produce M_n of up to 200,000, which at present represents an upper limit. If higher molecular weight material is prepared, then side reactions become significant and the "living" characteristics tend to be lost. However, if lower molecular weights are prepared, then polydispersities as low as 1.05 can be achieved.

In nitroxide-mediated polymerizations, the unimolecular-initiating species (P_n—X) decomposes to produce a small quantity of chains initiating radicals P_n• and the mediating, or persistent, radical X•. Some coupling of P• radicals occurs in the early stages of the reaction to produce an irreversibly terminated (P—P) species, thereby producing a small excess of X•. This excess is self-limiting as the formation of dormant chains is more efficient, and coupling reactions are negligible once the activation–deactivation equilibrium is established. The production of this small excess of X• gives rise to the all-important persistent radical effect, which controls the reaction.

Experiments have shown that the mediating radical X• is free to diffuse and does not remain close to the same chain end throughout the reaction. This then facilitates the introduction of terminal functional groups under mild conditions. These functionalized polymers are useful for subsequent reactions, such as block copolymer formation.

3.18 ATOM TRANSFER RADICAL POLYMERIZATION (ATRP)

Atom transfer radical polymerization (ATRP) is based on the well-established method of carbon–carbon bond formation by atom transfer radical addition used in organic synthesis.

A successful ATRP reaction depends on the reversible activation of a dormant species, such as an alkyl halide (R—X), by a transition metal halide catalyst (Mt^{z}Y/L). The metal halide catalyst abstracts the halogen atom (X) from the dormant species to form a radical R$^\bullet$ and an oxidized metal complex (XMt^{z+1}Y/L). Here, X and Y are halogens (Cl or Br), Mt is a transition metal (Cu, Ni, Pd, Rh, Ru, Mo), and L is a suitable ligand.

A typical ATRP reaction is shown below.

Initiation

$$\left(R\!\!-\!\!Cl\right) + Cu(I)Cl/(2bipy) \rightleftharpoons R^\bullet + Cu(II)Cl_2/(2bipy)$$

$$\Big\downarrow M$$

$$\left(P_m\!\!-\!\!Cl\right) + Cu(I)Cl/(2bipy) \rightleftharpoons P_n^\bullet + Cu(II)Cl_2/(2bipy)$$

Propagation

$$\left(P_n\!\!-\!\!Cl\right) + Cu(I)Cl/(2bipy) \underset{\textstyle\left(M\right)}{\rightleftharpoons} P_n^\bullet + Cu(II)Cl_2/(2bipy)$$

Here, the catalyst is a complex of Cu(I)Cl and the ligand 2,2 bipyridyl. This abstracts chlorine from an alkyl halide such as 1-phenyl ethyl chloride, and a reversible redox process is established where the Cu(II)Cl$_2$/(2bipy) species acts as the persistent or mediating radical shown in Equation 3.32.

ATRP reactions are very versatile with a high tolerance toward the presence of functional groups, such as the allyl, amino, epoxy, hydroxy, and vinyl groups, on both monomer and initiator. The reactions can be carried out either in bulk or in solution and also in heterogeneous systems. If solvents are used, chain transfer to solvent should be low, but the effect on the catalyst must be considered.

ATRP methods are successful when monomers such as styrenes, (meth)acrylates, (meth)acrylamides, and acrylonitrile are used but are less useful for monomers containing acid groups.

3.19 REVERSE ATRP

An alternative approach to initiating an ATRP is to use azobisisobutyronitrile (AIBN), a radical source used commonly in conventional radical polymerizations. By generating free radicals in the presence of a transition metal halide/ligand complex in its higher oxidation state, the required dynamic equilibrium can be generated:

Initiation

$$I_2 \longrightarrow 2\,I^\bullet$$

$$I^\bullet + X_2Mt^{z+1}/L \rightleftharpoons I\!\!-\!\!X + XMt^z/L$$

$$\Big\downarrow M$$

$$I\!\!-\!\!P_n^\bullet + X_2Mt^{z+1}/L \rightleftharpoons I\!\!-\!\!P_n\!\!-\!\!X + XMt^z/L$$

Propagation

$$\left(\begin{array}{c} \text{I—P}_{n}^{\bullet} \\ \text{M} \end{array}\right) + \text{X}_{2}\text{Mt}^{z+1}/\text{L} \rightleftharpoons \text{I—P}_{n}\text{—X} + \text{XMt}^{z}/\text{L}$$

Although initiation using azo compounds has been successful with a number of systems, the use of peroxides has been less successful.

3.20 DEGENERATIVE CHAIN TRANSFER REACTION (DT)

In degenerative chain transfer, the dormant species (P_n—X) is attacked by the propagating radical $P_m^{\bullet}$ to form a new dormant species (P_m—X) and $P_n^{\bullet}$. Two variations of this sample exchange reaction have been reported.

In the first group, X is simply transferred from radical to radical, i.e.,

$$P_n\text{—X} + P_m^{\bullet} \rightleftharpoons P_m\text{—X} + P_n^{\bullet}$$

as in

1. Iodine mediated polymerizations, where X = I; this has limited applicability.
2. Tellurium-mediated living radical polymerizations (TERP), where X is an organo tellurium species, such as $TeCH_3$. Polymerization of styrene and methyl methacrylate has been studied using AIBN initiator and $TeCH_3$, but results suggest that TERP involves both DT and DC mechanisms.

3.21 REVERSIBLE ADDITION FRAGMENTATION CHAIN TRANSFER (RAFT)

The second and much more important group of processes involves a conventional radical polymerization carried out in the presence of a compound that can act as a reversible chain transfer agent. This requirement is satisfied by dithio compounds (3) with the structure:

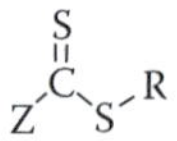

$$\underset{\text{(3)}}{\overset{\displaystyle \overset{S}{\underset{||}{}}}{\underset{Z}{\diagdown}}\overset{\displaystyle C}{}\underset{S}{\diagup}\overset{\displaystyle \diagup R}{}}$$

where Z = aryl, alkyl, SR′, OR′, $NR_2′$, and R is a good homolytic leaving group with respect to the polymer chain P_n, where radical $R^{\bullet}$ must be capable of initiating a polymerization reaction.

The process, which is known as reversible addition fragmentation chain transfer (RAFT), consists of the following steps:

Initiation

$$I \cdot \xrightarrow{\text{monomer}} P_n^\bullet$$

Addition–fragmentation

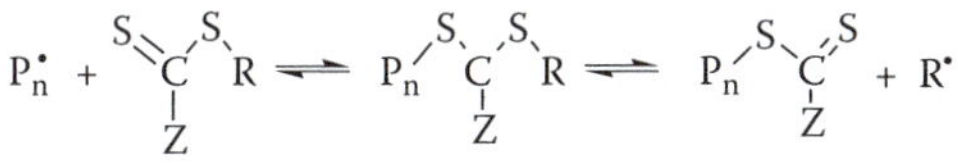

$$P_n^\bullet + \underset{Z}{\overset{S}{\underset{\|}{C}}}{\overset{S}{\diagdown}}R \rightleftharpoons P_n \overset{S}{\diagup}\underset{\underset{Z}{|}}{C}{\overset{S}{\diagdown}}R \rightleftharpoons P_n \overset{S}{\diagdown}\underset{\underset{Z}{|}}{C}{\overset{S}{\diagup}} + R^\bullet$$

Reinitiation

$$R^\bullet + \text{monomer (M)} \longrightarrow P_m^\bullet$$

Equilibration

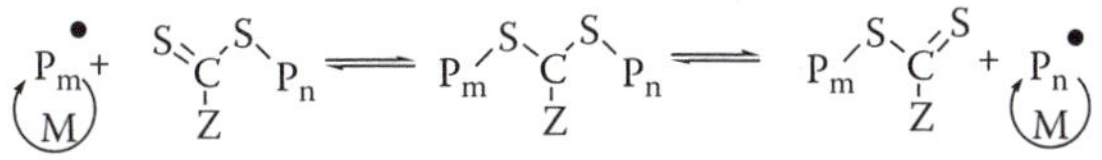

$$\underset{\textstyle\left(\!M\!\right)}{\overset{\bullet}{P_m}}{+} \underset{Z}{\overset{S}{\underset{\|}{C}}}{\overset{S}{\diagdown}}P_n \rightleftharpoons P_m \overset{S}{\diagup}\underset{\underset{Z}{|}}{C}{\overset{S}{\diagdown}}P_n \rightleftharpoons P_m \overset{S}{\diagdown}\underset{\underset{Z}{|}}{C}{\overset{S}{\diagup}} + \underset{\textstyle\left(\!M\!\right)}{\overset{\bullet}{P_n}}$$

of which the fundamental one is the reversible trapping of the majority of the propagating species into a dormant thiocarbonyl compound, thereby reducing the possibility of bimolecular termination reactions.

This approach can be used for a wide range of monomers and produces controlled molecular weight polymers with low polydispersities. When the dithiocarbonates are used in the polymerization of styrene and (meth)acrylates, samples with molecular weights of up to $M_n \sim 84{,}000$ and polydispersities in the range 1.14 to 1.84 are produced.

If the symmetrical trithiocarbonates (4)

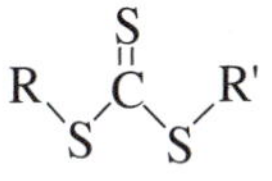

$$R \diagdown \underset{S}{\overset{\overset{\textstyle S}{\|}}{C}} \diagup R'$$

(4)

$R = CH_3$, CH_2Ph, $CH(CH_3)Ph$

$R' = CH_2Ph$, $C(CH_3)_2CN$, $CH(Ph)COOH$, $CH(CH_3)Ph$

are used, an extra dimension is introduced. Depending on the choice of R and R', compounds with either one or two good leaving groups can be prepared, and as the ability of the group to leave increases with the bulk and stability of the radical, CH_3 is a poor leaving group whereas $-C(CH_3)_2CN$ is a good leaving group. Thus, if the trithiocarbonate has one good and one poor leaving group, i.e., $R = CH_3$, $R' = C(CH_3)_2CN$ polymers with the structure

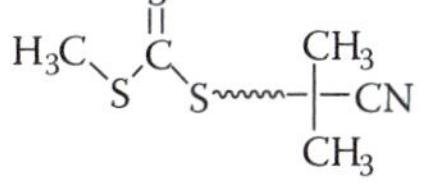

$$H_3C \diagdown \underset{S}{\overset{\overset{\textstyle S}{\|}}{C}} S \sim\sim\sim \overset{\textstyle CH_3}{\underset{\textstyle CH_3}{|}} - CN$$

will be obtained where the polymer chain has grown from one site only. If, on the other hand, $R = R' = CH_2Ph$, then chain growth occurs on both sites:

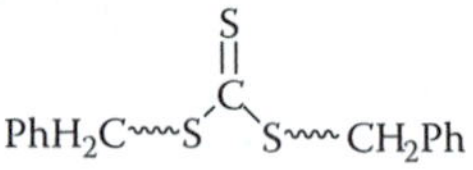

$$PhH_2C\text{\textasciitilde}S \underset{\underset{S}{\overset{\overset{S}{\parallel}}{C}}}{} S\text{\textasciitilde}CH_2Ph$$

In both cases the trithiocarbonate link can be cleaved using a mild nucleophile (e.g., ethylene diamine/THF) to produce (a) one or (b) two chains terminated by an SH group. Polymers with low polydispersities can be obtained when the monomers styrene (M_n up to 29,000; PDI of 1.06 to 1.19), methyl methacrylate (M_n up to 59,000; PDI of 1.13 to 1.17), and methyl acrylate (M_n up to 67,000; PDI of 1.09) are used. This reagent is also useful in the preparation of block copolymers.

Other suitable RAFT agents include dithiocarbamate derivatives in which the nitrogen atom is incorporated in an aromatic ring with structures (5) and (6).

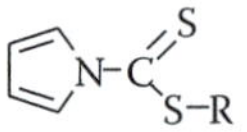

(5) 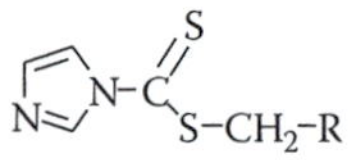   (6)

$$R = CH_2Ph, C(CH_3)_2CN$$

These have been used in the polymerization of styrene (M_n~15,000; PDI ~ 1.2), methyl methacrylate ($M_n \sim 51,000$; PDI = 1.19), and methyl acrylate ($M_n \sim 82,000$; PDI = 1.14).

One drawback in using the dithioesters, in particular, is that they are foul-smelling oily liquids that require careful handling. Improvements can be made, and dithioesters containing α-cyanobenzyl groups are essentially odorless (7).

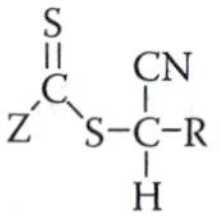

$$Z \underset{\underset{H}{\overset{\overset{S}{\parallel}}{C}}}{} S-\underset{H}{\overset{CN}{C}}-R \quad (7)$$

$$R = Ph; Ph\text{-}CH_3, Ph\text{-}F, Naphthyl, Dibenzyl$$

The use of these reagents does not detract from the efficiency of the RAFT process and M_n values of up to 33,000 with PDI values of 1.04 to 1.07 have been obtained for bulk and solution polymerization of styrene, methyl methacrylate, and n-butyl acrylate.

Xanthates (8) can also be employed, but the process is then called *macromolecular design via the interchange of xanthates* or MADIX.

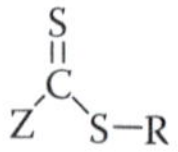

$$Z \underset{\overset{\parallel}{S}}{\overset{\overset{S}{\parallel}}{C}} S-R \quad (8)$$

$$Z = OC_2H_5, OCH_2CF_3 ; R = CH(CH_3)Ph, CH(CH_3)COOC_2H_5$$

These reagents have low-chain transfer constants (C_s = 0.6–3.5) in contrast to the dithioesters (C_s up to 6000), and this tends to produce polymers with broader molecular weight distributions. The advantages are that they are more easily synthesized and, more importantly, can be used in emulsion polymerizations, where other RAFT agents have failed.

3.22   CRP OF VINYL CHLORIDE

ATRP can be used to polymerize activated monomers, but has been unsuccessful with nonactivated monomers such as vinyl chloride. As PVC is an important commercial polymer, the production of controlled molecular weight and low polydispersity samples would be useful. Percec and co-workers have developed a method based on a combination of two processes: a single electron transfer step (SET) and degenerative chain transfer to control the propagation stage. The polymerization is initiated by mixing iodoform (CHI_3) with either zerovalent Cu(0), generated *in situ* by the disproportionation of Cu(I) reacting with a ligand L to produce Cu(0) and Cu(II)L, or sodium dithionite ($Na_2S_2O_4$), which dissociates in water to produce a radical ion $SO_2^{\bullet-}$. Reaction with either produces the radical $CHI_2^{\bullet}$ that initiates the polymerization of vinyl chloride to produce a telechelic PVC with a dichloroiodomethyl group at both ends. This can react further with $CHI_2^{\bullet}$ and set up the degenerative chain transfer reactions from growing chains to the initiator CHI_3, or dormant chains to $CHI_2^{\bullet}$.

Polyvinyl chloride with M_n = 2,000 to 12,000, PDI = 1.5, and no structural defects can be prepared in this way.

3.23   THE KINETICS OF CRP PROCESSES

The main features of CRP reactions can be summarized as follows: The polymerizations are started either by using a conventional radical initiator (or its equivalent) or by using a polymer adduct with an appropriate functional group. In the early stages, some of the initiator radicals terminate by combination and as excess of the mediating radical builds up, an activation–deactivation pseudoequilibrium is rapidly established, which favors the deactivation process, i.e., $k = (k_{act}/k_{deact})$ is small (10^{-8} to 10^{-10} M), thereby maintaining a high concentration of the dormant species and a low concentration of the propagating species. This minimizes the termination and chain transfer reactions and creates the conditions for a living polymerization, i.e., little or no termination, rapid initiation of a constant number of growing chains equal to the initiator concentration, and a steady increase in chain length throughout the reaction. This means that the degree of polymerization (x_n) can be calculated from Equation 3.34

$$x_n = \frac{\left[M\right]_o}{\left[I\right]_o} \times p \qquad (3.34)$$

where $[I]_o$ and $[M]_o$ are the initial concentrations of initiator (or transfer agent) and monomer, respectively, and p is the fractional monomer conversion, i.e.,

$$p = \left(\frac{[M]_o - [M]_t}{[M]_o} \right).$$

Analysis of the kinetics of CRP reactions is more complex than for conventional radical polymerizations; consequently, a detailed derivation of the basic equations will not be given here. The fundamental activation–deactivation pseudoequilibria that control the living characteristics of the various CRPs have been outlined in Equation 3.31 to Equation 3.33, and if the steady state is to be achieved rapidly, then the rates of activation and deactivation must be considerably larger than the rates of the initiation and termination reactions. In successful CRP reactions, the time taken to reach the equilibrium steady state is estimated to be in the range 1 to 100 ms.

The application of a steady-state analysis to these polymerization reactions leads to a series of rate equations that can be simplified for certain reaction conditions. If the rate of initiation is finite ($v_i > 0$), i.e., when a conventional radical initiator is used to start the reaction, and the concentration of the persistent radical $[X^\bullet]_o$ or $[AX^\bullet]_o$ is initially zero, then Equation 3.35 is obtained

$$\ln\left(\frac{[M]_o}{[M]_t}\right) = k_p \left(\frac{v_i}{k_t}\right)^{1/2} \cdot t \tag{3.35}$$

which shows that the conversion index ($[M]_o/[M]_t$) is first order with respect to time of reaction. Here, $[M]_o$ and $[M]_t$ are the concentrations of monomer at time 0 and t, respectively, and k_t is the rate constant for termination.

For CRP reactions that have no conventional initiator present, i.e., $v_i = 0$, two limiting cases can be considered.

1. If $[X^\bullet]_o$ or $[AX^\bullet]_o = 0$, Equation 3.36 is derived

$$\ln\left(\frac{[M]_o}{[M]_t}\right) = \frac{3}{2} k_p \left(\frac{K[I]_o}{3k_t}\right)^{1/3} \cdot t^{2/3} \tag{3.36}$$

where now the conversion index is a function of $t^{2/3}$. Here, $[I]_o$ is the initial concentration of the initiation adduct $[P\text{-}X]_o$, and k_p is the rate constant for propagation. This equation can also be derived using a power law kinetic analysis.

2. If $[X^\bullet]_o$ or $[AX^\bullet]_o$ are large, then the conversion index is again first order in t, Equation 3.37

$$\ln\left(\frac{[M]_o}{[M]_t}\right) = \left(\frac{k_p K [I]_o}{[X^\bullet]_o}\right) \cdot t \tag{3.37}$$

Stationary state kinetics can also be used to derive a relation for the polydispersity index (x_w/x_n) of the form

$$\left(\frac{x_w}{x_n}\right) = 1 + \frac{1}{x_n} + \left(\frac{2-p}{p}\right) \cdot \left(\frac{k_p [I]_o}{k_{deact}}\right) \tag{3.38}$$

where k_{deact} is the rate constant for the deactivation step in the various equilibria, $= k_c[X^\bullet]$ for NMP; $= k_{da}[AX^\bullet]$ for ATRP, and $k_{ex}[P\text{-}X]$ for RAFT processes (Equation 3.31 to Equation 3.33). As the term $(x_n)^{-1}$ will usually be small compared to the PDI value, it can be neglected. The minimum and best value possible for the PDI in a given CRP can then be obtained from this equation, if $p = 1$.

The development of molecular weights and polydispersities as a function of the conversion is exemplified in Figure 3.7 for the RAFT polymerization of methyl methacrylate in benzene solution, initiated by AIBN, with cumyl dithiobenzoate as the chain transfer agent.

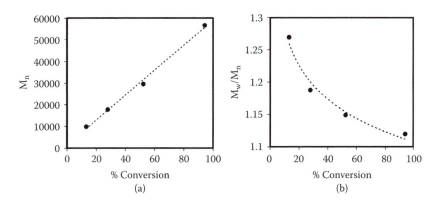

FIGURE 3.7 Evolution of (a) molecular weight (as polystyrene equivalents) and (b) polydispersity with conversion during polymerization of methyl methacrylate in benzene at 60°C initiated by AIBN, with cumyl dithiobenzoate as the chain transfer agent. (Adapted from Chiefari et al., *Macromolecules*, 31, 5559, 1998. Reproduced with permission of the American Chemical Society.)

3.24 APPLICATION TO EXPERIMENTAL DATA

Equation 3.35 and Equation 3.36 have been tested for all three major types of CRP process, and three examples will be used by way of illustration.

The nitroxide-mediated polymerization of styrene has been studied using a wide variety of nitroxides. If the phosphate nitroxide DEPN is used,

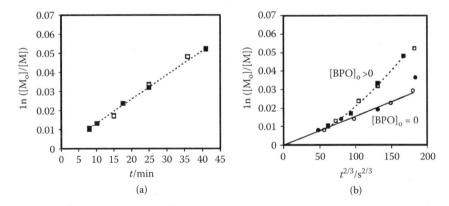

the value of $K = 6 \times 10^9$ M is larger than the TEMPO reaction, and so it takes longer to reach the stationary state. This allows one to test the power law behavior of the system, in addition to the steady-state analysis

Data for the polymerization of styrene, PS-DEPN (the polystyrene adduct) in the presence and absence of benzoyl peroxide (BPO) are shown in Figure 3.8(a) and Figure 3.8(b). The conversion index is first order with respect to t for systems containing BPO, in accord with Equation 3.35, but in the absence of BPO, a power-dependence, Equation 3.36, of $t^{2/3}$ is observed.

The basic kinetic features of ATRP are similar to NMP, and this can be demonstrated for the copper-mediated polymerization of styrene. The ATRP of styrene in t-butyl benzene, catalyzed by Cu(I)Br/L, where L is the ligand diheptyl bipyridine, and a (PS-Br) adduct in the presence or absence of a radical initiator 2,2′-azobis (2,4,4-tri methyl pentane) VR110 at 110°C, has been studied. The data are shown in Figure 3.9(a) and Figure 3.9(b). Again, it is seen that for the system containing VR110, steady-state kinetics apply, and the conversion index is first order in t [Figure 3.9(a)], but in the absence of the initiator power law kinetics are obeyed [Figure 3.9(b)].

FIGURE 3.8 Plots of $\ln([M]_0/[M])$ vs. (a) t and (b) $t^{2/3}$ for the styrene/PS-DEPN-(P$_0$-X)/BPO systems (80°C): $[P_0\text{-X}]_0 = 25$ mM; $[BPO]_0 = 4.7$ mM. The dotted line is the best-fit representation of the duplicated experiment. (From Yoshikawa, C. et al., *Macromolecules*, 35, 5801, 2002. With permission of the American Chemical Society.)

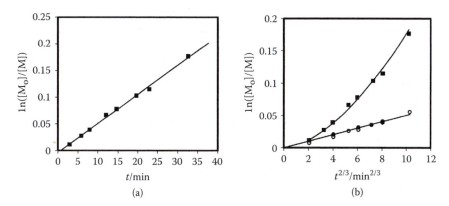

FIGURE 3.9 Plots of $\ln([M]_0/[M])$ vs. (a) t and (b) $t^{2/3}$ for the styrene/t-butylbenzene/PS-Br(P_0-X)/Cu(I)Br/dHbipy /(VR110) systems (110°C): $[P_0$-X$]_0 = 13$ mM; $[Cu(I)Br]_0 = 10$ mM; $[dHbipy]_0 = 30$ mM; $[VR110]_0 = 0$ (○,●) and 40 mM (■); [styrene]/[t-butylbenzene] = 50/50 v/v. The experiments are duplicated for $[VR110]_0 = 0$ (○,●). The line is the best-fit representation of the duplicated experiment. (From Yoshikawa, C. et al., *Macromolecules*, 36, 908, 2003. With permission of the American Chemical Society.)

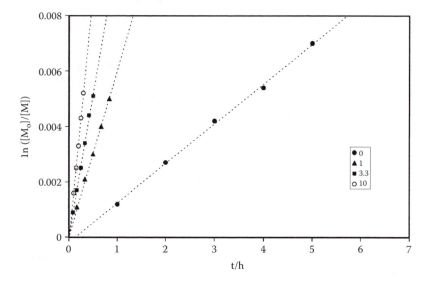

FIGURE 3.10 Plot of $\ln([M]_0/[M])$ vs. t (60°C) for the styrene/PS-SCSCH$_3$(P_0-X)/BPO system: $[P_0$-X$]_0 = 0.45$ mM; $[BPO]_0$ as indicated in the figure. (From Goto, A. et al., *Macromolecules*, 34, 402, 2001. With permission of the American Chemical Society.)

The situation is less clear-cut for RAFT systems. For a dithiocarbonate-mediated styrene polymerization studied by Goto and co-workers, the steady-state kinetic analysis applied both in the presence and absence of a BPO initiator (Figure 3.10). Similarly, for the solution polymerization of methyl methacrylate, mediated by dithioesters containing α-cyanobenzyl groups in the presence of AIBN initiator, pseudo-first-order plots were obtained although a significant induction period was detected.

In general, steady-state kinetics are successful in describing the majority of the CRP reactions when a conventional radical initiator is present in the reaction, but power law equations may be more appropriate when a conventional initiator is absent.

PROBLEMS

1. Draw structures of all possible configurational isomers resulting from the polymerization of 2-chlorobutadiene.
2. Draw repeat unit structures for the polymers obtained from:
    a. 1,4 polymerization of 2,3-dichlorobutadiene
    b. 1,2 addition of 2-methyl butadiene
    What polymer is formed by hydrogenation of (a)?
3. Explain why poly(vinyl alcohol) (PVA) cannot be prepared from its mono-mer. The synthetic route to this polymer involves polymerization of vinyl acetate followed by hydrolysis. Sketch the mechanism. Suggest a reason why PVA made from 100% hydrolyzed poly(vinyl acetate) is difficult to dissolve in water.
4. Write down the mechanism for the free-radical polymerization of vinyl-acetate initiated by 2,2′-azobis(2,4-dimethylpentanenitrile) (Vazo 52) in ethyl acetate including (a) initiation, (b) propagation, and (c) termination by disproportionation.
5. You are asked to select a suitable radical initiator for a polymerization process to be carried out at 50°C and which needs to be completed within 8 h. By considering that the initiator concentration must not fall below half its original value during the polymerization, select the most appro-priate initiator from the list below.

Initiator	T (°C)	$k_d \times 10^5$ (s^{-1})	E_a (kJ mol^{-1})
Benzoyl peroxide	80	2.5	124
tert-Butyl peroxide	130	2.48	142
Azobisbutyronitrile	50	0.20	129
2,2′-azobis-2,4-dimethyl valeronitrile	70	28.9	121

6. Overberger et al. (1949) studied the decomposition rates of a series of aliphatic azonitriles in toluene by measuring the volume of N_2 evolved as a function of time, at constant temperature. A set of V_t/V_∞ (where V_t and V_∞ are the volumes at time t, and t = ∞) data for the decomposition of the azo nitrile compound

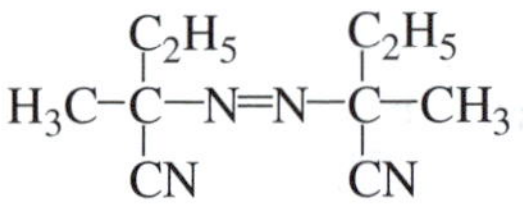

$$\begin{array}{ccc} C_2H_5 & & C_2H_5 \\ | & & | \\ H_3C-C-N=N-C-CH_3 \\ | & & | \\ CN & & CN \end{array}$$

at 80.2°C are given below. From these: (1) demonstrate that the decomposition reaction follows first order kinetics and (2) determine the rate of decomposition of the azo nitrile compound.

Time × 10⁻³ (s)	V_t/V_∞
2.86	0.213
5.10	0.381
6.33	0.429
7.35	0.493
9.18	0.577
10.82	0.647
12.04	0.687
13.16	0.733
15.82	0.786
17.04	0.802
18.57	0.831
20.51	0.870

7. The effect of α, α'-azobisisobutyronitrile initiator concentration [I] on the rate of polymerization of methyl methacrylate at 50°C has been studied by Arnett (1952). Use the following data to show that the termination reaction is second order with respect to the growing chains and initiation is first order with respect to the initiator concentration.

[I] × 10² (mol L⁻¹)	v_p × 10³ (mol L⁻¹ min⁻¹)	x_n
0.47	4.66	11090
0.62	5.54	11040
1.15	7.37	7250
1.27	7.92	8390
1.83	9.3	6150
2.56	11.4	8530
3.7	13.31	4880
5.21	15.6	3890
7.45	18.84	3000
10.77	22.45	2660
14.82	26.4	2810
21.06	31.61	2700

The number-average degree of polymerization, x_n, was also determined and data are listed in the table above. Use these results to establish a relationship between initiator concentration and x_n.

8. 15.45 g of methyl methacrylate were polymerized at 50°C for 55 min using 0.2096 g of α, α'-azobisisobutyronitrile (AIBN) as initiator, to give 1.6826 g of polymer. Radioactivity assay indicated that the polymer sample contained 7.07×10^{-4} g of azobisnitrile (Arnett and Peterson, 1952).

Considering that the first-order rate constant for the decomposition of
AIBN is given by:

$$\log k \,/\, \mathrm{min}^{-1} = \frac{-7019}{T} + 17.806,$$

calculate the initiator efficiency. The number-average molecular weight
of this polymer was found to be 390,000 from osmotic pressure measurements. Establish whether the termination reaction takes place by combination or disproportionation of growing chains.

9. Consider the kinetic scheme for free-radical polymerization. Demonstrate
   that when polymer chains terminate at sites located on the surface of the
   polymerization vessel:

$$RM_n^{\bullet} \overset{k_{\mathrm{wall}}}{\rightarrow} P$$

and this is the only mode of termination, then the overall rate of polymerization is given by:

$$v_{\mathrm{p}} = k_{\mathrm{pol}}\,[M][I]$$

where $k_{\mathrm{pol}} = 2\,k_{\mathrm{p}}\,k_{\mathrm{d}}/k_{\mathrm{wall}}$.

10. The ratio between the rates of termination by disproportionation and
    combination, $k_{\mathrm{t,d}}/k_{\mathrm{t,c}}$, has been evaluated by Berger (1975) for the poly-
    merization of styrene:

T (°C)	$k_{\mathrm{t,d}}/k_{\mathrm{t,c}}$
30	0.168
52	0.246
62	0.294
70	0.480
80	0.663

From these data, calculate the difference in activation energy between
disproportionation and combination, in the termination step.

11. Consider the free-radical polymerization of methyl methacrylate in toluene
    solution at 77°C, initiated by AIBN. When the initial monomer concentra-
    tion was 2.07 M and the initial AIBN concentration was 2×10^{-4} M, the
    initial rate of polymerization was determined to be $v_{\mathrm{p}} = 2.49 \times 10^{-3}$ M min^{-1}.
    a. Determine the initial rate of initiation, v_{i}, and $k_{\mathrm{p}}/(k_{\mathrm{t}})^{1/2}$ by considering
       that the rate constant for the decomposition of AIBN at 77°C is $k_{\mathrm{d}} = 5.7$
       $\times 10^{-3}$ min^{-1}, and that virtually all radicals are capable of initiating chains.

    b. Determine the initial kinetic chain length and the expected number average molecular weight assuming only termination mainly takes place by disproportionation.

    c. $E_p-0.5\ E_t$ gives a measure of the energy required to polymerize a given monomer. Estimate this for methyl methacrylate from your value of $k_p/(k_t)^{1/2}$ at 77°C and the value reported at 50°C by Arnett (1952), $k_p/(k_t)^{1/2} = 0.67$.

    d. Estimate the time required to polymerize 10% of the monomer.

12. In the polymerization of vinyl acetate with AIBN at 60°C Palit and Das (1954) measured the degree of polymerization in the presence of various chain transfer agents and at different [S]/[M] (i.e., solvent to monomer) ratios and intrinsic viscosities, [η]. A set of data is tabulated below.

Solvent	[S]/[M]	[η]
Ethyl acetate	1.076	0.980
	1.614	0.880
	2.421	0.680
	6.456	0.500
Methyl *iso*-butyl ketone	0.492	0.300
	0.738	0.240
	1.108	0.190
	2.954	0.095

Calculate the chain transfer constants, C_S. You will need to make use of the relationship:

$$\log x_n = 3.4655 + 1.45 \log [\eta]$$

13. In the polymerization of methyl methacrylate initiated by phenyl ethyl chloride (PECl), in the presence of Cu(I)Br/dinonyl bipyridine complex, the following data were obtained:

[PECl] (M)	[MMA] (M)	t (min)
0.10	9.0	0
0.08	7.2	40
0.05	4.5	118
0.03	2.7	200
0.01	0.9	380

Use these results to demonstrate that the reaction has the characteristics of a living radical polymerization, and identify the type of CRP.

14. In the thermally initiated bulk polymerization of styrene (density = 0.906 g cm⁻³) in the presence of dithiocarbamate chain transfer agent (CTA), the following data were obtained:

CTA	Percentage Conversion
2.5	30
2.5	50
2.5	70
2.5	90
2.0	50
1.5	50
1.0	50

a Which living polymerization method does this system represent?

b. Calculate the theoretical molecular weight for each of the reactions.

c. Do the data comply with any of the criteria for living polymerization?

REFERENCES

Arnett, L.M., *J. Am. Chem. Soc.* 74, 2027, 1952.

Arnett, L.M. and Peterson, J.H., *J. Am. Chem. Soc.* 74, 2031, 1952.

Berger, K.C., *Makromol. Chem.*, 176, 3575, 1975.

Cardenas, J.N. and O'Driscoll, K.F., *J. Polym. Sci. Polym. Chem. Ed.*, 14, 883, 1976.

Chiefari et al., *Macromolecules*, 31, 5559, 1998.

Dainton, F.S. and Ivin, K.J., *Q. Rev.*, 12, 61, 1958.

Davis, T.P., O'Driscoll, K.F., Piton, M.C., and Winnik, M.A., *Macromolecules*, 22, 2785, 1989.

Goto, A. et al., *Macromolecules*, 34, 402, 2001.

Melville, H.W., *J. Chem. Soc.*, 247, 1947.

Olaj, O.F., Bitai, I., and Hinkelmann, F., *Makromol. Chem.*, 188, 1689, 1987.

Overberger et al., *J. Am. Chem. Soc.*, 71, 2661, 1949.

Palit, S.R. and Das, S.K., *Proc. R. Soc. London* 226A, 82, 1954.

Schulz, G.V., *Chem. Ber.*, 80, 232, 1947.

Yoshikawa, C. et al., *Macromolecules*, 35, 5801, 2002.

BIBLIOGRAPHY

Allcock, H.R. and Lampe, F.W., *Contemporary Polymer Chemistry*, Prentice-Hall, 1981.

Allen, G. and Bevington, J.C., Eds., *Comprehensive Polymer Science*, Vol. 3, 4, Pergamon Press, 1989.

Alter, H. and Jenkins, A.D., Chain-reaction polymerization, in *Encyclopedia of Polymer Science and Technology*, Interscience Publishers, 1965.

Bamford, C.H. and Tipper, C.F.H., Eds., *Free Radical Polymerization, Comprehensive Chemical Kinetics*, Vol. 14A, Elsevier, 1976.

Billmeyer, F.W., *Textbook of Polymer Science*, 3rd ed., John Wiley and Sons, 1984.

Blackley, D.C., *Emulsion Polymerization,* John Wiley and Sons, 1975.

Bouton, T.C., Henderson, J.N., and Bevington, J.C., Eds., *Polymerization Reactors and Processes*, ACS Symposium Series 104, 1979.

Flory, P.J., *Principles of Polymer Chemistry*, Cornell University Press, 1953, chap. 4.

Ham, G.E., *Vinyl Polymerization*, Vol. I, Marcel Dekker, 1967.

Jenkins, A.D., The reactivity of polymer radicals, in *Advances in Free Radical Chemistry*, Vol. 2, Logos Press, 1967.

Lenz, R.W., *Organic Chemistry of Synthetic High Polymers*, Interscience Publishers, 1967, chap. 9–11.

Margerison, D. and East, G.C., *Introduction to Polymer Chemistry*, Pergamon Press, 1967, chap. 4.

Napper, D.H., *Polymer Stabilization of Colloidal Dispersions*, Academic Press, 1984.

Odian, G., *Principles of Polymerization*, 4th ed., John Wiley and Sons, 2004.

Rempp, P. and Merrill, E.W., *Polymer Synthesis*, Hüthig and Wepf Verlag, 1986.

Smith, D.A., *Addition Polymers,* Butterworths, 1968, chap. 2.

4 Ionic Polymerization

4.1 GENERAL CHARACTERISTICS

Radical-initiated polymerizations are generally nonspecific, but this is not true for ionic initiators because the formation and stabilization of a carbonium ion or carbanion depend largely on the nature of the group R in the vinyl monomer $CH_2=CHR$. For this reason, cationic initiation is usually limited to monomers with electron-donating groups that help stabilize the delocalization of the positive charge in the π-orbitals of the double bond. Anionic initiators require electron-withdrawing substituents ($-CN, -COOH, -CH=CH_2$, etc.) to promote the formation of a stable carbanion, and when there is a combination of both mesomeric and inductive effects, the stability is greatly enhanced.

As these ions are associated with a counterion or gegen ion, the solvent has a profound influence. Chain propagation will depend significantly on the separation of the two ions, and this separation will also control the mode of entry of an adding monomer. Also, the gegen ion itself can influence both the rate and stereochemical course of the reaction. Although polar and highly solvating media are obvious choices for ionic polymerizations, many cannot be used because they react with and negate the ionic initiators. This is true of the hydroxyl solvents, and even ketones will form stable complexes with the initiator to the detriment of the reaction. As solvents of much lower dielectric constant have to be used, the resulting close proximity of the gegen ion to the chain end requires that the propagating species be treated as an *ion pair*, but even in low-polarity media such as methylene chloride, ether, tetrahydrofuran (THF), nitrobenzene, etc., the ion pair separation can vary sufficiently for the effects to be distinguishable.

Ionic-initiated polymerizations are much more complex than radical reactions. When the chain carrier is ionic, the reaction rates are rapid, difficult to reproduce, and yield high-molar-mass material at low temperatures by mechanisms that are often difficult to define.

Complications in the kinetic analysis can arise from cocatalyst effects in which small quantities of an inorganic compound, such as water, will have an unexpectedly large influence on the polymerization rate.

Initiation of an ionic polymerization can occur in one of four ways, all of which involve essentially the loss or gain of an electron (e^-) by the monomer to produce an ion or radical ion:

1. $M + I^+ \rightarrow MI^+$ Cationic
2. $M + I^- \rightarrow MI^-$ Anionic
3. $M + e^- \rightarrow {}^\bullet M^-$ Anionic
4. $M - e^- \rightarrow {}^\bullet M^+$ Cationic (charge transfer)

4.2 CATIONIC POLYMERIZATION

Ionic polymerizations proceed by a chain mechanism and can be dealt with under the general headings that were used for the radical reactions: initiation, propagation, and termination. A common type of cationic initiation reaction is that represented in (1), where I^+ is typically a strong Lewis acid. These electrophilic initiators are classified into three groups: (a) classical protonic acids or acid surfaces — HCl, H_2SO_4, $HClO_4$; (b) Lewis acids or Friedel–Crafts catalysts — BF_3, $AlCl_3$, $TiCl_4$, $SnCl_4$; and (c) carbenium ion salts.

The most important initiators are the Lewis acids MX_n, but they are not particularly active alone and require a cocatalyst SH to act as a proton donor. In general, first we have an ionization process

$$MX_n + SH \rightleftharpoons [SMX_n]^- \; H^+$$

followed by an initiation mechanism that is probably the two-step process

(i) $\quad \begin{array}{c} H \\ \backslash \\ /C=C\backslash \\ H \end{array} \begin{array}{c} R_1 \\ \\ R_2 \end{array} + H^+[SMX_n]^- \longrightarrow \left[\begin{array}{c} H \quad H \quad R_1 \\ \backslash \quad \vdots \quad / \\ C \dot= C \\ / \quad \quad \backslash \\ H \quad \quad R_2 \end{array} \right]^+ [SMX_n]^-$

(ii) $\qquad\qquad\qquad\qquad\qquad\qquad \rightleftharpoons \left[\begin{array}{c} H \quad R_1 \\ | \quad | \\ H-C-C^+ \\ | \quad | \\ H \quad R_2 \end{array} \right] [SMX_n]^-$

where step one is the rapid formation of a π-complex, and step two is a slow intramolecular rearrangement. Although the need for a cocatalyst is recognized, it is often difficult to demonstrate, and a useful reaction that serves this purpose is the polymerization of isobutylene. This reaction proceeds rapidly when trace quantities of water are present but remains dormant under anhydrous conditions. The active catalyst–cocatalyst species required to promote this reaction is

$$BF_3 + H_2O \rightleftharpoons H^+[BF_3OH]^-$$

and the complex reacts with the monomer to produce a carbenium ion chain carrier that exists as an ion pair with $[BF_3OH]^-$.

$$H^+ [BF_3OH]^- + (CH_3)_2C=CH_2 \longrightarrow (CH_3)_3C^+ [BF_3OH]^-$$

The type of cocatalyst also influences the polymerization rate because the activity of the initiator complex depends on how readily it can transfer a proton to the monomer. If the polymerization of isobutylene is initiated by $SnCl_4$, the acid strength of the cocatalyst governs the rate, which decreases in the cocatalyst order: acetic acid > nitroethane > phenol > water.

Other types of initiator are less important; thus, strong acids protonate the double bond of a vinyl monomer

$$HA + H_2C{=}CR_1R_2 \longrightarrow A^- \; CH_3C^+ \, R_1R_2$$

whereas iodine initiates polymerization with the ion pair

$$2I_2 \longrightarrow I^+ I_3^-$$

which forms a stable π-complex with olefins such as styrene and vinyl ethers.

$$H_2C{=}CHR + I^+ I_3^- \longrightarrow ICH_2{-}\overset{+}{\underset{R}{C}}HI_3^-$$

High-energy radiation is also thought to produce cationic initiation, but this may lead to fragmentation, and a mixture of free-radical and cationic centers.

4.3 PROPAGATION BY CATIONIC CHAIN CARRIERS

Chain growth takes place through the repeated addition of a monomer in a head-to-tail manner, to the carbenium ion, with retention of the ionic character throughout.

$$CH_3{-}\overset{R_1}{\underset{R_2}{C^+}}[SMX_n]^- + nCH_2{=}CR_1R_2 \xrightarrow{k_p} CH_3[CR_1CR_2CH_2{-}]_{\overline{n}}\overset{R_1}{\underset{R_2}{C^+}}[SMX_n]^-$$

The mechanism depends on the *counterion*, the *solvent*, the *temperature*, and the *type of monomer*. Reactions can be extremely rapid when strong acid initiators such as BF_3 are used, and produce long-chain polymer at low temperatures. Rates tend to be slower when the weaker acid initiators are used, and a polymerization with $SnCl_4$ may take several days. Useful reaction temperatures are in the range 170 to 190 K, and both molar mass and reaction rate decrease as the temperature is raised.

Propagation also depends greatly on the position and type of the gegen ion associated with the chain carrier. The position of the gegen ion can be altered by varying the dielectric constant of the solvent, and large changes in k_p can be obtained as shown in Table 4.1 for a perchloric-acid-initiated polymerization of styrene in several media.

It has been suggested that the various stages of the ionization producing carbenium ions can be represented as

$$RX \rightleftharpoons R^+ X^- \rightleftharpoons R^+/X^- \rightleftharpoons R^+ + X^-$$

| | | solvent separated | |
| covalent | intimate ion pair | ion pair | free ions |

TABLE 4.1
Cationic Polymerization of Styrene in Media of Varying
Dielectric Constant ε

Solvent	ε	Catalyst	$\dfrac{k_p}{dm^3mol^{-1}s^{-1}}$
CCl_4	2.3	$HClO_4$	0.0012
$CCl_4 + (CH_2Cl)_2(40/60)$	5.16	$HClO_4$	0.40
$CCl_4 + (CH_2Cl)_2(20/80)$	7.0	$HClO_4$	3.20
$(CH_2Cl)_2$	9.72	$HClO_4$	17.0
$(CH_2Cl)_2$	9.72	$TiCl_4/H_2O$	6.0
$(CH_2Cl)_2$	9.72	I_2	0.003

The increasing polarity of the solvent alters the distance between the ions from an intimate pair, through a solvent separated pair to a state of complete dissociation. As free ions propagate faster than a tight ion pair, the increase in free ion concentration with change in dielectric constant is reflected in an increase in k_p. The separation of the ions also lowers the steric restrictions to the incoming monomer, so that free ions exert little stereoregulation on the propagation, and too great a separation may even hinder reactions that are assisted by coordination of the monomer with the metal in the gegen ion. As a first approximation, it can be stated that as the dielectric constant of the medium increases, there is a linear increase in the polymer chain length and an exponential increase in the reaction rate, but in some cases the bulk dielectric of the medium may not determine the effect of the solvent on an ion in its immediate environment. This leads to deviations from the simple picture. The nature of the gegen ion affects the polymerization rate. Larger and less tightly bound ions lead to larger values of k_p; hence, a decrease in k_p is observed as the initiator changes from $HClO_4$ to $TiCl_4.H_2O$ to I_2, for the reaction of styrene in 1,2-dichloroethane.

4.4 TERMINATION

The termination reaction in a cationic polymerization is less well defined than for the radical reactions, but is thought to take place either by a unimolecular rearrangement of the ion pair

$$\sim\!\!\!CH_2\overset{R_1}{\underset{R_2}{C^+}}[SMX_n]^- \longrightarrow \sim\!\!\!CH\!\!=\!\!CR_1R_2 + H^+[SMX_n]^-$$

or through a bimolecular transfer reaction with a monomer

$$\sim\!\!\!CH_2\overset{R_1}{\underset{R_2}{C^+}}[SMX_n]^- + H_2C\!\!=\!\!\overset{R_1}{\underset{R_2}{C}} \longrightarrow \sim\!\!\!CH\!\!=\!\!CR_1R_2 + CH_3\overset{R_1}{\underset{R_2}{C^+}}[SMX_n]^-$$

The first involves hydrogen abstraction from the growing chain to regenerate the catalyst–cocatalyst complex, whereas the second reforms a monomer–initiator complex, thereby ensuring that the kinetic chain is not terminated by the reaction. In the unimolecular process, actual covalent combination of the active center with a catalyst–cocatalyst complex fragment may occur, giving two inactive species. This serves to terminate the kinetic chain and reduce the initiator complex and, as such, is a more effective route to reaction termination.

4.5 GENERAL KINETIC SCHEME

Many cationic polymerizations are both rapid and heterogeneous, which makes the formulation of a rigorous kinetic scheme extremely difficult. At best, a scheme of general validity can be deduced, but this should not be applied indiscriminately. Following the steady-state approach outlined for radical reactions, the rate of initiation v_i of a cationic reaction is proportional to the catalyst–cocatalyst concentration c and the monomer concentration [M].

$$v_i = k_i c [M] \tag{4.1}$$

Termination can be taken as a first-order process in contrast to the free-radical mechanisms, and

$$v_i = k_t [M^+] \tag{4.2}$$

Under steady-state conditions, $v_i = v_t$ and

$$[M^+] = k_i c [M] / k_t. \tag{4.3}$$

This gives an overall polymerization rate v_p of

$$v_p = k_p [M] \, [M^+] = (k_p k_i / k_t) c [M]^2 \tag{4.4}$$

and a chain length of

$$x_n = v_p / v_t = (k_p / k_t)[M], \tag{4.5}$$

if termination, rather than transfer, is the dominant process. When chain transfer is significant,

$$x_n = k_p / k_{tr}. \tag{4.6}$$

Although not universally applicable, this scheme gives an adequate description of the polymerization of styrene by $SnCl_4$ in ethylene dichloride at 298 K.

4.6 ENERGETICS OF CATIONIC POLYMERIZATION

Having established a kinetic scheme, an explanation for the increase in overall rate with decreasing temperature is possible. The rate is proportional to $(k_i k_p / k_t)$, so the overall activation energy E is given by

$$E = E_i + E_p - E_t, \tag{4.7}$$

and for the chain length

$$E_x = E_p - E_t. \tag{4.8}$$

Propagation in a cationic polymerization requires the approach of an ion to an uncharged molecule in a relatively nonpolar medium, and as this is an operation with a low activation energy, E_p is much less than E_i, E_t, or E_{tr}. Consequently, E is normally in the range -40 to $+60$ kJ mol^{-1}, and when it is negative, the rather unexpected increase in k_p is obtained with decreasing temperature. It should be noted, however, that not all cationic polymerizations have negative activation energies; the polymerizations of styrene by trichloroacetic acid in nitromethane and by 1,2-dichloroethylene have E equal to $+57.8$ kJ mol^{-1} and $+33.6$ kJ mol^{-1}, respectively.

The chain length, on the other hand, will always decrease as the reaction temperature rises because E_t is always greater than E_p.

4.7 TELECHELIC POLYMERS VIA CATIONIC POLYMERIZATION

A telechelic polymer is defined as a relatively low-molar-mass species ($M_n \leq 20,000$), with functional end groups that can be used for further reaction to synthesize block copolymers or for network formation. Cationic polymerization methods can be used to prepare these functionalized polymers using the *initiator-transfer*, or "Inifer," technique perfected by Kennedy. If the initiating catalyst–cocatalyst system is prepared from a Lewis acid and an alkyl or aryl halide, i.e.,

$$BCl_3 + RCl \rightleftharpoons [R^{\oplus} BCl_4^{\ominus}]$$

and a monomer such as isobutene is added, then the resulting polymer can undergo a transfer reaction with RCl to produce a halogen-terminated chain and a regenerated initiating species.

$$[R^{\oplus} BCl_4^{\ominus}] + n H_2C = \underset{\underset{CH_3}{|}}{\overset{\overset{CH_3}{|}}{C}} \longrightarrow R^{\sim\sim}CH_2 \underset{\underset{CH_3}{|}}{\overset{\overset{CH_3}{|}}{C^{\oplus}}} BCl_4^{\ominus}$$

$$\Big\downarrow RCl$$

$$R^{\sim\sim}CH_2 \underset{\underset{CH_3}{|}}{\overset{\overset{CH_3}{|}}{C}} -Cl \quad + \quad [R^{\oplus} BCl_4^{\ominus}]$$

The terminal chlorine can then be converted by subsequent reactions into other useful functional groups.

$$>C=CH_2; \ -CH_2OH; \ -NCO; \ -\underset{\underset{CH_2SO_3H}{|}}{C}=CH_2$$

Bifunctional telomers can also be prepared, and these functional polymers can be used to form block copolymers using coupling reactions.

4.8 CATIONIC RING OPENING POLYMERIZATION

Cyclic monomers such as lactones, lactams, cyclic amines, and cyclic ethers can be encouraged to undergo ring-opening reactions under the influence of cationic initiators to form linear polymers. The tendency to ring-open depends on the ring size, and the primary driving force in small rings is the relief of ring strain. The main reasons for the presence of ring strain are bond angle distortion, conformational strain, and nonbonded interactions in the ring. The heat of polymerization, ΔH_p, is a good indication of the magnitude of this quantity. Table 4.2 lists some representative values for cyclic ethers, and Table 4.3 shows a range of cyclic monomers that are susceptible to cationic ring-opening polymerization reactions. The data in Table 4.2 indicate that

TABLE 4.2
Dependence of ΔH_p on the Ring Size for Cyclic Ethers[a]

Monomer	Ring Size	$-\Delta H_p$(kJ mol^{-1})
Ethylene oxide (oxirane)	3	94.5
Trimethylene oxide (oxetane)	4	81
Tetrahydrofuran (oxolane)	5	15
Tetrahydropyran (oxane)	6	~0
1,4-Dioxane	6	~0
Hexamethylene oxide (oxepane)	7	33.5[b]

[a] Sawada, H., *J. Macromol. Sci. Rev. Macromol. Chem.*, C5(1), 151, 1970.
[b] Busfield, W.K., Lee, R.M., and Merigold, D., *Makromol. Chem.*, 156, 183, 1972.

TABLE 4.3
Heterocycles That Can Be Polymerized Cationically

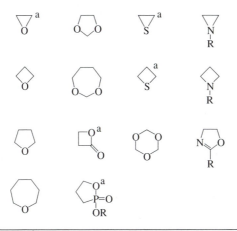

[a] These compounds can also be polymerized anionically.

oxane and 1,4 dioxane do not polymerize, presumably because of the stability of the six-membered ring, but it should be noted that six-membered lactones, lactams, and trioxane can form polymers. Monomers with rings larger than six tend to have lower ring strains, and if they can polymerize they may also have low ceiling temperatures.

Two general mechanisms for the chain propagation in cationic ring-opening polymerizations have been suggested.

One mechanism involves the primary interaction of the catalyst system with the monomer, forming an onium ion that acts as the initiating species. Propagation is then predominantly an S_N2-type substitution reaction.

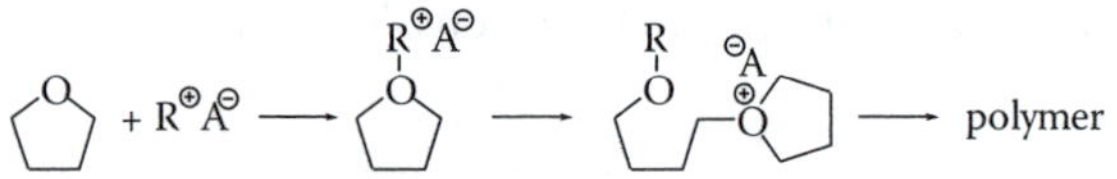

An alternative mechanism requires ring cleavage by the catalyst to form an ionic species. This is followed by attack of another monomer, with ring opening and regeneration of the active site.

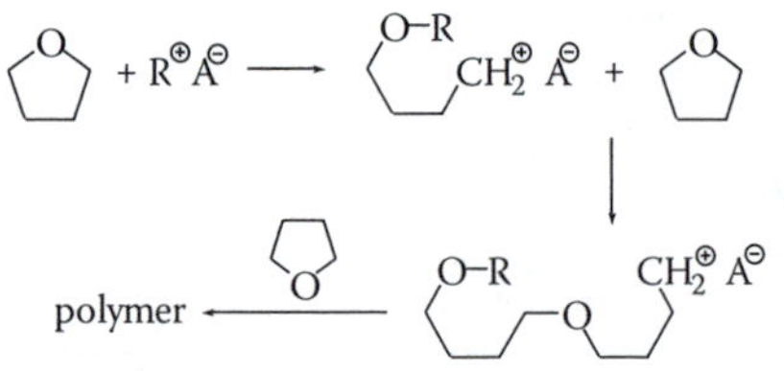

The reactions can produce commercially useful polymers, but because of the low ceiling temperatures and the tendency to degrade by an "unzipping" mechanism, end-capping of the molecules may be desirable. Thus, polyformaldehyde is an engineering plastic, prepared from 1,3,5-trioxane using boron trifluoride etherate as the initiator and can be stabilized by acetylating the terminal hydroxyl group.

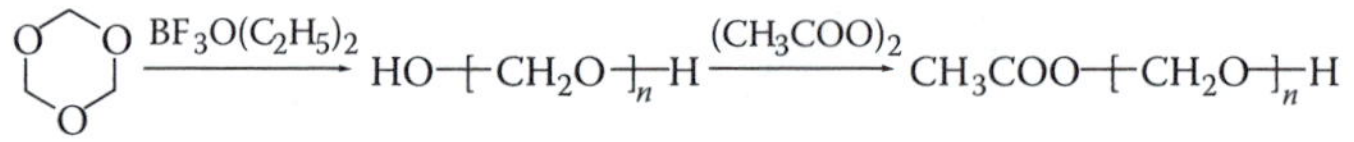

Cyclic ethers are not the only heterocycles that can be polymerized. Poly(ethylene imine) can be prepared by the ring opening of aziridine, which can be initiated by protonic acids, followed by nucleophilic attack of the monomer to give the dimer. Further addition of the monomer can lead to linear polymer formation, but the dimer

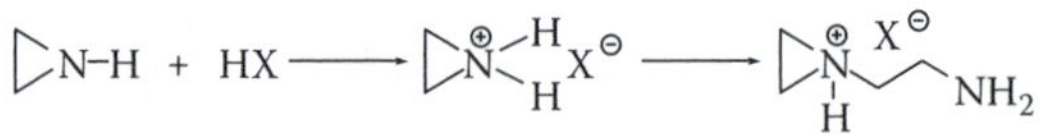

can also transfer a proton to another amino group and form an uncharged dimer. Continued chain growth leads to the formation of secondary amine functions that can react with an arizidinium ion, followed by proton transfer to form branched points in the chain.

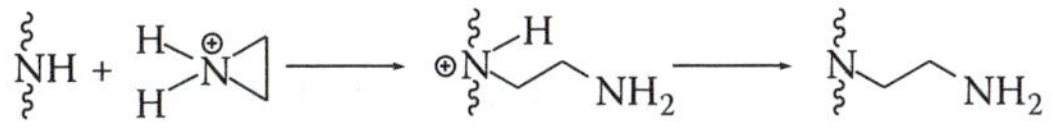

Consequently 20 to 30% branching can be present in the final product.

A linear form of poly(ethylene imine) LPEI can be prepared by using a protecting group.

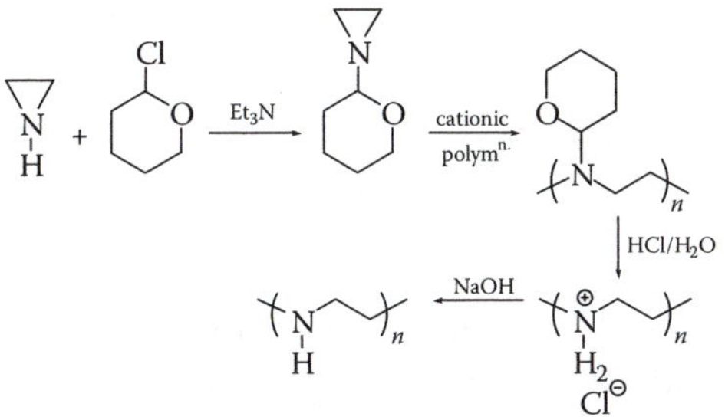

N-(α-tetrahydropyranyl)aziridine is prepared and polymerized cationically to give the poly(iminoether), which is then hydrolyzed with aqueous acid to give the polymer salt. On neutralizing, LPEI is formed.

4.9 STABLE CARBOCATIONS

Many of the uncertainties inherent in Friedel–Crafts catalyst–cocatalyst systems can be removed if stable, well-defined initiators are used. Bawn and co-workers have made use of triphenyl methyl and tropylium salts of the general formula $Ph_3C^+X^-$ and $C_7H_7^+X^-$, where X^- is a stable anion such as ClO_4^-, $SbCl_6^-$, and PF_6^-.

Initiation occurs by one of three mechanisms:

1. Direct addition: $I^+ + CH_2 = CHR \rightarrow ICH_2 - \overset{+}{C}HR$
2. Hydride extraction: $I^+ + CH_2 = CHR \rightarrow IH + CH \overset{+}{=} CH - R$
3. Electron transfer: $I^+ + CH_2 = CHR \rightarrow I^\bullet + {}^\bullet[CH_2 = CHR]^+$

The reaction of trityl hexafluorophosphate and THF has been shown to proceed without evidence of a termination reaction, and a "living" cationic system can be obtained. The reaction takes place below room temperature.

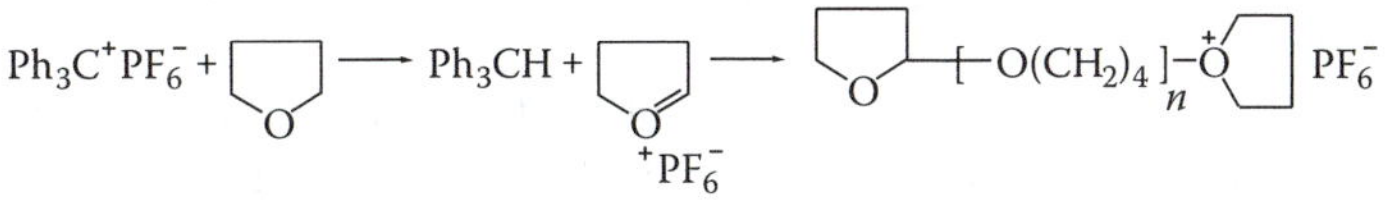

The effect of the counterion is a noticeable factor in the elimination of the termination reaction, and neither $SbCl_6^-$ nor any other anion studied has proved as good as PF_6^-.

It has been reported that when alkyl vinyl ethers are polymerized using an (HI/I_2) initiating system, living polymers are produced. The reaction involves the addition of HI to a solution of the monomer in a nonpolar solvent at low temperatures. This produces an inert adduct, but no polymer formation. However, a rapid reaction takes place when I_2 is introduced, which is much faster than is usually observed by simple I_2 initiation, and polymers with narrow molar mass distribution can be prepared.

$$H_2C{=}CH \xrightarrow{\;HI\;} HCH_2CHI \xrightarrow{\;I_2\;} \overset{\delta^+}{H}CH_2{-}\overset{\delta^-}{C}HI\cdots I_2$$
$$\underset{OR}{} \qquad \underset{OR}{} \qquad \underset{OR}{}$$

$$H{+}H_2CCH{\xrightarrow{}_n}\overset{\delta^+}{C}H_2\overset{\delta^-}{C}HI\cdots I_2 \longleftarrow$$

"living" polymer

The living character of the system is thought to arise from the stability of the propagating species, which suppresses any tendency toward termination or chain transfer reactions.

Other similar nonterminating systems have been identified, but the influence of the anion on the capability of the system in producing living polymers varies from monomer to monomer.

4.10 ANIONIC POLYMERIZATION

The polymerization of monomers with strong electronegative groups — acrylonitrile, vinyl chloride, styrene, and methyl methacrylate — can be initiated by either mechanism (2) or (3) of Section 4.1.

In (2), an ionic or ionogenic molecule is required that is capable of adding the anion to the vinyl double bond and so creating a carbanion.

$$CX \longrightarrow C^+ + X^-$$
$$X^- + M \longrightarrow MX^-$$

The gegen ion C^+ may be inorganic or organic, and typical initiators include KNH_2, n-butyl lithium, and Grignard reagents (alkyl magnesium bromides).

If the monomer has a strong electron-withdrawing group, then only a weakly positive initiator (Grignard) will be required for polymerization, but when the side group is phenyl or the electronegativity is low, a highly electropositive metal initiator, such as a lithium compound, is needed.

Mechanism (3) is the direct transfer of an electron from a donor to the monomer to form a radical anion. This can be accomplished by means of an alkali metal, and Na or K can initiate the polymerization of butadiene and methacrylonitrile; the latter reaction is carried out in liquid ammonia at 198 K.

$$Na + H_2C{=}\underset{\underset{CN}{|}}{\overset{\overset{CH_3}{|}}{C}} \longrightarrow Na^+ + \left[{}^\bullet CH_2{-}\underset{\underset{CN}{|}}{\overset{\overset{CH_3}{|}}{C}}{}^- \right]$$

The anionic reactions have characteristics similar in many ways to the cationic polymerizations. In general, they are rapid at low temperatures but are slower and less sensitive to changes in temperature than the cationic reactions. Reaction rates depend on the dielectric constant of the solvent, the resonance stability of the carbanion, the electronegativity of the initiator, and the degree of solvation of the gegen ion. Many anionic polymerizations have no formal termination step but are sensitive to traces of impurities, and as carbanions are quickly neutralized by small quantities of water, alcohol, carbon dioxide, and oxygen, these are effective terminating agents. This underlines the need for rather rigorous experimental procedures to exclude impurities, when anionic polymerizations are being studied.

4.11 LIVING POLYMERS

One of the first anionic reactions studied in detail was the polymerization of styrene in liquid ammonia, with potassium amide as initiator, reported by Higginson and Wooding (1952). The reaction scheme proposed for this contains no formal termination step, and if all the impurities that are liable to react with the carbanions are excluded from the system, propagation should continue until all monomer has been consumed, leaving the carbanion intact and still active. This means that if more monomer could be introduced, the active end would continue growing unless inadvertently terminated. These active polycarbanions were first referred to as *living polymers* by Szwarc (1968).

One of the first living polymer systems studied was the polymerization of styrene initiated by sodium naphthalene. The initiator is formed by adding sodium to a solution of naphthalene in an inert solvent, tetrahydrofuran.

The sodium dissolves to form an addition compound and, by transferring an electron, produces the green naphthalene anion radical. Addition of styrene to the system leads to electron transfer from the naphthyl radical to the monomer to form a red styryl radical anion.

It is thought that a dianion is finally formed which is capable of propagating from both ends.

$$Na^{+\,-}[Ph\ddot{C}HCH_2CH_2\dot{C}H^{\cdot}Ph]^-\, Na^+$$

Note that the absence of both a termination and a transfer reaction means that if no accidental termination by impurity occurs, the chains will remain active indefinitely.

The validity of this assumption has been demonstrated (1) by adding more styrene to the living polystyryl carbanions, and (2) by adding another monomer such as isoprene, to form a block copolymer.

The lack of a formal termination reaction in living anionic systems can be put to good use, particularly in the preparation of block copolymers as well as the more unusual star-shaped and comb-shaped structures, as will be seen in Chapter 5. Block copolymer synthesis can be effected in several ways, either by direct initiation of a new block sequence on addition of a second monomer, or by coupling reactions involving the living anionic chain ends. As there are monomers that are not reactive under anionic conditions, use of a coupling reaction between preformed blocks with functional terminal units is an alternative method. These functionalized blocks can be prepared by deliberately adding a compound at the end of the reaction that terminates the chain and is itself incorporated to form a useful end group. Thus, hydroxyl groups can be inserted by reaction with a lactone or oxirane, and the addition of water will protonate the anion to form the primary alcohol group.

$$\sim CH_2-\overset{\underset{\displaystyle R}{|}}{C}H^{\ominus} + \triangle_O \longrightarrow \sim CH_2\overset{\underset{\displaystyle R}{|}}{C}HCH_2CH_2O^{\ominus} \xrightarrow{H^{\oplus}} \sim CH_2OH$$

Carboxylic acid functions can be formed using carbon dioxide,

$$\sim CH_2-\overset{\underset{\displaystyle R}{|}}{C}H^{\ominus} + CO_2 \longrightarrow \sim \overset{\underset{\displaystyle R}{|}}{C}H-C\overset{O}{\underset{O^{\ominus}}{<}} \xrightarrow{H^{\oplus}} \sim \overset{\underset{\displaystyle R}{|}}{C}HCOOH$$

whereas phosgene in excess produces a terminal acid chloride.

$$\sim CH_2-\overset{\underset{\displaystyle R}{|}}{C}H^{\ominus} + COCl_2 \longrightarrow \sim CH_2-\overset{\underset{\displaystyle R}{|}}{C}H-C\overset{O}{\underset{Cl}{<}}$$

Ketones and amide functional groups can also be introduced.

4.12 KINETICS AND MOLAR MASS DISTRIBUTION IN LIVING ANIONIC SYSTEMS

It can be assumed that in a system designed specifically to produce a living polymer, the initiator is completely dissociated in the medium before monomer is added. Under those circumstances, free-ion propagation of the chain should occur when monomer is brought in contact with the initiator, and all chains will begin to grow at approximately the same time. This should lead to a polymer sample with a very narrow distribution of molar masses and, as we shall see, a Poisson-type distribution is expected. If the initial concentration of the initiator is [GA], then the initiating steps are

$$GA \rightarrow G^{\oplus} + A^{\ominus}$$

$$G^{\oplus} + A^{\ominus} + M \rightarrow AM_1^{\ominus} + G^{\oplus}$$

(4.9)

and chain growth will start at $[AM_1^{\ominus}] = [GA]$ centers. The rate of propagation is then

$$-\frac{d[M]}{dt} = v_p = k_p[AM^{\ominus}][M] = k_p[GA][M]$$

(4.10)

and as this is a first-order rate equation with respect to the monomer concentration, it can be integrated to give

$$[M] = [M]_0 \exp(-k_p[GA]t)$$

(4.11)

where $[M]_0$ is the monomer concentration at $t = 0$.

The kinetic chain length $(\bar{v})$ at any time during the reaction is then

$$\bar{v} = ([M]_0 - [M])/[GA]$$

(4.12)

and substituting Equation 4.11 gives

$$\bar{v} = \frac{[M]_0}{[GA]}\{1 - \exp(-k_p[GA]t)\}$$

(4.13)

demonstrating that after all the monomer is consumed, i.e., $(t \rightarrow \infty)$, then

$$\bar{v} = [M]_0/[GA] = x_n$$

when there is no termination reaction.

It can be shown that this type of polymerization leads to a Poisson distribution of chain lengths by considering the following steps.

The rate of addition of the second monomer to the active center

$$AM_1^{\ominus} + M \rightarrow AM_2^{\ominus}$$

is (always assuming the presence of $G^{\oplus}$)

$$-\frac{d[AM_1^{\ominus}]}{dt} = k_p[AM_1^{\ominus}][M] = k_p[AM_1^{\ominus}][M]_0 \exp(-k_p[GA])t)$$

(4.14)

which on integration, remembering that $[AM_1^\ominus] = [GA]$ at $t = 0$, is

$$[AM_1^\ominus] = [GA] \exp \left\{ \frac{-[M]_0}{[GA]}[1 - \exp (k_p[GA]t)] \right\} \tag{4.15}$$

or in a simplified form using Equation 4.13, this becomes

$$[AM_1^\ominus] = [GA] \exp (-\bar{v}) \tag{4.16}$$

Consider next the addition of the third monomer

$$[AM_2^\ominus] + M \rightarrow AM_3^\ominus$$

and the rate of change in concentration of species $[AM_2^\ominus]$,

$$\frac{d[AM_2^\ominus]}{dt} = k_p[AM_1^\ominus][M] - k_p[AM_2^\ominus][M]$$

or, again using Equation 4.11 and Equation 4.16,

$$\frac{d[AM_2^\ominus]}{dt} = k_p[M]_0 \exp (-k_p[GA]t)\{[GA] \exp (-\bar{v}) - [AM_2^\ominus]\} \tag{4.17}$$

This can be simplified by differentiating Equation 4.13 with respect to t

$$d\bar{v} = k_p[M]_0 \exp (-k_p[GA])t) dt \tag{4.18}$$

and substituting to give

$$\frac{d[AM_2^\ominus]}{d\bar{v}} = [GA] \cdot \bar{v} \cdot \exp (-\bar{v}) \tag{4.19}$$

Generalization of this analysis for $n - 1$ additions of monomer to the first active site $[AM^\ominus]$ to give an n-mer chain leads to

$$[AM_2^\ominus] = [GA] \cdot \bar{v}^{n-1} \cdot \frac{\exp (-\bar{v})}{(n-1)!} \tag{4.20}$$

This can be expressed as the number fraction N_n/N, where N_n is the number of chains with degree of polymerization n, and N is the total number of chains

$$\frac{N_n}{N} = \frac{[AM_n^{\ominus}]}{[GA]} = \frac{\bar{v}^{n-1} \exp{(-\bar{v})}}{(n-1)!} \tag{4.21}$$

which is the form of a Poisson distribution.

The number fraction distribution is shown in Figure 4.1 for $\bar{v} = 50$, for which the heterogeneity index

$$\frac{M_w}{M_n} = 1 + \frac{\bar{v}}{(\bar{v}+1)^2} \tag{4.22}$$

can be calculated from Equation 4.22 to be 1.02.

Also shown for comparison is the distribution curve for a radical-initiated polymerization for a similar kinetic chain length, calculated for termination by combination.

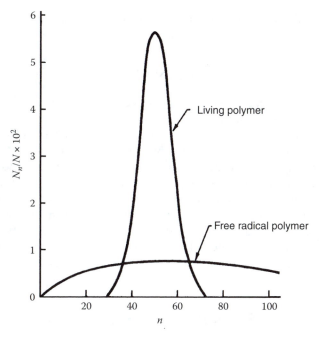

FIGURE 4.1 Number distribution curves for a polymer prepared by anionic living polymerization and, for comparison, that generated by free-radical polymerization with termination by combination (both curves calculated for $\bar{v} = 50$).

4.13 METAL ALKYL INITIATORS

The organolithium derivatives, such as *n*-butyl lithium, are particular members of this group of electron-deficient initiators. In general, the initiation involves addition to the double bond of the monomer

$$RLi + H_2C=CHR_1 \longrightarrow RCH_2-\overset{\overset{\displaystyle H}{|}}{\underset{\underset{\displaystyle R_1}{|}}{C}}{}^- Li^+$$

and propagation is then

$$RCH_2\overset{\overset{\displaystyle H}{|}}{\underset{\underset{\displaystyle R_1}{|}}{C}}{}^- Li^+ + nH_2C=CHR_1 \longrightarrow R\left(CH_2-CHR_1\right)_n CH_2\overset{\overset{\displaystyle H}{|}}{\underset{\underset{\displaystyle R_1}{|}}{C}}{}^- Li^+$$

Kinetic analysis of the reactions shows that the initiation is not a simple function of the basicity of R, however, owing to the characteristic tendency for organolithium compounds to associate and form tetramers or hexamers. The kinetics are usually complicated by this feature, which is solvent dependent and, consequently, fractional reaction orders are commonplace.

The alkyl lithiums have proved commercially useful in diene polymerization, and some steric control over the polymerization can be obtained.

4.14 SOLVENT AND GEGEN ION EFFECTS

Both the solvent and gegen ion have a pronounced influence on the rates of anionic polymerizations. The polymerization rate generally increases with increasing polarity of the solvent; for example, $k_p = 2.0$ dm^3 mol^{-1} s^{-1} for the anionic polymerization of styrene in benzene, but $k_p = 3800$ dm^3 mol^{-1} s^{-1} when the solvent is 1,2-dimethoxyethane. Unfortunately, the dielectric constant is not a useful guide to polarity or solvating power in these systems, as $k_p = 550$ dm^3 mol^{-1} s^{-1} when the solvent is changed to THF, whose dielectric constant ε is higher than ε for 1,2-dimethoxyethane.

The influence of the gegen ion on the polymerization of styrene in THF at 298 K is shown in Figure 4.2, compiled from data obtained by Szwarc (1968). Clearly, the smaller Li$^+$ ions can be solvated to a greater extent than the larger ones, and the decreasing rate reflects the increasing tendency for ion pairs to be the active species rather than free ions as the solvating power of the solvent deteriorates.

4.15 ANIONIC RING-OPENING POLYMERIZATION

The ring-opening polymerization of oxiranes, thiiranes, and thietanes can be initiated by both cationic and anionic methods, but there are some heterocyclic compounds such as lactones and lactams that are more suited to the anionic technique.

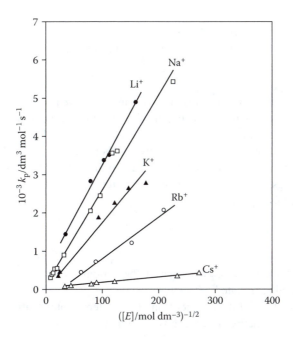

FIGURE 4.2 Behavior of the experimental propagation rate constant k_p as a function of the concentration [E] of "living ends" for various salts of living polystyrene in tetrahydrofuran at 298 K. (From Szwarc, M., *Carbanions, Living Polymers and Electron Transfer Processes*, Interscience Publishers, 1968. With permission.)

Polyethylene oxide is readily prepared by the reaction of ethylene oxide with the potassium salt of an alcohol, and the chain is terminated by a transfer reaction to excess alcohol present.

$$\text{(epoxide)} + K^{\oplus}OR^{\ominus} \longrightarrow ROCH_2CH_2O^{\ominus}\ K^{\oplus}$$
$$\Big\downarrow \text{monomer}$$
$$\sim\!\!\sim CH_2CH_2OH + K^{\oplus}OR^{\ominus} \xleftarrow{\ ROH\ } \sim\!\!\sim(CH_2CH_2O)_{\!n}CH_2CH_2O^{\ominus}\ K^{\oplus}$$

Lactones can be used to prepare polyesters, but the ring size is an important consideration, e.g., the five-membered ring γ-butyrolactone will not polymerize, whereas the six-membered ring δ-valerolactone reacts.

$$\text{(δ-valerolactone)} \longrightarrow \left(\!(CH_2)_4\!-\!\overset{\overset{\displaystyle O}{\displaystyle \|}}{C}\!-\!O\right)_{\!n}$$

Certain lactams will undergo ring-opening polymerizations to give polyamides, and nylon-6 can be prepared from the water-catalyzed reaction of caprolactam.

$$H_2C-C\overset{H_2}{\underset{C}{}}\overset{O}{\underset{}{}} \quad\longrightarrow\quad +(CH_2)_5-\overset{}{\underset{O}{C}}-\overset{}{\underset{H}{N}}\overset{}{\xrightarrow{}}_n$$

Care must be taken with this reaction, and an alternative procedure is to use a two-component catalyst system by reacting the lactam with a base to produce an activated monomer. This then reacts with a promoter such as the acyl lactam, which initiates the ring-opening growth of the linear polymer. For a series of cyclic lactams, the reaction rates are a function of ring size and are in the order of 8- > 7- > 11- >> 5- or 6-membered rings. These are important commercial processes, and nylon-4 is also prepared using this type of reaction.

PROBLEMS

1. Write down the first two steps of the cationic polymerization of vinyl methyl ether initiated with HF.

2. NMR analysis of the product from the low-temperature cationic polymerization of 4-methyl-1-pentene showed that both isobutyl and methyl side groups were present. Explain this finding.

3. The structures of the polymers resulting from the cationic polymerization of 4-methyl-1-pentene with an $AlCl_3$ catalyst in C_2H_5Cl solvent at −78, −50, and +5°C were investigated by ^{13}C nuclear magnetic resonance spectroscopy (Mizuno and Kawachi, 1992). The main product of the reaction is structurally similar to an ethylene isobutylene alternating copolymer.
   a. Propose a mechanism for the cationic polymerization.
   b. Identify other possible products of this reaction.

4. The kinetics of the living polymerization of isobutylene using 5-*tert*-butyl-1,3-*bis*(2-chloro-2-propyl)benzene (*t*-Bu-*m*-DCC)/TiCl$_4$ initiating system in 2,4-dimethyl pyridine were determined by gravimetrically monitored polymer yield as a function of reaction time (Storey and Choate, 1997). For this reaction to be classified as "living," how would you expect:
   a. ln $[M_o]/[M]$ to vary with time?
   b. M_n to vary with degree of conversion?
   A set of apparent rate constant values, k_{app}, were obtained from ln $[M_o]/[M]$ vs. time plots, and a set of k_{app} values for different *t*-Bu-*m*-DCC concentrations (all other parameters being constant) were collected. From these, estimate the kinetic order of the reaction with respect to [*t*-Bu-*m*-DCC].

[*t*-Bu-*m*-DCC] × 10^3/M	k_{app} × 10^4/s^{-1}
0.40	3.4
0.60	5.2
0.80	7.0
1.10	9.6
1.30	12

5. The polymerization of isobutylene (IB) has been carried out using a 1,4-*bis*(1-chloro-1-methylethyl)benzene (DiCumCl)/BCl$_3$ initiating system in chloroform at 75°C. An incremental monomer addition method was used, and data are summarized below, where W$_p$ is the weight of polymer and M$_n$ is the number average molecular weight. If [DiCumCl] = 4.6 × 10^{-3} mol l^{-1} and V$_o$ = 25 ml, establish whether the system shows characteristics of a living polymerization.

W$_p$ (g)	M$_n$ × 10^{-3} (g mol^{-1})
0.1404	2.5
0.2871	3.5
0.4211	4.4
0.5720	5.0
0.6773	5.8
0.7626	6.4
0.8739	7.3

6. From the definition of number average, derive an expression for M_n as a function of the kinetic chain length, $\bar{v}$, assuming a Poisson distribution. In doing so, remember that:

$$e^x = \sum_{i=0}^{\infty} \frac{x^i}{i} = \sum_{i=0}^{\infty} \frac{x^{i-1}}{(i-1)!}$$

and

$$i\,\bar{v}^{i-1} = \mathrm{d}\,(\bar{v}^{i})/\mathrm{d}\bar{v}$$

though $e^{-\bar{v}}$ does not depend on i.
Because:

$$N_w = \frac{1 + 3\bar{v} + \bar{v}^2}{1 + \bar{v}}$$

demonstrate that

$$\frac{N_w}{N_n} = 1 + \frac{\bar{v}}{\left(1 + \bar{v}\right)^2}$$

7. One of the first anionic reactions to be studied was the polymerization of styrene in liquid ammonia (Higginson and Wooding, 1952). This type of anionic polymerization consists of a two-step initiation process:

$$KNH_2 \rightarrow K^+ + :NH_2^-$$

$$:NH_2^- + M \xrightarrow{k_1} H_2N-M^-$$

involving the dissociation of the potassium amide into the constituent ions followed by addition of the anion to M to generate an active chain carrier. By considering that this latter step is slow relative to the first and thus rate determining, and that chain termination occurs by transfer to solvent:

$$H_2N(M)_n M^- + NH_3 \xrightarrow{k_1} H_2N(M)_n MH + NH_2^-$$

derive an expression for the rate of polymerization and kinetic chain length. Given that the activation energy for the transfer process is larger than that for propagation and that the overall activation energy is +38 kJ mol⁻¹, predict how chain length and reaction rate will vary with increasing temperature.

8. In the anionic polymerization of styrene in liquid ammonia catalyzed by KNH_2 (Higginson and Wooding, 1952) considered in the previous problem, it was found that plots of 1/[styrene] vs. time were linear. What is the order of the reaction with respect to [styrene]? Is this consistent with the derivation in Problem 7?

9. Design a synthetic route for the preparation of poly(styrene-*block*-methylmethacrylate).

10. Suggest three possible routes to prepare poly(styrene-*block*-butadiene-*block*-styrene) by living anionic polymerization.

11. Consider the anionic polymerization of isoprene with butyllithium. What molar concentration of initiator would be required to obtain a sample with molecular weight 136,000 g mol⁻¹ if the isoprene concentration is 2 M? What is the expected polydispersity of this sample?

12. In anionic polymerization, which of these two monomers is more reactive: methyl methacrylate or methacrylonitrile?

REFERENCES

Bhattacharyya, D.N., Lee, C.L., Smid, J., and Szwarc, M., *J. Phys. Chem.*, 69, 612, 1965.
Bywater, S., Polymerization initiated by lithium and its compounds, *Adv. Polym. Sci.*, 4, 66 1965.
Bywater, S. and Worsfold, D.J., *J. Phys. Chem.*, 70, 162 1966.
Higginson, W.C.E. and Wooding, N.S., *J. Chem. Soc.*, 760, 1952.
Kennedy, J.P. and Lenger, A.W., Recent advances in cationic polymerization, *Adv. Polym. Sci.*, 3, 508, 1965.
Mizuno, A. and Kawachi, H., *Polymer*, 33, 57, 1992.
Storey, R.F. and Choate, K.R., *Macromolecules*, 30, 4799, 1997.
Szwarc, M., *Carbanions, Living Polymers and Electron Transfer Processes*, Interscience Publishers, 1968.

BIBLIOGRAPHY

Allen, G. and Bevington, J.C., Eds., *Comprehensive Polymer Science,* Vol. 3, Pergamon Press, 1989.

Goethals, E.J., Ed., *Cationic Polymerization and Related Processes*, Academic Press, 1984.

Hogen-Esch, T.E. and Smid, J., Eds., *Recent Advances in Anionic Polymerization*, Elsevier Science Publishing, 1987.

Hsieh, H.L. and Quirk, R.P., *Recent Anionic Polymerization: Principles and Practical Applications,* Marcel Dekker, 1996.

Ivin, K.J. and Saegusa, T., Eds., *Ring Opening Polymerization*, Elsevier Applied Science Publishers, 1984.

Kennedy, J.P. and Marechal, E., *Carbocationic Polymerization*, Krieger Publishing Company, 1990.

Lenz, R.W., *Organic Chemistry of Synthetic High Polymers*, Interscience Publishers, 1967, chap. 13, 14.

Morton, M., *Anionic Polymerization: Principles and Practice*, Academic Press, 1983.

Plesch, P.H., *The Chemistry of Cationic Polymerization*, Pergamon Press, 1963.

Puskas, J.E. and Kaszas, G., Carbocationic polymerization, in *Encyclopedia of Polymer Science and Technology*, John Wiley and Sons, 2003.

Quirk, R.P., Anionic polymerization, in *Encyclopedia of Polymer Science and Technology*, John Wiley and Sons, 2002.

Smith, D.A., *Addition Polymers*, Butterworths, 1968, chap. 3.

Szwarc, M., *Carbanions, Living Polymers and Electron Transfer Processes*, Interscience Publishers, 1968.

5 Linear Copolymers and Other Architectures

5.1 GENERAL CHARACTERISTICS

In the addition reactions considered in the previous chapters, the emphasis has been on the formation of a polymer from only one type of monomer. Often, it is found that these homopolymers have widely differing properties and one might think that by using physical mixtures of various types, a combination of all the desirable properties would be obtained in the resulting material. Unfortunately, this is not always so; instead, it is more likely that the poorer qualities of each become exaggerated in the mixture.

An alternative approach is to try to synthesize chains containing more than one monomer and examine the behavior of the product. By choosing two (or perhaps more) suitable monomers, A and B, chains incorporating both can be prepared using free radical or ionic initiators, and many of the products exhibit the better qualities of the parent homopolymers. This is known as *copolymerization*.

Even in the simplest case, that of copolymerization involving two monomers, a variety of structures can be obtained, and five important types exist:

1. *Statistical copolymers:* These are formed when irregular propagation occurs and the two units enter the chain in a statistical fashion, i.e., ⌇⌇ABBAAAABAABBBA⌇⌇. This is the most commonly encountered structure.

2. *Alternating copolymers:* These are obtained when equimolar quantities of two monomers are distributed in a regular alternating fashion in the chain ⌇⌇ABABABA⌇⌇. Many of the step-growth polymers formed by the condensation of two (A—A), (B—B) type monomers could be considered alternating copolymers, but these are commonly treated as homopolymers with the repeat unit corresponding to the dimeric residue.

3. *Block copolymers:* Instead of having a mixed distribution of the two units, the copolymers may contain long sequences of one monomer joined to another sequence or block of the second. This produces a linear copolymer of the form AA ⌇⌇ AABBB⌇⌇B, i.e., an {A} {B} block, but other combinations are possible.

4. *Graft copolymers:* A nonlinear or branched block copolymer is formed by attaching chains of one monomer to the main chain of another homopolymer.

5. *Stereoblock copolymers:* Finally, a very special structure can be formed from one monomer in which the distinguishing feature is the tacticity of each block, i.e.,

$$\text{\textsterling}(-CH_2.\underset{X\ H}{C}-)_m\text{\textsterling}(-CH_2.\underset{H\ X}{C}-)_n\text{\textsterling}$$

In general, block and graft copolymers possess the properties of both homopolymers, whereas the random and alternating structures have characteristics that are more of a compromise between the extremes.

It soon becomes obvious that the factors influencing the course of even simple copolymerizations are much more complex than those in a homopolymerization. For example, attempts to polymerize styrene and vinyl acetate result in copolymers containing only 1 to 2% of vinyl acetate, whereas a small quantity of styrene will tend to inhibit the free-radical polymerization of vinyl acetate. At the other extreme, two monomers such as maleic anhydride and stilbene are extremely difficult to polymerize separately, but form copolymers with relative ease.

5.2 COMPOSITION DRIFT

It was realized by Staudinger, as early as 1930, that when two monomers copolymerize, the tendency of each monomer to enter the chain can differ markedly. He found that if an equimolar mixture of vinyl acetate and vinyl chloride was copolymerized, the chemical composition of the product varied throughout the reaction, and that the ratio of chloride to acetate in the copolymers changed from 9:3 to 7:3 to 5:3 to 5:7.

This phenomenon, known as *composition drift*, is a feature of many copolymerizations and has been attributed to the greater reactivity of one of the monomers in the mixture. Consequently, in a copolymerization, it is necessary to distinguish between the composition of a copolymer being formed at any one time in the reaction and the overall composition of the polymer formed at a given degree of conversion.

Two major questions arise that must be answered if the criteria controlling copolymerizations are to be formulated:

1. Can the composition of the copolymer be predicted when it is prepared from the restricted conversion of a mixture of two monomers?
2. Can one predict the behavior of two monomers that have never reacted before?

To answer the first question, we must explore the relative reactivity of one monomer to another, whereas an attempt to answer the second is embodied in the Q–e scheme.

5.3 THE COPOLYMER EQUATION

To begin to answer the first question, we must establish a suitable kinetic scheme. The following group of homo- and heteropolymerization reactions were proposed by Dostal in 1936 for a radical copolymerization between two monomers M_1 and M_2 and, ultimately, extended and formalized by a number of workers who established a practical equation from the reactions:

$$\sim\!\!\sim M_1^\bullet + M_1 \xrightarrow{\;k_{11}\;} \sim\!\!\sim M_1^\bullet \tag{5.1a}$$

$$\sim\!\!\sim M_1^\bullet + M_2 \xrightarrow{\;k_{12}\;} \sim\!\!\sim M_2^\bullet \tag{5.1b}$$

$$\sim\!\!\sim M_2^\bullet + M_2 \xrightarrow{\;k_{22}\;} \sim\!\!\sim M_2^\bullet \tag{5.1c}$$

$$\sim\!\!\sim M_2^\bullet + M_1 \xrightarrow{\;k_{21}\;} \sim\!\!\sim M_1^\bullet \tag{5.1d}$$

where k_{11} and k_{22} are the rate constants for the *self-propagating* reactions and k_{12} and k_{21} are the corresponding *cross-propagation* rate constants.

Under steady-state conditions, and assuming that the radical reactivity is independent of chain length and depends only on the nature of the terminal unit, the rate of consumption of M_1 from the initial reaction mixture is then

$$-d[M_1]/dt = k_{11}[M_1][M_1^\bullet] + k_{21}[M_1][M_2^\bullet], \tag{5.2}$$

and M_2 by

$$-d[M_2]/dt = k_{22}[M_2][M_2^\bullet] + k_{12}[M_2][M_1^\bullet]. \tag{5.3}$$

The *copolymer equation* can then be obtained by dividing Equation 5.2 by Equation 5.3 and assuming that $k_{21}[M_1][M_2^\bullet] = k_{12}[M_2][M_1^\bullet]$ for steady-state conditions, so that

$$d[M_1]/d[M_2] = ([M_1]/[M_2])\{(r_1[M_1] + [M_2])/([M_1] + r_2[M_2])\}, \tag{5.4}$$

where $k_{11}/k_{12} = r_1$, and $k_{22}/k_{21} = r_2$.

The quantities r_1 and r_2 are the *relative reactivity ratios*, defined more generally as the ratio of the reactivity of the propagating species with its own monomer to the reactivity of the propagating species with the other monomer.

5.4 MONOMER REACTIVITY RATIOS

The copolymer equation provides a means of calculating the amount of each monomer incorporated in the chain from a given reaction mixture or feed when the reactivity ratios are known. It shows that if monomer M_1 is more reactive than M_2, then M_1 will enter the copolymer more rapidly; consequently, the feed becomes progressively poorer in M_1, and composition drift occurs. The equation is then an "instantaneous" expression, which relates only to the feed composition at any given time.

As r_1 and r_2 are obviously the factors that control the composition of the copolymer, one must obtain reliable values of r for each pair of monomers (comonomers) if the copolymerization is to be completely understood and controlled. This can be achieved by analyzing the composition of the copolymer formed from a number of comonomer mixtures with various $[M_1]/[M_2]$ ratios, at low (5 to 10%) conversion (where monomer reactivities do not differ greatly).

If we now define F_1 and F_2 as the mole fractions of monomers M_1 and M_2 being added to the growing chain at any given time, and f_1 and f_2 as the corresponding mole fractions of the monomers in the feed mixture, then the copolymer equation can be written as

$$F_1 = (r_1 f_1^2 + f_1 f_2)/(r_1 f_1^2 + 2 f_1 f_2 + r_2 f_2^2) \qquad (5.5)$$

This can be rearranged and simplified further if

$$F = (F_1/F_2) \text{ and } f = (f_1/f_2)$$

to give

$$\{f(1-F)/F\} = r_2 - (f^2/F)r_1$$

which is the linear form of Equation 5.5 proposed by Finemann and Ross.

A plot of $\{f(1-F)/F\}$ against (f^2/F) should then be linear and yield r_1 from the slope and r_2 from the intercept. Several other linear forms have been suggested for the determination of the reactivity ratios, but it is now much easier to estimate r_1 and r_2 from a nonlinear least-squares fit to the composition data.

Some representative values of r_1 and r_2 for a number of comonomers are shown in Table 5.1. These are seen to differ widely.

5.5 REACTIVITY RATIOS AND COPOLYMER STRUCTURE

It is obvious, from the wide-ranging values of reactivity ratios shown in Table 5.1, that the structure of the copolymer will also be a function of r_1 and r_2.

Several types of copolymeric structure can be obtained, as shown in Section 5.1, and the influence of monomer reactivity ratios can be illustrated by examining plots of the "instantaneous" copolymer composition F_1 against the "instantaneous" monomer composition in the feed f_1, for various combinations of r_1 and r_2.

Consider first the unusual case when $r_1 \approx r_2 \approx 1$. This situation arises when little or no preference for either monomer is shown by the polymer radical, i.e., $k_{11} \approx k_{12}$ and $k_{22} \approx k_{21}$, and copolymerization is then entirely random. Under these conditions, $F_1 = f_1$, and this is represented by curve I in Figure 5.1. As this plot is reminiscent of corresponding plots for an ideal system of two liquids, the copolymers formed under these conditions, and indeed any copolymer in which the product $(r_1 r_2)$ is unity, are called *ideal* copolymers. Completely random copolymers are formed from the comonomer pairs: tetrafluoroethylene + monochlorotrifluoroethylene; isoprene + butadiene; and vinyl acetate + isopropenyl acetate. However, if $r_1 > 1$ and $r_2 < 1$

TABLE 5.1

Some Reactivity Ratios r_1 and r_2 for Free Radical Initiated Copolymerizations

M_1	M_2	r_1	r_2	r_1r_2
Acrylonitrile	Acrylamide	0.87	1.37	1.17
	Butadiene	2.0	0.1	0.2
	Methyl acrylate	0.84	0.83	0.70
	Styrene	0.01	0.40	0.004
	Vinyl acetate	6.0	0.07	0.42
Butadiene	Methyl methacrylate	0.70	0.32	0.22
	Styrene	1.40	0.78	1.1
Ethylene	Propylene	17.8	0.065	1.17
Maleic anhydride	Acrylonitrile	0	6	0
	Methyl acrylate	0.02	3.5	0.07
Methyl methacrylate	Vinyl acetate	22.2	0.07	1.55
	Vinyl chloride	10	0.1	1.0
Styrene	p-Fluorostyrene	1.5	0.7	1.05
	α-Methylstyrene	2.3	0.38	0.87
	Vinyl acetate	55	0.01	0.55
	2-Vinyl pyridine	0.55	1.14	0.63
Tetrafluoroethylene	Monochlorotrifluoroethylene	1.0	1.0	1.0
Vinyl chloride	Vinyl acetate	1.35	0.65	0.88
	Vinylidene chloride	0.5	0.001	0.0005

or vice versa, but $r_1r_2 = 1$, there will be composition drift of the kind shown in Figure 5.2, and when the differences between r_1 and r_2 become large, departure from ideal conditions is significant. The curve for $r_1 = 5.0$ and $r_2 = 0.2$ clearly shows that M_1 enters the copolymer more frequently than M_2, and random copolymers become increasingly difficult to prepare.

Values of (r_1r_2) are, however, more likely to be above or below unity, and curve II of Figure 5.1 represents the nearly ideal pair acrylamide + acrylonitrile, for which $r_1r_2 = 1.17$. This shows the slight deviation from ideal copolymerization and illustrates the use of the curve as a guide to the composition drift, which can be expected when $r_1 \neq r_2$.

In systems where r_1 and r_2 are both less than unity, copolymerization is favored and only short sequences of M_1 and M_2 tend to form. In the extreme case, when k_{11} and k_{22} are zero, $r_1 = r_2 = 0$, and a regular alternating (1:1) copolymer is formed; this is represented by curve III of Figure 5.1. Strictly alternating copolymers can be prepared from the comonomers maleic anhydride + styrene, fumaronitrile + α-methyl styrene, and others; however, these are rather special cases, and more generally, there is a greater likelihood for the system to lie in the range $0 < r_1r_2 < 1$. Thus, the closer the product (r_1r_2) is to zero, the greater is the tendency for M_1 and M_2 to alternate in the chain. The copolymer composition plots for these types of system are sigmoidal

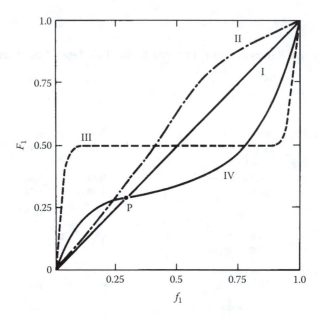

FIGURE 5.1 Variation of F_1 with f_1 for copolymerizations which are: I, completely random; II, almost ideal (i.e., $r_1r_2 = 1.17$); III, regular alternating; IV, intermediate between alternating and random (i.e., $0 < r_1r_2 < 1$), i.e., statistical.

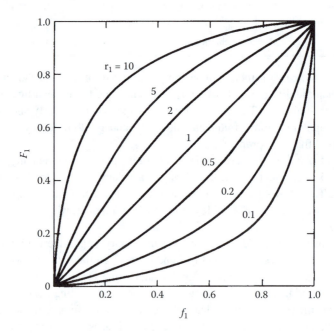

FIGURE 5.2 Plot of the mole fraction F_1 of comonomer 1 in the copolymer as a function of the mole fraction f_1 of comonomer 1 in the feed for copolymerizations in which $r_1r_2 = 1$, showing compositional variations for several indicated values of r_1.

(curve IV) and cross the ideal line at a point P. At this point $F_1 = f_1$, and P indicates the *azeotropic copolymer composition*. This is an important feature of the system, as it represents a feed composition that will produce a copolymer of constant composition throughout the whole reaction, without having to make adjustments to the feed. This type of copolymerization, in which no composition drift is observed, is known as *azeotropic* copolymerization, and the critical composition f_{1C} required to obtain the necessary conditions can be calculated from

$$f_{1C} = (1 - r_2)/\{2 - (r_1 + r_2)\}. \tag{5.6}$$

When r_1 and r_2 are greater than unity, i.e., $r_1 r_2 \gg 1$, conditions favoring long sequences or blocks of each monomer in the copolymer are obtained, and, in extreme cases, homopolymer formation may predominate.

5.6 MONOMER REACTIVITIES AND CHAIN INITIATION

Monomer reactivities have been found to be essentially independent of the free-radical process used (e.g., bulk, emulsion) but can be affected tremendously for the same pair of monomers if the chain carrier is changed.

For example, monomer reactivity ratios for styrene and methyl methacrylate in a free-radical copolymerization are $r_1 = 0.5$, $r_2 = 0.44$. This represents a statistical copolymerization. Contrast this with the anionic reaction, where $r_1 = 0.12$ and $r_2 = 6.4$, or the cationic reaction where $r_1 = 10.5$ and $r_2 = 0.1$. Obviously, the propagation rates are no longer similar, and this is represented in Figure 5.3, where it can be seen that the anionic technique produces a copolymer rich in methyl methacrylate, whereas the cationic system leads to a copolymer with a high styrene content.

This illustration merely accentuates the need to answer the questions, why do the values of r_1 and r_2 differ so widely and why does r for a given monomer change when the comonomer is changed?

5.7 INFLUENCE OF STRUCTURAL EFFECTS ON MONOMER REACTIVITY RATIOS

The propagation rates in ionic polymerizations are influenced by the polarity of the monomers; in free-radical reactions, the relative reactivity of the monomers can be correlated with resonance stability, polarity, and steric effects; we shall consider only radical copolymerizations.

5.7.1 RESONANCE EFFECTS

The reactivity of a free radical is known to depend on the nature of the groups in the vicinity of the radical. If, in a vinyl monomer ($CH_2=CHR$), the group R is capable of aiding delocalization of the radical, the radical stability will increase, and some

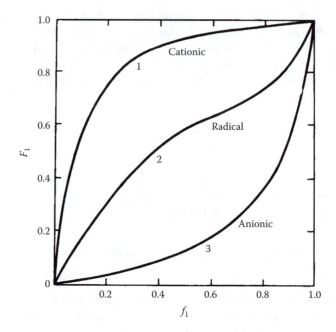

FIGURE 5.3 Copolymerization of styrene and methyl methacrylate initiated by 1, SnCl₄; 2, benzoyl peroxide; and 3, sodium in liquid ammonia, showing the vast differences in the dependence of F_1 on f_1 for the various types of initiator, in which component 1 is styrene. (From Pepper, D.C., *Q. Rev.*, 8, 88, 1954. With permission.)

of the more common substituents can be arranged in order of increasing electron withdrawal:

$$C_6H_5 > -CH=CH_2 > -\underset{O}{\overset{\parallel}{C}}-CH_3 > C \equiv N > -\underset{O}{\overset{\parallel}{C}}-OR > Cl > R > -O-\underset{O}{\overset{\parallel}{C}}-CH_3$$

Thus, styrene ($R = C_6H_5$) has a radical whose resonance stabilization is high (84 kJ mol⁻¹), whereas vinyl acetate

$$R = (O-\underset{O}{\overset{\parallel}{C}}-CH_3)$$

has a very unstable radical.

As a reactive monomer forms a stable free radical, the radical reactivity will be the reverse order of the groups above. This means that monomers containing conjugated systems (styrene, butadiene, acrylates, acrylonitriles, etc.) will be highly reactive monomers but will form stable, and so relatively unreactive, radicals. Conversely, unconjugated monomers (ethylene, vinyl halides, vinyl acetate, etc.) are relatively unreactive toward free radicals but will form unstable and highly reactive adducts.

The suppression of radical reactivity toward a monomer is also found to be a stronger effect than the corresponding enhancement of monomer reactivity. This is true for styrene, whose radical is about 10^3 times less reactive toward a given monomer than the vinyl acetate radical, but the styrene monomer is only 50 times more reactive toward a given radical than the vinyl acetate monomer.

We can now see why styrene and vinyl acetate are such a poor comonomer pair. The copolymerization requires that the stable styrene radical react with the unreactive vinyl acetate monomer, but this is such a slow process that the styrene tends to homopolymerize.

Broadly speaking, an efficient copolymerization tends to take place when the comonomers are either both reactive or both relatively unreactive, but not when one is reactive and the other unreactive. As with most generalizations, this is rather an extreme statement and cannot be treated too rigorously, especially when one realizes that resonance is not the only factor contributing to copolymerization behavior, and that both steric and polar effects have to be considered.

5.7.2 POLAR EFFECTS

It has been observed that strongly alternating copolymers are formed when comonomers with widely differing polarities are reacted together. The polarity is again determined by the side group. Thus, electron-withdrawing substituents, e.g., $-COOR$, $-CN$, $-COCH_3$, all decrease the electron density of the double bond in a vinyl monomer relative to ethylene, whereas electron-donating groups, e.g., $-CH_3$, $-OR$, $-OCOCH_3$, increase the electron density. Hence, acrylonitrile forms statistical copolymers with methyl vinyl ketone ($r_1 r_2 = 1.1$), whereas copolymerization of acrylonitrile with vinyl ethers leads to alternating structures ($r_1 r_2 \approx 0.0004$).

Polar forces also help to overcome steric hindrance. Neither maleic anhydride nor diethyl fumarate will form homopolymers, but both will react with styrene, stilbene, and vinyl ethers to form alternating copolymers because of the strong polar interaction. For example, the reaction between stilbene and maleic anhydride is

5.8 THE Q–e SCHEME

All these factors contribute to the rate of copolymerization, but in a manner that makes it difficult to distinguish the magnitude of each effect.

Attempts to correlate copolymerization tendencies are thus mainly on a semi-empirical footing and must be treated as useful approximations rather than rigorous relations. A generally useful scheme was proposed by Alfrey and Price, who denoted the reactivities or resonance effects of monomers by a quantity Q and radicals by P, whereas the polar properties were assigned a factor e, which is assumed to be the same for both a monomer and its radical.

An expression for the rate constant of the cross-propagation reaction can then be derived as

$$k_{12} = P_1 Q_2 \exp(-e_1 e_2), \qquad (5.7)$$

where P_1 relates to the radical $M_1^\bullet$ and Q_2 to the monomer M_2. This has been called the Q–e scheme and can be used to predict monomer reactivity ratios by extending the treatment to give the relations for r_1 and r_2:

$$r_1 = (k_{11}/k_{12}) = (Q_1/Q_2)\exp\{-e_1(e_1 - e_2)\}, \qquad (5.8)$$

$$r_2 = (k_{22}/k_{21}) = (Q_2/Q_1)\exp\{-e_2(e_2 - e_1)\}, \qquad (5.9)$$

and

$$r_1 r_2 = \exp\{-(e_1 - e_2)^2\}. \qquad (5.10)$$

By choosing arbitrary reference values for styrene of $Q = 1.0$ and $e = -0.8$, a table of relative values of Q and e for monomers can be compiled (see Table 5.2).

On doing this, one finds that for substituents capable of conjugating with the double bond, $Q > 0.5$, whereas for groups such as Cl, OR, and alkyl, $Q < 0.1$, thereby reflecting the assumption that Q is a measure of resonance stabilization.

The values of e are also informative; for instance, maleic anhydride with two strong electron-attracting side groups has $e = +1.5$, indicating an electropositive double bond. This leads to a repulsion of other maleic anhydride molecules, and so no homopolymerization takes place. Similarly, isobutylene has $e = -1.1$, and repulsion of like monomers is again a strong possibility. Copolymerization of oppositely charged monomers, however, should take place readily.

Although the scheme suffers from the disadvantages that steric effects are ignored, that the use of the same value of e for both monomer and radical is a

TABLE 5.2
Selected Values of Q and e for Monomers

Monomer	Q	e
Styrene (reference)	1.0	−0.8
Acrylonitrile	0.60	1.20
1,3-Butadiene	2.39	−1.05
Isobutylene	0.033	−0.96
Ethylene	0.015	−0.20
Isoprene	3.33	−1.22
Maleic anhydride	0.23	2.25
Methyl methacrylate	0.74	0.40
α-Methyl styrene	0.98	−1.27
Propylene	0.002	−0.78
Vinyl acetate	0.026	−0.25
Vinyl chloride	0.044	0.20

doubtful assumption, and that monomers other than monosubstituted ethylenes do not fit in very well, it has proved useful in a qualitative way and should be accepted for what it is: a useful approximation.

The equation is similar to the Hammett equation, which correlates monomer reactivity with structure, but the Hammett treatment is limited to substituted aromatic compounds.

5.9 ALTERNATING COPOLYMERS

The factors that control the entry of monomers into a chain in a strictly alternating sequence are combinations of strong polar and steric effects. Thus, a powerful electron donor such as SO_2 can react "spontaneously" with an electron acceptor such as bicyclo(2.2.1)hept-2-ene even at temperatures as low as 230 K to form a (1:1) alternating copolymer, i.e.,

$$\text{(structure)} + SO_2 \longrightarrow \text{(structure)} \qquad (5.11)$$

In most cases, the spontaneity is absent but in the presence of a radical initiator, maleic anhydride, which is a powerful electron acceptor, reacts readily with a wide range of donor molecules (e.g., styrene, vinyl acetate, vinyl ethers) to produce copolymers with a strong tendency to form alternating structures. The perfection of the alternating sequence will depend on the relative strengths of the donor–acceptor pairs, and as this becomes weaker, statistical-copolymer formation becomes more likely.

Strong acceptor molecules are usually vinyl compounds with a cyano or a carbonyl group conjugated with the double bond. The alternating tendency in systems involving this type of monomer can be enhanced by adding a Lewis acid that can complex with the acceptor and in so doing reduce the electron density on the double bond. The formation of these complexes can alter the characteristics of the acceptor molecule quite markedly; thus, uncomplexed methyl methacrylate has $Q = 0.74$ and $e = 0.4$, but when complexed with $ZnCl_2$, these change to $Q = 26.2$ and $e = 4.2$.

The acceptor–Lewis acid complex reacts with conjugated donor molecules (e.g., styrene) under quite mild conditions to produce highly alternating structures. However, much stronger conditions are required for nonconjugated donor molecules (ethylene, propylene, vinyl acetate, etc.), where it is necessary to use alkyl aluminum sesquichloride as the Lewis acid at a temperature of 195 K. It has been postulated that a ternary molecular complex is formed

$$\begin{matrix} \text{H} & \text{CH}_2 \\ \text{R–C} & \text{C–C}{\equiv}\text{N} \\ \text{H}_3\text{C} & \text{H} \quad \text{EtAlCl}_2 \end{matrix} \longrightarrow \left[\begin{matrix} \text{H} \\ \text{R–C}^{\oplus} \; {}^{\ominus}\text{CH} \\ \text{H}_3\text{C} \cdot \; \cdot\text{C–C}{\equiv}\text{N} \\ \text{H} \quad \text{EtAlCl}_2 \end{matrix} \right] \qquad (5.12)$$

$$\downarrow$$

$$+\!\!\left(\text{CH}_2\text{CHR–CH}_2\text{CHCN}\right)_{\overline{n}}$$

that can then undergo polymerization to form the alternating copolymer. This could be an oversimplification and need not be the mechanism for every system.

A more specialized reaction involves the use of a Ziegler catalyst, formed from vanadium and titanium halides complexed with alkyl aluminum compounds, to synthesize alternating copolymers from propylene and dienes, but the mechanism is now different.

A novel type of spontaneous alternating copolymerization, developed by Saegusa, leads to copolymer formation via a zwitterion, in which both the propagating ion and the gegen ion are situated at opposite ends of the chain. In general, an electrophilic monomer (M_E) interacts with a nucleophilic monomer (M_N) in the absence of catalyst to form a dimeric dipolar species.

$$M_N + M_E \rightarrow {}^{\oplus}M_N - M_E^{\ominus} \tag{5.13}$$

Polymerization then proceeds with the retention of the separated charges to form a "living" polymer.

$$n({}^{\oplus}M_N - M_E^{\ominus}) \rightarrow {}^{\oplus}M_N - (M_E - M_N)_{n-1} - M_E^{\ominus} \tag{5.14}$$

The majority of the M_N monomers are heterocyclic

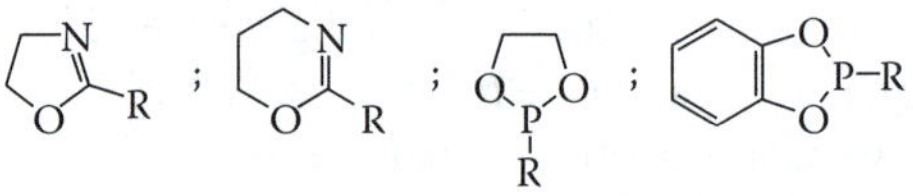

R = H, CH$_3$, Ph, OPh

whereas M_E monomers exist in a greater variety of forms, including both heterocyclic and acrylic types, e.g.,

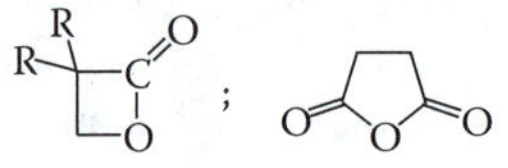

$$H_2C=CXCOOH, H_2C=CHCOOR$$
$$X = H, CH_3, halogen$$

Typical reactions occurring between M_N and M_E monomers are:

1. 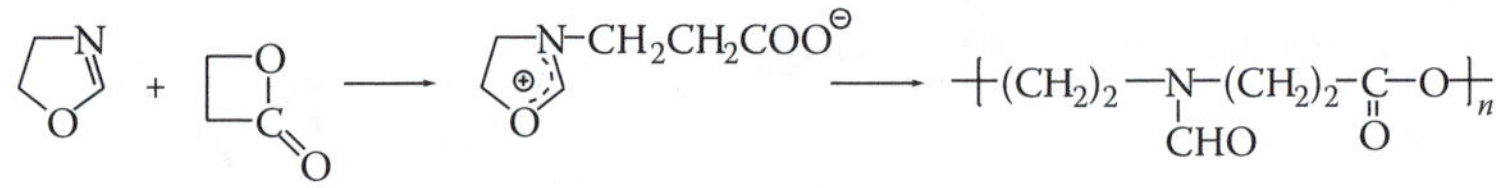 $\quad$ (5.15)

2. 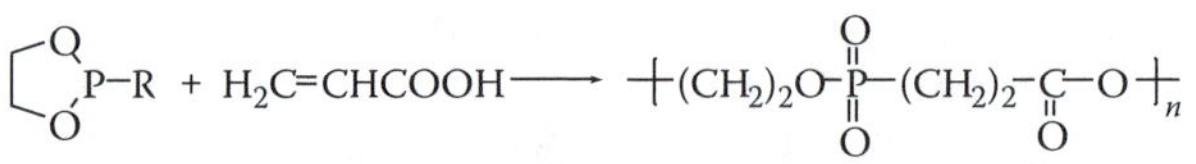 $\quad$ (5.16)

but in the majority of cases, only low-molecular-weight polymers (<10,000) are produced.

Comonomers that have a natural tendency to spontaneous alternation, i.e., an electron-poor and an electron-rich pair such as styrene or substituted maleimides, can be copolymerized using ATRP with CuBr/bpy catalysts. For other comonomer pairs, e.g., styrene and methyl methacrylate, the RAFT process has been used successfully to prepare alternating structures if a Lewis acid such as diethyl aluminum chloride (Et_2AlCl) is added and complexed with the MMA. Subsequent addition of a mixture of AIBN, styrene, and cumyl dithiobenzoate, with heating to 60°C, produces poly(St-alt-MMA) with molecular weights of up to 20,000 and PDI < 1.3. As the alternating copolymer formed retains its active functionality, it can act as a macro chain-transfer agent (CTA), and further addition of styrene will produce the block copolymer poly(St-alt-MMA)-b-polystyrene. Because of the highly reactive nature of the alkyl aluminum halides, the first part of the reaction is carried out in a glove box and under dry, oxygen-free conditions, but the macro CTA formed can be isolated and stored safely for future reactions.

5.10 BLOCK COPOLYMER SYNTHESIS

Block copolymers (Figure 5.4) can be synthesized by sequential addition reactions using: (1) ionic initiators in which an active site is kept "alive" on the end of the initial block, which is then capable of initiating chain growth of a second monomer on the end of the first chain; (2) coupling of different blocks with functional terminal units, either directly or through a reaction involving a small intermediate molecule; and (3) bifunctional radical initiators in which a second potentially active site is incorporated at one end of the first chain grown, which can initiate, at a later stage, a new chain from the macroradical produced.

Ionic reactions are particularly successful in preparing well-defined block copolymers by making use of the observation that there is no easily discernible termination

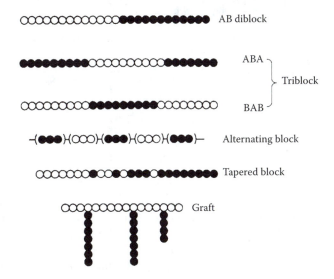

FIGURE 5.4 Block and graft copolymers.

step and, if kept free from impurities, the living carbanionic end groups can be used to initiate the polymerization of a second monomer.

The main limitation to the method is that the anion of one monomer must be able to initiate the polymerization of a second monomer, and this may not always be the case. Thus, polystyryl lithium can initiate the polymerization of methyl methacrylate to give an (A − B) diblock but, because of its relatively low nucleophilicity, the methyl methacrylate anion cannot initiate styrene propagation. Best results are achieved when two monomers of high electrophilicity are used, e.g., styrene (St) with butadiene (Bd) or isoprene and (A − B − A) triblocks can be formed as shown in Equation 5.17a and Equation 5.17b.

$$(PSt)_x^- Li^+ + yBd \longrightarrow (PSt)_x (PBd)_y^- Li^+ \tag{5.17a}$$

$$(PSt)_x (PBd)_y^- Li^+ + zSt \longrightarrow (PSt)_x (PBd)_y (PSt)_z^- Li^+ \tag{5.17b}$$

$$(PSt)_w (PBd)_x^- Li^+ + ClCH_2Cl + Li^+ \, ^-(PBd)_y(PSt)_z$$

$$\tag{5.18}$$

$$\longrightarrow (PSt)_w (PBD)_{x+y}(PSt)_z + 2LiCl$$

The triblock can also be prepared by coupling the two carbanions using an organic dihalide (Equation 5.18), and other coupling agents such as phosgene or dichlorodimethylsilane are equally effective. This method can also be used to prepare radial blocks with multifunctional compounds, as illustrated with silicon tetrachloride.

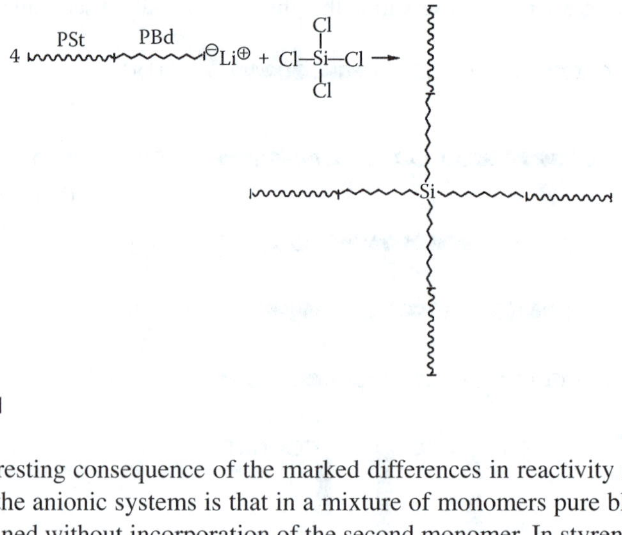

SCHEME 5.1

An interesting consequence of the marked differences in reactivity ratios found in some of the anionic systems is that in a mixture of monomers pure blocks of one can be obtained without incorporation of the second monomer. In styrene–butadiene mixtures, the latter reacts most rapidly and can be almost completely polymerized before the styrene begins to react. As the butadiene becomes depleted, styrene is incorporated progressively until it is the only monomer left, and the remaining chain grown is purely polystyrene. This produces a "tapered" diblock copolymer.

Triblock copolymers can be constructed if a bifunctional initiator is generated, as when sodium naphthalene is used with styrene or α-methylstyrene. Radical anions are formed, which combine to give a dianion, and growth can then take place from both ends. Addition of a second monomer then yields a triblock structure.

5.10.1 TRANSFORMATION REACTIONS

Potentially, there are greater numbers of monomers that are suitable for cationic polymerization than for anionic, but the cationic method is less successful in block copolymer synthesis because, in many systems, the existence of a living carbocationic species is doubtful. Consequently, the involvement of carbocations in block copolymer synthesis tends to be limited to mixed reactions, e.g., the coupling of poly(tetrahydro-furan) cations with polystyryl anions to give an (A – B) diblock (Equation 5.19).

$$(5.19)$$

A more versatile approach is to use a transformation reaction in which one type of active terminal species is converted into a second type. Two general reactions have been identified: (1) a terminal unit anion-cation transformation by a two-electron oxidation process and (2) carbanion to free-radical conversion, which is a one-electron oxidation step.

In the anion-cation transformation reaction, the anionically generated living polymer chain is end-capped with a halide, producing a chain that can be isolated for subsequent reaction. This can be used to initiate a cationic polymerization of a suitable monomer by activating the end with a silver or lithium salt according to the general scheme shown in Equation 5.20a to Equation 5.20c.

$$\text{www}M_1^- \, Li^+ + BrRBr \longrightarrow \text{www}M_1RBr + LiBr \qquad (5.20a)$$

$$\text{www}M_1RBr + Ag^+Y^- \longrightarrow \text{www}M_1R^+Y^- + AgBr \qquad (5.20b)$$

$$\text{www}M_1R^+Y^- + nM_2 \longrightarrow \text{www}M_1\text{www}M_2^+Y^- \qquad (5.20c)$$

Halides may not always be the best terminating agents, and Grignard reagents have been used for this purpose with much greater success.

The reverse cation-anion transformation is also feasible and involves the end-capping of the carbenium ion with a species capable of further reaction with an alkyllithium.

$$\sim\!\!\sim\!\! M_1^+ Y^- + RNH_2 \longrightarrow \sim\!\!\sim\!\! M_1 NRH + HY \tag{5.21a}$$

$$\sim\!\!\sim\!\! M_1 NRH + R'Li \longrightarrow \sim\!\!\sim\!\! M_1 NR^- Li^+ + R'H \tag{5.21b}$$

$$\sim\!\!\sim\!\! M_1 NR^- Li^+ + nM_2 \longrightarrow \sim\!\!\sim\!\! M_1 N \sim\!\!\sim\!\! M_2^- Li^+ \tag{5.21c}$$

Anion-radical transformations can be effected in a number of ways, but one must always begin with the carbanion-terminated chain.

1. This chain can be end-capped with a halide perester (Equation 5.22), which provides a chain with a potential radical-forming site at one end. Thermal decomposition of this group in a second-stage reaction, in the presence of another monomer, generates an alkoxy macroradical from which to grow the second block, but also produces a second radical fragment likely to produce some homopolymer as a contaminant

$$\sim\!\!\sim\!\! M_1^{\ominus} Li^{\oplus} + XR\!-\!\underset{\underset{O}{\|}}{C}\!-\!O\!-\!O\!-\!\underset{\underset{O}{\|}}{C}\!-\!RX \longrightarrow$$

$$\sim\!\!\sim\!\! M_1 R\!-\!\underset{\underset{O}{\|}}{C}\!-\!O\!-\!O\!-\!\underset{\underset{O}{\|}}{C}\!-\!RX \xrightarrow[nM_2]{heat} \tag{5.22}$$

$$\sim\!\!\sim\!\! M_1 R\!-\!\underset{\underset{O}{\|}}{C}\!-\!O\!\sim\!\! M_2^{\bullet} + XR\!-\!\underset{\underset{O}{\|}}{C}\!-\!O\!\sim\!\! M_2^{\bullet}$$

2. An alternative route involves end-capping to produce a terminal hydroxyl followed by reaction with trichloroacetyl isocyanate (Equation 5.23). This new, reactive end group can be used to initiate the growth of a second block via the photoreduction method proposed by Bamford in which magnesium or rhenium carbonyls are excited by UV or visible radiation and extract a chloride atom from the terminal unit, thereby creating a radical site. As only one radical is formed, this is a much "cleaner" reaction compared with (1); however, block lengths are more difficult to control in both these radical reactions, and the exact structure of the product formed can depend on the mechanism of the termination reaction.

$$\sim\!\!\sim\!\! M_1 OH + OCN\!-\!\underset{\underset{O}{\|}}{C}\!-\!CCl_3 \longrightarrow \sim\!\!\sim\!\! M_1\!-\!O\!-\!\underset{\underset{O}{\|}}{C}\!-\!\overset{H}{\underset{}{N}}\!-\!\underset{\underset{O}{\|}}{C}\!-\!CCl_3$$

$$\sim\!\!\sim\!\! M_1\!-\!O\!-\!\underset{\underset{O}{\|}}{C}\!-\!\overset{H}{\underset{}{N}}\!-\!\underset{\underset{O}{\|}}{C}\!-\!CCl_3 + Mn_2(CO)_{10} \xrightarrow{hv}$$

$$\sim\!\!\sim\!\! M_1\!-\!O\!-\!\underset{\underset{O}{\|}}{C}\!-\!\overset{H}{\underset{}{N}}\!-\!\underset{\underset{O}{\|}}{C}\!-\!CCl_2^{\bullet} + n\,M_2 \longrightarrow \tag{5.23}$$

$$\sim\!\!\sim\!\! M_1\!-\!O\!-\!\underset{\underset{O}{\|}}{C}\!-\!\overset{H}{\underset{}{N}}\!-\!\underset{\underset{O}{\|}}{C}\!-\!CCl_2\!\sim\!\! M_2^{\bullet}$$

5.10.1.1 Cationic to CRP

The carbocationic polymerization of styrene using $SnCl_4$ results in Cl-terminated PS chains. These can be isolated, purified, and in the presence of a copper catalyst such as $CuCl/(dNbpy)_2$, further polymerization of MA or MMA can be initiated to form diblock copolymers with polydispersities of 1.2 and 1.6, respectively.

A transformation from cationic ring-opening polymerization (CROP) to ATRP, or vice versa, can also be effected, as shown below.

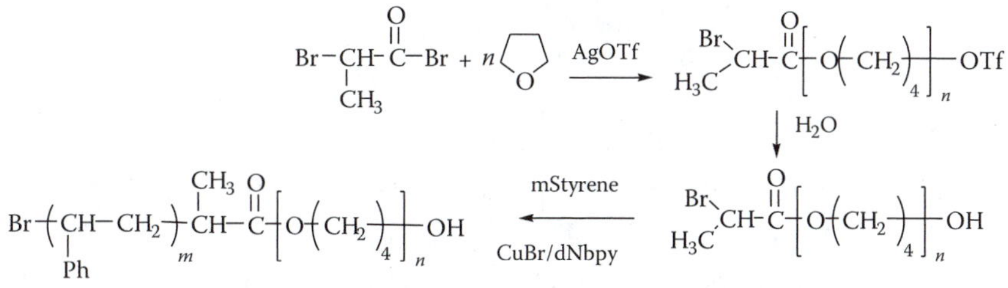

Here, the cationic ring-opening polymerization of tetrahydrofuran (THF) using silver triflate (AgOTf) in the presence of 2-bromopropionylbromide produces a Br-terminated PTHF. Block copolymers can then be prepared using S, MMA, and MA, employing a $CuBr/(dNbpy)_2$ catalyst.

ABA triblocks can be prepared by forming a bifunctional PTHF, which is then reacted with a sodium OTEMPO as shown.

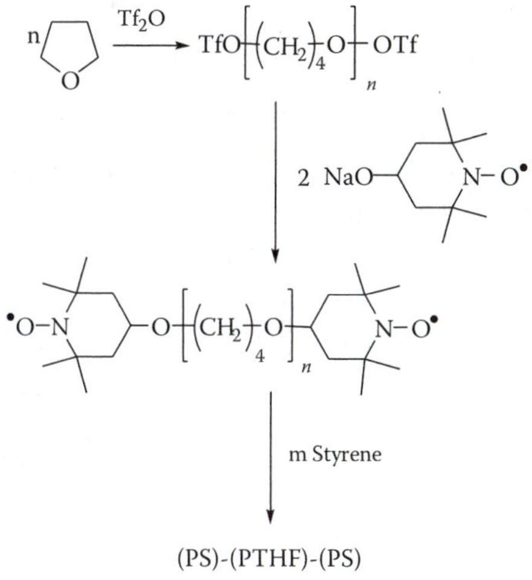

(PS)-(PTHF)-(PS)

Further reaction with styrene produces triblock ABA copolymers.

Dual initiators, i.e., 2-bromo(3,3'-diethoxypropyl)-2-methylpropionate (BrDEP), with two different initiating functions, produce unusual diblocks such as poly(vinyl methyl ether)-b-poly(acrylic acid) as shown.

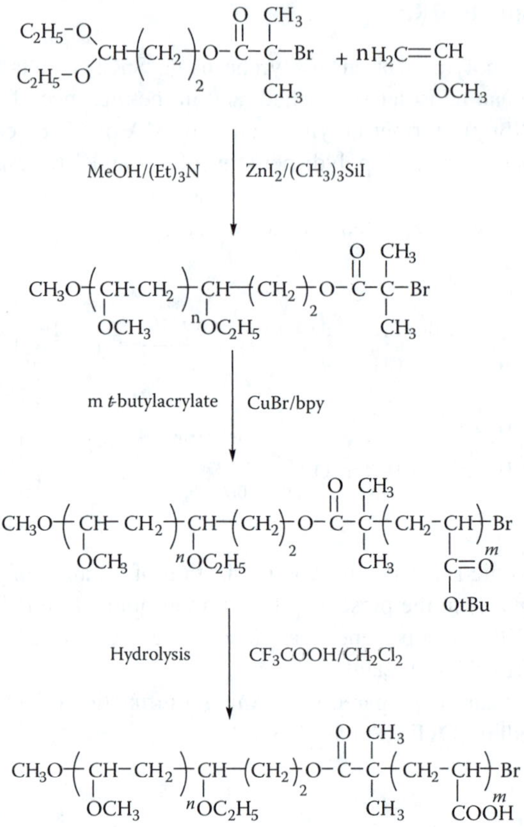

$$C_2H_5\text{-}O \diagdown CH\text{-}(CH_2)_2\text{-}O\text{-}\overset{O}{\overset{\|}{C}}\text{-}\overset{CH_3}{\underset{CH_3}{\overset{|}{C}}}\text{-}Br \quad + \quad n\,H_2C\text{=}CH\underset{OCH_3}{|}$$

MeOH/(Et)$_3$N | ZnI$_2$/(CH$_3$)$_3$SiI

$$CH_3O\text{-}\left(\underset{OCH_3}{\overset{|}{CH}}\text{-}CH_2\right)_n\text{-}CH\text{-}(CH_2)_2\text{-}O\text{-}\overset{O}{\overset{\|}{C}}\text{-}\overset{CH_3}{\underset{CH_3}{\overset{|}{C}}}\text{-}Br$$

m *t*-butylacrylate | CuBr/bpy

$$CH_3O\text{-}\left(\underset{OCH_3}{\overset{|}{CH}}\text{-}CH_2\right)_n\text{-}CH\text{-}(CH_2)_2\text{-}O\text{-}\overset{O}{\overset{\|}{C}}\text{-}\overset{CH_3}{\underset{CH_3}{\overset{|}{C}}}\text{-}\left(CH_2\text{-}\underset{\underset{OtBu}{\overset{|}{C=O}}}{\overset{|}{CH}}\right)_m\text{-}Br$$

Hydrolysis | CF$_3$COOH/CH$_2$Cl$_2$

$$CH_3O\text{-}\left(\underset{OCH_3}{\overset{|}{CH}}\text{-}CH_2\right)_n\text{-}CH\text{-}(CH_2)_2\text{-}O\text{-}\overset{O}{\overset{\|}{C}}\text{-}\overset{CH_3}{\underset{CH_3}{\overset{|}{C}}}\text{-}\left(CH_2\text{-}\underset{COOH}{\overset{|}{CH}}\right)_m\text{-}Br$$

5.10.1.2 Anionic to CRP

The anionic polymerization of butadiene (BD) with *sec*-butyl lithium to form living polybutadiene lithium can be terminated by reaction with a nitroxide, which can then be used to mediate a subsequent styrene polymerization, as shown.

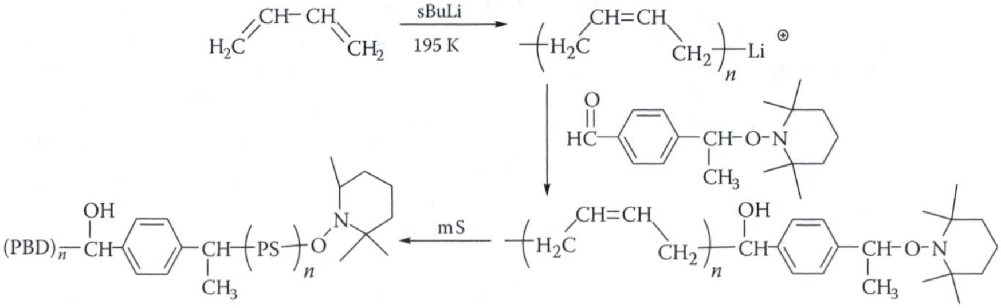

The ring-opening polymerization of ε-caprolactone (ECL) using AlEt$_3$ in tribromoethanol produces a Br-terminated poly(caprolactone), e.g.,

$$H \left(O \left(CH_2 \right)_5 \overset{\text{O}}{\underset{\text{O}}{C}} \right)_n OCBr_3.$$

This can act as a macroinitiator that can be used to polymerize MMA, using a nickel bromide complex catalyst ($NiBr_2(PPh_3)_2$) thereby yielding a diblock copolymer (PECL-b-PMMA).

5.10.1.3 ROMP to ATRP

The ROMP of norbornene, using a molybdenum alkylidene catalyst, followed by end-capping with *p*-(bromo methyl) benzaldehyde, produces a Br-terminated polynorbornene (1). This can be used in an ATRP reaction to initiate block copolymer formation

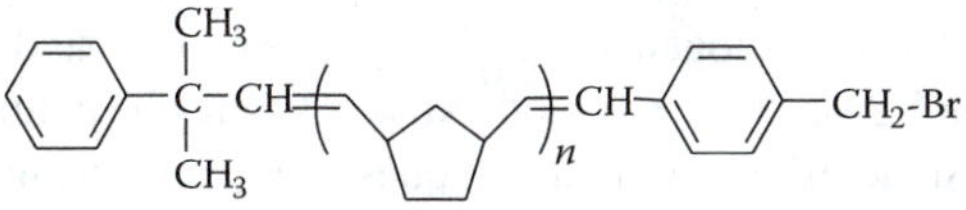

(1)

on addition of styrene or methyl acrylate as the comonomers. Dicyclopentadiene can also be used in place of the norbornene. Halogen-terminated bifunctional polybutadiene has been prepared using 1,5 cyclooctadiene (COD) initiated by a ruthenium catalyst, and this can be used in an ATRP reaction to make ABA triblocks with styrene, but methyl methacrylate was less successful. The metathesis ring-opening reaction of 1,5-cyclocotadiene produces a polybutadiene (PBD) with perfect 1,4-microstructure and is a superior way of producing PS-*b*-PBD-*b*-PS triblocks.

5.10.1.4 Step-Growth ATRP

Bifunctional polysulfones can be prepared by step-growth polymerization and end-capped using 2-bromopropionyl bromide (2). Triblocks are then formed using ATRP with a CuBr/(dNbpy)$_2$ catalyst and either styrene or i-butyl acrylate monomers.

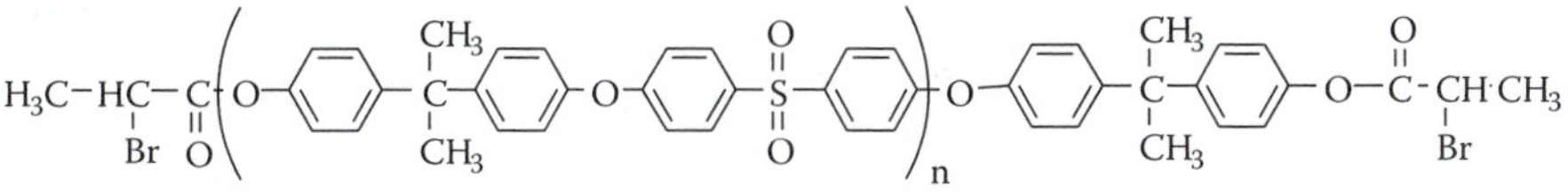

(2)

  Dihydroxy polyester blocks can be functionalized in a similar manner and can act as ATRP macroinitiators to form ABA triblock copolymers. Other blocks such as poly(phenylene vinylene) have been functionalized with TEMPO moieties and used to initiate S polymerization, giving AB diblocks.

5.10.2 COUPLING REACTIONS

It is clear from the foregoing that polymer chains can be synthesized with functional groups in the α or the ω position, or both. If two different types of block are functionalized, they can be linked together to form copolymers.

Anionic polymerizations can be terminated by addition of another molecule, which will introduce an ω-functional group in the chain. Excess carbon dioxide or cyclic anhydrides lead to terminal carboxylic groups, whereas addition of excess phosgene produces an acid chloride function. Similarly, isocyanates generate ω-amide functions, and lactones yield ω-hydroxyl groups.

The "Inifer" process developed by Kennedy can be used to functionalize vinyl monomers via a cationic route by initiating a polymerization with an alkyl halide–boron trichloride mixture $\{R^+BCl_4^-\}$. The termination by transfer to an alkyl halide leaves a halide-terminated polymer. This can be transformed to a hydroxyl terminal unit via the sequence: (1) dehydrohalogenation, (2) hydroboration, and (3) oxidation and hydrolysis (Equation 5.24). These ω-functional blocks may be coupled to form diblock copolymers using standard reaction techniques, e.g., diisocyanate will couple ω-hydroxy and ω-amine blocks together. Direct reactions can also occur, and ω-acid chlorides combine readily with ω-hydroxy units.

$$\text{\~\~CH}_2\text{-}\overset{\overset{\displaystyle CH_3}{|}}{\underset{\underset{\displaystyle CH_3}{|}}{C^{\oplus}}} BCl_4^{\ominus} \;+\; RCl \;\longrightarrow\; \text{\~\~CH}_2\text{-}\overset{\overset{\displaystyle CH_3}{|}}{\underset{\underset{\displaystyle CH_3}{|}}{C}}\text{-}Cl \;+\; R^{\oplus}\,BCl_4^{\ominus}$$

$$(5.24)$$

$$\text{\~\~CH}_2\text{-}\overset{\overset{\displaystyle CH_3}{|}}{\underset{\underset{\displaystyle CH_3}{|}}{C}}\text{-}Cl \quad\overset{(i)\quad(ii)\quad(iii)}{\longrightarrow}\quad \text{\~\~CH}_2\text{-}\overset{}{\underset{\underset{\displaystyle CH_3}{|}}{CH}}\text{-}CH_2OH$$

Macroazonitriles can be employed, and structures based on (3) in Equation 5.25 with either diol, acid, or acid chloride terminal functions are preferred. Functionalized chains can be linked to the azo compounds, after which the azo group can be decomposed thermally to produce radical sites for further chain growth.

$$Cl\text{-}\underset{\underset{\displaystyle O}{\|}}{C}\text{-}R\text{-}N{=}N\text{-}R\text{-}\underset{\underset{\displaystyle O}{\|}}{C}\text{-}Cl \;+\; 2OH\text{\~\~} \;\longrightarrow$$
$$\qquad\qquad (3)$$

$$(5.25)$$

$$\text{\~\~}O\text{-}\underset{\underset{\displaystyle O}{\|}}{C}\text{-}R\text{-}N{=}N\text{-}R\text{-}\underset{\underset{\displaystyle O}{\|}}{C}\text{-}O\text{\~\~}$$

Both polar and nonpolar blocks can be incorporated using coupling strategies. Polystyrene-b-poly(ethyleneglycol) can be prepared by NMP.

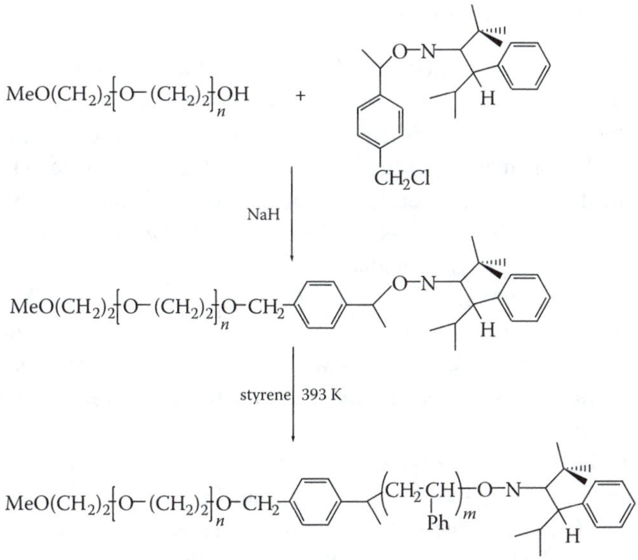

A similar strategy can be employed using RAFT techniques and starting materials like (4)

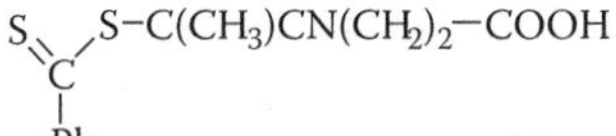

$$S=C(S-C(CH_3)CN(CH_2)_2-COOH)$$

(4)

using dicyclohexyl carbodiimide as the coupling again. Alternatively, a bifunctional RAFT agent (5)

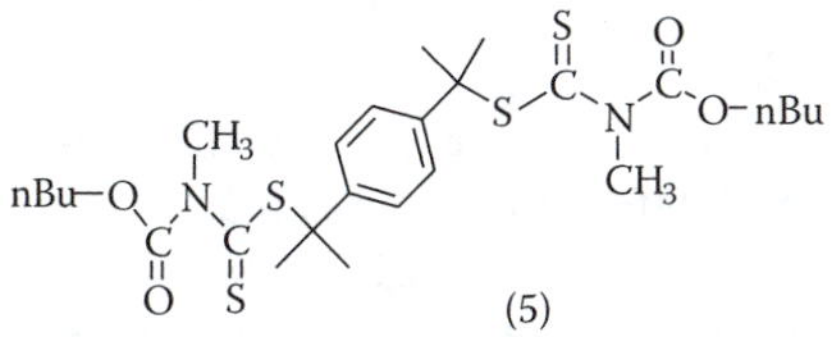

(5)

S-(1,4 phenylene *bis*(propane-2,2-diyl)*bis*(*N*,*N'* butoxy carbonyl-methyl dithio carbamate) can be used to form blocks with monohydroxyl end capped poly(ethylene-co-butene), and further addition of *n*-butyl acrylate produces an ABA triblock.

Hydroxy-terminated PEG is readily transformed into ATRP macroinitiators by conversion of the hydroxyl groups to terminal halide units followed by reaction with a transition metal/ligand complex, and used to form diblocks with several monomers, including the sodium salt of methacrylic acid.

5.10.3 USE OF CRP METHODS

As with all living polymerizations, the formation of block copolymers is possible if the active chain end of one polymer block can initiate the polymerization of a second monomer. This may mean that when block copolymers are prepared, the sequence of monomer addition may be critical. In this respect, the CRP techniques are no different from the other ionic reactions, but they do have one specific advantage. Because many of the CRP processes do not reach 100% conversion, it is best to separate and purify the polymer adduct formed in the first step, which can then be stored and used as a macroinitiator for the growth of the second block, with no loss of activity. The use of such a macroinitiator helps to control the subsequent reaction more effectively and produces products with very low polydispersities, because for the macroinitiator, diffusion and reactivity are decreased, thereby minimizing radical–radical coupling.

Several approaches can be used to prepare block copolymers, one example being sequential addition. Vinyl-block copolymers can be prepared in the conventional way by polymerization of one monomer with the subsequent addition of a second monomer. Thus, in the first step, the block formed can be a polymer adduct, examples of which are

NMP

ATRP

RAFT

Although block copolymers such as polystyrene-*b*-poly(butyl acrylate) (PS-*b*-PBA) can be prepared by NMP, the reaction is much more successful if the (PBA) alkoxyamine-terminated block is used to initiate the styrene polymerization than vice versa. However, a PS macroinitiator can be used to prepare well-defined diblocks with isoprene as the second monomer, i.e., (PS-*b*-PI), using NMP techniques and 2,2,5 trimethyl-3-(1-phenyl ethoxy)-4-phenyl-3-azahexane (TMPAH) as the nitroxide mediator.

The TEMPO method tends to be successful only when styrene-based monomers are part of the system, but are usually unsuccessful when acrylates, and in particular methacrylates, are used. The newer nitroxides can be more successful with acrylates, but methyacrylates still present a problem when preparing block copolymers with nitroxide-mediated reactions.

In ATRP, the nature of the halogen end group can play an important role. The initiation of MMA polymerization by a Cl-terminated poly(methyl acrylate) macroinitiator is poor, but is much faster if a Br-terminated poly(methyl acrylate) is used. In general, efficient block copolymer formation occurs when the rate of the cross-propagation reaction is comparable to the rate of propagation of the second monomer. The following order has been suggested:

acrylonitrile (AN)> alkyl methacrylates (AMA)> styrene (S) ≈ alkyl acrylates (AA).

Hence, the insensitivity to addition sequence for styrene and (AA) monomers makes it relatively easy to form (PS-b-PAA), (PS-b-PAA-b-PS), and (PAA-b-PS-b-PAA) di- and triblock copolymers. However, when AN and MMA are involved, the poly(acrylonitrile) block must be formed first, i.e., second monomer addition should be in the order shown in the preceding text. Unlike NMP methods, the use of ATRP presents no major problems when preparing block copolymers containing acrylates and methacrylates. Triblock (ABC) copolymers can be prepared by starting with a bromine-terminated polystyrene macroinitiator to polymerize t-butyl acrylate, followed by methyl acrylate or methyl methacrylate using a CuBr/pentamethyl diethylene triamine catalyst.

In RAFT block copolymerizations, the main requirement to produce AB diblocks with a low polydispersity index PDI is that the dithioester macroinitiator has a high chain transfer constant in the polymerization reaction producing the second block. In other words, in the equilibrium step

$$ X^{\bullet} + \underset{Z}{\overset{S}{\underset{|}{C}}}\overset{Y}{\diagup} \;\rightleftharpoons\; X^{\cdot S}\underset{\underset{Z}{|}}{C}\diagdown^{S}Y \;\rightleftharpoons\; X^{\cdot S}\underset{\underset{Z}{|}}{C}\diagup^{S} + Y^{\bullet} $$

the leaving group ability of the propagating radical Y• should be greater than, or at least comparable to, that of X•. This means that if block copolymers of MMA with styrene or MA are to be prepared, then the PMMA block should be formed first, as PS• and PMA• radicals are poor leaving groups when compared with the PMMA• radical. Examples of block copolymers formed in this way are shown in Table 5.3

TABLE 5.3
Molecular Weight Data for AB Diblock Copolymers

Monomer A	M_n	Monomer B	M_n	Solvent
Styrene	20,300	N,N-dimethyl acrylamide	43,000	Benzene
Styrene	20,300	p-Methylstyrene	25,400	Benzene
Butyl acrylate	33,600	Acrylic acid	52,400	DMF
Methyl acrylate	24,100	Ethyl acrylate	30,900	Benzene
Methyl methacrylate	3,230	Methacrylic acid	4,720	DMF
Methyl methacrylate	17,400	Styrene	35,000	Bulk
Benzyl methacrylate	1,800	2-(dimethylamino)ethyl methacrylate	3,500	Ethyl acetate
Benzyl methacrylate	1,800	Methacrylic acid	2,400	DMF
Ethylene oxide	750	Styrene	7,800	Bulk
Ethylene oxide	750	styrene	10,800	Benzene

Note: DMF = N,N-dimethyl formamide.

from the data reported by Moad and Rizzardo. RAFT is often considered to be the most versatile of the CRPs as polar and nonpolar monomers can be used in solution and aqueous dispersion.

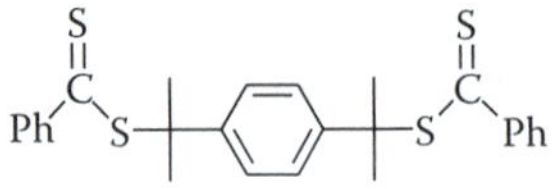

$$\text{Ph}-\underset{\underset{S}{\overset{S}{\|}}}{C}-S-\left[\underset{}{\bigcirc}\right]-S-\underset{\underset{S}{\overset{S}{\|}}}{C}-\text{Ph} \qquad (6)$$

Triblock copolymers can be prepared by using either the bifunctional chain transfer agent or the trithiocarbonates.

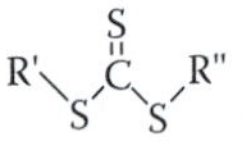

$$R'\diagdown_{S}\overset{\overset{S}{\overset{\|}{C}}}{\diagup}_{S}\diagdown R''$$

R' = CH$_3$, CH$_2$Ph, CH(CH$_3$)Ph

R" = CH$_2$Ph, C(CH$_3$)$_2$CN, CH(Ph)COOH, CH(CH$_3$)Ph

$$(7)$$

As mentioned in Chapter 3, the use of R' and R" groups with good or poor leaving abilities can make (7) either mono- or bifunctional. Hence, for (6) with R'=R"=CH$_2$Ph, a triblock copolymer can be prepared:

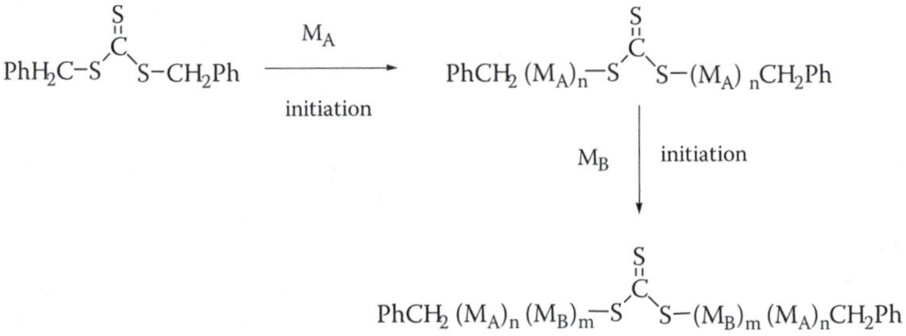

$$\text{PhH}_2\text{C-S}\overset{\overset{S}{\overset{\|}{C}}}{\diagup}\text{S-CH}_2\text{Ph} \quad \xrightarrow[\text{initiation}]{M_A} \quad \text{PhCH}_2(M_A)_n\text{-S}\overset{\overset{S}{\overset{\|}{C}}}{\diagup}\text{S-}(M_A)_n\text{CH}_2\text{Ph}$$

$$\Big\downarrow M_B \quad \text{initiation}$$

$$\text{PhCH}_2(M_A)_n(M_B)_m\text{-S}\overset{\overset{S}{\overset{\|}{C}}}{\diagup}\text{S-}(M_B)_m(M_A)_n\text{CH}_2\text{Ph}$$

Treatment of this ABA triblock copolymer with a mild nucleophile will cleave the chain at the trithiocarbonate group to produce two AB diblocks each with a SH terminal unit. If, however, R'=CH$_3$ and R"=C(CH$_3$)$_2$CN, then CH$_3$ is a poor leaving group, so only diblocks are formed, e.g.,

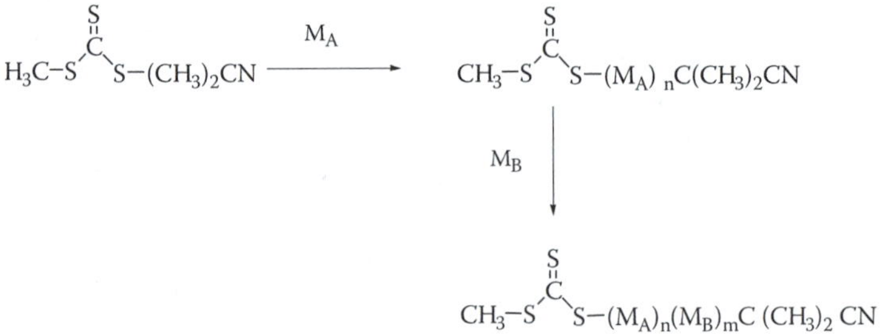

$$\text{H}_3\text{C-S}\overset{\overset{S}{\overset{\|}{C}}}{\diagup}\text{S-}(\text{CH}_3)_2\text{CN} \quad \xrightarrow{M_A} \quad \text{CH}_3\text{-S}\overset{\overset{S}{\overset{\|}{C}}}{\diagup}\text{S-}(M_A)_n\text{C}(\text{CH}_3)_2\text{CN}$$

$$\Big\downarrow M_B$$

$$\text{CH}_3\text{-S}\overset{\overset{S}{\overset{\|}{C}}}{\diagup}\text{S-}(M_A)_n(M_B)_m\text{C}(\text{CH}_3)_2\text{CN}$$

Hydrophilic monomers can also be used to form block copolymers with MMA. In particular, dimethyl amino ethyl methacrylate (DMAEMA) and 2-hydroxy ethyl methacrylate (HEMA) form block copolymers when initiated by chloro-terminated PMMA blocks using ATRP methods. 4-Vinyl pyridine (4VP) can also be initiated to give (PMMA-*b*-P4VP) diblocks.

The RAFT process has been used to prepare a variety of block copolymers, as shown in Table 5.3; for example, DMAMEA and ethylene oxide can form diblocks with benzyl methacrylate, but more important, acrylic acid and methacrylic acid can be used, which has so far been impossible employing ATRP.

5.11 GRAFT COPOLYMER SYNTHESIS

In the synthesis of graft copolymers (Figure 5.4), the sites at which the second and subsequent blocks are attached to the first are no longer terminal units but are at positions along the backbone of the first chain. There are three main techniques for preparing graft copolymers: (1) grafting "from"; (2) grafting "onto"; and (3) via macromonomers.

The grafting "from" procedure requires active sites to be created on the polymer chain capable of initiating the growth of other chain branches comprising a second monomer. Free-radical sites can be formed by direct or mutual radiation with γ-rays of a polymer in the presence of the second monomer. This is a simple method but can also lead to homopolymer formation.

Pre-irradiation of a polymer in the presence of oxygen leads to formation of relatively stable peroxy groups on the polymer, and this allows the polymer to be isolated and stored for further reaction. Polymers prepared in this way can be heated in the presence of a second monomer: the peroxy groups decompose to produce radical sites, and the grafting process can take place. This method has been used to prepare poly(styrene-*graft*-acrylonitrile). The free-radical approach can be used in other ways. Graft copolymers are formed when a chain transfer to preformed polymer can be effected with a second monomer present in the reaction mixture, but this depends on the radical source, e.g., methyl acrylate can be grafted to natural rubber when benzoyl peroxide is the initiator but is much less likely to do so with azobisisobutyronitrile. The effectiveness of this grafting "from" technique is a function of the reactivity and polarity of the radical site and the monomer. Alternatively, radical-forming sites can be introduced into the chain backbone by *in situ* modification of some monomer units or by copolymerization. Thus, a polymer with trihalide groups pendant to the chain can be activated in the presence of a second monomer, thereby forming a graft rather than a block copolymer. This type of reaction may also lead to a cross-linked structure if termination of the radical by combination predominates.

$$(5.26)$$

By altering the number of active sites on the backbone, the number and distribution of grafted chains can be controlled. The lengths of each graft will depend on both the rate of initiation and the monomer concentration, but the mechanism by which termination of the growing radical takes place will determine the ratio of branches to cross-links in the system. If this is exclusively by combination, then the occurrence of cross-linking will be high. However, the network formation can be modified by addition of a chain transfer agent, which will produce a mixture of branches and cross-links but also some homopolymers. In such systems, the amount of chain transfer agent added will determine the ratio of branches to cross-links.

A similar mixture of structures, but with little of the contaminating homopolymer, will be obtained when a second monomer is used whose radicals terminate partly by disproportionation and partly by combination.

Finally, photodegradation of pendant ketones results in the formation of radical sites capable of initiating a graft, but like many of the other radical techniques, there is also a tendency for homopolymerization to occur.

Anionic sites suitable for grafting "from" reactions can be introduced by metallation, involving the complexation of a hydrocarbon polymer by organolithium compounds. The reaction is assisted by complexing the lithium first with tetramethylenediamine, which acts as a solvating base. Aromatic chlorine is readily exchanged for lithium, which then acts as an initiator for the anionic polymerization of suitable monomers.

$$2 \sim\!\!\sim M_1^{\ominus} + COCl_2 \longrightarrow \sim\!\!\sim (M_1)\!\sim\!\!\sim\underset{\underset{O}{\|}}{C}\!\sim\!\!\sim (M_1)\!\sim\!\!\sim \tag{5.27a}$$

$$\sim\!\!\sim (M_1)\!\sim\!\!\sim\underset{\underset{O}{\|}}{C}\!\sim\!\!\sim (M_1)\!\sim\!\!\sim + \sim\!\!\sim M_2^{\ominus} \longrightarrow \sim\!\!\sim (M_1)\!\sim\!\!\sim\underset{\underset{(M_2)}{\overset{OH}{|}}}{C}\!\sim\!\!\sim (M_1)\!\sim\!\!\sim \tag{5.27b}$$

Grafting "onto" methods involve having sites on the main chain, which can be attacked by a growing second chain, thereby linking the two by covalent bonding. In anionic polymerizations, the linking of two chains by phosgene creates a polymeric ketone, which can react with a second chain. Other electrophilic functional groups are effective in this reaction, e.g., ester, nitrile, anhydride, etc., and can be used as grafting sites by growing carbanions such as the polystyryl ion. Many of these grafting techniques produce random branching along the primary chain, also a distribution of branch lengths. A more controlled grafting procedure can be used to produce regular comb-branch structures, and this is achieved either by polymerization of macromonomers or by using a polymer-analogous reaction on a suitable backbone.

In the first case, macromonomers can be prepared by functionalizing a short chain with a vinyl unit. A typical reaction is shown in Equation 5.28, and other methods have been reported. Polymerization of these monomers produces a well-defined graft structure with branches located at regular intervals along the chain. If the starting macromonomers have a uniform length, then the branches will also be regular, but mixed lengths can also be prepared. Copolymerization with another monomer will alter the regularity of the branching points but will maintain the uniformity of branch length.

$$\text{\raise1pt\hbox{$\leadsto$}CH-OH} + \text{H}_2\text{C}=\underset{\underset{\text{COCl}}{|}}{\text{CH}} \longrightarrow \text{H}_2\text{C}=\text{CH}-\underset{\underset{\text{O}}{\|}}{\text{C}}-\underset{\underset{\text{R}}{|}}{\text{CH}}\leadsto \tag{5.28}$$

Poly(acid chloride)s have been used for polymer-analogous reactions, and ω-functionalized units can be condensed at these sites to produce structures similar to those obtained using macromonomers.

5.12 STATISTICAL AND GRADIENT COPOLYMERS

The preparation of statistical copolymers using anionic or cationic living polymerizations is difficult to control, and these methods are rarely used for this purpose. Controlled free-radical techniques are much more versatile and also have a distinct advantage over conventional free-radical copolymerizations. In the latter, chains are initiated and terminated constantly throughout the course of the reaction, and because of the differing reactivities of the monomers, "composition drift" occurs during the course of the copolymerization. If the reaction is allowed to proceed to high conversions, a mixture of copolymer chains with differing compositions is obtained. When CRP techniques are used, all the chains are initiated at the same time and grow at approximately the same rate; hence, the change in the instantaneous comonomer concentration is the same for all the growing chains. This means that the statistical copolymers produced at the end of the reaction have approximately the same chemical composition, even though the monomer sequences along the chains may differ. This produces a much more homogeneous product whose composition can be changed by starting with different ratios of the comonomers.

As the difference between the reactivity ratios increases, gradient, or tapered, copolymers will tend to form, and in extreme cases, even block copolymer formation may be expected. In gradient copolymers, there is a progressive dilution of monomer A by monomer B as the chain grows, so that the end product will resemble a block copolymer with a broad compositional interface between the blocks. This contrasts with conventional block copolymers in which the change from one block to the next is sharply defined. The broader interface imparts interesting properties to the gradient copolymers, which exhibit different thermal behavior and special interfacial properties that make them particularly useful as compatibilizers in polymer blend preparation.

Two methods can be used to prepare gradient copolymers:

1. Spontaneous gradient formation using the simultaneous reaction of monomer pairs with sufficiently differing reactivity ratios
2. Forced gradient copolymerization using continuous feed or semibatch techniques

Statistical and gradient copolymers have been prepared successfully using NMP and ATRP in bulk, solution, and mini-emulsion, particularly when styrene is used as a monomer with acrylates, methacrylates, acrylonitrile, and substituted styrenes as comonomers. A number of BPO/TEMPO-mediated statistical polymerizations have been reported, but in some cases, control can only be effected if the initial comonomer feed is rich in styrene.

Gradient copolymers of styrene with MA and MMA have been prepared by ATRP using CuCl and CuBr complexed with bipyridine as catalyst. Similarly, gradient copolymers of MMA and MA and styrene with AN could be formed using copper-based ATRP. Forced gradients were also achieved by adding the second monomer gradually throughout the polymerization, and the composition of the gradient can also be controlled by changing the rate of addition of the second monomer. This change in gradient alters the mechanical behavior of the resulting products.

5.13 COMPLEX MOLECULAR ARCHITECTURES

Hyperbranched molecules can be prepared using condensation polymerization methods and AB$_2$-type molecules typified by the hydroxydiacids. However, a "self-condensing" unimolecular initiator, such as

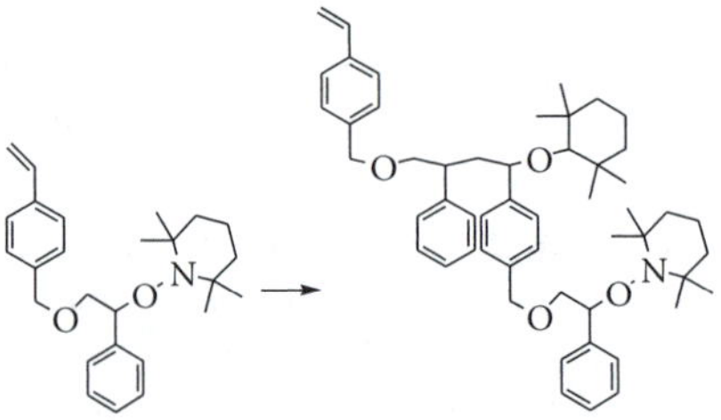

(8)

can act like an AB$_2$ monomer and combine with itself to form a highly branched molecule, using controlled radical techniques. More regular, star-shaped (multiarm) molecules have been prepared using living anionic polymerizations, and CRP methods have now been developed to prepare similar structures.

Use of a multifunctional core, e.g., (9) has been used to prepare star-shaped molecules by applying a RAFT process with styrene and acrylate monomers.

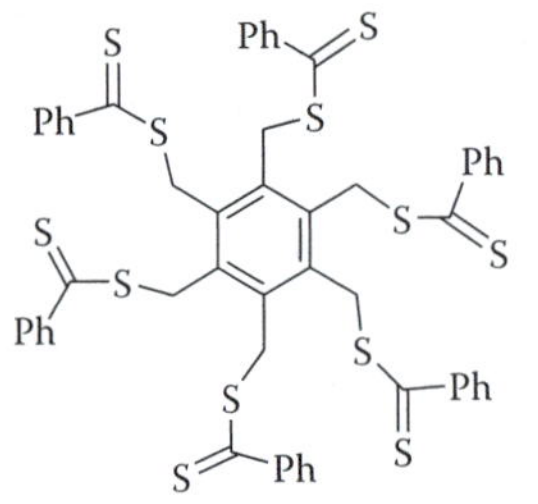

(9)

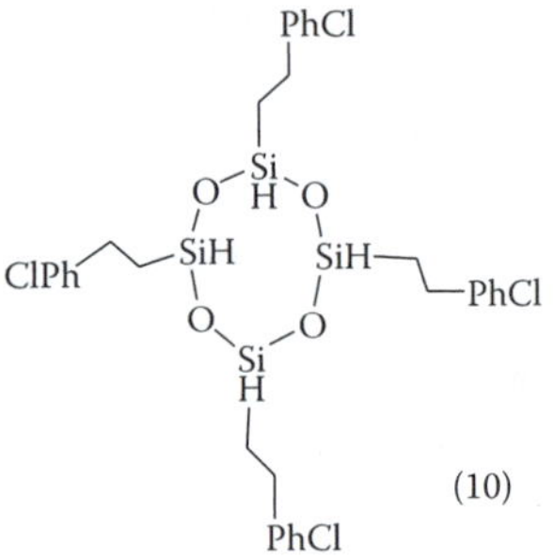

(10)

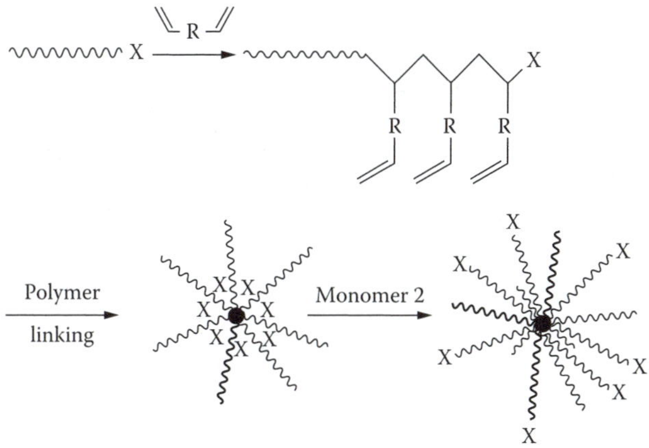

FIGURE 5.5 Schematic representation of the growth of the poly(*t*-butyl acrylate) star by the arm-first approach. (From Matyjaszewski, K., *Polym. Int.*, 52, 1559, 2003. With permission of ACS.)

This approach involves growth of the side arms outward from the core molecule, but an alternative is to prepare monofunctional living chains that are then coupled with a multifunctional core molecule. Other core molecules can be used that are modified inorganic structures, such as cyclotetrasiloxanes (10) and cyclotriphosphazenes. These can be used as the initiating core for ATR polymerization of styrene or acrylate chains to form four- or six-armed star molecules.

Linear polymer chains with an active terminal unit can be prepared using ATRP and NMP. These can be isolated, stored, and used later to couple with multifunctional core molecules. The use of divinyl benzene as the core is illustrated in Figure 5.5.

Multiarm polymers (11) can be prepared that still have the reactive functional groups (Z) close to the core. As these are still active, they can be used as sites to initiate the growth of more arms by adding either the same monomer used to prepare (11) or a second monomer to produce mikto-arm star polymers, in which the arms have different chemical structures. Thus, an active ended poly(*t*-butyl acrylate), prepared by ATRP, can be coupled with divinyl benzene to form a multiarm star polymer. This structure can be converted to a mikto-arm star polymer by reacting the living ends still present with *n*-butyl acrylate, and so propagate poly(*n*-butyl acrylate) chains from the core outward.

5.14 DENDRIMERS

The synthesis of multiarmed and hyperbranched polymers uses techniques that lead to products with variable molecular weights and, consequently, relatively broad molecular weight distributions. Most polymer chemists would prefer to exercise greater control over these variables and continue to strive to develop methods of preparing polymers with precise structures, narrow or monodisperse molecular weight distributions, and controlled growth. These goals have largely been achieved in the synthesis of "spherical dendrimers," which are highly branched, approximately spherical or globular macromolecules. A dendrimer is usually synthesized in a careful stepwise fashion with branches that increase exponentially on moving from a central multifunctional core to the surface of the molecule. The final structures have a common three-dimensional architecture, comprising multifunctional core molecule, several levels of branches, and end groups on the surface. The molecular weight will depend on how many branching layers are incorporated, and the end product should theoretically be monodisperse.

There are two main synthetic routes used to prepare dendrimers: (1) divergent growth and (2) convergent growth.

5.14.1 Divergent Growth

The divergent method is the original approach, developed by Tomalia. The reaction begins with a multifunctional core molecule from which the branches are grown outward by adding the branching units in an iterative, two-step process. The first step involves the reaction of an (A-B) molecule with the core, in which group A readily reacts with the core molecule and group B is dormant. In the second step, a monomer (C-D) is used in which C reacts with group B, leaving dormant D groups as the terminal units on each branch. This was called the *first generation* by Tomalia, but is now more commonly referred to as *generation* 0. Group D is normally bifunctional (or trifunctional) and is one that can react with group A. Thus, further branching can be developed using the two-step synthetic sequence to produce the "next generation." Repetition of this process builds up the dendrimer, generation by generation, until a critical branched state is reached in which the surface of the dendrimer is so tightly packed that further growth is restricted because of lack of space.

The method can be illustrated by the preparation of a poly(amido amine), (PAMAM). These are synthesized using either trifunctional ammonia (NH_3) or tetrafunctional ethylene diamine (EDA), $NH_2CH_2CH_2NH_2$, as the core and EDA and methylacrylate (MA) as the branch monomers. The first step is the Michael addition of MA to ammonia or EDA followed by the amidation of each carboxymethyl group using EDA.

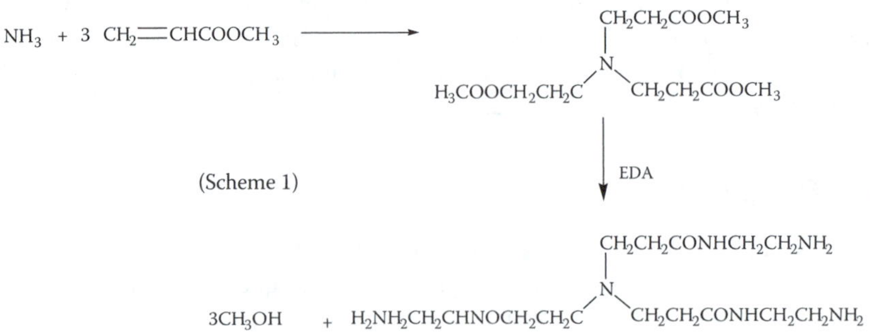

$$NH_3 + 3\ CH_2{=}CHCOOCH_3 \longrightarrow$$

$$\begin{array}{c} CH_2CH_2COOCH_3 \\ | \\ N \\ H_3COOCH_2CH_2C \quad CH_2CH_2COOCH_3 \end{array}$$

(Scheme 1) $\downarrow$ EDA

$$\begin{array}{c} CH_2CH_2CONHCH_2CH_2NH_2 \\ | \\ N \\ 3CH_3OH \quad + \quad H_2NH_2CH_2CHNOCH_2CH_2C \quad CH_2CH_2CONHCH_2CH_2NH_2 \end{array}$$

(12)

The product (12) can be isolated and purified, after which the two-step reaction is repeated to produce the next generation (which will now have six end groups). By repeating these steps, a molecule with up to ten generations can be prepared.

Although an idealized structure can be drawn for the dendritic molecule, the greater the amount of growth, the greater the probability of defects. These include incomplete addition reactions, due to steric hindrance, loop formation, retro-Michael reactions, and other side reactions.

5.14.2 CONVERGENT GROWTH

In an attempt to overcome some of the problems encountered in the divergent method, convergent synthetic routes were developed by Hawker and Frechet. In this approach, the synthesis starts at the outside of the molecule and progresses inward, toward the core. The method is best described with reference to the original Frechet synthesis, shown below.

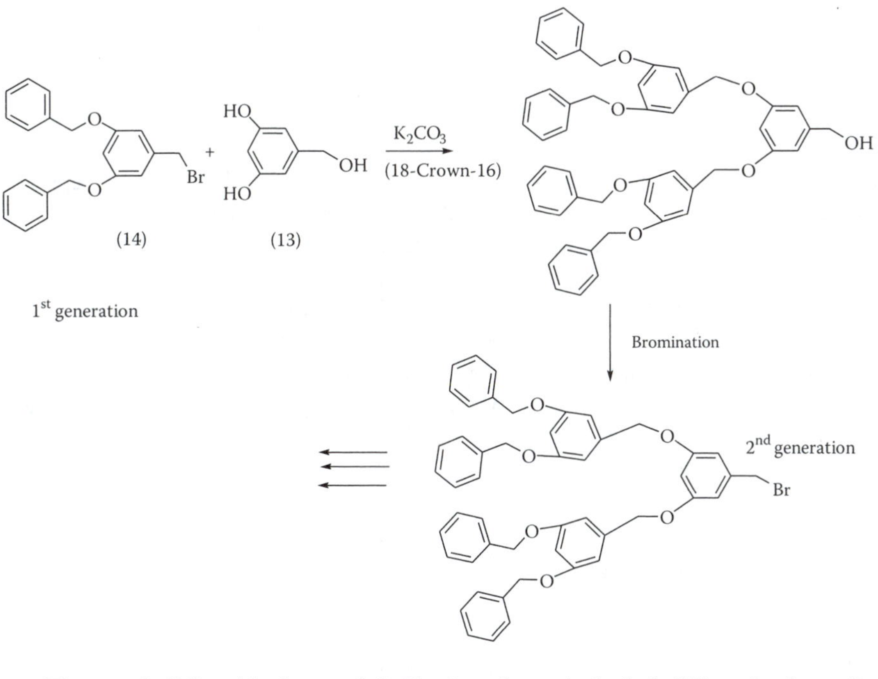

The two building blocks are 3,5-dihydroxybenzyl alcohol (13) and a benzylic bromide such as (14), which is referred to as the first-generation molecule. In the first of a two-step synthesis, compound (13) reacts with two molecules of (14) to produce a second-generation wedge with four terminal phenyl units and a passive primary alcohol as the focal point. The latter is activated in the second step by bromination, and this product can be isolated and purified before proceeding to the next stage. The two synthetic steps are then repeated using (14) to link two of the second-generation wedges to form a third-generation wedge with eight terminal units and one focal point, which is brominated, separated, purified, and is ready for the next stage. Repetition of the process can produce dendritic wedge-shaped molecules up to the sixth generation. The dendrimer is assembled finally by reaction with a multifunctional core molecule, in this case 1,1,1,tris(4-hydroxyphenyl) ethane.

This is a versatile approach that allows the formation of unsymmetrical structures in which wedges with different chemical compositions, i.e., different end groups, can be attached to the same core molecule.

5.14.3 DENDRIMER MOLECULAR WEIGHT

The molecular weight of the dendrimer, generated by a divergent approach, can be calculated theoretically, but there is a certain amount of confusion in the literature, partly brought about by differences in the numbering of the generations. Taking the synthesis of PAMAM (using NH_3 as the core) as an illustration, structure (12) was originally labeled generation 1, but tables of molecular weights in the literature refer to this as generation 0. Tomalia proposed Equation 5.29 for the calculation of the molecular weight MW as

$$MW = M_c + N_c \left\{ M_R \left[\frac{N_R^G - 1}{N_R - 1} \right] + M_t N_R \right\} \qquad (5.29)$$

a function of the generation number G (leading to a G = 1, 2, 3,4, ... series). Here, M_c, M_R, and M_t are the molecular weights of the core, monomer repeat unit, and terminal group, respectively; N_c is the number of branches at the core unit (core multiplicity); and N_R is the number of branches on each monomer unit (repeat-unit multiplicity). This was later revised, and the term (N_b^{G+1}) replaced N_R^G, in which N_b is the multiplicity of the branch unit. Calculation of MW then requires differentiation between the core, the monomer repeat units, and the terminal group and can lead to confusion and miscalculation. A simplified form of Equation 5.29 has been proposed by Majoros, which removed these ambiguities and makes the calculation of MW much easier.

The form proposed is

$$MW = MW_{core} + \left(MW_m - MW_E \right) \cdot N_c \left[\frac{N_r^{g+1} - 1}{N_r - 1} \right] \qquad (5.30)$$

where N_c is the core multiplicity, N_r is the multiplicity of the repeat unit $(MW_m - MW_E)$, g is the generation number, which, because a geometric series is used in the growth calculation, means the labeling starts at 0, i.e., g = 0, 1, 2, 3, ...

By calculating the monomer repeat unit as $(MW_m - MW_E)$, in which MW_m is the sum of the MW of each monomer unit and MW_E is the MW of the molecule eliminated during the reaction (see Scheme 5.1 in Section 5.10), there is no need to identify the terminal unit. This will, of course, also depend on the type of reaction involved in activating the branch unit. For example, in the preparation of poly (propylene imine) dendrimers, a hydrogenation reaction is used to activate the branch unit, i.e., $RCH_2CH_2CN \xrightarrow{4H} RCH_2CH_2CH_2NH_2$. So the monomer repeat unit would then be $(MW_m + 4H)$.

The use of Equation 5.30 can be illustrated using PAMAM as an example. If EDA is used as the core molecule, then $MW_{core} = 60$ and $N_c = 4$. The monomers in the branch unit are EDA and MA, and CH_3OH is the eliminated molecule so the monomer repeat unit (EDA + MA – CH_3OH) is (60 + 86 – 32) = 114 and $N_r = 2$.

Substitution in Equation 5.30 for $g = 0$ gives

$$MW = 60 + (114) \cdot 4 \left[\frac{2^{0+1} - 1}{2 - 1} \right] = 516 \tag{5.31}$$

and for $g = 4$

$$MW = 60 + (114) \cdot 4 \left[\frac{2^5 - 1}{2 - 1} \right] = 14,136 \tag{5.32}$$

The calculations then yield the values shown in Table 5.4 for g up to 10.

If the degree of polymerization DP is required, then

$$DP = N_c \left[\frac{N_r^{g+1} - 1}{N_r - 1} \right] \tag{5.33}$$

and the number of terminal units Z is given by

$$Z = N_c N_r^g \tag{5.34}$$

5.14.4 PROPERTIES OF DENDRIMERS

In solution, dendrimers form close-packed, globular, or spherical shapes when the molecular weight is high enough, unlike the expanded flexible coil structures of linear polymers. This has a significant effect on the rheological properties, e.g., the

TABLE 5.4
PAMAM Using EDA Core

Generation	Molar Mass	Number of Terminal Groups
0	516	4
1	1,428	8
2	3,252	16
3	6,900	32
4	14,196	64
5	28,788	128
6	57,972	256
7	116,340	512
8	233,076	1,024
9	466,548	2,048
10	933,492	4,096

viscosity of higher-molecular-weight dendrimers in solution is lower than that of an equivalent molecular weight linear polymer. This is molecular weight dependent as the intrinsic viscosity has been shown to pass through a maximum and then decline when measured as a function of molecular weight.

The nature of the groups on the surface also greatly affects the solubility behavior. Thus, hydrophilic groups enhance solubility in polar liquids. If convergent-growth methods are used, then it is possible to construct dendrimers in which segments with polar surface groups are combined with wedges having nonpolar surface groups (Figure 5.6).

These can form monolayers at aqueous–organic liquid interfaces.

5.14.5 APPLICATIONS OF DENDRIMERS

Although a high-molecular-weight dendrimer has a very tightly packed surface layer, there are fairly large internal cavities (see Figure 5.7). These have been used to trap

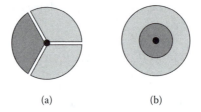

(a)   (b)

FIGURE 5.6 Schematic representation of dendritic copolymers with different polarities: (a) segment-block dendrimer and (b) layer-block dendrimer. (From Klajnert, B. and Bryszewska, M., *Acta Biochimica Polonica*, 48, 199, 2001. With permission.)

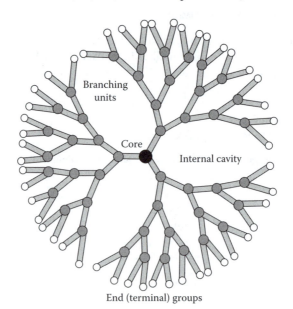

FIGURE 5.7 Schematic illustration of a fourth-generation dendrimer. (From Klajnert, B. and Bryszewska, M., *Acta Biochimica Polonica*, 48, 199, 2001. With permission.)

small molecules, a feature first explored by Meier. Small molecules, such as Rose Bengal and p-nitrobenzoic acid, can be held in the cavities of a fourth-generation poly(propylene imine) dendrimer, with 64 surface branches, which form a "dendritic box." The box can be "sealed" by reacting the surface amino groups with L-(phenyl alanine) and the molecules subsequently liberated by hydrolysis of this outer layer. Similar systems are now being actively investigated for use in drug delivery in the human body.

Release systems have also been developed using azobenzene groups on the surface, which can be "opened" and "closed" via a photochemical-stimulated cis–trans isomeric change reaction. Other applications being explored for dendritic structures include light harvesting, use as contrast agents for magnetic resonance techniques, use for *in vitro* diagnostics, application in catalysis, and as carriers in gene therapy.

PROBLEMS

1. In a certain copolymer system, the reactivity ratios are $r_1 = 0.001$ and $r_2 = 0.002$. Sketch the copolymer composition diagram expected for this pair of monomers.

2. Mayo et al. (1948) have reported reactivity ratios for the copolymerization of vinyl acetate (reactivity ratio $= r_2$) with eight representative monomers (corresponding reactivity ratio $= r_1$).

Monomer, M_1	r_1	r_2
Acrylonitrile	4.05	0.61
Trichloroethylene	0.01	0.66
Styrene	55	0.01
Methyl acrylate	9	0.1
Vinyl chloride	1.68	0.23
Vinyl ethyl ether	0	3.0
Vinyl bromide	4.5	0.35
Methyl methacrylate	20	0.15

a. In each case, calculate the composition of the copolymer that would be formed at low conversion from equimolar mixtures of the two monomers.

b. What is the comonomer feed composition required to form a copolymer with 50% vinyl acetate units?

c. Identify those pairs of monomers that display azeotropic conditions, and calculate the azeotropic copolymer composition.

d. Which system is closest to an ideal copolymerization?

3. Determine the composition of the copolymer obtained when an equimolar mixture of styrene (S) and methyl methacrylate (MMA) undergoes:

a. cationic copolymerization initiated by $SnCl_4$ (the reactivity ratios are $r_S = 10.5$ and $r_{MMA} = 0.1$)

b. anionic copolymerization initiated by sodium in liquid ammonia ($r_S = 0.5$ and $r_{MMA} = 0.44$)

4. Suggest conditions for the synthesis of:
   a. random copolymers of styrene and 4-vinyl phenol
   b. block copolymers of styrene and 4-vinyl phenol, i.e., polystyrene-*b*-poly(4-vinyl phenol) by ATRP
   c. gradient copolymers of styrene and 4-vinyl phenol.
5. Poly(propylene imine) dendrimers are prepared by a divergent route, using 1,4-diamino butane as the core molecule and acrylonitrile ($CH_2=CHCN$) as the monomer branch unit. Growth occurs via the two-stage process: (a) a double Michael addition of acrylonitrile to the core followed by (b) hydrogenation under pressure to the nitrile group to a primary amine.
   i. Draw the structure for generation 0.
   ii. Calculate the molecular weight and number of end groups for generation-three poly(propylene imine) dendrimer.
6. Calculate the molecular weight and number of end groups from generation 0 to generation 6 for a poly (amido amine) prepared using NH_3 as core molecule and ethylene diamine and methyl acrylate as monomers.

REFERENCES

Chong, Y.K., Le, T.P.T., Moad, G., Rizzardo, E., and Thang, S.H., *Macromolecules*, 32, 2071, 1999.
Estes, G.M., Cooper, S.L., and Tobolsky, A.B., Block copolymers, *Rev. Macromol. Chem.*, 5–2, 167, 1970.
Hawker, C.J., Bosman, A.W., and Harth, E., *Chem. Rev.*, 101, 3661, 2001.
Klajnert, B. and Bryszewska, M., *Acta Biochimica Polonica*, 48, 199, 2001.
Matyjaszewski, K., *Polym. Int.*, 52, 1559, 2003.
Matyjaszewski, K. and Xia, J., *Chem. Rev.*, 101, 2921, 2001.
Mayo, F.R., Walling, C., Lewis, F.M., and Hulse, W.F., *J. Am. Chem. Soc.*, 70, 1523, 1948.
Pepper, D.C., *Q. Rev.*, 8, 88, 1954.

BIBLIOGRAPHY

Allen, G. and Bevington, J.C., Eds., *Comprehensive Polymer Science*, Vol. 3, 4, Pergamon Press, 1989.
Allport, D.G. and Janes, W.H., Eds., *Block Copolymers*, Applied Science Publishers, 1973.
Cowie, J.M.G., *Alternating Copolymers*, Plenum Press, 1985.
Folkes, M.J., Eds., *Processing, Structure and Properties of Block Copolymers*, Elsevier Applied Science Publishers, 1985.
Noshay, A. and McGrath, J.E., *Block Copolymers: Overview and Critical Survey*, Academic Press, 1977.
Odian, G., *Principles of Polymerization*, 4th ed., John Wiley and Sons, 2004.
Rempp, P. and Merrill, E.W., *Polymer Synthesis*, Hüthig and Wepf, 1986.

6 Polymer Stereochemistry

The physical behavior of a polymer depends not only on the general chemical composition but also on the more subtle differences in microstructure. As it is now possible to exercise a larger degree of control over the synthesis of specific structures, it is prudent at this point to elaborate on the types of microstructural variations encountered before discussing how each can be produced. Several kinds of isomerism or microstructural variations can be identified, and these are grouped under four main headings: architectural, orientational, configurational, and geometric.

6.1 ARCHITECTURE

Differences here include branching, network formation, and polymers derived from isomeric monomers, for example, poly(ethylene oxide), I, poly(vinyl alcohol), II, and polyacetaldehyde, III, in which the chemical composition of the monomer units is the same, but the atomic arrangement is different in each case. This makes a considerable difference to the physical properties of the polymers, e.g., the glass transition temperature T_g of structure I is 206 K, for II $T_g = 358$ K, and for III $T_g = 243$ K.

$$
\begin{array}{ccc}
\text{I} & \text{II} & \text{III} \\[4pt]
+\!CH_2CH_2\!-\!O\!+_n & \left(CH_2\!-\!\underset{\underset{OH}{|}}{CH}\right)_n & \left(\underset{\underset{CH_3}{|}}{CH}\!-\!O\right)_n
\end{array}
$$

6.2 ORIENTATION

When a radical attacks an asymmetric vinyl monomer, two modes of addition are possible:

$$
R^{\bullet} + H_2C{=}\underset{\underset{X}{|}}{CH} \longrightarrow
\begin{cases}
\!-\!RCH_2\underset{\underset{X}{|}}{CH}^{\bullet} & \text{I} \\[10pt]
\!-\!R\underset{\underset{X}{|}}{CH}CH_2^{\bullet} & \text{II}
\end{cases}
$$

This leads to the configuration of the monomer unit in the chain being either head-to-tail

$$
\text{w}CH_2\!-\!\underset{\underset{X}{|}}{CH}\!-\!CH_2\!-\!\underset{\underset{X}{|}}{CH}\!-\!CH_2\!-\!\underset{\underset{X}{|}}{CH}\!-\!CH_2\!-\!\underset{\underset{X}{|}}{CH}\text{w}
$$

III

if route I is favored, or a chain containing a proportion of head-to-head, tail-to-tail structure IV if route II is followed.

$$\text{wCH}_2\text{-CH-CH-CH}_2\text{-CH}_2\text{-CH-CH-CH}_2\text{-CH}_2\text{-CHw}$$
$$\qquad\; X \;\; X \qquad\qquad\quad X \;\; X \qquad\qquad\quad X$$

head-head   tail-tail

IV

The actual mode of addition depends on two factors: the stability of the product and the possible steric hindrance to the approach of R· caused by a large group X in the molecule. The reaction in route I is highly favored because first, there is a greater possibility of resonance stabilization of this structure by interaction between group X and the unpaired electron on the adjacent α-carbon atom, and second, this direction of radical attack is least impeded by the substituent X. The preferred structure is then the head-to-tail orientation (III), and whereas the alternative structure IV may occur occasionally in the chain, especially when termination by combination predominates, the existence of an exclusively head-to-head orientation is unlikely, unless synthesized by a special route.

Experimental evidence supports the predominance of structure III in the majority of polymers; the most notable exceptions are poly(vinylidene fluoride) with 4 to 6% and poly(vinyl fluoride) with 25 to 32% head-to-head links detected by NMR studies. The presence of head-to-tail structures can be demonstrated in a number of ways, and the general principle is illustrated using poly(vinyl chloride) as an example. Treatment of this polymer with zinc dust in dioxan solution leads to elimination of chlorine, which can proceed by two mechanisms.

(a) $\text{wH}_2\text{C-CH-CH}_2\text{-CH-CH}_2\text{w} \longrightarrow \text{wCH}_2\text{-CH}_2\text{CH-CH}_2\text{w} + \text{ZnCl}_2$
              Cl   Cl   C
                                                              H$_2$

(b) $\text{wH}_2\text{C-CH-CH-CH}_2\text{w} \longrightarrow \text{wCH}_2\text{-CH=CH-CH}_2\text{w} + \text{ZnCl}_2$
              Cl   Cl

Statistical analysis of chlorine loss via route (a) indicates that only 86.4% of the chlorine will react due to the fact that, as elimination is a random process, about 13.6% of the chlorine atoms become isolated during the reaction and will remain in the chain. Elimination by mechanism (b) results in total removal of chlorine. Analysis of poly(vinyl chloride) after treatment with zinc dust showed 84 to 86% chlorine elimination, and this figure remained constant even after prolonged heating of the reaction mixture. This leads one to the conclusion that the polymer is almost entirely in the head-to-tail orientation.

6.3 CONFIGURATION

It has long since been recognized that when an asymmetric vinyl monomer $CH_2=CHX$ is polymerized, every tertiary carbon atom in the chain can be regarded as a chiral center by virtue of the fact that m and n are not normally equal in any chain. Under these circumstances, the two possible configurations shown [(i) and (ii)] can only be interconverted by breaking a bond.

(i)   (ii)

No real progress in the preparation of distinguishable stereoisomers was made until the advent of the Ziegler–Natta catalysts, which will be discussed later. Since then, the study of stereoregular polymers has expanded rapidly in a vigorous and exciting manner, helped greatly by the application of NMR to the accurate characterization of the microstructure, but before examining these topics, a brief outline of the nomenclature is required.

If every tertiary carbon atom in the chain is asymmetric, one might expect the polymer to exhibit optical activity. Normally, homoatomic carbon chains show no optical activity because two long chains constitute part of the group variations; as these become longer (and more alike) in relation to the chiral center, the optical activity decreases to a vanishingly small value. Vinyl polymers derived from $(CH_2=C^*XY)$ monomers fall into this category as they are centrosymmetric relative to the main chain, and the tertiary carbons are then only pseudo-asymmetric.

This is not true for heteroatomic chains $\sim CH_2C^*HX \cdot O \sim$ such as where C^* is a true asymmetric center, and these polymers are optically active. In this case, an absolute configuration can be assigned, using preferably the Cahn–Ingold–Prelog system, referring either to the R-(rectus) or to the S-(sinister) form.

The two forms (i) and (ii) can be distinguished by arbitrarily assigning them d- or l-configurations, which have nothing to do with optical activity and merely refer to the group X being positioned either below or above the chain in a planar projection.

There are then three distinctive distributions of the d-and l-forms among the units in a chain, and these decide the chain tacticity.

6.3.1 MONOTACTIC POLYMERS

a. The *isotactic* form: When a polymer, in the all trans zigzag conformation, is viewed along the bonds comprising the chain backbone, then if each asymmetric chain atom has its substituents in the same steric order, the polymer is said to be *isotactic*. In other words, the arrangement of the substituent groups is either all d or all l.

b. *Syndiotactic:* A chain is termed *syndiotactic* when observation along the main chain shows the opposite configuration around each successive asymmetric center in the chain.

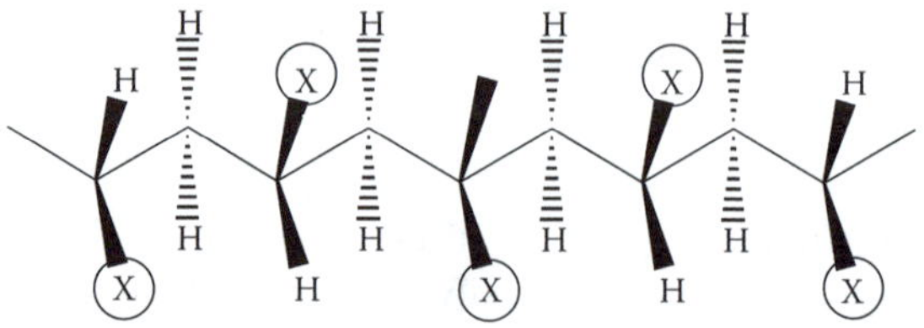

c. *Atactic:* When the stereochemistry of the tertiary carbons in the chain is random, the polymer is said to be atactic or heterotactic.

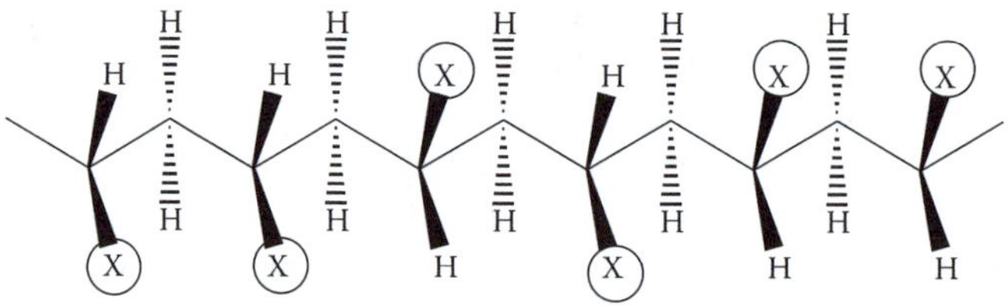

It is often easier to obtain a clear picture of the spatial arrangement of the chains by referring to a Newman projection, and these are shown for comparison in Figure 6.1(a,b,c).

6.3.2 DITACTIC POLYMERS

A more complicated picture emerges when the polymerization of 1,2-disubstituted ethylenes (CHR=CHR′) is considered because now each carbon atom in the chain becomes a chiral center. The resulting ditactic structures are illustrated in Figure 6.1(d,e,f). Two isotactic structures are obtained, the *erythro*, in which all the carbon atoms have the same configuration, and the *threo*, in which the configuration alternates. Only one disyndiotactic structure is possible. The differences arise from the stereochemistry of the starting material; if the monomer is cis-substituted the *threo* form is obtained, whereas a trans monomer leads to the *erythro* structure.

6.3.3 POLYETHERS

When the spacing between the asymmetric centers increases, as in the heteroatomic polymer poly(propylene oxide), the isotactic and syndiotactic structures become less easily recognized in planar projection. Using an extended zigzag structure, the *isotactic* form now has its substituents alternating across the plane containing the main chain bonds.

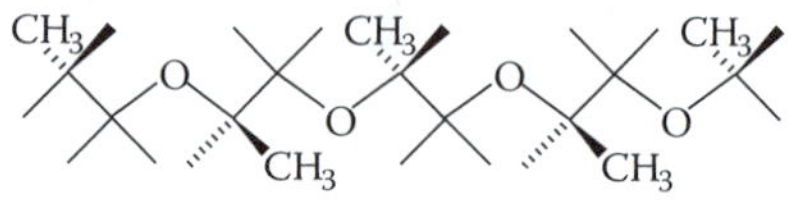

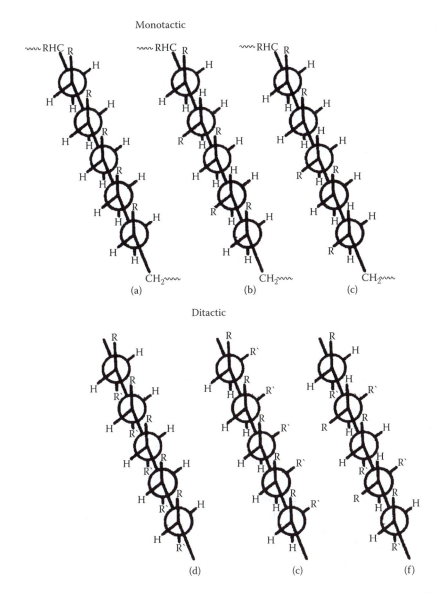

FIGURE 6.1 Newman projections of various stereoregular forms: (a) isotactic, (b) syndio-tactic, (c) atactic, (d) erythro-di-isotactic, (e) threo-di-isotactic, and (f) di-syndiotactic.

The reverse is true for the *syndiotactic* chain in which the substituents are all located on one side.

6.4 GEOMETRIC ISOMERISM

In addition to the configurational isomerism encountered in polymers derived from asymmetric olefins, geometric isomerism is obtained when conjugated dienes are polymerized, e.g., $(CH_2=CX-CH=CH_2)$. Chain growth from monomers of this type can proceed in a number of ways, illustrated conveniently by 2-methyl-1,3-butadiene (isoprene). Addition can take place either through a 1,2-mechanism or a 3,4-mechanism, both of which could lead to isotactic, syndiotactic, or atactic structures, or by a 1,4-mode leaving the site of unsaturation in the chain.

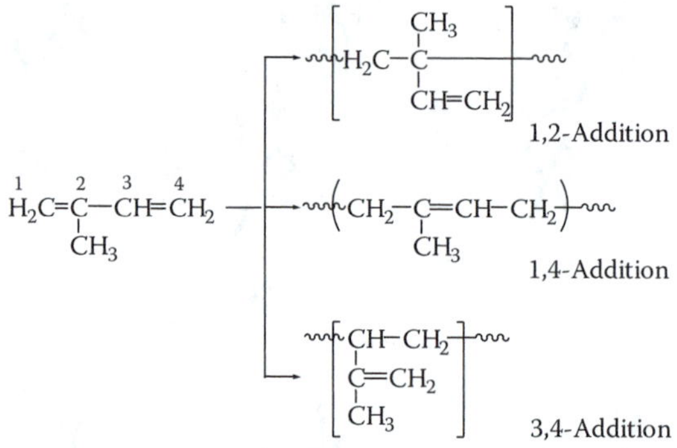

This means that the 1,4 polymer can exist in the cis or trans form or a mixture of both.

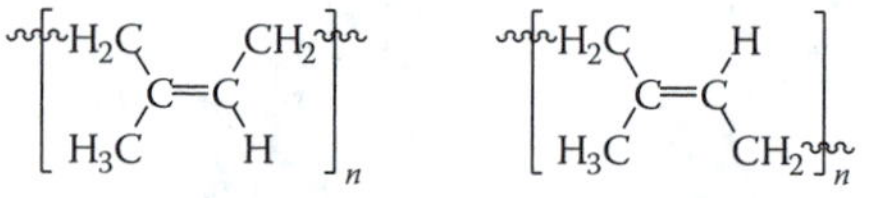

In theory, it is possible to synthesize eight distinguishable stereochemical forms or mixtures of these. For a symmetrical monomer such as 1,3-butadiene $(CH_2=CH-CH=CH_2)$ the 1,2- and 3,4-additions are indistinguishable, and the possible number of stereoforms diminishes accordingly.

Additional variations are possible when 1,4-disubstituted dienes are considered, $XCH=CH-CH=CHY$. For $Y = H$

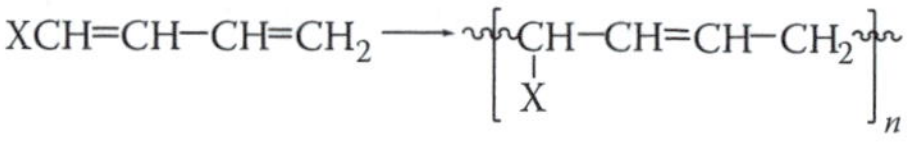

both cis and trans isomerism is possible together with isotactic, syndiotactic, and atactic placements for the group X when the addition is 1,4. When $Y \neq H$ then *threo* and *erythro* forms are also possible in the 1,4-polymer. The name *tritactic* has been suggested for polymers prepared from monomers with different X and Y groups.

6.5 CONFORMATION OF STEREOREGULAR POLYMERS

Many of the stereoregular polymers prepared are highly crystalline, and the tendency to form ordered structures increases as the stereoregularity becomes more pronounced. We shall see later that crystalline order is usually associated with regular symmetrical polymer structures, whereas the asymmetric monomers form highly unsymmetrical chains. Some other factors must aid crystallite formation.

A stable form of polyethylene is the all trans zigzag form in which it crystallizes. An extended zigzag pattern becomes untenable, however, for an isotactic polymer with a bulky substituent because the distance between the substituent centers in this conformation is only 0.254 nm. Obviously, the low energy form for an isotactic species must be attained by placing the substituents in staggered positions of maximum separation, and this is achieved when bond rotation generates a helix. One particular helical form is shown for polypropylene in Figure 6.2. Working from carbon 1, we have the following sequence: 1 and 4 are trans to each other (*t*), carbons 2 and 5 are *gauche* (*g*), 3 and 6 are trans, 4 and 7 are *gauche*, and so on. Carbon 1 repeats at carbon 7; hence, the helix is threefold with three monomer residues constituting one complete turn. In a shorthand notation, this is a 3_1-helix with a *tgtgtg* conformation, and departure from this pattern to a *ttgg* sequence would simply lead the chain back on itself. This type of helix can also be built up on a triangular

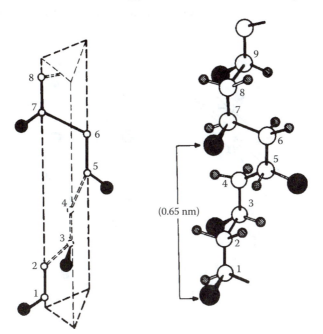

FIGURE 6.2 A 3_1-helix formed by a poly α-olefin, in this case polypropylene. This structure is also seen to fit a triangular template.

template, which can be used as a simple model to demonstrate the structure. A helix generated in this way using rotations of 120° should result in an identity period of 0.62 nm. Polypropylene has a period of 0.65 nm, and the structure may be generated with equal ease either as a left- or right-handed helix. A helix turning in a clockwise direction, when the chain is viewed along its axis, is said to be a right-handed helix; if anticlockwise, it is left-handed.

The syndiotactic configuration is much more suitable for the extended zigzag form as the substituents are already staggered for convenient packing on either side of the chain, but a twofold helix can also be generated by adopting a *ttgg* sequence, which has been identified in syndiotactic polypropylene. A square mandrel can be used to demonstrate this structure (Figure 6.3). The type of helix formed depends largely on the size of the substituent, and a number of these are shown in Figure 6.4. As the helix is a regular-ordered structure, it can be arranged compactly in a three-dimensional close structure with relative ease, which explains how the unsymmetrical chain monomer can be accommodated in a crystalline polymer structure. Highly crystalline samples are obtained when the polymer is sufficiently stereoregular to enable it to form significantly long helical or regular zigzag sections for ordered chain arrangement to take place.

The automatic identification of crystallinity with stereoregularity should be avoided; however, as they are not necessarily synonymous, and whereas highly stereoregular polymers tend to be crystalline, the existence of any polymer in a crystalline state does not automatically mean the sample is markedly stereoregular.

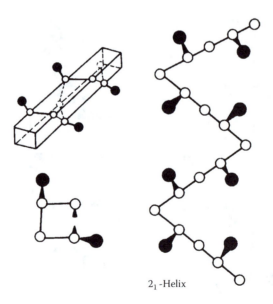

2_1-Helix

FIGURE 6.3 A poly α-olefin in the syndiotactic configuration showing the *ttgg* sequence along the chain, and the twofold helix, which fits a square template.

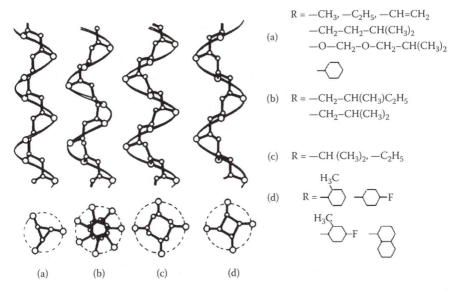

$R = -CH_3, -C_2H_5, -CH=CH_2$
$-CH_2-CH_2-CH(CH_3)_2$ (a)
$-O-CH_2-O-CH_2-CH(CH_3)_2$

(b) $R = -CH_2-CH(CH_3)C_2H_5$
$-CH_2-CH(CH_3)_2$

(c) $R = -CH (CH_3)_2, -C_2H_5$

(d) $R =$

(a) (b) (c) (d)

FIGURE 6.4 A diagrammatic representation of various other ordered helical structures adopted by isotactic polymers. (From Natta, G. and Corradini, P., *Rubber Chem. Technol.*, 33, 703, 1960. With permission.)

6.6 FACTORS INFLUENCING STEREOREGULATION

The low-pressure polymerization of ethylene, reported by Ziegler in 1955, signaled the emergence of a new phase in polymer science. The catalyst, prepared from $TiCl_4$ and $(C_2H_5)_3Al$, was heterogeneous, and it was subsequently demonstrated by Natta (1961) that coordination catalysts of this type could be used to exercise control over the stereoregular structure of the polymer. Initially, it was thought that only heterogeneous catalysis would lead to stereoregular polymers, but we now know that this is untrue and that stereoregulation can be effected under specific rigorously defined conditions, regardless of the solubility of the catalyst system.

If stereoregulation is simply the control of the mode of entry of a monomer unit to a growing chain, examination of the factors influencing this addition should provide an understanding of how to exert such control.

Free-radical initiation can usually be thought of as generating a chain by a Bernoulli-trial propagation, in which the orientation of the incoming monomer is unaffected by the stereostructure of the polymer. It can then add on in one of the two ways in which the active end is assumed to be a planar sp^2 hybrid, and the configuration of the adding monomer is finally determined only when another monomer adds on to it in the next step. In other words, this addition leads to an isotactic or syndiotactic placement of the pseudo-asymmetric center 1 with respect to 2. When the chain carrier is a free species, i.e., a radical, the stereoregularity of the polymer is a function of the relative rates of the two methods of addition, and this is governed by the temperature. Consideration of the relative magnitudes of the enthalpy and entropy of activation for isotactic and syndiotactic placements shows that whereas the differences are small, the syndiotactic structure is favored. This is, of course, aided by the greater steric hindrance and repulsions experienced by the substituents in the isotactic configuration and will vary in extent with the nature of the group X. Thus, for a free-radical polymerization at 373 K, the fraction of syndiotactic placements is 0.73 for methyl methacrylate monomer but only 0.51 when vinyl chloride is used. A decrease in the polymerization temperature increases the tendency toward syndiotactic placements, but as radical reactions are normally high-temperature processes, atactic structures predominate. Low-temperature free-radical propagation has been found to produce syndiotactic polymers from the polar monomers, isopropyl and cyclohexyl acrylate, and methyl methacrylate.

The same general principles apply for freely propagating ionic chain carriers, but if coordination between the monomer and the active end takes place, the stereoregulation is altered. The configuration of the monomer is then influenced by the stereochemistry of the growing end, and the possible number of ways the monomer can join the chain is in excess of two. These coordination catalysts include the Ziegler–Natta type as the largest group, and others such as butyl lithium, phenyl magnesium bromide, and boron trifluoride etherate. The resulting polymer is normally isotactic, although some cases exist in which highly syndiotactic polymers are obtained (see Table 6.1).

The orienting stage in coordination polymerization can be pictured as being multicentered, with the monomer position governed by coordination with the gegen-ion

---

TABLE 6.1
Polymerizations Using Coordination Catalysts Where Quoted Tacticities Are >90%

Monomer	Catalyst	Structure
Isobutyl vinyl ether	$BF_3(C_2H_5)_2O$ in propane at 213 K	Isotactic
Methyl acrylate	C_6H_5MgBr or $n\text{-}C_4H_9Li$ in toluene at 253 K	Isotactic
Propylene	$TiCl_4 + (C_2H_5)_3Al$ in heptane at 323 K	Isotactic
Propylene	$VCl_4 + Al(i\text{-}C_4H_9)_2Cl$ in anisole or toluene at 195 K	Syndiotactic

and the propagating chain end. As the gegen-ion will tend to repel the substituent X on the incoming monomer, it is forced to approach in a way that leads to predominantly isotactic placement. If coordination plays a major role in determining the configuration of the incoming monomer, then the greater the coordinating power the more regular the resulting polymer should be, but the nature of the monomer is also important. Polar monomers (the acrylates and vinyl ethers) are capable of taking an active part in the coordination process and will only require catalysts with moderate powers of orientation, but nonpolar monomers such as the α-olefins will require stronger coordinating catalysts to maintain the required degree of stereoregulation in the addition process. In extreme cases, the heterogeneous Ziegler–Natta catalysts are required, in which severe restrictions are imposed on the method of monomer approach to the growing chain end, and these must be used for the nonpolar monomers, which yield only atactic polymers with homogeneous catalyst systems.

6.7 HOMOGENEOUS STEREOSPECIFIC CATIONIC POLYMERIZATIONS

An example of this type of reaction is provided by the alkyl vinyl ethers ($CH_2=CHOR$). Isobutyl vinyl ether was the first monomer studied, which produced a stereoregular polymer using a $BF_3 + (C_2H_5)_2O$ catalyst and will be used as the illustrative monomer. A homogeneous-stereospecific polymerization can be carried out in toluene at 195 K using such soluble complexes as $(C_2H_5)_2TiCl_2AlCl_2$ or $(C_2H_5)_2TiCl_2Al(C_2H_5)Cl$, or, if a suitable choice of mixed solvents is made, a homogeneous system with $BF_3 + (C_2H_5)_2O$ can be obtained, which is capable of producing the isotactic polymer.

The mechanism proposed by Bawn and Ledwith (1962) postulates the existence of an sp^3 configuration for the terminal carbon in the growing chain due to the attendant gegen ion, and especially in low-dielectric solvents. They also point out that the structure of the alkyl vinyl ethers, with the exception of the ethyl and isopropyl members, will be subject to steric shielding of one side of the double bond, i.e.,

This blocks one mode of double-bond opening and assists stereoregulation. This conclusion is supported by the lack of any crystalline polymer in the product when the ethyl and isopropyl groups are used where no blocking is possible.

The formation of a loose six-membered ring is thought to stabilize the growing carbonium ion in the reaction so that the only route for monomer approach is past the counterion.

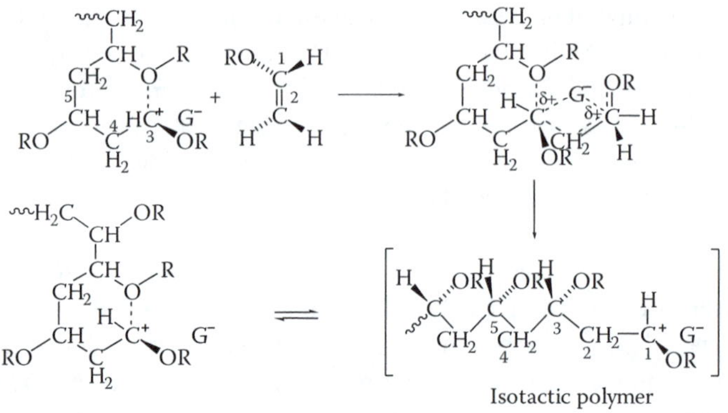

Isotactic polymer

A four-centered cyclic-transition state is involved in the propagation stage leading to the insertion of a monomer unit between the catalyst and the chain end, with subsequent regeneration of the cyclic structure. An alternative transition state, proposed by Cram and Kopecky, has a similar but more rigid structure.

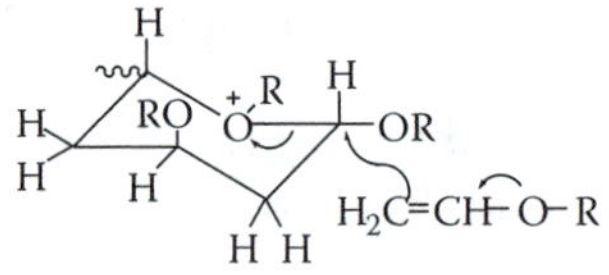

Both mechanisms ignore the nature of the catalyst forming the gegen-ion, but obviously, as this will act as a template for the attacking monomer, it will exert an influence on the rate of reaction and the type of stereoregularity imposed. The most probable configuration is isotactic because of the tendency for the gegen-ion to repel the substituent group of the incoming monomer.

6.8 HOMOGENEOUS STEREOSELECTIVE ANIONIC POLYMERIZATIONS

The various factors influencing the stereoregularity, when the propagating chain end is a carbanion, are conveniently highlighted in a study of the polymerization of methyl methacrylate by organolithium catalysts.

The propagating chain end in an anionic reaction initiated by a reagent such as *n*-butyl lithium can be thought of as existing in one of the following states, analogous to carbonium ion formation.

$$RLi \rightleftharpoons R^{\ominus} Li^{\oplus} \rightleftharpoons R^{\ominus} //Li^{\oplus} \rightleftharpoons R^{\ominus} Li^{\oplus}$$

covalent   contact ion   solvent separated   free ions
pair   ion pair

The extent of the separation will depend on the polarity of the reaction medium and in nonpolar-hydrocarbon solvents, such as toluene, covalent molecules, or contact ion pairs, are most likely to exist. With increasing solvent polarity, there is a greater tendency to solvate the ions, eventually producing free ion for strictly anionic polymerizations. These lead to conditions similar to a free-radical polymerization in which the stereoregulation is reduced and syndiotactic placements are favored at low-temperatures.

The effects of solvent and temperature are manifest in the polymerization of methyl methacrylate with *n*-butyl lithium at 243 K in a series of mixed solvents prepared from toluene and dimethoxyethane (DME). The NMR spectra of the products indicate the compositions in Table 6.2 and reveal that a predominantly isotactic material is produced in a low-polarity medium, but that this becomes highly syndiotactic as the solvating power of the medium increases.

An additional point emerges from this; higher syndiotactic contents are obtained when the Lewis-base strength of the solvent increases, and this factor probably accounts for the efficiency of the ether in this system. When the catalyst is 9-fluorenyl lithium, the reaction of methyl methacrylate at 195 K in toluene leads to isotactic polymer, whereas a change of solvent to tetrahydrofuran results in a syndiotactic product.

Stereoregulation is also altered by the nature of the solvent when Grignard reagents and alkali metal alkyls are used as initiators. In toluene, for example, the isotactic placements in the chain decrease as reagents change from Li to Na to K.

If general conclusions can be drawn from the behavior of methyl methacrylate, it appears that stereoregulation in anionic polymerizations, involving either polar monomers or monomers with bulky substituents, will lead to predominantly syndiotactic polymers when a free, dissociated ion occurs at the propagating end. This is because it is the most stable form arising from a minimization of steric and repulsive forces. If, however, some strongly regulating kinetic mechanism is available, for instance monomer + gegen-ion coordination, then the less favorable isotactic placement occurs.

In an attempt to explain the mechanism of stereoregulation in systems catalyzed by lithium alkyls, Bawn and Ledwith (1962) proposed that the penultimate residue plays an important role in the addition. A loose cyclic intermediate is formed when the Li+

TABLE 6.2
The Effect of Mixed Solvent Composition on the Tacticity of Poly(methyl methacrylate) Initiated by *n*-Butyl Lithium at 243 K

Toluene/DME	Isotactic	Heterotactic	Syndiotactic
100/0	0.59	0.23	0.18
64/36	0.38	0.27	0.35
38/62	0.24	0.32	0.44
2/98	0.16	0.29	0.55
0/100[a]	0.07	0.24	0.69

Note: The mole fractions of the various configurations are given.

[a] Measured at 203 K.

counterion coordinates with the carbonyl of the penultimate unit and with the terminal unit in a resonance enolic structure V.

This can be represented alternatively as a transition state VI similar to that in an S_N2 reaction. With one side of the Li^+ shielded, monomer approach is restricted, and the path of least resistance places the α-methyl group on the incoming monomer in a trans position, relative to the α-methyl group on the carbanion, during the π-complex formation. Addition then proceeds through a series of bond exchanges as the carbanion joins the monomer methylene group. The carbonyl group of the monomer coordinates with the ion by replacing the interaction with the previous penultimate group, and the cyclic intermediate is regenerated.

The steric restriction imposed by the α-methyl group aids the formation of an isotactic polymer, and in its absence (i.e., methyl acrylate), there is a reduced probability of isotactic placements. A compensating feature in the higher acrylates arises from the shielding of one side of the monomer by the bulkiness of the ester group. In the branched homologues, isopropyl and t-butyl acrylate, π-bonding with the Li^+ ion is forced to take place on one side of the monomer only, thereby enhancing the formation of isotactic polymer quite markedly.

As this and other mechanisms all postulate the existence of structures stabilized by intramolecular solvation, the addition of Lewis-bases or polar solvents should disrupt the required template and encourage conventional anionic propagation by free ions. This automatically reduces the probability of an isotactic placement occurring.

6.9 HOMOGENEOUS DIENE POLYMERIZATION

The principles applied in the previous section to essentially polar monomers can be extended to the stereoregular polymerization of dienes by alkali metals and metal alkyls. We have already seen that the cis-trans isomerism presents a variety of possible structures for the polydiene to adopt and complicates the preparation of a sample containing only one form rather than a mixture. Thus polyisoprene may contain units in the 1,2 or 3,4, or cis-1,4 or trans-1,4 configuration without even considering the tacticity of the 1,2 or 3,4 monomer sequences in the chain.

Most work has centered on the preparation of a particular form of geometric isomer because the type and distribution of each isomeric form in the chain has a profound influence on the mechanical and physical properties of the sample. The original discovery that metallic lithium in a hydrocarbon solvent catalyzed the production of an all *cis*-1,4 polyisoprene stimulated interest in this area and quickly raised two points that must be satisfied if a suitable mechanism is to be postulated.

1. Lithium and lithium–alkyl catalysts produce highly specific stereostructures, but when replaced by Na or K, this effect diminishes.
2. Stereospecific polymerization takes place in the bulk state or in hydrocarbon solvents, but the addition of a polar solvent leads to drastic changes.

To explain these features, the following mechanism has been put forward. Initiation produces a "Schlenk" adduct VII.

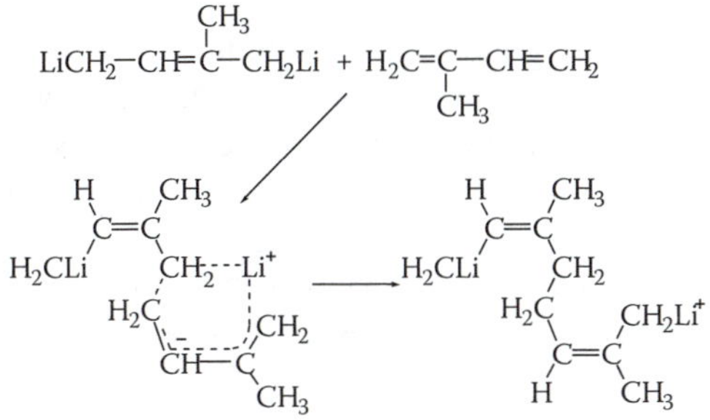

$$LiCH_2-CH=C-CH_2Li \; + \; H_2C=C-CH=CH_2$$

with CH_3 substituents on the central carbons

VII   *cis*-1,4

The lithium ion then forms a chelate complex with the isoprene monomer locking it into a cis-configuration, which is maintained during the addition reaction. This type of complex is suitable when a small ion like Li^+ is used but will be disrupted by the larger gegen-ions Na^+ and K^+, thereby allowing freer approach of the reactants. The presence of ethers also alters the stereospecificity by competing for the Li^+ and altering the spatial arrangement of the chelating pattern.

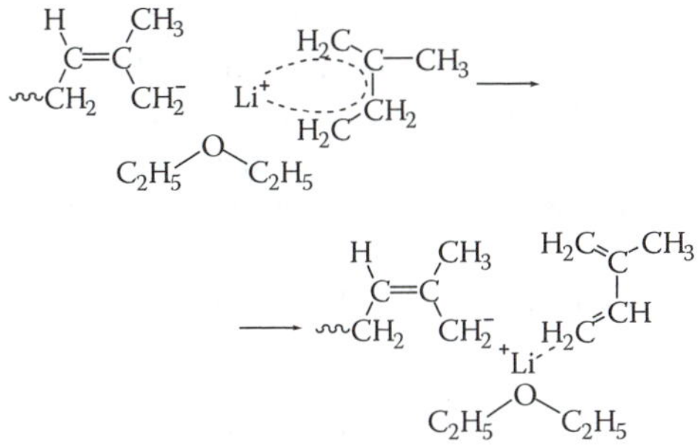

The monomer can then enter in a random fashion. The absence of significant 1,2- or 3,4-addition is thought to be caused by the shielding of carbon 3 in the transition state. However, all such proposals remain speculative.

6.10 SUMMARY

We can now summarize a few important points dealt with so far. Three factors influence stereoregularity during chain propagation:

1. *Steric factors*, which force the unit into a spatial arrangement determined by the size and position of the substituents already in the chain.
2. *Polar factors* because solvents that allow contact ion pairs favor isotactic placements, but pairs separated by heavy solvation (free ions) lead to syndiotactic structures.
3. *Coordination* because if the end group of a growing chain has a planar (sp^2) configuration with no established parity, such as that found in free-radical or free-ion propagation, then the configuration of this unit is established only during the course of addition of an incoming monomer. Normally, this will result in a syndiotactic placement with respect to the penultimate unit. Otherwise, coordination occurs between the gegen-ion, the incoming polar monomer, and the end or penultimate unit.

For polar monomers, the soluble catalysts can produce isotactic structures, but for nonpolar monomers, the homogeneous catalysts lead mainly to atactic or syndiotactic polymers, and a heterogeneous catalyst is required for isotactic placements to occur. These will be discussed in the next chapter.

PROBLEMS

1. Draw the structures of the polymers generated by 1,4 addition via anionic polymerization starting from:
   a. Methyl 2-methyl-2,4-hexadienoate (MMHd)
   b. Methyl 2,4-dimethyl-2,4-pentadienoate (MDMPd)
   c. Hydrogenation of poly(MDMPd) and poly(MMHd)
      What is the difference between these two structures? What type of copolymer is produced? (Hirabayashi et al., 2000)
2. Head-to-head polymers cannot be made directly, and a number of indirect approaches have been developed for their synthesis. For each of the methods in the following list, write down a synthetic scheme and name the polymer that is formed:
   a. 1,4-Polymerization of 2,3-dimethylbutadiene-1,3 (anionic) followed by hydrogenation
   b. Cationic polymerization of hexadiene-2,4 followed by hydrogenation
   c. Free-radical polymerization of 2,3-diphenylbutadiene-1,3 followed by hydrogenation
   d. Alternating polymerization of maleic anhydride and ethylene followed by esterification with methanol
   e. Chlorination of *cis*-1,4-polybutadiene

3. Which of the following monomers may be prepared in different isomeric forms?
    a. Vinylidene fluoride
    b. Vinyl chloride
    c. Ethyl acrylate
    d. Cyclopentene
4. Draw all possible configurational and stereo isomers resulting from polymerization of chloroprene.
5. Explain why for vinyl monomers of type CHR=CHR' two diisotactic isomers are possible but only one disyndiotactic structure is possible.
6. Draw a section of a poly(2-pentene) chain in the *threo*-diisotactic configuration.
7. Draw structures of isotactic and syndiotactic polyacetaldehyde.
8. Explain why polytetrafluoroethylene (PTFE) (van der Waals diameter of fluorine atoms = 270 pm) chains adopt a helical conformation in the crystal.
9. Consider the low-temperature cationic polymerization of vinyl ethers and discuss the effect of solvent on polymer stereochemistry.

REFERENCES

Bawn, C.E.H. and Ledwith, A., *Q. Rev.,* 16, 361, 1962.
Hirabayashi, T., Yamamoto, H., Kojima, T., Takasu, A., and Inai, Y., *Macromolecules*, 33, 4304, 2000.
Natta, G. and Corradini, P., *Rubber Chem. Technol.,* 33, 703, 1960.
Vogl, O., *J. Macromol. Sci. Chem.,* A21, 1725, 1984.

BIBLIOGRAPHY

Cooper, W., Stereospecific polymerization, in *Progress in High Polymers,* Vol. I, Academic Press, 1961.
Goodman, M., Concepts of polymer stereochemistry, *Topics in Stereochemistry,* Vol. 2, Wiley-Interscience, 1967.
Ketley, A.D., *The Stereochemistry of Macromolecules,* Vol. I–III, Edward Arnold, 1968.
Natta, G., Precisely constructed polymers, *Sci. Am.,* 205, 33, 1961.
Schildknecht, G.E., Stereoregular polymers, in *Encyclopedia of Chemistry*, Reinhold Publishing, 1966.
Seymour, R.B., *Introduction to Polymer Chemistry,* McGraw-Hill, 1971, chap. 6.

7 Polymerization Reactions Initiated by Metal Catalysts and Transfer Reactions

7.1 POLYMERIZATION USING ZIEGLER–NATTA CATALYSTS

Stereoregular polymerizations carried out in homogeneous systems, using essentially polar monomers, whose ability to coordinate with the catalyst complex imposes a stereospecific mechanism on the addition, have been dealt with in Chapter 6. As the polarity of the monomer decreases, however, the ability to control the configuration of the incoming monomer decreases, and atactic polymers result.

One of the most significant advances in synthetic polymer science in the 1950s was the discovery by Ziegler that, in the presence of a catalyst prepared from aluminum alkyl compounds used in conjunction with a transition metal halide, ethylene could be polymerized at ambient temperatures and atmospheric pressure. Ziegler found that this reaction produced a linear, highly crystalline polymer (now manufactured and marketed as high-density polyethylene, HDPE) as opposed to the extensively branched, less crystalline polyethylene produced by radical-initiated, high-pressure processes (low-density polyethylene, LDPE). The work was developed further by Natta and his co-workers, who found that semicrystalline linear polymers from propylene, 1-butene, and several other α-olefins could be prepared using similar catalyst systems. It was also demonstrated that the crystallinity in these polymers was enhanced by their highly stereoregular structure, and this paved the way for the preparation of stereoregular α-olefins.

In all cases the systems were heterogeneous and the active initiators are now known by the general name *Ziegler–Natta catalysts*. This encompasses a vast number of substances prepared from different combinations of organometallic compounds where the metal comes from the main Groups I, II, or III and is combined with the halide or ester of a transition metal (Groups IV to VIII). Table 7.1 contains a number of the common components of the Ziegler–Natta catalysts, but this list is far from exhaustive.

These catalysts tend to control two features, (1) the rate and (2) the specificity of the reaction, but this varies from reaction to reaction, and only a judicious choice of catalyst can effect control over both of these aspects.

TABLE 7.1

Components of Ziegler–Natta Catalysts

Metal Alkyl or Aryl	Transition Metal Compounds
$(C_2H_5)_3Al$	$TiCl_4$; $TiBr_3$
$(C_2H_5)_2AlCl$	$TiCl_3$; VCl_3
$(C_2H_5)AlCl_2$	VCl_4; $(C_5H_5)_2TiCl_2$
$(i\text{-}C_4H_9)_3Al$	$(CH_3COCHCOCH_3)_3V$
$(C_2H_5)_2Be$	$Ti(OC_4H_9)_4$
$(C_2H_5)_2Mg$	$Ti(OH)_4$; $VOCl_3$
$(C_4H_9)Li$	$MoCl_5$; $CrCl_3$
$(C_2H_5)_2Zn$	$ZrCl_4$
$(C_2H_5)_4Pb$	$CuCl$
$((C_6H_5)_2N)_3Al$	WCl_6
C_6H_5MgBr	$MnCl_2$
$(C_2H_5)_4AlLi$	NiO

Unfortunately, the insolubility of the catalyst poses the problems that the kinetics are hard to reproduce and the reaction mechanisms are difficult to formulate with real confidence. This means that the choice of a suitable catalyst for a system is somewhat empirical and based very much on trial and error, until optimum conditions are established.

It is useful to remember that both heterogeneous and homogeneous catalysts exist in the Ziegler–Natta group, but the latter only yield atactic or occasionally syndiotactic polymers from nonpolar monomers. As only the heterogeneous Ziegler–Natta catalysts produce isotactic poly-α-olefins, these have received the most attention. The interest in this type of system has been immense, as evidenced by the vast quantity of published material, and it was most fitting that both Ziegler and Natta were recognized for their work by being awarded jointly the Nobel Prize for chemistry in 1963.

A wide range of nonpolar monomers can be polymerized using Ziegler–Natta catalysts, and monomer reactivity decreases with increasing steric hindrance about the double bond.

7.2 NATURE OF THE CATALYST

Frequently, the product of Ziegler–Natta polymerization is sterically impure and can be preferentially extracted to give two products: a highly crystalline stereoregular fraction and an amorphous atactic one. This may be attributed to the size of the catalyst particles as stereoregularity is enhanced by having large particles, whereas a finely divided catalyst tends to produce an amorphous polymer.

The crystal form of the catalyst is also important, and the violet α, γ, and δ forms of $TiCl_3$ produce a greater quantity of isotactic polypropylene when combined with an aluminum alkyl than the brown β-structure. As the active sites for heterogeneous polymerizations are believed to be situated on the crystal surfaces, the structure is all-important. In the layered structure of α-TiCl$_3$, where every third Ti^{3+} ion in the lattice is missing, a number of Cl vacancies occur on the surface to maintain electrical neutrality in the crystal. The Ti^{3+} on the surface is then only 5-coordinated, leaving a vacant d-orbital, $\square$, and an active site is created when an alkyl group replaces a chloride ion to form TiRCl$_4$ $\square$. (See diagrams in Section 7.4.)

In β-TiCl$_3$, the linear chains form bundles in which some Ti ions are surrounded by five Cl$^-$ ions and some by only four Cl$^-$. This means that the steric control at the sites with two vacancies is now less rigid and stereoregulation is much poorer.

Catalyst composition affects both stereoregulation and polymer yield. Thus, Ti^{3+} is a more active producer of isotactic polypropylene than Ti^{4+} or Ti^{2+}, although an increase in the length of the associating alkyl group decreases the efficiency of stereoregular placements. Varying the transition metal and the associated aluminum compounds in the catalysts also influences the nature of the product.

7.3 NATURE OF ACTIVE CENTERS

Most of the experimental evidence points to propagation taking place at a carbon-to-transition metal bond with the active center being anionic in character. Free-radical reactions are considered to be nonexistent in the Ziegler–Natta systems because neither chain transfer nor catalyst consumption occurs. The active centers also live longer than radicals and resemble "living" polymer systems in many ways, one being that block copolymers can be produced by feeding two monomers alternatively into the system.

Although a number of reaction mechanisms have been suggested, two are worth considering in detail. These are based on the view that the active centers are localized rather than migrating and that the α-olefin is complexed at the transition metal center prior to incorporation into the chain; i.e., growth is always from the metal end of the growing chain.

The active species are then considered to be either *bimetallic* or *monometallic*.

7.4 BIMETALLIC MECHANISM

Natta and his associates have postulated a mechanism involving chain propagation from an active center formed by the chemisorption of an electropositive metal alkyl of small ionic radius on the cocatalyst surface. This yields an electron-deficient bridge complex such as I, and chain growth then emanates from the C–Al bond.

It is suggested that the nucleophilic olefin forms a π-complex with the ion of the transition metal, and following a partial ionization of the alkyl bridge, the monomer is included in a six-membered ring transition state. The monomer is then incorporated into the growing chain between the Al and C, allowing regeneration of the complex.

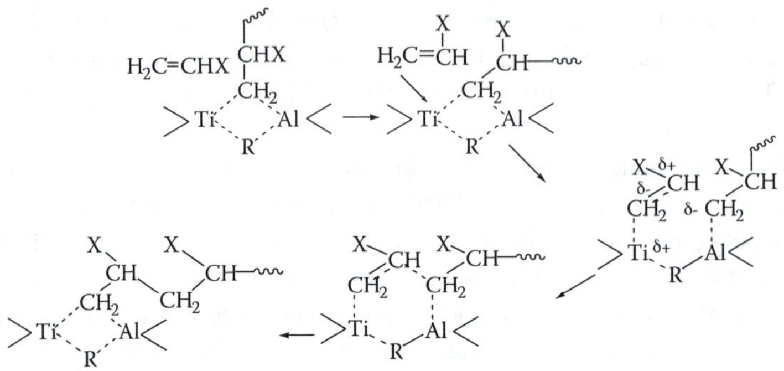

Although limited experimental evidence does lend support to this concept, major objections have been voiced by Ziegler, who is of the opinion that as dimeric aluminum alkyls are inefficient catalysts in the "Aufbau" reaction, the Ti–Al complex is not likely to be the effective catalytic agent. Other more recent work also favors the second and simpler alternative, the monometallic mechanism.

7.5 MONOMETALLIC MECHANISM

Majority opinion now favors the concept that the d-orbitals in the transition element are the main source of catalytic activity and that chain growth occurs at the titanium–alkyl bond. The ideas now presented are predominantly those of Cossee and Arlman, and will be developed using propylene as the monomer.

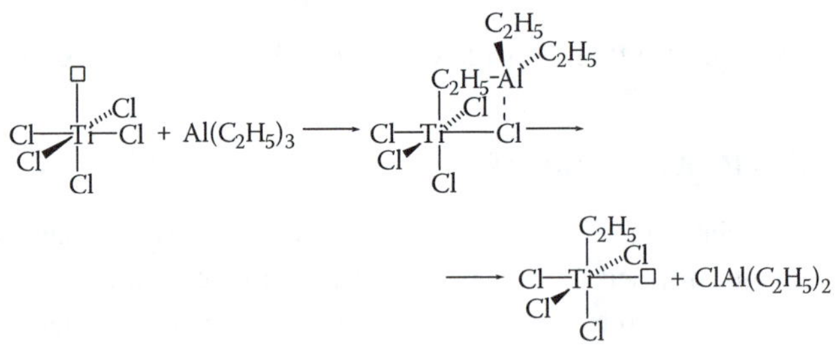

The first stage is the formation of the active center, illustrated here using α-TiCl$_3$ as the catalyst. The suggestion is that alkylation of the 5-coordinated Ti^{3+} ion takes place by an exchange mechanism after chemisorption of the aluminum alkyl on the surface of the TiCl$_3$ crystal. The four chloride ions remaining are the ones firmly embedded on the lattice, and the vacant site is now ready to accommodate the incoming monomer unit. The reaction is confined to the crystal surface, and the active complex is purely a surface phenomenon in heterogeneous systems.

The attacking monomer is essentially nonpolar but forms a π-complex with the titanium at the vacant d-orbital. A diagram of a section of the complex shows that the propylene molecule is not much bigger than a chloride ion and, consequently, the double bond can be placed adjacent to the Ti ion and practically as close as the halide. After insertion of the monomer between the Ti–C bond, the polymer chain then migrates back into its original position ready for a further complexing reaction.

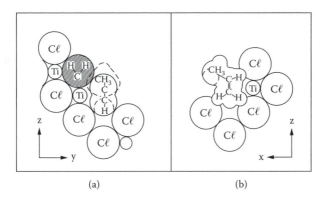

The reactivity of the active center is attributed primarily to the presence of d-orbitals in the transition metal. The initial state of the active center shows that the $=CH_2$ group will be capable of considerable distortion from its equilibrium position, because of the availability of adjacent d-orbitals. Complexing (b) takes place when the π-bonding orbitals of the olefin overlap with the vacant $d_{x^2-y^2}$ orbital of the Ti^{3+} while at the same time the π*-antibonding orbitals can overlap the d_{yz} orbitals of the Ti^{3+}. Formation of the transition state is aided by the ability of the $=CH_2$ group to migrate by partial overlap with the d_{z^2}, d_{yz}, and π*-orbitals (see Figures 7.1 and 7.2).

FIGURE 7.1 Cross-sectional diagrams of the propylene–catalyst complex through (a) the y–z plane and (b) the x–z plane of the octahedral structure. (Adapted from Bawn, C.E.H. and Ledwith, A., *Q. Rev.*, 16, 361, 1962.)

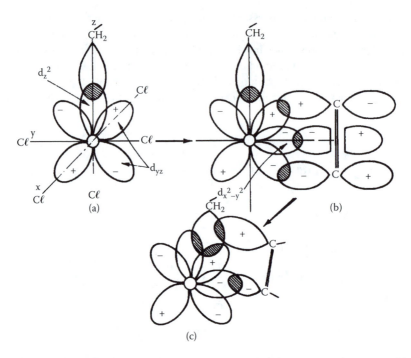

FIGURE 7.2 Representation of the relevant orbital overlap in (a) the active center, (b) the titanium–olefin complex, and (c) the transition state.

The main features of the monometallic mechanism are as follows: (1) an octahedral vacancy on the Ti^{3+} is available to complex the olefin, (2) the presence of an alkyl to transition metal bond at this site is required, and (3) the growing polymer chain is always attached to the transition metal.

7.6 STEREOREGULATION

To obtain a stereoregular polymer, the chemisorption of the monomer on the catalyst surface must be controlled so that the orientation of the incoming monomer is always the same. Examination of models shows that a molecule such as propylene will fit into the catalyst surface in only one way if a position of closest approach of the double bond to the Ti^{3+} ion is to be achieved. This places the $=CH_2$ group pointing into the lattice, and for steric reasons the orientation of the $-CH_3$ group to one side is preferred. This determines the configuration of the monomer during the complexing stage and is always the same. Repeated absorption of the monomer in this orientation, prior to reaction, leads to an isotactic polymer.

For the Cossee–Arlman mechanism to operate, migration of the vacant site back to its original position is necessary; otherwise, an alternating position is offered to the chemisorbed monomer and a syndiotactic polymer would result. This implies that the tacticity of the polymer formed depends essentially on the rates of both the alkyl shift and the migration. As both of these will slow down when the temperature is decreased, formation of syndiotactic polymer should be favored at low temperatures; syndiotactic polypropylene can in fact be obtained at 203 K.

7.7 RING-OPENING METATHESIS POLYMERIZATION (ROMP)

The Ziegler catalysts are not the only group of complexes capable of promoting polymerizations. There is a growing interest in *olefin metathesis reactions*, particularly those involving the transition-metal-catalyzed, ring-opening polymerizations of cycloalkenes and bicycloalkenes. The reactions can be used to produce linear chains containing unsaturated sites and, in the case of bicycloalkanes, ring structures are also incorporated. The development of catalysts capable of sustaining living polymer systems has led to the ability to prepare block copolymer structures or functionalized telomers; statistical copolymers have also been synthesized by these routes.

The majority of catalyst systems are based on transition metal compounds of tungsten, molybdenum, rhenium, ruthenium, and titanium carbene complexes. Among the most widely used are the tungsten halides, WCl_6, WF_6, and $WOCl_4$ but, in general, these and other catalysts may also require activation by a cocatalyst that is an organometallic compound or a Lewis acid. Thus $[WCl_6:(C_2H_5)_2AlCl]$, $[TiCl_4:(C_2H_5)_3Al]$, $[RuCl_3(hydrate):C_2H_5OH]$, and the general group $M(CHR)$ $(Ar)(OR)_2$, where M = Mo or W, R = alkyl, and Ar = 2,6-diisopropylphenyl, are all active catalyst–cocatalyst combinations. Many more exist, and the structure of the polymer formed can depend on the ratio of the catalyst to cocatalyst in the mixture. Metal carbene and metallacyclic complexes that have extended the range of monomer derivatives capable of undergoing ROMP have now been discovered. For example, using titanocene compounds such as Cp_2TiCl_2 (Cp = η^5–C_5H_5), active metallacyclobutane derivatives can be generated and used to form stable, living polymer systems that can be stored for some considerable time while retaining their activity.

The types of monomer susceptible to metathesis polymerization reactions are limited, however, and those that are most suitable are strained ring structures. Thermodynamic considerations account for the lack of reaction with six-membered rings, and strain-free cyclohexene derivatives or conjugated cyclodienes tend to be excluded. Newer catalysts that can polymerize polar monomers are now available. Cyclic monomers contain functional groups, –OH, –COOH, –COOR, –CONH$_2$, and –NH$_2$, can be used with the newer catalysts that have been developed.

7.8 MONOCYCLIC MONOMERS

Rapid polymerization reactions are observed when highly strained cycloalkene rings are used, an example being the conversion of cyclopentene into polypentenamer using $(WCl_6:Al_2Et_3Cl_3)$ as catalyst.

The material is a linear polymer retaining the double bond in the chain; this means that there is the possibility of cis–trans geometric isomerism, which will affect the properties of the product. Polypentenamer is a useful elastomer and can be prepared with either a high trans content ($T_g = 183$ K, $T_m = 293$ K) or a high cis content ($T_g = 159$ K, $T_m = 232$ K); e.g., 99% cis content is obtained with $MoCl_5/AlEt_3$ as the catalyst where there are equimolar ratios of Mo:Al.

Both structures have a glass transition temperature T_g, which is lower than that of natural rubber, although the T_g for the cis-polymer is the lowest recorded so far for a hydrocarbon elastomer. Variation of the catalyst composition can be used to control the cis–trans content of the chain. For molar ratios (Al:W) of up to 2:1, a predominantly cis-polypentenamer is formed, but as the molar ratio is raised, increasing amounts of the trans placements are found in the chain, and at 6:1 this amounts to ≈90% trans content.

Chain propagation is thought to proceed by coordination of the carbon–carbon double bond of the monomer at the vacant site of a transition metal carbene to form a π-complex. The unstable metallocyclobutane produced is then opened to reform the vacant site, and in so doing, incorporates the monomer in the chain.

This is known as the *nonpairwise mechanism*.

Other monocyclic monomers will undergo ring-opening reactions: cyclobutene and 1-methyl cyclobutene produce polybutadiene and cis-polyisoprene, respectively; but this synthetic route cannot compete with established methods for preparing synthetic elastomers. Reactions are slower with the larger rings, as the strain decreases, but a useful elastomer with a high trans content and well-developed crystallinity (such as the product sold under the trade name Vestenamer) can be prepared from cis-cyclooctene (the trans compound is unreactive).

$(T_g = 208$ K; $T_m = 358$ K)

7.9 BICYCLO- AND TRICYCLOMONOMERS

In general, bi- and tricyclic monomers can be polymerized more easily than the monocyclic monomers, and this is reflected in the wider range of catalytic systems that can be used.

Norbornene, or bicyclo[2.2.1]hept-2-ene, is one of the better-known monomers that can be subjected to ring-opening polymerization by metathesis catalysts. This monomer, or its derivatives, produces polymers with both a ring and a site of unsaturation in the main chain repeat unit.

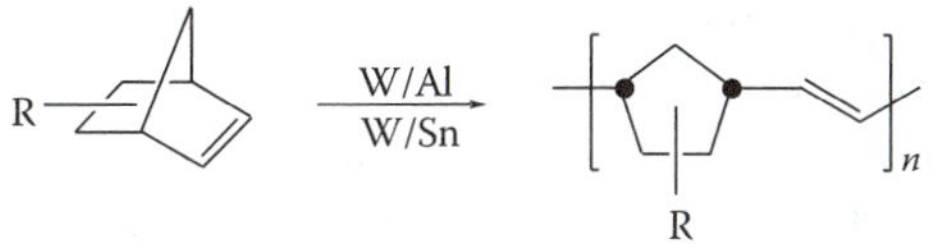

Such polymers can have both cis and trans structures, and a polynorbornene with a high trans content ($T_g = 308$ K, $T_m > 440$ K), prepared using $MoCl_5$ or $RuCl_3$ catalysts, is sold as the product Norsorex. In addition, the chain contains the chiral centers (shown as —•—), and other stereoregular structures are now possible. If the chirality of the centers on either side of the double bond is the same, then a *racemic* diad is formed, whereas if they have opposite chiralities, a *meso* diad results. The structures with *meso* diads are called *isotactic* placements and those with *racemic* diads are called *syndiotactic* placements.

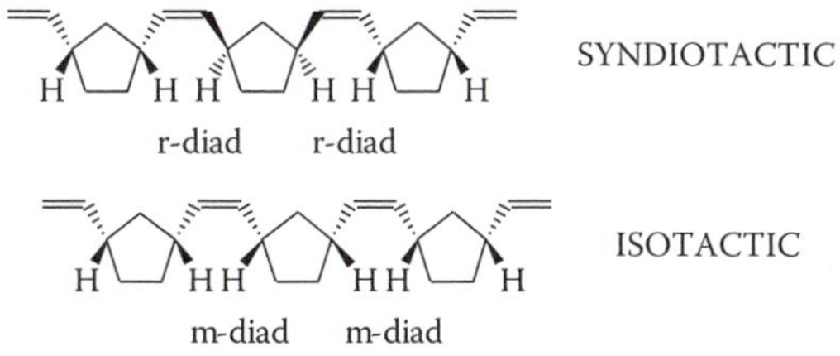

Successful polymerization of heterobicyclic monomers depends on careful selection of the catalyst system, and only a few examples exist at present. Thus, 7-oxa-norbornene derivatives can ring-open when reacted with osmium or ruthenium catalysts.

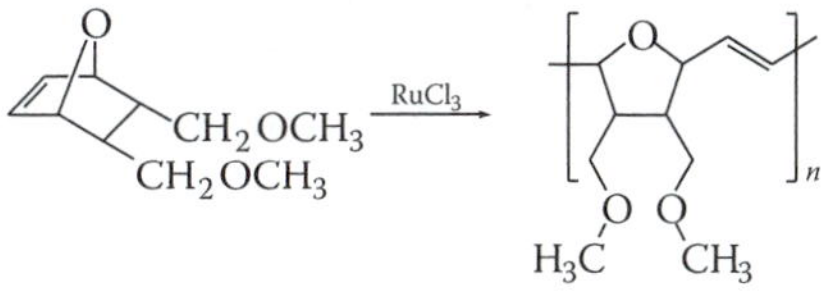

Tricyclo[5.2.1.0]dec-8-ene can be polymerized by several catalytic combinations to produce highly stereospecific polymers.

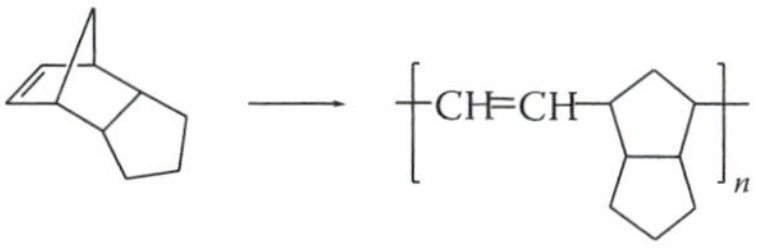

If MoCl$_5$ is used, a trans-polyalkenamer is formed; when ReCl$_5$ is the catalyst, the cis-polymer is produced; and if the monomer is reacted in the presence of WCl$_6$, a mixed cis–trans structure is obtained.

7.10 COPOLYALKENAMERS

The copolymerization of various pairs of cycloalkenes can be carried out in the presence of ROMP catalytic systems. The structures obtained are predominantly statistical because the differences in ring size or ring substituents, which will always exist, ensure that the reactivities of the monomers are sufficiently dissimilar to make uniform alternating copolymers virtually impossible to prepare.

The copolymerization of cyclopentene and cycloheptene in the presence of (WCl$_6$:Et$_2$AlCl) and benzoyl peroxide leads to a copolyalkenamer with 75% trans placements and approximately 80% of the pentenamer units in the chain. Mono- and bicycloalkenes can also copolymerize when tungsten or molybdenum catalysts are used; hence, cyclopentene and norbornene form the elastomeric structure shown:

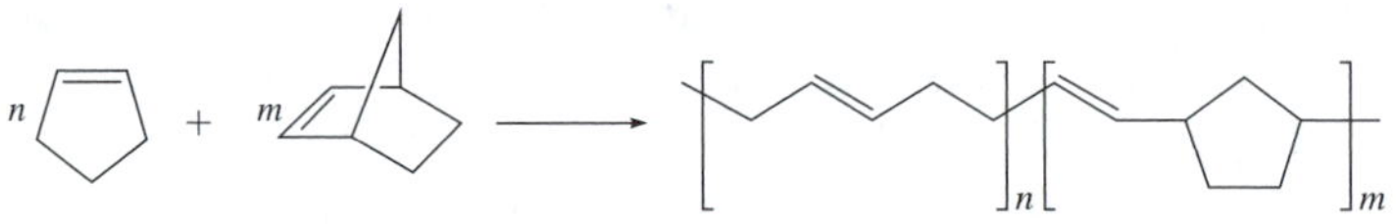

Similarly, substituted norbornenes can be used to produce copolymers with differing pendant groups.

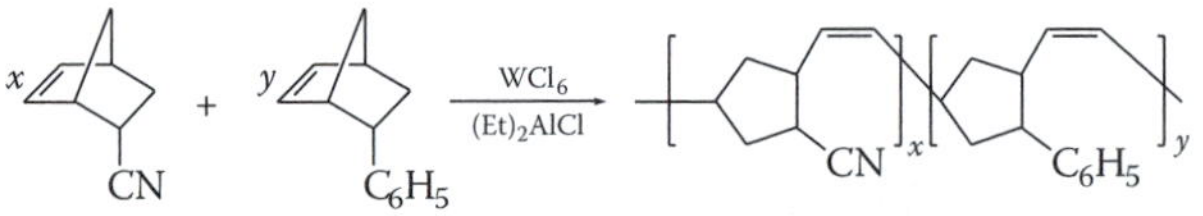

7.11 LIVING SYSTEMS

The development of novel titanium carbene complexes by Grubbs has opened up a route to "living" polymer systems, using coordinating polymerizations as opposed to those derived from ionic initiators, which can be used to form block copolymers or produce chains with a functionalized end group. The initiating species are formed by the reaction of norbornene with a titanocyclobutane derived from 3,3-dimethyl cycloprene

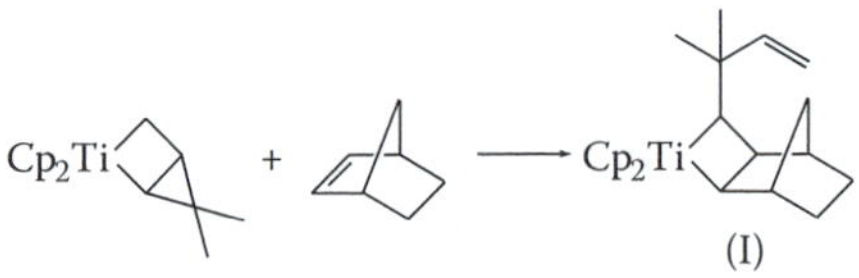

(I)

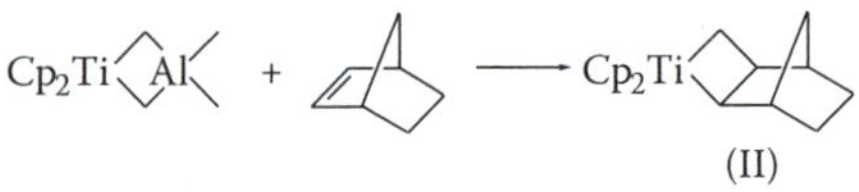

(II)

These can be used to prepare narrow molar mass distribution polymers, and as no termination reaction can be detected, they have all the features of a living polymer system; for example, a plot of percentage conversion of monomer against the molar masses of the polymers formed is a straight line passing through zero, which is a good test of a living system.

The stable living polymer chain derived from norbornene is

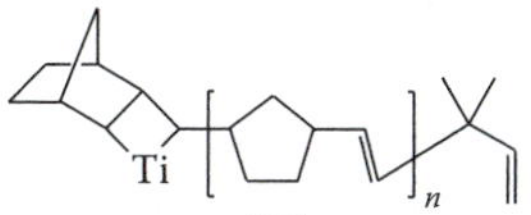

(III)

Heating (III) in the presence of another monomer leads to diblock copolymer formation

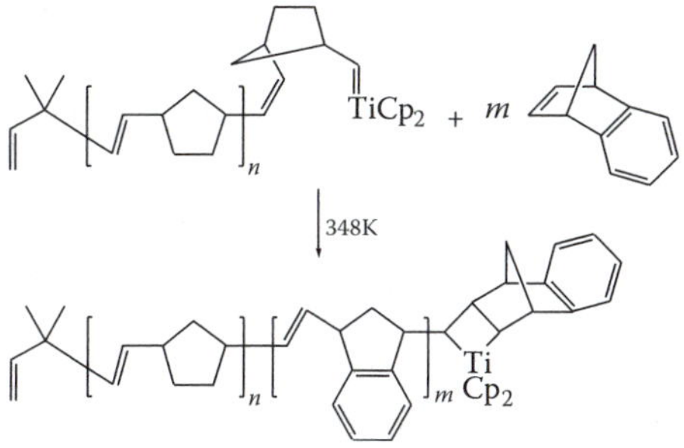

and by making a third addition to the reaction mixture of another monomer (or norbornene again), a triblock will be formed.

The chains can be terminated by heating with a reagent that will react with the carbene, and aldehydes or ketones will undergo a Wittig-type reaction to form a terminal olefin unit.

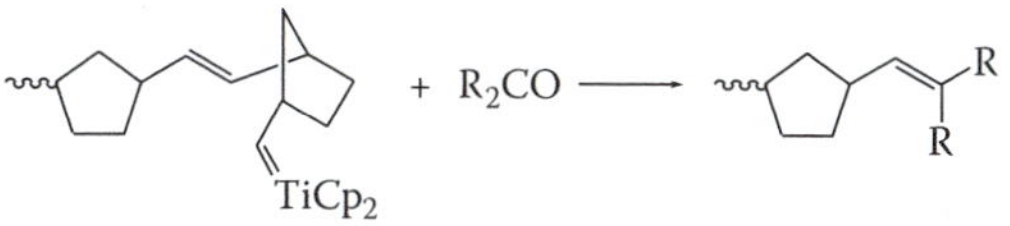

The end-capping reaction can be used to functionalize the terminal group and provide a telomer for further graft or block copolymer synthesis by coupling. Alternatively, it can be designed to give a group capable of initiating a different polymerization technique by a transfer reaction. For example, end-capping with phthalaldehyde gives a chain with a terminal aldehyde group that can be used to initiate an aldol GTP reaction (see Section 7.13) and form a diblock copolymer

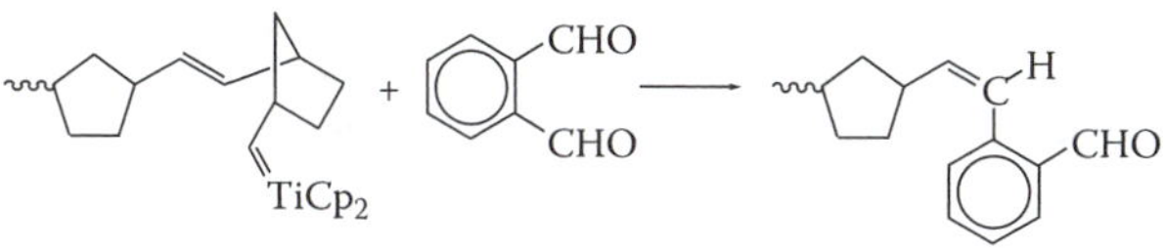

7.12   GROUP TRANSFER POLYMERIZATION (GTP)

The use of conventional anionic polymerization methods to produce living polymers from acrylic or methacrylic monomers has not met with much success. In 1983, however, the discovery of group transfer polymerization by workers at DuPont rectified the situation. The method developed is a new type of reaction leading to living polymers, which can be used for polar monomers, particularly derivatives of acrylic and methacrylic acids.

The reaction is a sequential Michael addition of an organosilicon compound to α,β-unsaturated esters, ketones, nitriles, and carboxamides. Chain propagation proceeds by transfer of the silyl group from the silyl ketene acetal catalyst to the monomer with the generation of a new ketene acetal, and if inadvertent termination is avoided, repeated addition of the monomer leads to a living polymer.[1] The typical reaction shown as follows uses the catalyst 1-methoxy-1-trimethylsiloxy-2-methyl prop-1-ene, but a cocatalyst is required, and this is either an anionic species or a Lewis acid.

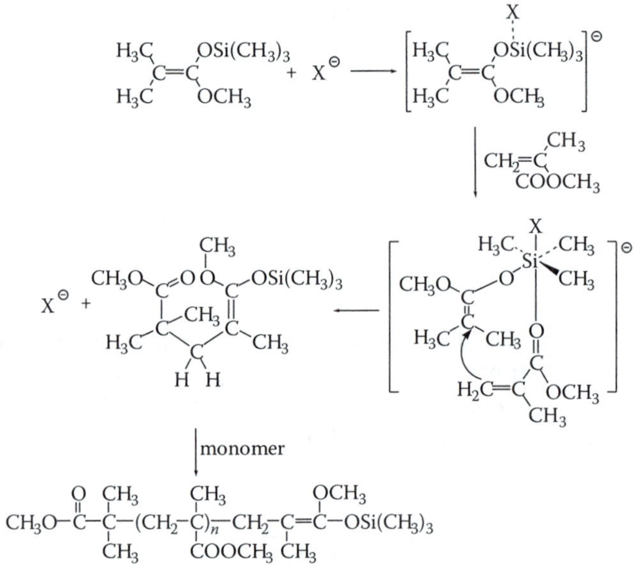

A catalytic complex forms first, to which the monomer is added sequentially, with the trimethyl silyl group transferring to the incoming monomer. A chain with the living ketene silyl acetal group is then produced, which is capable of further reaction if more monomer is made available.

Effective anionic cocatalysts are bifluoride ions; e.g., tris(dimethyl amino) sulfonium bifluoride $[(Me_2N)_3S \cdot HF_2]$ or azides (TAS N_3) and cyanides (TASCN), where TAS = $(Me_2N)_3S$. Lewis acids, zinc halides, and dialkyl aluminum chloride can also act as cocatalysts.

---

[1] Recent work suggests that GTP is not truly a living polymerization reaction and that there is a slow termination reaction involving cyclization. Consequently, block copolymer formation may depend on the rate of addition of a second monomer and the rate of polymerization.

Monomers containing active hydrogen (e.g., acids and hydroxy compounds) are not suitable for GTP, but acrylates, acrylonitrile, and *N,N*-dimethyl acrylamide can be polarized quite readily. The reaction is very susceptible to impurities, and all reagents and solvents must be scrupulously dried. Although both low- and high-temperature polymerization reactions are possible, a range of 270 to 320 K is preferred.

As with other living systems, GTP lends itself to the preparation of block copolymers using, for example, monomers such as methylmethacrylate (MMA) and butyl methacrylate (BMA).

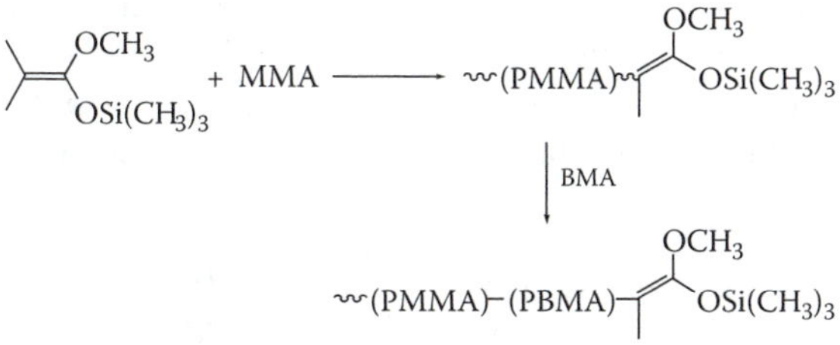

The block lengths can be controlled by altering the monomer-to-initiator molar ratio.

The formation of telechelic polymers is also possible. Modification of the initiator with a trimethyl silyl group that can subsequently be removed by treatment with a methanol/Bu_4NF mixture gives a hydroxy-terminated chain.

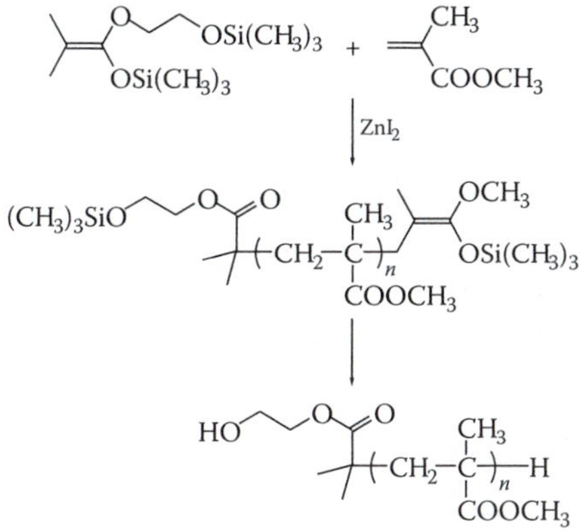

If $(Me)_2C=C(OSiMe_3)_2$ is used as the initiating species, a carboxyl-terminated chain can be obtained. In either case, coupling reactions between these functional chains will lead to block copolymer formation.

7.13 ALDOL GROUP TRANSFER POLYMERIZATION

A related technique called *aldol GTP* makes use of the reaction of an aldehyde with a silyl vinyl ether. The cocatalysts used are again Lewis acids, and if used with *t*-butyl dimethyl silyl vinyl ether, the combination will polymerize aldehydes:

$$RCHO + H_2C=CHOSi(CH_3)_2Bu^t \xrightarrow[\text{acid}]{\text{Lewis}} RCHCH_2CHO$$

$$\underset{OSi(CH_3)_2Bu^t}{|}$$

$$\Big| \text{monomer}$$

$$\underset{OSi(CH_3)_2Bu^t}{|}$$

$$R-CH(CH_2-CH)_n-CH_2CHO \longleftarrow R-CH(CH_2-CH)_n-CH_2CHO$$
$$\underset{OH}{|} \qquad \underset{OH}{|} \qquad\qquad\qquad \underset{OSi(CH_3)_2Bu^t}{|}$$

Hydrolysis of the trialkyl silyl groups gives poly(vinyl alcohol). Block copolymers in which one block is poly(vinyl alcohol) can be synthesized if a telechelic with an aldehyde terminal group is used to initiate the aldol GTP; structures such as poly(styrene-*block*-vinyl alcohol) can be prepared in this way. Alternatively, as silyl ketene acetals can react with aldehydes, block structures can be formed by a coupling process.

$$\sim\!\!\big(CH_2-CH\big)_n-CH_2CHO \;+\; \underset{R_3SiO}{\overset{H_3CO}{}}\!\!C=C\overset{/}{\underset{CH_2}{}}\!\!\overset{H_3C}{\underset{COOCH_3}{\big(\overset{|}{C}-CH_2\big)_m}}$$
$$\underset{OSiR_3}{|}$$

$$HF_2^{\ominus}\Big| THF$$

$$\sim\!\!\big(CH_2-CH\big)_{n+1}\!\!\big(\overset{H_3C}{\underset{COOCH_3}{\overset{|}{C}-CH_2}}\big)_{m+1}$$
$$\underset{OSiR_3}{|}$$

$$\Big|$$

$$\sim\!\!\big(CH_2-CH\big)_{n+1}\!\!\big(\overset{H_3C}{\underset{COOCH_3}{\overset{|}{C}-CH_2}}\big)_{m+1}$$
$$\underset{OH}{|}$$

During this reaction, cleavage of the silyl groups by fluoride ion in the presence of methanol forms the vinyl alcohol block.

7.14 METALLOCENE CATALYSTS

Polymers with relatively high levels of stereoregularity can be synthesized using heterogeneous Ziegler–Natta catalysts, but the multicentered nature of the reactive sites makes the reaction difficult to control.

The development of homogeneous metallocene catalysts has improved this considerably, and it is now possible to design polyolefins, in particular, with a desired microstructure and the corresponding material properties, by changing the catalyst structure. The first homogeneous Ziegler–Natta catalyst discovered was *bis*(cyclopentadienyl)titanium chloride, Cp_2TiCl_2, used in conjunction with an alkyl aluminum compound acting as a cocatalyst. Here, Cp represents the cyclopentadienyl rings that are π-bonded to the metal atom. This was not a particularly effective catalyst system although it was found that the addition of small quantities of water improved its performance because of the formation of alkyl aluminoxanes. These take part in the creation and stabilization of the active centers by alkylating the halogens on the metallocene to form the active site and reacting with impurities. Much improved metallocene catalyst systems have now been developed, which usually consist of one or two cyclopentadienyl (Cp) or substituted cyclopentadienyl (Cp') rings π-bonded to a group 4B metal; e.g., Zr, Ti, Hf. In the overwhelming majority of systems, the rings are linked by a bridging unit, which imparts a rigidity to the structure.

$B = \{CH_2CH_2\}$; $(CH_3)_2Si=$; $(CH_3)_2C=$

Y = halogen or alkyl group

R = H, alkyl or fused ring system

M = transition metal from groups 4B, 5B or 6B

Two categories of metallocene catalysts are now recognized: (1) metallocenes combined with aluminoxanes and (2) cationic metallocenes, where the transition metal is positively charged and stabilized by an anion that can be displaced by an incoming monomer during the polymerization reaction.

7.14.1 METALLOCENE/ALUMINOXANE CATALYSTS

Of all of the alkyl aluminoxanes, methyl aluminoxane (MAO) appears to be the most effective cocatalyst when combined with a metallocene. If stereoregulation is not required (e.g., polymerization of ethylene), an unbridged, achiral catalyst combination such as *bis*(cyclopentadienyl) zirconium dichloride (Cp_2ZrCl_2)/MAO can be used. This produces narrow molecular weight distribution polyethylenes (PE) at high polymerization rates; for example, at 8-bar pressure and 388 K, 40,000 kg of PE are produced per (g Zr)$^{-1}$ h^{-1}. The speed of polymerization can be increased by altering the Al/Zr ratio from 1,070 to 460,000, and the molecular weight can be controlled by changing the polymerization temperature. Lowering the reaction temperature from 373 K to 273 K increases the molecular weight from 10^3 to 10^6. These achiral catalysts are also very effective in producing copolymers with a narrower chemical composition and narrower molecular weight distributions than the heterogeneous Ziegler–Natta systems. Thus, Cp_2ZrCl_2/MAO is effective in the production of ethylene/1-butene, ethylene/propylene copolymers and terpolymers of ethylene/1-butene/1-decene.

7.14.2 STEREOREGULATION

The achiral catalysts such as Cp_2ZrCl_2/MAO can be used to polymerize propylene, but they produce only atactic polymer. They are unbridged, have C_{2V} symmetry, and are unsuitable for preparing stereoregular polymers. Much greater success is achieved by using chiral ansa-metallocenes with MAO. These have two Cp units locked into a chiral arrangement by a bridging group and can have C_2, C_S, or C_1 symmetry, depending on the type of Cp' ring system and nature of the bridging group. The bridge imparts the necessary rigidity to the metallocene and prevents rotation about the coordination axis. Two classic examples of chiral ansa-metallocene are the Brintzinger catalysts:

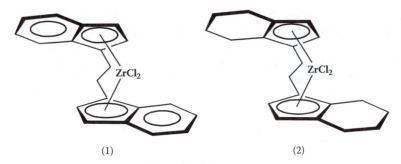

(1)   (2)

They are (1) ethylene(*bis*-indenyl)zirconium dichloride Et(Ind)$_2$ZrCl$_2$ and (2) ethylene(*bis*-tetrahydroindenyl) zirconium dichloride Et(H$_4$Ind)$_2$ZrCl$_2$. Both have C$_2$ symmetry and can be produced in a racemic mixture of d and l forms, with very low concentrations of the meso form that will produce only atactic polymer. The structures (1) and (2) when combined with MAO favor the production of isotactic polyolefins. The Et(Ind)$_2$ ZrCl$_2$/MAO system is the more active of the two, and when used to catalyze the polymerization of propylene at low temperature (273 to 293 K), produces highly (83%) isotactic polypropylene with high molecular weights and low polydispersities (1.9 to 2.6). Increasing the temperature of polymerization will increase the rate of polymerization but at the expense of decreasing the molecular weight, molecular weight distribution, and stereoregularity. At higher temperatures, the catalysts become more flexible, and the control over stereoregular addition of monomers during polymerization reaction is reduced.

Small changes in a metallocene structure can have a significant effect on the molecular weight and rate of polymerization of polypropylene. This can be illustrated using (CH$_3$)$_2$Si(Ind)$_2$ZrCl$_2$ (3).

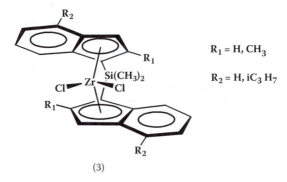

$R_1 = H, CH_3$

$R_2 = H, iC_3 H_7$

(3)

The catalyst (3) combined with MAO when used in the bulk polymerization of propylene at 323 K yields a 94% isotactic polypropylene of M$_w$ = 6 × 10^4. Modification of this structure by introducing a methyl group in each indenyl ring system next to the bridging unit (i.e., R$_1$ = CH$_3$, R$_2$ = H) gives a catalyst that at 323 K will produce 97% isotactic polypropylene with M$_w$ = 3.4 × 10^5 at a slightly reduced rate. Further modification by introducing the isopropyl group (R = CH$_3$, R$_2$ = iC$_3$H$_7$) produces a catalyst that yields a 98% isotactic polypropylene with M$_w$ = 4.6 × 10^5 but at twice the rate of the previous two catalysts.

Syndiotactic polypropylene can be prepared by changing to a catalyst with C_S symmetry such as isopropylidene (cyclopentadienyl fluorenyl) zirconium dichloride, iPr(CpFlu)ZrCl$_2$ (4). This is not a chiral structure, but it does have an internal symmetry plane that favors syndiotactic placement along the polymer chain.

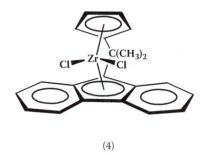

(4)

The Zr can also be replaced by Hf, and with MAO these catalysts produce polypropylene with a very high level of syndiotacticity, high molecular weight, and low polydispersity at high polymerization rates.

Polypropylene with thermoplastic elastomeric properties can be prepared using Et(Me$_4$Cp)(Ind)TiCl$_2$/MAO. This catalyst, which has C_1 symmetry, or an unbridged catalyst (e.g., (2-PhInd)$_2$ ZrCl$_2$) that can rotate between chiral and achiral forms, produces stereoblock polypropylenes with alternating blocks of isotactic and atactic placements along the chain. The isotactic blocks tend to crystallize and act as physical cross-linking sites between the atactic sequences to produce a material with elastomeric properties that is thermoreversibly cross-linked. To produce a stereoblock structure, the catalyst must have two different reactive states: one that can stereoregulate the monomer, the other being nonstereospecific. The *bis*(2-phenyl, indenyl) zirconium dichloride shown as follows can oscillate between the chiral C$_2$ form and the achiral C$_{2v}$ stereoconformation:

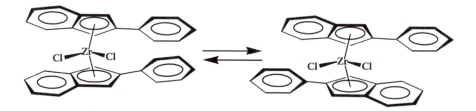

These states then interchange during the course of the chain-growth reaction to produce sequences of isotactic and atactic placements.

Small structural changes in the catalysts can produce even more remarkable results. Substitution of a methyl group in the cyclopentadienyl ring of iPr(CpFlu)ZrCl$_2$ led to the production of hemiisotactic polypropylene. In this structure, every second stereogenic carbon has an isotactic configuration, and the centers in between have atactic placements. Thus, there is order and disorder along the chain in a well-defined manner.

hemi-isotactic

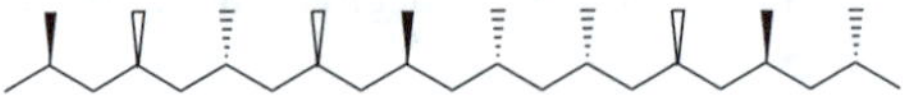

hemi-syndiotactic

An unusual feature of the chiral metallocenes, such as structures (1) and (2), is their ability to catalyze the polymerization of cycloalkenes without opening the ring. Isotactic, high-melting-point cycloalkenes can be prepared from cyclopentene, cyclobutene, and norbornene, using $Et(Ind)_2ZrCl_2/MAO$ as the catalyst system.

7.14.3 CATIONIC METALLOCENES

Catalyst systems that do not require the presence of MAO can be prepared by forming a cationic complex of the transition metal, which is associated with a stabilizing anion that can be easily displaced by an incoming monomer during chain growth.

Two examples show how these can be prepared:

(a) $Cp_2Zr(CH_3)_2 + B(C_6F_5)_3 \longrightarrow Cp_2Zr^{\oplus}CH_3 + (C_6F_5)B(CH_3)^{\ominus}$

$Et(Ind)_2Zr(CH_3)_2 + Ph_3C^{\oplus}B(C_6F_5)^{\ominus}$

(b) $\downarrow$

$Et(Ind)_2Zr^{\oplus}CH_3 + B(C_6F_5)^{\ominus} + Ph_3CCH_3$

The system $Et(Ind)_2Zr^{\oplus}CH_3.B(C_6F_5)^{\ominus}$ has been used to produce isotactic polypropylene at 218 K. A cationic catalyst such as $Cp_2Zr^{\oplus} Me_2/B(C_6F_5)_3^{\ominus}$ can be used to produce high-molecular-weight polyisobutylene and isoprene at 243 K.

7.14.4 MECHANISM OF STEREOREGULATION

Several methods have been proposed to explain the stereoregulation mechanism in metallocene catalysis. Most now agree that the active species is cationic, the insertion of the monomer is between the transition metal and growing chains, and that the steric structure of the catalyst determines the orientation of the incoming monomer. Without giving detailed analyses of all the models, the essential features can be summarized as follows.

The MAO methylates replace the halogens on the metallocene to form the active site:

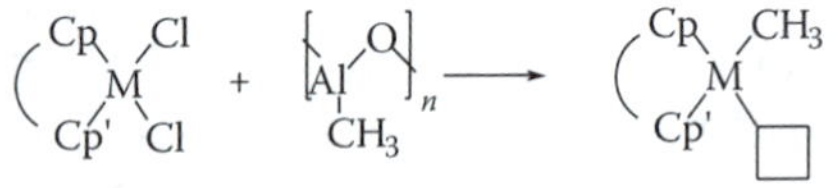

This is a vacant coordination site that will bind the incoming monomer during the chain propagation. The polymerization process is thought to be essentially the same

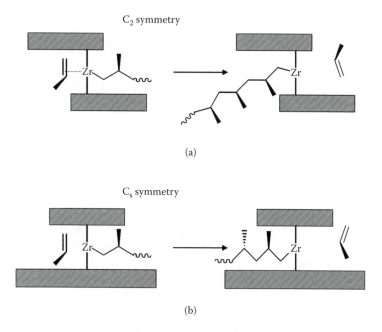

(a)

(b)

FIGURE 7.3 Stereoregulation mechanism of metallocene catalysts.

as the monometallic mechanism described earlier. The coordinated monomer is added between the metal and the polymer chain by a cis migratory insertion process. For monomers such as propylene this process is influenced by both stereoselectivity and regioselectivity. The simple schematic is shown in Figure 7.3.

In Figure 7.3(a), the metallocene has C_2 symmetry, i.e., Et(Ind)$_2$ZrCl$_2$ (1); the binding sites are equivalent and each successive monomer coordinates with the same facial selectivity, which is the most favorable sterically (chiral-site stereocontrol). This leads to a predominance of isotactic placements.

In Figure 7.3(b), the metallocene has C_S symmetry, i.e., iPr(C$_p$ Flu)ZrCl$_2$ (4), and each successive incoming monomer can present an alternating facial selectivity and so produces mainly syndiotactic placements along the chain. Although this is not a particularly sophisticated model, it serves to illustrate how the steric structure of the catalyst can influence the steric placement of the incoming monomer.

7.15   CONCLUDING REMARKS

Polyolefins are extremely important commercially. In 2005, the annual demand for polyethylene was 20×10^6 tons, whereas the average yearly consumption of all types of polypropylene polymers was approximately 4 kg for every human on the planet. As control of the polymer structure makes the recycling process easier, it is easily seen that metallocenes play a major role in improving this situation and help meet the demand for more precisely structured materials.

PROBLEMS

1. Would you expect stereoregularity of propylene to be higher when using $AlEt_3 + \beta TiCl_3$, $AlEt_3 + \alpha TiCl_3$, or $AlEt_3 + TiCl_4$? Explain.

2. Place the following monomers in order of decreasing reactivity with Ziegler–Natta catalysts.

$$CH_2{=}CH(CH_3)$$
$$CH_2{=}CHCH(CH_3)_2$$
$$CH_2{=}CH\ C(CH_3)_3$$

3. Draw all possible isomers resulting from polymerization of isoprene. Explain why polymerization of isoprene with Ziegler–Natta catalysts leads to *trans*-1,4, *cis*-1,4 or 3,4 isomers but no 1,2 product has been observed.

4. What are the products of the ROMP of 1-methyl-1,5-cyclooctadiene and *cis*-, *trans*-cyclodeca-1,5-diene. To which copolymer structures are these equivalent?

5. Draw structures of each of the metallocene catalysts given below and indicate (a) type of symmetry and (b) stereochemistry of the polymer produced from polymerization of propene.

    (a)   *rac*-$Me_2Si(2Me-4-Ph-1-Ind)_2\ ZrCl_2$

    (b)   *meso*-$Me_2Si(4-Ph-1-Ind)_2\ ZrCl_2$

    (c) $Me_2C(Cp)(9-Flu)\ ZrCl_2$

REFERENCES

Bawn, C.E.H. and Ledwith, A., *Q. Rev.*, 16, 361, 1962.

Bazan, G., Schrock, R.R., Khosravi, E., Feast, W.J., and Gibson, V.C., *Polym. Commun.*, 9, 258, 1989.

Coates, G.W., *Chem. Rev.*, 100, 1223, 2000.

Gilliom, L.R. and Grubbs, R.H., *J. Am. Chem. Soc.*, 108, 733, 1986.

Resconi, L., Cavallo, L., Fait, A., and Piemontesi, F., *Chem. Rev.*, 100, 1253, 2000.

Webster, O.W., Hertler, W.R., Sogah, D.Y., Farnham, W.B., and Rajanbabu, T.V., *J. Macromol. Sci.-Chem.*, A21, 943, 1984.

BIBLIOGRAPHY

Allen, G. and Bevington, J.C., Eds., *Comprehensive Polymer Science: The Synthesis, Characterization, Reactions, and Applications of Polymers*, Vol. 4, Pergamon Press, 1989.

Boor, J., *Ziegler-Natta Catalysis and Polymerization*, Academic Press, 1979.

Dragutan, V., Balaban, A.T., and Dimonie, M., *Olefin Metathesis and Ring Opening Polymerization of Cyclo Olefins*, John Wiley, 1985.

Hiemenz, P.C. and Lodge, T., *Polymer Chemistry*, 2nd ed., CRC Press, 2007, chap. 5.

Ivin, K.J., *Olefin Metathesis*, Academic Press, 1983.

Kaminsky, W. and Sinn, H., Eds., *Transition Metals and Organometallics as Catalysts for Olefin Polymerization,* Springer-Verlag, 1988.

Keii, T., *Kinetics of Ziegler-Natta Polymerization*, Chapman and Hall, 1972.

Mark, H.F., *Encyclopedia of Polymer Science and Technology*, 12 volume set, 3rd ed., John Wiley and Sons, 2004.

Odian, G., *Principles of Polymerization*, 4th ed., John Wiley and Sons, 2004, chap. 8.

Quirk, R.P., Ed., *Transition Metal Catalysed Polymerizations,* Harwood Academic Press, 1983.

Stevens, M., *Polymer Chemistry: An Introduction*, 3rd ed., Oxford University Press, 1998.

8 Polymers in Solution

8.1 THERMODYNAMICS OF POLYMER SOLUTIONS

The interaction of long-chain molecules with liquids is of considerable interest from both a practical and theoretical viewpoint. For linear and branched polymers, liquids that will dissolve the polymer completely to form a homogeneous solution can usually be found, whereas cross-linked networks will only swell when in contact with compatible liquids. In this chapter, we shall deal with linear or branched polymers and treat the swelling of networks in Chapter 14.

When an amorphous polymer is mixed with a suitable solvent, it disperses in the solvent and behaves as though it too is a liquid. In a good solvent, classed as one that is highly compatible with the polymer, the liquid–polymer interactions expand the polymer coil from its unperturbed dimensions in proportion to the extent of these interactions. In a "poor" solvent, the interactions are fewer, and coil expansion or perturbation is restricted.

The fundamental thermodynamic equation used to describe these systems relates the Gibbs free-energy function G to the enthalpy H and entropy S; i.e., $G = H - TS$. A homogeneous solution is obtained when the Gibbs free energy of mixing $\Delta G^M < 0$, i.e., when the Gibbs free energy of the solution G_{12} is lower than the Gibbs functions of the components of the mixture G_1 and G_2.

$$\Delta G^M = G_{12} - (G_1 + G_2) \tag{8.1}$$

8.2 IDEAL MIXTURES OF SMALL MOLECULES

To understand the behavior of polymers in solution more fully, knowledge of the enthalpic and entropic contributions to ΔG^M is essential, and it is instructive to consider first mixtures of small molecules to establish some fundamental rules concerning ideal and nonideal behavior. Raoult's law is a useful starting point and defines an ideal solution as one in which the activity of each component in a mixture a_i is equal to its mole fraction x_i. This is valid only for components of comparable size and where the intermolecular forces acting between both like and unlike molecules are equal. The latter requirement means that component molecules of each species can interchange positions without altering the total energy of the system; i.e., $\Delta H^M = 0$ and, consequently, it only remains for the entropy contribution ΔS^M to be calculated.

For a system in a given state, the entropy is related to the number of distinguishable arrangements the components in that state can adopt and can be calculated from the Boltzmann law, $S = k \ln \Omega$, where Ω is the number of statistical microstates available to the system. We can begin by considering the mixing of N_1 molecules of component

1 with N_2 molecules of component 2, and this can be assumed to take place on a hypothetical lattice containing $(N_1 + N_2) = N_0$ cells of equal size. Although this formalism is not strictly necessary for the analysis, the arrangement of spherical molecules of equal size in the liquid state will, to the first near-neighbor approximation, be similar to a regular lattice structure, and so it is a useful structure to use as a framework for the mixing process.

The total number of possible ways in which the component molecules can be arranged on the lattice increases when mixing takes place and is equal to $(N_1 + N_2)!$ $= N_0!$, but as the interchanging of a molecule of component 1 with another molecule component 1, or component 2 with component 2 will be an indistinguishable process, the net number of distinguishable arrangements will be

$$\Omega = \frac{(N_1 + N_2)!}{N_1!N_2!} = \frac{N_0!}{\Pi N_i!} \tag{8.2}$$

The configurational (or combinatorial) entropy S_c can then be derived from the Boltzmann law and

$$S_c = k \ln \frac{N_0!}{N_1!N_2!} = k \left(\ln N_0! - \ln N_1! - \ln N_2! \right) \tag{8.3}$$

For large values of N_i, Stirling's approximation can be used to deal with the factorials; i.e., $\ln N! = N \ln N - N$, and Equation 8.3 becomes

$$S_c = k(N_0 \ln N_0 - N_0 - N_1 \ln N_1 + N_1 - N_2 \ln N_2 + N_2) \tag{8.4}$$

which on dividing by N_0 gives

$$S_c = -k \left[N_1 \ln \frac{N_1}{N_0} + N_2 \ln \frac{N_2}{N_0} \right] \tag{8.5}$$

If $x_i = (N_i/N_0)$, the mole fraction of component i, then

$$S_c = -k[N_1 \ln x_1 + N_2 \ln x_2] \tag{8.6}$$

For pure components, $x_i = 1$, and as ΔS^M, the change in entropy on mixing, is given by $(S_c - S_1 - S_2)$, we can write

$$S_c = \Delta S_{id}^M = -k \sum N_i \ln x_i$$

So for a two-component mixture,

$$\Delta S_{id}^M = -k[N_1 \ln x_1 + N_2 \ln x_2] \tag{8.7}$$

This expression is derived on the assumptions that (1) the volume change on mixing $\Delta V^M = 0$, (2) the molecules are all of equal size, (3) all possible arrangements have the same energy, $\Delta H^M = 0$, and (4) the motion of the components about their equilibrium positions remains unchanged on mixing. Thus, the free energy of mixing, ΔG^M, is

$$\Delta G^M = -T\Delta S^M = kT(N_1 \ln x_1 + N_2 \ln x_2) \tag{8.8}$$

which shows that mixing in ideal systems is an entropically driven, spontaneous process.

8.3 NONIDEAL SOLUTIONS

Any deviations from assumptions (1) to (4) will constitute a deviation from ideality — an ideal solution is a rare occurrence — and several more realistic types of solution can be identified:

1. *Athermal* solutions, where $\Delta H^M = 0$ but ΔS^M is not ideal
2. *Regular* solutions, where ΔS^M is ideal but $\Delta H^M \neq 0$
3. *Irregular* solutions, in which both ΔS^M and ΔH^M deviate from their ideal values

Polymer solutions tend to fall into category (3), and the nonideal behavior can be attributed not only to the existence of a finite heat of mixing but also to the large difference in size between the polymer and solvent molecules. The polymer chain can be regarded as a series of small segments covalently bonded together, and it is the effect of this chain connectivity that leads to deviations from an ideal entropy of mixing. The effect of connectivity can be assessed by calculating the entropy change associated with the different number of ways of arranging polymer chains and solvent molecules on a lattice and, as will be demonstrated, this differs from that calculated for the ideal solution. This is embodied in the theory developed by Flory and Huggins but still represents only the combinatorial contribution, whereas there are other (noncombinatorial) contributions to the entropy that come from the interaction of the polymer with the solvent and are much harder to quantify. Nevertheless, the Flory–Huggins theory forms the cornerstone of polymer solution thermodynamics and is worth considering further.

8.4 FLORY–HUGGINS THEORY: ENTROPY OF MIXING

The dissolution of a polymer in a solvent can be regarded as a two-stage process. The polymer exists initially in the solid state in which it is restricted to only one of the many conformations that are available to it as a free isolated molecule. On passing into the liquid solution, the chain achieves relative freedom and can now change rapidly among a multitude of possible equi-energetic conformations, dictated partly by the chain flexibility and partly by the interactions with the solvent.

Flory and Huggins considered that formation of the solution depends on (1) the transfer of the polymer chain from a pure, perfectly ordered state to a state of disorder, which has the necessary freedom to allow the chain to be placed randomly on a lattice, and (2) the mixing process of the flexible chains with solvent molecules (Figure 8.1).

The formalism of the lattice was used for convenience to calculate the combinatorial entropy of mixing according to the method outlined in Section 8.2 for small molecules, including the same starting assumptions and restrictions.

Consider a polymer chain consisting of r covalently bonded segments whose size is the same as the solvent molecules, i.e., $r = (V_2/V_1)$, where V_i is the molar volume of component i. To calculate the number of ways this chain can be added to a lattice, the necessary restriction imposed is that the segments must occupy r contiguous sites on the lattice because of the connectivity. The problem is to examine the mixing of N_1 solvent molecules with N_2 monodisperse polymer molecules comprising r segments; we can begin by adding i polymer molecules to an empty lattice with a total number of cells N_0.

$$N_0 = (N_1 + rN_2) \tag{8.9}$$

Thus, the number of vacant cells left, which can accommodate the next $(i + 1)$ molecule, is given by

$$(N_0 - ri) \tag{8.10}$$

The $(i + 1)$ molecule can now be placed on the lattice, segment by segment, bearing in mind the restrictions imposed, i.e., the connectivity of the segments, which requires the placing of each segment in a cell adjoining the preceding one. This in turn will depend on the availability of a suitable vacancy. The first segment can be placed in any empty cell, but the second segment is restricted to the immediate near neighbors surrounding the first (see Figure 8.1). This can be given by the coordination number of the lattice z, but we must also know if a cell in the coordination shell is empty. If we let p_i be the probability that an adjacent cell is vacant, then, to a reasonable approximation, this can be equated with the fraction of cells occupied by i polymer chains on the lattice, i.e.,

$$p_i = (N_0 - ri)/N_0 \tag{8.11}$$

which is valid for large values of z. So the expected number of empty cells available for the second segment is zp_i, and having removed one more vacant cell from the immediate vicinity, the third and each succeeding segment will have $(z - 1)p_i$ empty cells to choose from. The total number of ways in which the $(i + 1)$ molecule can be placed on the lattice is then

$$\omega_{(i+1)} = (N_0 - ri)z(z-1)^{r-2}[(N_0 - ri)/N_0]^{r-1} = (N_0 - ri)^r[(z-1)/N_0]^{r-1} \tag{8.12}$$

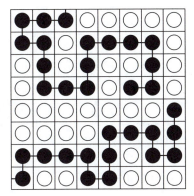

FIGURE 8.1 Placement of polymer chains and solvent molecules on a lattice as required by the Flory–Huggins theory.

This gives the set of possible ways in which the $(i + 1)$ molecule can be accommodated on the lattice. The total number of ways for all N_2 molecules to be placed can then be obtained from the product of all possible ways, i.e.,

$$\omega_1 \omega_2 \dots \omega_i \dots \omega_{N_2} = \prod_{i=1}^{N_2} \omega_i$$

The polymer molecules are all identical and so, by analogy with Equation 8.2, the total number of distinguishable ways of adding N_2 polymer molecules is

$$\Omega_p = \prod_{i=1}^{N_2} \omega_i / N_2! \tag{8.13}$$

Substituting for ω_i gives

$$\omega_p = \frac{1}{N_2!} \prod_{i=1}^{N_2} \{ [N_0 - r(i-1)]^r [(z-1)/N_0]^{r-1} \}$$

$$= \left(\frac{1}{N_2!} \right) \left(\frac{z-1}{N_0} \right)^{N_2(r-1)} \prod_{i=1}^{N_2} \{ N_0 - r(i-1) \}^r \tag{8.14}$$

To evaluate the product term, we can multiply and divide by r

$$\prod_{i=1}^{N_2} \{ N_0 - r(i-1) \}^r = r^{(N_2 r)} \prod_{i=1}^{N_2} \left\{ \frac{N_0}{r} - i + 1 \right\}^r \tag{8.15}$$

This can be converted into the more convenient factorial form by remembering that the product

$$\left(\frac{N_0}{r}+1-1\right)^r \left(\frac{N_0}{r}+1-2\right)^r \left(\frac{N_0}{r}+1-3\right)^r \cdots \left(\frac{N_0}{r}+1-N_2\right)^r \qquad (8.16)$$

is equivalent to

$$\left\{\frac{(N_0/r)!}{(N_0/r-N_2)!}\right\}^r = \left\{\frac{(N_0/r)!}{(N_1/r)!}\right\}^r \qquad (8.17)$$

and so Equation 8.14 can be written as

$$\Omega_p = \left(\frac{1}{N_2!}\right) r^{N_2 r} \left\{\frac{(N_0/r)!}{(N_1/r)!}\right\}^r \left[\frac{z-1}{N_0}\right]^{N_2(r-1)} \qquad (8.18)$$

The remaining empty cells on the lattice can now be filled by solvent molecules, but as there is only one distinguishable way in which this can be done, $\Omega_s = 1$, there is no further contribution to Ω_p and the entropy of the system. The latter can now be calculated from the Boltzmann equation. The factorials can again be approximated using Stirling's relation and, although this requires considerable manipulation, which will be omitted here, it can eventually be shown that

$$S^M/k = \ln\Omega_p = -N_1 \ln\left(\frac{N_1}{N_0}\right) - N_2 \ln\left(\frac{N_2}{N_0}\right) + N_2\left\{(r-1)\ln(z-1)-(r-1)\right\} \qquad (8.19)$$

To convert this into a form that will allow us to express this in the correct site fraction form, we can add and subtract $N_2 \ln R$ on the R.H.S. of Equation 8.19 to give

$$S^M/k = -N_1 \ln\left[\frac{N_1}{N_1+rN_2}\right] - N_2 \ln\left[\frac{rN_2}{N_1+rN_2}\right] + N_2\left\{(r-1)\ln\frac{(z-1)}{e}+\ln r\right\} \qquad (8.20)$$

For the pure solvent, $N_2 = 0$ and the entropy $S_1 = 0$. Similarly, the entropy of the pure polymer S_2 can be obtained for $N_1 = 0$, which gives

$$S_2 = N_2\left\{(r-1)\ln\frac{(z-1)}{e}+\ln r\right\} \qquad (8.21)$$

Equation 8.21 then represents the entropy associated with the disordered or amorphous polymer on the lattice in the absence of solvent.

It follows that the entropy change on mixing disordered polymer and solvent is

$$\Delta S^M = S^M - S_1 - S_2 = -k\left\{N_1 \ln \phi_1 + N_2 \ln \phi_2\right\} \qquad (8.22)$$

where ϕ_i, the volume fraction, can replace the site fraction if it is considered that the number of sites occupied by the polymer and solvent is proportional to their respective volumes.

Equation 8.22 is the expression for the combinatorial entropy of mixing of an athermal polymer solution, and comparison with Equation 8.7 shows that they are similar in form except for the fact that now the volume fraction is found to be the most convenient way of expressing the entropy change rather than the mole fraction used for small molecules. This change arises from the differences in size between the components, which would normally mean mole fractions close to unity for the solvent, especially when dilute solutions are being studied.

We can gain a further understanding of how the size of the polymer chain affects the magnitude of ΔS^M and why it differs from ΔS_{id}^M (Equation 8.7) by recasting Equation 8.22 in terms of the volume fractions $\phi_i = (n_i V_i / V)$, where n_i and V_i represent the number of moles and the volume of component i, respectively, and V is the total volume.

As V_i can conveniently be expressed as a function of a reference volume V_0 such that $V_i = r_i V_0$ and assuming that, without introducing significant error, r can be equated with the degree of polymerization for the polymer, then

$$\Delta S^M = -\frac{RV}{V_0}\left[\frac{\phi_1}{r_1}\ln \phi_1 + \frac{\phi_2}{r_2}\ln \phi_2\right] \qquad (8.23)$$

If the volume fraction form is retained, then for a simple liquid mixture, $r_1 = r_2 = 1$, but for a polymer solution, $r_2 \gg 1$ and the last term in Equation 8.23 will be smaller than the equivalent term calculated for small molecules. Consequently, ΔS^M per mole of lattice sites (or equivalent volume) will be very much less than ΔS_{id}^M, and the contribution of the combinatorial entropy to the mixing process in a polymer solution is not as large as that for solutions of small molecules when calculated in terms of volume fractions and expressed as per mole of sites.

8.5 ENTHALPY CHANGE ON MIXING

The derivation of ΔS^M from the lattice theory has been made on the assumption that no heat or energy change occurs on mixing. This is an uncommon situation as experimental experience suggests that the energy change is finite. We can make use of regular solution theory to obtain an expression for ΔH^M where this change in energy is assumed to arise from the formation of new solvent–polymer (1–2) contacts on mixing, which replace some of the (1–1) and (2–2) contacts present in the pure solvent, and the pure polymer components, respectively. This can be represented by a quasi-chemical process

$$\tfrac{1}{2}(1-1) + \tfrac{1}{2}(2-2) \to (1-2) \qquad (8.24)$$

where the formation of a solvent–polymer contact requires first the breaking of (1–1) and (2–2) contacts, and can be expressed as an interchange energy $\Delta\varepsilon_{12}$ per contact, given by

$$\Delta U^M = \Delta\varepsilon_{12} = \varepsilon_{12} - \tfrac{1}{2}(\varepsilon_{11} + \varepsilon_{22}) \tag{8.25}$$

Here, ε_{ii} and ε_{ij} are the contact energies for each species. The energy of mixing ΔU^M can be replaced by ΔH^M if no volume change takes place on mixing, and for q new contacts formed in solution

$$\Delta H^M = q\Delta\varepsilon_{12} \tag{8.26}$$

The number of contacts can be estimated from the lattice model by assuming that the probability of having a lattice cell occupied by a solvent molecule is simply the volume fraction ϕ_1. This means that each polymer molecule will be surrounded by (ϕ_1 rz) solvent molecules, and for N_2 polymer molecules

$$\Delta H^M = N_2\phi_1\ rz\ \Delta\varepsilon_{12} = N_1\phi_2\ z\ \Delta\varepsilon_{12} \tag{8.27}$$

considering the definition of ϕ_2, i.e., $rN_2\phi_1 = N_1\phi_2$.

This equation is the van Laar expression derived for regular solutions and shows that this approach can be applied to polymer systems. To eliminate z, a dimensionless parameter (χ_1) per solvent molecule is defined as

$$kT\chi_1 = z\ \Delta\varepsilon_{12} \tag{8.28}$$

which is the difference in energy between a solvent molecule when it is immersed in pure polymer and when in pure solvent. It can also be expressed in the alternative form $RT\chi_1 = BV_1$, where B is now an interaction density.

The final expression is

$$\Delta H^M = kT\chi_1 N_1\phi_2 \tag{8.29}$$

and the interaction parameter χ_1 is an important feature of polymer solution theory, which will be met with frequently. The Flory–Huggins interaction parameter is zero for athermal solutions, positive for endothermic, and negative for exothermic mixing.

8.6 FREE ENERGY OF MIXING

Having calculated the entropy and enthalpy contributions to mixing, these can now be combined to give the expression for the free energy of mixing, $\Delta G^M = \Delta H^M - T\Delta S^M$ as

$$\Delta G^M = kT\ \underbrace{[N_1 \ln\phi_1 + N_2 \ln\phi_2}_{\text{Combinatorial term}} + \underbrace{N_1\phi_2\chi_1]}_{\substack{\text{Contact}\\\text{dissimilarity}}} \tag{8.30}$$

It is more useful to express Equation 8.30 in terms of the chemical potentials of the pure solvent (μ_1^0), by differentiating the expression with respect to the number of solvent molecules, N_1, to obtain the partial molar Gibbs free energy of dilution (after multiplying by Avogadro's number),

$$\frac{\partial \Delta G^M}{\partial N_1} = (\mu_1 - \mu_1^0) = RT\left[\ln(1-\phi_2) + \left(1-\frac{1}{r}\right)\phi_2 + \chi_1\phi_2^2\right] \tag{8.31}$$

This could also be carried out for the polymer (N_2), but as it makes no difference which one is taken (both having started from ΔG^M), Equation 8.31 is more convenient to use. Although this expression is not strictly valid for the dilute solution regime, it can be converted into a structure that is extremely informative about deviations from ideal solution behavior encountered when measuring the molar mass by techniques such as osmotic pressure. If the logarithmic term is expanded using a Taylor series,

$$\ln(1-\phi_2) = -\phi_2 - \phi_2^2/2 - \phi_2^3/3\,.....$$

but truncated after the squared term, assuming ϕ_2 is small, then

$$(\mu_1 - \mu_1^0) = -RT[(\phi_2/r) + (\tfrac{1}{2} - \chi_1)\phi_2^2] \tag{8.32}$$

This can be modified by remembering that $r = (V_2/V_1)$ and $\phi_2 = c_2\bar{v}_2$, where $\bar{v}_2$ is the partial specific volume of the polymer. This can be related to the polymer molecular weight M_2 through $\bar{v}_2 = (V_2/M_2)$, so that $(\phi_2/r) = c_2V_1/M_2$ and, finally,

$$(\mu_1 - \mu_1^0) = -RT\left[\frac{c_2V_1}{M_2} + \bar{v}_2^2(\tfrac{1}{2} - \chi_1)c_2^2\right] \tag{8.33}$$

8.7 LIMITATIONS OF THE FLORY–HUGGINS THEORY

The simple lattice theory does not describe the behavior of dilute polymer solutions particularly well because of the following invalid simplifications in the theoretical treatment: (1) it was assumed that the segment-locating process is purely statistical, but this would only be true if $\Delta\varepsilon_{12}$ was zero; (2) the treatment assumed that the flexibility of the chain is unaltered on passing into the solution from the solid state — this limits the calculation of ΔS^M to the combinatorial contribution only and neglects any contribution from continual flexing of the chain in solution; (3) any possible specific solvent–polymer interactions that might lead to orientation of the solvent molecules in the vicinity of the polymer chain are neglected; i.e., polar solutions may be inadequately catered to by this theory; (4) a uniform density of lattice site occupation is assumed, but this will only apply to relatively concentrated solutions; and (5) the parameter χ_1 is often concentration dependent, but this is ignored. It is now accepted that a noncombinatorial entropy contribution arises from the formation of new (1–2) contacts in the mixture, which change the vibrational frequencies of the two components; i.e., assumption (4) in Section 8.2 must be relaxed. This can be allowed for by recognizing that χ_1 is actually a free-energy parameter comprising entropic χ_H and enthalpic χ_S contributions such that $\chi_1 = \chi_H + \chi_S$. These are defined by

$$\chi_H = - T(d\chi_1/dT) \quad \text{and} \quad \chi_S = d(T\chi_1)/dT \quad (= - \Delta S/k)$$

Experiments tend to show that the major contribution comes from the χ_S component, indicating that there is a large decrease in entropy (noncombinatorial), which is acting against the dissolution process of a polymer in a solvent.

In spite of much justifiable criticism, the Flory–Huggins theory can still generate considerable interest because of the limited success that can be claimed for it in relation to phase equilibria studies.

8.8 PHASE EQUILIBRIA

The Flory–Huggins theory can be used to predict the equilibrium behavior of two liquid phases when both contain amorphous polymer and one or even two solvents.

Consider a two-component system consisting of a liquid (1) that is a poor solvent for a polymer (2). Complete miscibility occurs when the Gibbs free energy of mixing is less than the Gibbs free energies of the components, and the solution maintains its homogeneity only as long as ΔG^M remains less than the Gibbs free energy of any two possible coexisting phases.

The situation is represented by curve T_4 in Figure 8.2. The miscibility of this type of system is observed to be strongly temperature dependent, and as T decreases the solution separates into two phases. Thus, at any temperature, say, T_1, the Gibbs free energy of any mixture, composition x_2''' in the composition range x_2' to x_2'', is higher than either of the two coexisting phases whose compositions are x_2' and x_2'', and phase separation takes place. The compositions of the two phases x_2' and x_2'' do not correspond to the two minima but are measured from the points of contact of the double tangent AB with the Gibbs free-energy curve. The same is true of other temperatures lying below T_c (T_1 to T_3), and the inflexion points can be joined to bound an area representing the heterogeneous two-phase system, where there is limited solubility of component 2 in 1 and vice versa. This is called a *cloud-point curve*.

As the temperature is increased, the limits of this two-phase coexistence contract, until eventually they coalesce to produce a homogeneous, one-phase mixture at T_c, the *critical solution temperature*. This is sometimes referred to as the *critical con-solute point*.

In general, we can say that if the free-energy–composition curve has a shape that allows a tangent to touch it at two points, phase separation will occur.

The critical solution temperature is an important quantity and can be accurately defined in terms of the chemical potential. It represents the point at which the inflexion points on the curve merge, and so it is the temperature where the first, second, and third derivatives of the Gibbs free energy with respect to mole fraction are zero.

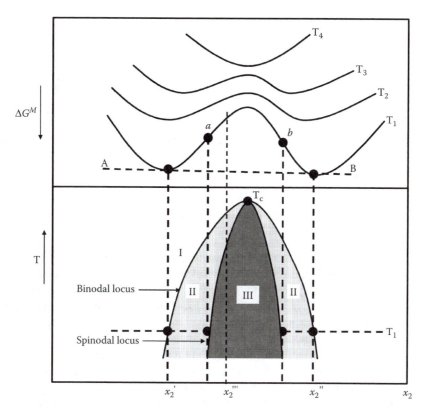

FIGURE 8.2 Schematic diagram of the Gibbs free energy of mixing ΔG^M as a function of the mole fraction x_2 of solute (top half), showing the transition from a system miscible in all proportions at a temperature T_4 through the critical temperature T_c, to partially miscible systems at temperatures T_3 to T_1. The contact points for the common tangents drawn to the minima are shown projected onto the temperature—mole fraction plane to form the binodal (cloud-point) curve, whereas projection of the inflexion points forms the spinodal curve. The lower part of the diagram indicates the one-phase stable region I, the metastable region II, and the unstable region III.

$$\partial(\Delta G^M)/\partial x_2 = \partial^2(\Delta G)/\partial x_2^2 = \partial^3(\Delta G^M)/\partial x_2^3 = 0 \qquad (8.34)$$

It is also true that the partial molar Gibbs free energies of each component are equal at this point, and it emerges that the conditions for incipient phase separation are

$$\partial\mu_1/\partial\phi_2 = \partial^2\mu_1/\partial\phi_2^2 = \partial^3\mu_1/\partial\phi_2^3 = 0 \qquad (8.35)$$

By remembering that $\Delta G_1 = (\mu_1 - \mu_1^0)$, application of these criteria for equilibrium to Equation 8.31 leads to the first derivative of that equation

$$(1 - \phi_{2,c})^{-1} - (1 - 1/x_n) - 2\phi_{2,c}\,\chi_{1,c} = 0 \qquad (8.36)$$

whereas the second derivative is

$$(1 - \phi_{2,c})^{-2} - 2\chi_{1,c} = 0 \tag{8.37}$$

where the subscript c denotes critical conditions. The critical composition at which phase separation is first detected is then

$$\phi_{2,c} = 1/(1 + x_n^{1/2}) \approx 1/x_n^{1/2} \tag{8.38}$$

and

$$\chi_{1,c} = \tfrac{1}{2} + 1/x_n^{1/2} + 1/2x_n \tag{8.39}$$

which indicates that $\chi_{1,c} = 0.5$ at an infinitely large chain length.

The interaction parameter χ_1 is a useful measure of the solvent power. Poor solvents have values of χ_1 close to 0.5, whereas an improvement in solvent power lowers χ_1. Generally, a variation from 0.5 to -1.0 can be observed, although for many synthetic polymer solutions the range is 0.6 to 0.3. A linear temperature dependence of the general form $\chi_1 = a + b/T$ is also predicted for χ_1, which suggests that as the temperature increases, the solvating power of the liquid should increase.

The relationship between chain length and solvent power as expressed by Equation 8.39 is illustrated in Figure 8.3. The implication is that if χ_1 can be carefully controlled, conditions could be attained that, for a polydisperse sample, would allow a given molecular species to precipitate, while leaving larger or smaller molecules in solution. This process is known as *fractionation*.

Experimentally, a polymer sample can be fractionated in a variety of ways; two in common use are (1) addition of a nonsolvent to a polymer solution and (2) lowering the temperature of the solution.

In the first method, the control of χ_1 is effected by adding a nonsolvent to the polymer solution. If the addition is slow, χ_1 increases gradually until the critical value for large molecules is reached. This first causes precipitation of the longest chains, which can be separated from the shorter chains that remain in solution. Successive additions of small quantities of nonsolvent to the solution allow a series of fractions of steadily decreasing molar mass to be separated.

In the second method, χ_1 is varied by altering the temperature, with similar results. For both techniques, it is useful to dissolve the polymer initially in a poor solvent with a large χ_1 value. This ensures that only small quantities of nonsolvent are required to precipitate the polymer in method 1 and that the temperature changes required in method 2 are small.

8.9 FLORY–KRIGBAUM THEORY

To overcome the limitations of the lattice theory resulting from the discontinuous nature of a dilute polymer solution, Flory and Krigbaum discarded the idea of a

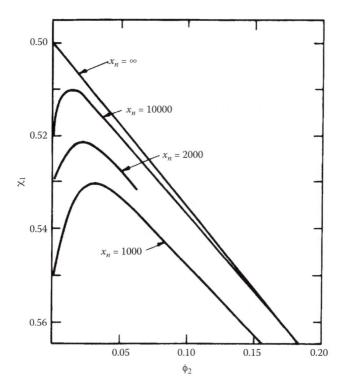

FIGURE 8.3 Variation of χ_1 with volume fraction ϕ_2 of the polymer in solution, showing the effect of changing chain length x_n.

uniform distribution of chain segments in the liquid. Instead, they considered the solution to be composed of areas containing polymer separated by the solvent. In these areas, the polymer segments were assumed to possess a Gaussian distribution about the center of mass, but even with this distribution the chain segments still occupy a finite volume from which all other chain segments are excluded. It is within this excluded volume that the long-range interactions originate, as discussed more fully in Chapter 10.

Flory and Krigbaum defined an enthalpy (κ_1) parameter and an entropy of dilution (ψ_1) parameter such that the thermodynamic functions used to describe these long-range effects are given in terms of the excess partial molar quantities

$$\Delta H_1^E = RT\kappa_1\phi_2^2 \tag{8.40}$$

$$\Delta S_1^E = R\psi_1\phi_2^2 \tag{8.41}$$

From Equation 8.33 it can be seen that the excess free energy of dilution is

$$(\mu_1 - \mu_1^0)^E = \Delta G_1^E = -RT(\tfrac{1}{2} - \chi_1)\phi_2^2 \tag{8.42}$$

Combination of these nonideal terms then yields

$$(\tfrac{1}{2} - \chi_1) = (\psi_1 - \kappa_1) \tag{8.43}$$

We will show in Chapter 9 that a relationship exists between the second virial coefficient B and the interaction parameter χ_1

$$B = RT \frac{\overline{v}_2^{\,-2}}{V_1} (\tfrac{1}{2} - \chi_1) \tag{8.44}$$

When $B = 0$ and $\chi_1 = \tfrac{1}{2}$, the solution appears to behave as though it were ideal. The point at which this occurs is known as the *Flory* or *theta* point and is in some ways analogous to the Boyle point for a nonideal gas. Under these conditions,

$$\psi_1 = \kappa_1, \text{ that is, } \Delta H_1^E = T \Delta S_1^E$$

The temperature at which these conditions are obtained is the Flory or theta temperature Θ, conveniently defined as $\Theta = T\kappa_1/\psi_1$. This tells us that Θ will only have a meaningful value when ψ_1 and κ_1 have the same sign.

Substitution in Equation 8.42 followed by rearrangement gives

$$(\mu_1 - \mu_1^0)^E = -RT\psi_1 \left(1 - \frac{\Theta}{T}\right)\phi_2^2 \tag{8.45}$$

and shows that deviations from ideal behavior vanish when $T = \Theta$.

The theta temperature is a well-defined state of the polymer solution at which the excluded volume effects are eliminated and the polymer coil is in an unperturbed condition (see Chapter 10). Above the theta temperature, expansion of the coil takes place, caused by interactions with the solvent, whereas below Θ the polymer segments attract one another, the excluded volume is negative, the coils tend to collapse, and eventual phase separation occurs.

8.10 LOCATION OF THE THETA TEMPERATURE

The theta temperature of a polymer–solvent system can be measured from phase separation studies. The value of $\chi_{1,c}$ at the critical concentration is related to the chain length of the polymer by Equation 8.39, and substitution in Equation 8.45 leads to

$$\psi_1(\Theta/T_c - 1) = 1/x_n^{1/2} + 1/2x_n \tag{8.46}$$

where now we have replaced r with the equivalent degree of polymerization x_n. Rearrangement gives

$$1/T_c = (1/\Theta)\{1 + (1/\psi_1)(1/x_n^{1/2} + 1/2x_n)\} \tag{8.47}$$

Remembering that $x_n = (M\bar{v}_2/V_1)$, where M and $\bar{v}_2$ are the molar mass and partial specific volume of the polymer, respectively, and V_1 is the molar volume of the solvent, the equation states that the critical temperature is a function of M and the value of T_c at infinite M is the theta temperature for the system.

Precipitation data for several systems have proved the validity of Equation 8.47. Linear plots are obtained with a positive slope from which the entropy parameter ψ_1 can be calculated, as shown in Figure 8.4. Typical values are shown in Table 8.1, but ψ_1 values measured for systems such as polystyrene–cyclohexane have been found to be almost ten times larger than those derived from other methods of measurement. This appears to arise from the assumption in the Flory–Huggins theory that χ_1 is concentration independent and improved values of ψ_1 are obtained when this is rectified.

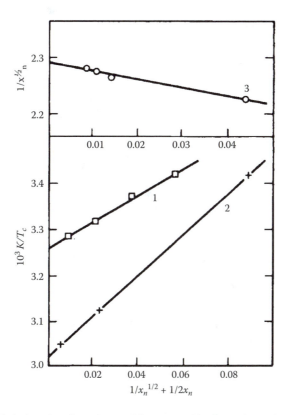

FIGURE 8.4 Chain length x_n dependence of the upper critical consolute solution temperature T_c for (1) polystyrene in cyclohexane and (2) polyisobutylene in di-isobutyl ketone (data from Schultz, A.R. and Flory, P.J., *J. Am. Chem. Soc.*, 74, 4760, 1952), and the lower critical solution temperature for (3) polyoctene-1 in *n*-pentane (data from Kinsinger, J.B. and Ballard, L.E., *Polym. Lett.*, 2, 879, 1964).

TABLE 8.1

Theta Temperatures and Entropy Parameters for Some Polymer–Solvent Systems

Polymer	Solvent	Θ (K)	ψ_1
1. Polystyrene	Cyclohexane	307.2	1.056
2. Polyethylene	Nitrobenzene	503	1.090
3. Polyisobutene	Diisobutyl ketone	333.1	0.653
4. Poly(methylmethacrylate)	4-Heptanone	305	0.610
5. Poly(acrylic acid)	Dioxan	302.2	−0.310
6. Polymethacrylonitrile	Butanone	279	−0.630

Note: Values have been derived using Equation 8.47; Θ=theta temperature, ψ = entropy of dilution.

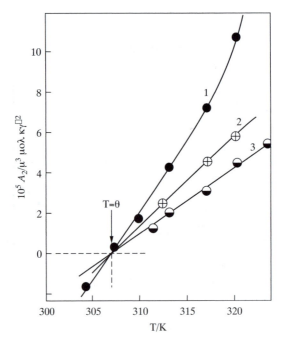

FIGURE 8.5 Location of the theta temperature Θ for poly(α-methyl styrene) in cyclohexane. Values of A_2 are measured for (1) $M_n = 8.6 \times 10^4$ g mol^{-1}, (2) $M_n = 3.8 \times 10^5$ g mol^{-1}, and (3) $M_n = 1.5 \times 10^6$ g mol^{-1}.

The theta temperature, calculated from Equation 8.47 for each system, is in good agreement with that measured from the temperature variation of the second virial coefficient $A_2 (= B/RT$, see Chapter 9). Curves of A_2, measured at various temperatures in the vicinity of Θ, are constructed as a function of temperature for one or more molar masses as shown in Figure 8.5. Intersection of the curves with the T-axis occurs when $A_2 = 0$ and $T = \Theta$. The curves for each molar mass of the same polymer should all intersect at $T = \Theta$.

8.11 LOWER CRITICAL SOLUTION TEMPERATURES

So far we have been concerned with nonpolar solutions of amorphous polymers, whose solubility is increased with rising temperature, because the additional thermal motion helps to decrease attractive forces between like molecules and encourages energetically less favorable contacts. The phase diagram for such a system, when the solvent is poor, is depicted by area A in Figure 8.6, where the critical temperature T_c occurs near the maximum of the cloud-point curve and is often referred to as the *upper critical solution temperature* (UCST). This behavior follows from that depicted in Figure 8.2.

For nonpolar systems ΔS^M is normally positive but weighted heavily by T, and so solubility depends mainly on the magnitude of ΔH^M, which is normally endothermic (positive). Consequently, as T decreases, ΔG^M eventually becomes positive and phase separation takes place.

Values of Θ and ψ_1, in Table 8.1, show that for systems 1 to 4 the entropy parameter is positive, as expected, but for poly(acrylic acid) in dioxan and polymethacrylonitrile in butanone, ψ_1 is negative at the theta temperature. As $\psi_1 = \kappa_1$ when $T = \Theta$, the enthalpy is also negative for these systems. This means that systems 5 and 6 exhibit an unusual decrease in solubility as temperature rises, and the cloud-point curve is now inverted as in area B. The corresponding critical temperature is located at the minimum of the miscibility curve and is known as the *lower critical solution temperature* (LCST).

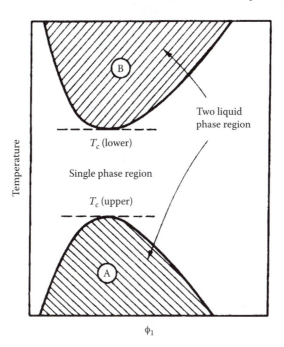

FIGURE 8.6 Schematic diagram of the two types of phase boundaries commonly encountered in polymer solutions: (A) the two-phase region characterized by the upper critical solution temperature and (B) the two-phase region giving the lower critical solution temperature, with a single-phase region lying between the two.

In systems 5 and 6, this phenomenon is a result of hydrogen-bond formation between the polymer and solvent, which enhances the solubility. As hydrogen bonds are thermally labile, a rise in T reduces the number of bonds and causes eventual phase separation. In solutions, which are stabilized in this way by secondary bonding, the LCST usually appears below the boiling temperature of the solvent, but it has been found experimentally that an LCST can be detected in nonpolar systems when these are examined at temperatures approaching the critical temperature of the solvent. Polyisobutylene in a series of n-alkanes, polystyrene in methyl acetate and cyclohexane, and cellulose acetate in acetone all exhibit LCSTs.

The separation of polymer–solvent systems into two phases as the temperature increases is now recognized to be a characteristic feature of all polymer solutions. This presents a problem of interpretation within the framework of regular solution theory, as the accepted form of χ_1 predicts a monotonic change with temperature and is incapable of dealing with two critical consolute points.

The problem of how to accommodate, in a theoretical framework, the existence of two miscibility gaps requires a new approach, and a more elaborate treatment by Prigogine and co-workers encompasses the difference in size between the components of a mixture, which cannot be ignored for polymer solutions. They replaced the rigid lattice model used by Flory and Huggins, which is valid only at absolute zero, with a flexible lattice whose cells change in volume with temperature and pressure. This allowed them to include in their theory dissimilarities in free volume between polymer and solvent together with the corresponding interactions. The same approach was extended by both Patterson and Flory to deal specifically with polymer systems.

The most important of the new parameters is the so-called structural effect, which is related to the number of degrees of freedom "$3c$" that a molecule possesses, divided by the number of external contacts q. This structural factor (c/q) is a measure of the number of external degrees of freedom per segment and changes with the length of the component. Thus, the ratio decreases as a liquid becomes increasingly polymeric.

The expansion and free volume can then be characterized by the ratio of the thermal energy arising from the external degrees of freedom available to the component, $U_{thermal}$, and the interaction energy between neighboring nonbonded segments, $U_{cohesive}$, which will oppose the thermal energy effects, i.e.,

$$\frac{U_{thermal}}{U_{cohesive}} = \left(\frac{ckT}{q} \right) \cdot \frac{1}{\varepsilon^*} \tag{8.48}$$

where ε^* is the characteristic cohesive energy per contact.

For convenience q may be replaced by r, the number of chain segments, although q will actually be less than r because some of the external contacts are used in forming the covalent bonds in the chain.

Free-volume dissimilarities become increasingly important as the size of one component increases with respect to the second, as in polymer solutions, and when these differences are sufficiently large, phase separation can be observed at the LCST.

The differences in expansivity can be accounted for if the interaction parameter is now expressed as

$$\chi = -(U_1/RT)v^2 + (C_{p1}/2R)\tau^2 \tag{8.49}$$

where the first term reflects the interchange energy on forming contacts of unlike type and includes segment size differences, whereas the second term is the new "structural" contribution coming from free-volume changes on mixing a dense polymer with an expanded solvent. This can be represented schematically as in Figure 8.7.

The first term in Equation 8.49, shown by curve 1 in Figure 8.7, is merely an expression of the Flory–Huggins theory where χ decreases constantly with rising temperature, but now inclusion of the new free-volume term, shown by curve 2, modifies the behavior of χ. The second term gains in importance as the expansivities of the two components become increasingly divergent with temperature, and the net effect is to increase χ again until it once more attains its critical value at high temperature. The LCST that results is then a consequence of these free-volume differences and is an entropically controlled phenomenon.

This can be illustrated in the following ways. In terms of the flexible lattice model, one can imagine the polymer and liquid lattices expanding at different rates until a temperature is reached at which the highly expanded liquid lattice can no longer be distorted sufficiently to accommodate the less-expanded polymer lattice and form a solution; i.e., the loss in entropy during the distortion becomes so large and unfavorable that phase separation (LCST) takes place. Alternatively, a polymer solution can be thought of as a system formed by the condensation of solvent into a polymer. As the temperature increases, the entropy loss incurred during condensation becomes greater until eventually it is so unfavorable that condensation in the polymer is impossible and phase separation takes place. Neither picture is particularly rigorous, but they serve to emphasize the fact that the LCST is an entropically controlled phenomenon.

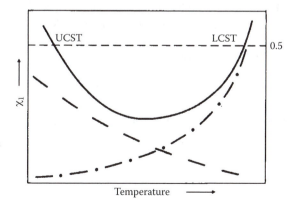

FIGURE 8.7 Schematic diagram of χ_1 as a function of temperature, showing the composite curve of 1 (– –), first term in Equation 8.49; and 2 (– • –), the free-volume contribution from the second term in Equation 8.49, which results in the observation of the LCST.

8.12 SOLUBILITY AND THE COHESIVE ENERGY DENSITY

Solvent–polymer compatibility problems are repeatedly encountered in industry. For example, in situations requiring the selection of elastomers for use as hosepipes or gaskets, the correct choice of elastomer is of prime importance, as contact with highly compatible fluids may cause serious swelling and impair the operation of the system. Wrong selection can have far-reaching consequences; the initial choice of an elastomer for the seals in the landing gear of DC-8 aircraft resulted in serious jamming because the seals became swollen when in contact with the hydraulic fluid. This almost led to grounding of the plane, but replacement with an incompatible elastomer made from ethylene–propylene copolymer rectified the fault.

To avoid such problems, a technologist may wish to have at his or her disposal a rough guide to aid the selection of solvents for a polymer or to assess the extent of polymer–liquid interaction other than those already described. Here, use can be made of a semiempirical approach suggested by Hildebrand and based on the premise that "like dissolves like." The treatment involves relating the enthalpy of mixing to the cohesive energy density (E/V) and defines a solubility parameter $\delta = (E/V)^{1/2}$, where E is the molar energy of vaporization and V is the molar volume of the component. The proposed relation for the heat of mixing of two nonpolar components

$$\Delta H^{M} = V^{M}(\delta_1 - \delta_2)^2 \phi_1 \phi_2 \tag{8.50}$$

shows that ΔH^M is small for mixtures with similar solubility parameters, and this indicates compatibility.

Values of the solubility parameter for simple liquids can be readily calculated from the enthalpy of vaporization. The same method cannot be used for a polymer, and one must resort to comparative techniques. Usually, δ for a polymer is established by finding the solvent that will produce maximum swelling of a network or the largest value of the limiting viscosity number, as both indicate maximum compatibility. The polymer is then assigned a similar value of δ (see Table 8.2). Alternatively, Small and Hoy have tabulated a series of group molar attraction constants from which a good estimate of δ for most polymers can be made.

TABLE 8.2
Average Solubility Parameters for Some Common Polymers

Polymer	$\delta/(J\ cm^{-3})^{1/2}$	Polymer	$\delta/(J\ cm^{-3})^{1/2}$
Polytetrafluoroethylene	12.7	Polystyrene	18.7
Polyisobutylene	16.3	Poly(methylmethacrylate)	19.0
Polyethylene	16.4	Poly(vinyl acetate)	19.2
Polyisoprene	16.7	Poly(vinyl chloride)	20.7
Polybutadiene	17.1	Nylon 6.6	27.8
Polypropylene	17.8	Poly(acrylonitrile)	28.8

TABLE 8.3
Group Contributions to F

Group	Small	Hoy
$-CH_3$	438	303.4
$-CH_2$	272	269.0
$-\overset{\overset{H}{\mid}}{\underset{\mid}{C}}-$	57	176.0
$-\overset{\mid}{\underset{\mid}{C}}-$	−190	65.5
$-CH(CH_3)$	495	(479.4)
$-C(CH_3)_2$	686	(672.3)
$-\overset{\overset{H}{\mid}}{C}=\overset{\overset{H}{\mid}}{C}-$	454	497.4
$-\overset{\mid}{C}=\overset{\overset{H}{\mid}}{C}-$	266	421.5
$-C(CH_3)=CH-$	(704)	(724.9)
Cyclopentyl	—	1295.1
Cyclohexyl	—	1473.3
Phenyl	1504	1392.4
p-Phenylene	1346	1442.2
$-F$	(250)	84.5
$-Cl$	552	419.6
$-Br$	696	527.7
$-I$	870	—
$-CN$	839	725.5
$-CHCN-$	(896)	(901.5)
$-OH$	—	462.0
$-O-$	143	235.3
$-CO-$	563	538.1
$-COOH$	—	(1000.1)
$-COO-$	634	668.2
$-O-\overset{\overset{O}{\parallel}}{C}-O-$	—	(903.5)
$-\overset{\overset{O}{\parallel}}{C}-O-\overset{\overset{O}{\parallel}}{C}-$	—	1160.7
$-\overset{\overset{O}{\parallel}}{C}-\overset{\overset{H}{\mid}}{N}-$	—	(906.4)
$-O-\overset{\overset{O}{\parallel}}{C}-\overset{\overset{H}{\mid}}{N}-$	—	(1036.5)
$-S$	460	428.4

The suggested group contributions are shown in Table 8.3, and the solubility parameter for a polymer can be estimated from the sum of the various molar attraction constants F for the groups that make up the repeat unit; i.e.,

$$\delta = (\Sigma F)/V = (\Sigma F)\rho/M_0.$$

Here, V is the molar volume of the repeat unit whose molar mass is M_0, and ρ is the polymer density.

Thus, for poly(methyl methacrylate) with $M_0 = 100.1$ and $\rho = 1.19$ g cm^{-3}, we have, using the Hoy values,

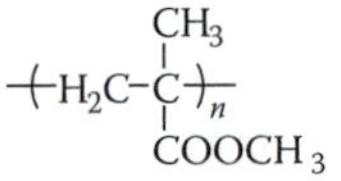

Group	F
2($-$CH$_3$)	2(303.4)
>CH$_2$	269
$-$COO$-$	668.2
$-\overset{\textstyle |}{\underset{\textstyle |}{C}}-$	65.5
	$\Sigma 1609.5$

$$\cfrac{}{}\left(\text{H}_2\text{C}-\overset{\textstyle \text{CH}_3}{\underset{\textstyle \text{COOCH}_3}{\text{C}}}\right)_n$$

Therefore

$$\delta = (1609.5)(1.19)/100.1$$
$$= 19.13 \ (\text{J cm}^{-3})^{1/2}$$

For a more complex polyhydroxyether of bisphenol A structure and with $\rho = 1.15$ g cm^{-3}:

Group	F
2(p-phenylene)	2(1442)
2($-$CH$_3$)	2(303)
2(> CH$_2$)	2(269)
2(Ether oxygen)	2(235.3)
$-$OH	462
$-\overset{\textstyle |}{\text{CH}}$	176
$-\overset{\textstyle |}{\underset{\textstyle |}{\text{C}}}-$	65.5
	$\Sigma 5202$

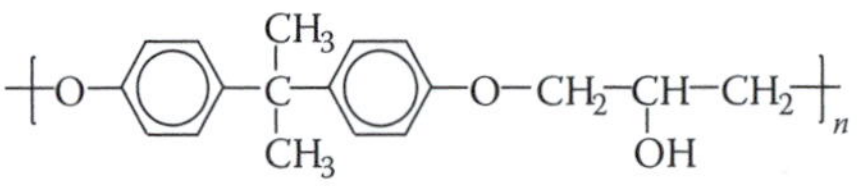

$$\left(\text{O}-\!\!\!\bigcirc\!\!\!-\overset{\textstyle \text{CH}_3}{\underset{\textstyle \text{CH}_3}{\text{C}}}-\!\!\!\bigcirc\!\!\!-\text{O}-\text{CH}_2-\overset{}{\underset{\textstyle \text{OH}}{\text{CH}}}-\text{CH}_2\right)_n$$

$$\delta = (5202)(1.16)/268 = 22.32 \ (\text{J cm}^{-3})^{1/2}$$

Both estimates are within 10% of experimentally determined values.

Attempts to correlate δ with χ_1 from the Flory–Huggins equation have met with limited success because of the unjustifiable assumptions made in the derivation. It is now believed, however, that χ_1 is not an enthalpy parameter but a free-energy parameter, and a relation of the form (cf. Section 8.7)

$$\chi_1 = 1/z + (V_1/RT)(\delta_1 - \delta_2)^2 \tag{8.51}$$

has improved the correlation. Here, the relation $1/z = \chi_s$ is supposed to compensate for the lack of a noncombinatorial entropy contribution in the Flory–Huggins treatment. Unfortunately, solubility is not a simple process, and secondary bonding may play an important role in determining component interactions.

More detailed approaches have made use of Hansen solubility parameters (HSP), based on the realization that the heat of vaporization consists of several components arising from van der Waals dispersion forces (D), dipole–dipole interaction (P), and hydrogen bonding (H). The total cohesive energy density is then the sum of the three contributions, and so the overall solubility parameter is

$$\delta = (\delta_D^2 + \delta_P^2 + \delta_H^2)^{1/2}$$

8.13 POLYMER–POLYMER MIXTURES

In the constant search for new materials with improved performance, the idea of mixing two or more different polymers to form new substances having a combination of all the attributes of the components is deceptively attractive; deceptively, because in practice it is rarely accomplished and only in a few cases have polymer blends or mixtures achieved industrial importance. The main reason is that most common polymers do not mix with one another to form homogeneous, one-phase solutions or blends, and an explanation for this is to be found in the thermodynamics of solutions, which have been outlined in the previous sections.

As we have seen, when two liquids or a liquid and a polymer are mixed, the formation of a homogeneous, one-phase solution is assisted mainly by the large favorable gain in combinatorial entropy. This entropic contribution is progressively reduced when one or both components increase in size, and the reason for this becomes obvious on inspection of Equation 8.23. When r_1 and r_2 both increase, then ΔS^M becomes smaller; consequently, attempts to mix two high-molar-mass polymer samples will receive little assistance from this function and must depend increasingly on a favorable (negative) heat of mixing embodied in the χ parameter. This loss of entropy can be conveniently illustrated using the simple lattice model shown in Figure 8.8. Here, a 10×10 lattice, containing 50 white and 50 black units randomly mixed (a), will result in approximately 10^{30} possible different arrangements of the units on mixing. If these white units are now connected to other white units, and black to black (b), to form five equal chains of each color with $r_1 = r_2 = 10$, then the number of possible arrangements of these chains decreases to about 10^3. Thus, as r_1 and r_2 approach infinity, ΔS^M will become negligible, and the free energy of mixing will become essentially dependent on ΔH^M, which now has to be either very small or negative.

The heat of mixing for the majority of polymer (1)–polymer (2) pairs tends to be endothermic and can be approximated by reference to the solubility parameters using Equation 8.50. This can be written as

$$\chi_{12} = \frac{V_0}{RT}(\delta_1 - \delta_2)^2 \tag{8.52}$$

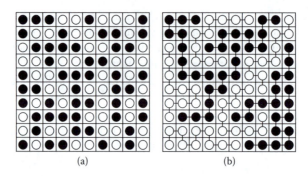

(a)   (b)

FIGURE 8.8 (a) A 10×10 lattice containing 50 black and 50 white spheres distributed randomly. (b) The same lattice with five white and five black polymer chains placed on the lattice. Each chain has $x_n = 10$. The number of possible arrangements falls from $\Omega = 10^{30}$ for (a) to $\Omega \approx 10^3$ for (b). Hence, $S = k \ln \Omega$ will also decrease dramatically.

where the reference volume normally assumes a value of $100 \text{ cm}^3 \text{ mol}^{-1}$. The critical value for χ_{12} can be estimated from

$$(\chi_{12})_c = \frac{1}{2}\left[\frac{1}{x_1^{1/2}} + \frac{1}{x_2^{1/2}}\right]^2 \tag{8.53}$$

where x_i is the degree of polymerization, related to the actual degree of polymerization x_n, and the reference volume by

$$x_i = x_n(V_R/V_0)$$

with V_R the molar volume of the repeat unit. The critical values for χ_{12}, above which the two polymers will phase-separate, calculated for various mixtures with $x_1 = x_2$, are shown in Table 8.4 along with the corresponding differences in δ. This shows that for mixing to take place between high-molecular-weight components, the solubility

---

TABLE 8.4
Critical Values for χ_{12} Derived for Various Chain Lengths, with Necessary Differences in Solubility Parameters

$x_1 = x_2$	$(\chi_{12})_c$	$(\delta_1 - \delta_2)_c$
50	0.0400	0.49
100	0.0200	0.35
200	0.0100	0.25
500	0.0040	0.15
700	0.0028	0.13
1000	0.0020	0.11
2000	0.0010	0.08
5000	0.0005	0.05

parameters would have to be virtually identical. This limits the number of possible combinations such that only a few examples exist in this category. These include polystyrene/poly(α-methyl styrene) below $M \approx 70,000$ and the polyacrylates mixed with the corresponding poly vinyl esters, e.g.,

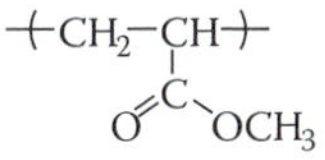

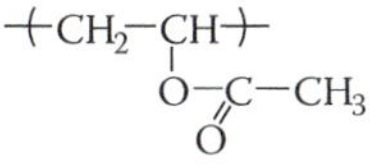

The situation changes if ΔH^M is negative as this will encourage mixing, and the search for binary polymer blends that are miscible has focused on combinations in which specific intermolecular interactions, such as hydrogen bonds, dipole–dipole interactions, ion–dipole interactions, or charge transfer complex formation, can exist between the component polymers. A substantial number of miscible blends have now been discovered using this principle, and it is possible to identify certain groups or repeat units that when incorporated in polymer chains tend to enter into these intermolecular interactions and enhance the miscibility.

A short selection of some of these complementary groups is given in Table 8.5, where a polymer containing groups or composed of units from column 1 will tend to form miscible blends with polymers containing groups or composed of units from column 2. Thus, it is believed that polystyrene forms miscible blends with poly(vinyl methylether) and poly(phenylene oxide)s (examples 1 and 2, respectively) because of interactions between the π-electrons of the phenyl rings and the lone pairs of the ether oxygens. Similarly, it has been suggested that a weak hydrogen bond, which is strong enough to induce miscibility, can form between the carbonyl unit of poly(methyl methacrylate) and the α-hydrogen of poly(vinyl chloride) (example 3 where $R = CH_3$); i.e.,

$$>C=O\ldots\ldots H-C-Cl$$

Much stronger hydrogen bonding interactions can be obtained if units such as

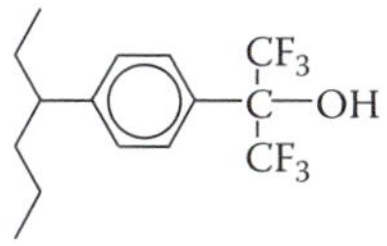

or sites for ion–dipole interactions such as

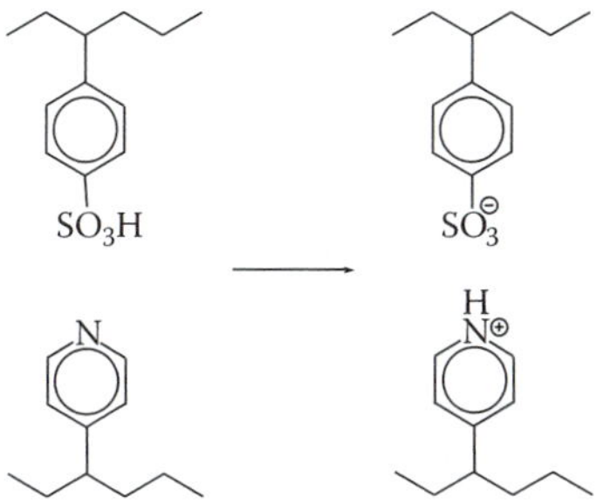

TABLE 8.5
Complementary Groups and Repeat Units Found in Miscible Binary Polymer Blends

Group 1	Group 2
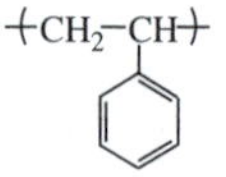 1. $+CH_2-CH+$ (phenyl)	$+CH_2-CH+$, $O-CH_3$
2. $+CH_2-CH+$ (phenyl) 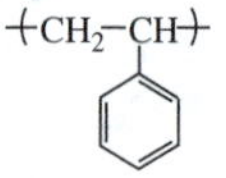 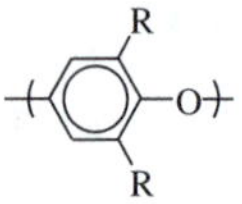	$+$ (aromatic ring with R, R) $-O+$
3. $+CH_2-CR+$, $O=C-OCH_3$	$+CH_2-CF_2+$
4. $+CH_2-CR+$, $O=C-OCH_3$	$+CH_2-CH+$, Cl
5. $+R_1-O-C-R_2-C-O+$, O O	$+CH_2-CH+$, Cl
6. $+CH_2-CH+$, $O-C-CH_3$, O	ONO_2
7. $+CH_2-CH+$, $O=C-O-CH_3$	ONO_2
8. (aromatic)-O-(aromatic)-$\overset{O}{\underset{O}{S}}$-	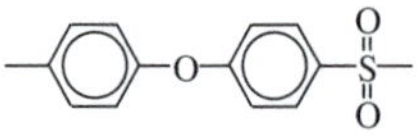 $+CH_2-CH_2-O+$

can be built into chains, and even in relatively small amounts these can transform immiscible pairs into totally miscible blends.

Many of these blends undergo quite rapid demixing as the temperature is raised, and an LCST phase boundary can be located above the glass transition temperature of the blend. There has been considerable debate through the years about the origins of the lower critical phase separation phenomenon in polymer blends. Various possible causes have been proposed: (1) free-volume dissimilarities, which may become unfavorable to mixing on increasing the temperature, (2) unfavorable entropy contributions arising from nonrandom mixing, or (3) temperature-dependent heat of mixing resulting from specific intermolecular interactions, which dissociate on heating, and favors miscibility at lower temperatures.

Although the latter seems the most likely cause in many blends where specific interactions have been identified, miscible blends can also be obtained when certain statistical copolymers are mixed with either a homopolymer, or another copolymer, in which no such interactions have been located. Thus, poly(styrene-*stat*-acrylonitrile) will form miscible blends with poly(methyl methacrylate) if the composition of the copolymer lies in the range of 10 to 39 wt% acrylonitrile. This range of compositions is called the *miscibility window* and has been reported to be present in other systems. The drive toward formation of a miscible solid solution in these cases is believed to arise when large repulsive interactions exist between the monomer units A and B constituting the copolymer; on mixing with a polymer (C), the number of these unfavorable A–B contacts are reduced by forming less repulsive A–C or B–C contacts, and a miscible blend results. Many of these blends also exhibit an LCST. Thus, the driving force toward lower critical phase separation in polymer–polymer solutions may depend on the system or may be a combination of the effects 1 to 3.

8.14 KINETICS OF PHASE SEPARATION

The phase diagram for a partially miscible polymer system exhibiting UCST behavior has been reported in Figure 8.2. Three different regions can be identified: a stable region (I), a metastable region (II) located between the bimodal and spinodal lines, and an unstable region (III). Lowering the temperature into either region II or III will cause phase separation, but it will take place according to different mechanisms.

Whereas in the unstable region phase separation takes place spontaneously by a process known as *spinodal decomposition* (SD), in the metastable region, for phase separation to take place there is a need to overcome an energy barrier. The latter phase separation mechanism, called *nucleation and growth* (NG), is initiated by local concentration fluctuations: the system phase separates if there is a fluctuation in concentration large enough to overcome the potential barrier to demixing. The result is a spheroidal domain structure whose domain size increases with time. On the contrary, spinodal decomposition occurs when the mixture becomes unstable as a result of temperature or pressure changes, and this phase separation process leads to a very characteristic interconnected structure (Figure 8.9). During the early stages of the SD phase separation, the amplitude of the wavelike composition fluctuations increases with time, whereas its wavelength remains constant.

The kinetics of the phase separation process have been studied experimentally and the results compared to the mean field approach developed by Cahn and Hilliard for metals, according to which the early stages of the phase separation process are given by

$$\frac{\partial \phi}{\partial t} = M \left[\left(\frac{\partial^2 G}{\partial \phi^2} \right) \nabla^2 \phi - 2K \nabla^4 \phi \right]^2 \qquad (8.54)$$

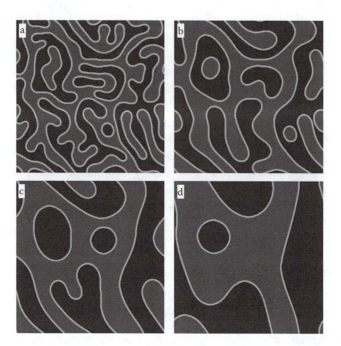

FIGURE 8.9 Spinodal decomposition process modeled using the Cahn–Hilliard equation. Time and length scales are dimensionless. Number of time steps following the quench are (a) 10,000, (b) 40,000, (c) 160,000, and (d) 640,000. Courtesy of Dr. Nigel Clarke (University of Durham, U.K.).

where ϕ is the composition, G represents the Gibbs free energy, M is the mobility, and K is the energy coefficient. This predicts the development of a characteristic domain size, which remains constant in the early stages but grows in amplitude. Figure 8.9 shows the development of the characteristic SD interconnected structure, modeled through Equation 8.54.

PROBLEMS

1. Using Equation 8.8 as your starting point, find an expression for the molar Gibbs free energy of mixing for an ideal solution, ΔG_m^M. Hence, calculate and plot $\Delta G_m^M/RT$ at regular intervals of mole fraction. Repeat these calculations for a *regular solution*, for which

$$\Delta H_m^M = RT\chi_{12}x_1x_2$$

with $\chi_{12} = 4$, 2.3, and 1.6 and discuss the effect of decreasing χ_{12} values on $\Delta G_m^M/RT$.

2. Construct a graph of the molar entropy of mixing $\Delta S_m^M/R$ vs. volume fraction for a polymer solution where the polymer degree of polymerization $r_2 = 10$ and 1000. You should do this by plotting $\Delta S_m^M/R$ values starting

from Equation 8.23 at regular volume fraction intervals. Comment on the effect that molecular weight has on the entropy of mixing.

3. Consider the molar Gibbs free energy of mixing for a polymer solution, ΔG_m^M. Evaluate ΔG_m^M for a polymer with degree of polymerization $r_2 = 1000$ at a polymer concentration $\phi_2 = 0.2$, when:
   a. $\Delta H_m^M = 0$
   b. $\chi_1 = 1$
   c. $\chi_1 = 0.1$

4. By differentiating the Flory–Huggins equation, demonstrate that the partial Gibbs free energy of mixing $\partial \Delta G^M / \partial n_1$ is given by Equation 8.31.

5. Calculate the cohesive energy density and the solubility parameter of toluene from the following data:
   a. molar enthalpy of vaporization $\Delta H_m^v = 37.99\,\text{kJ mol}^{-1}$
   b. density $= 0.862$ g cm^{-3}
   Hint: You will need to convert the enthalpy of vaporization to energy of vaporization, ΔU_m^v. Explain why the CED is defined in terms of the latter rather than the former quantity.

6. The CED of a polymer cannot be obtained directly from vaporization measurements. Among the experimental methods that can be used to determine δ, measurements of the degree of swelling of elastomers in liquids have proved very useful. Bristow and Watson (1958) carried out a series of swelling measurements on vulcanized natural and synthetic rubbers, in a wide range of solvents; χ values were obtained from the equilibrium degree of swelling for the specific rubber/solvent system, using rubber theory. Solubility parameters were derived from Equation 8.51:

$$\chi_1 = \beta + (V_1 / RT)(\delta_1 - \delta_2)^2,$$

By rearranging this equation, demonstrate that a plot of

$$\frac{\delta_1^2}{RT} - \frac{\chi}{V_1}$$

vs. the solubility parameter of the solvent, δ_1, yields the polymer solubility parameter δ_2 from the slope (assume that the parameter β is independent of the solubility parameters). Using this method, the authors reported $\delta_2 = 16.6$ (J cm^{-3})$^{1/2}$ for natural rubber. Compare this value with that obtained from group additivity (density $= 0.91$ g cm^{-3}).

7. The same authors as in Problem 6 measured the degree of swelling of three butadiene/acrylonitrile copolymers vulcanized and reported their solubility parameters as follows:

% Butadiene in Copolymer	$\delta/\text{J}^{1/2}$ cm$^{-3/2}$
82	17.8
70	20.2
61	21.0

Discuss the trend between δ values.

How would you expect the extent of swelling to vary for the different co-polymers in acetonitrile ($\delta = 24.13$ $J^{1/2}$ $cm^{-3/2}$) and in cyclohexane ($\delta = 16.6$ $J^{1/2}$ $cm^{-3/2}$)?

8. Shultz and Flory (1952), after measuring the critical temperature for precipitation of PIB fractions in di-isobutyl ketone, reported $\Theta = 331.1$ K and $\psi_1 = 0.653$ (Table 8.1). From these, calculate:

   a. The expected critical temperature for precipitation of PIB fractions with molecular weight equal to 22,700, 285,000, and 6,000,000.

   b. The critical concentration and interaction parameter for incipient phase separation of the same PIB fractions as in item a.

9. Calculate the interaction parameter between bisphenol A polycarbonate (BAPC) and hydroxyl styrene (OHS) units, $\chi_{BAPC\text{-}OHS}$, from the following data and the literature value of the solubility parameter of styrene (S), $\delta = 18.2$ $J^{1/2}$ $cm^{-3/2}$:

   a. $\chi_{S\text{-}OHS} = 0.25$

   b. $\chi_{BAPC\text{-}S} = 0.103$

10. A styrene butadiene copolymer whose solubility parameter is $\delta = 16.6$ $J^{1/2}$ $cm^{-3/2}$ is found to be insoluble in both pentane ($\delta = 14.5$ $J^{1/2}$ $cm^{-3/2}$) and ethyl acetate ($\delta = 18.6$ $J^{1/2}$ $cm^{-3/2}$) but soluble in a mixture of these two solvents. Explain.

11. Saeki et al. (1973) measured a set of upper and lower critical solution temperatures for the polystyrene/methylcyclohexane system at various molecular weights. Demonstrate that the data listed as follows for both UCST and LCST behavior follow the expected trend with molecular weight. (Note that r should be taken as the ratio of the molar volumes of the polymer and the solvent, and it can be assumed to be independent of the temperature.) What is the difference between the theta temperature for UCST and LCST?

$M_w \times 10^{-4}$ (g mol^{-1})	UCST (K)	LCST (K)
3.7	309.7	518.8
9.72	321.8	505.9
20.0	327.4	499.9
40.0	332.7	494.6
67.0	334.5	492.3
270.0	339.6	488.4

12. The interaction parameter of poly(methyl methacrylate) (PMMA) in toluene at 300 K is reported to be 0.450. Calculate the free energy resulting from mixing a PMMA sample with molecular weight 50,000 g mol^{-1} in toluene at 10% volume fraction (density of PMMA = 1.13 g cm^{-3}, density of toluene = 0.862 g cm^{-3}, and total volume of solution = 1 cm^3).

13. The interaction parameter of polyethyl adipate in ethyl acetate at 393 K is reported to be 0.55. At this temperature, what is the limiting value of molecular weight for which you expect the system to be miscible?

14. The temperature dependence of the interaction parameter for the polystyrene/ polyisoprene (PS/PI) blend is expressed by

$$\chi = A + B/T$$

with $A = 7.85 \times 10^{-3}$ and $B = 17.6$ K. What is the limiting molecular weight for a symmetric PS/PI blend to be miscible at 423 K?

15. Polymer interactions can be described in terms of temperature and composition-dependent energy density parameter, B_{12}. For a blend of two statistical copolymers, A_xB_{1-x} and C_yD_{1-y}, the blend interaction energy density is given by

$$B_{blend} = xyB_{AC} + (1-x)yB_{BC} + x(1-y)B_{AD}$$

$$+ (1-x)(1-y)B_{BD} - x(1-x)B_{AB} - y(1-y)B_{CD}$$

where x and y define the volume fraction compositions. Starting from this equation, derive an expression for B_{blend} for the simpler system of a homopolymer A mixed with a copolymer C_yD_{1-y}.

REFERENCES

Bristow, G.M. and Watson, W.F., *Trans. Faraday Soc.*, 54, 1731, 1958.
Kinsinger, J.B. and Ballard, L.E., *Polym. Lett.*, 2, 879, 1964.
Saeki, S., Kuwahara, N., Konno, S., and Kaneko, M., *Macromolecules*, 6, 246, 1973.
Schultz, A.R. and Flory, P.J., *J. Am. Chem. Soc.*, 74, 4760, 1952.

BIBLIOGRAPHY

Allen, G. and Bevington, J.C., Eds., *Comprehensive Polymer Science*, Vol. 2, Pergamon Press, 1989.
Billmeyer, F.W., *Textbook of Polymer Science*, John Wiley and Sons, 1985.
Coleman, M.M., Graf, J.F., and Painter, P., *Specific Interactions and the Miscibility of Polymer Blends*, Technomic Publishing, 1991.
Flory, P.J., *Principles of Polymer Chemistry*, Cornell University Press, 1953, chap. 12, 13.
Hansen, C., *Hansen Solubility Parameters — A User's Handbook*, CRC Press, 1999.
Hildebrand, J.H. and Scott, R.L., *Regular Solutions*, Prentice-Hall, 1962.
Koningsveld, R., Onclin, M.H., and Kleintjens, L.A., in *Polymer Compatibility and Incompatibility*, Harwood Academic Publishers, 1982.
Kurata, M., *Thermodynamics of Polymer solutions*, Gordon and Breach, 1982.
Morawetz, H., *Macromolecules in Solution*, 2nd ed., John Wiley and Sons, 1975.
Olabisi, O., Robeson, L.M., and Shaw, M.T., *Polymer-Polymer Miscibility*, Academic Press, 1969.
Paul, D.R. and Bucknall, C.B., Eds., *Polymer Blends: Formulation and Performance*, Vol. 1, 2, John Wiley and Sons, 2000.
Paul, D.R. and Newman, S., Eds., *Polymer Blends*, Vol. 1, 2, Academic Press, 1978.

Tompa, H., *Polymer Solutions*, Butterworths, 1956.
Van Krevelen, D.W., *Properties of Polymers*, 3rd ed., Elsevier, 1990.
Yamakawa, H., *Modern Theory of Polymer Solutions*, Gordon and Breach, 1982.

9 Polymer Characterization — Molar Masses

9.1 INTRODUCTION

Many of the distinctive properties of polymers are a consequence of the long chain lengths, which are reflected in the large molar masses of these substances. Although such large molar masses are now taken for granted, it was difficult in 1920 to believe and accept that these values were real and not just caused by the aggregation of much smaller molecules. Values of the order of 10^6 g mol^{-1} are now accepted without question, but the accuracy of the measurements is much lower than for simple molecules. This is not surprising, especially when polymer samples exhibit polydispersity, and the molar mass is, at best, an average dependent on the particular method of measurement used. Estimation of the molar mass of a polymer is of considerable importance, as the chain length can be a controlling factor in determining solubility, elasticity, fiber-forming capacity, tear strength, and impact strength in many polymers.

The methods used to determine the molar mass M are either relative or absolute. Relative methods require calibration with samples of known M and include viscosity and gel permeation chromatography. The absolute methods are often classified by the type of average they yield, i.e., colligative techniques yield number averages and light scattering yields the weight average.

9.2 MOLAR MASSES, MOLECULAR WEIGHTS, AND SI UNITS

The dimensionless quantity *relative molecular mass* (molecular weight), defined as the average mass of the molecule divided by one twelfth the mass of an atom of the nuclide ^{12}C, is often used in polymer chemistry and called the *molecular weight*. In this book, the quantity molar mass is also used and the appropriate SI units are given.

9.3 NUMBER-AVERAGE MOLAR MASS M_n

Determination of the number-average molar mass M_n involves counting the total number of molecules, regardless of their shape or size, present in a unit mass of the polymer. The methods are conveniently grouped into three categories: end-group assay, thermodynamic, and transport methods.

9.4 END-GROUP ASSAY

This technique is of limited value and can only be used when the polymer has an end group amenable to analysis. It can be used to follow the progress of linear condensation reactions when an end group, such as a carboxyl, is present and can be titrated. It is used to detect amino end groups in nylons dissolved in *m*-cresol, by titration with methanolic perchloric acid solution, and can be applied to vinyl polymers if an initiator fragment, perhaps containing halogen, is attached to the end of the chain.

The sensitivity of the method decreases rapidly as the chain length increases and the number of end groups drops. A practical upper limit might reach an M_n of about 15,000 g mol^{-1}.

9.5 COLLIGATIVE PROPERTIES OF SOLUTIONS

Because chemical methods are rather limited, the most widely used techniques for measuring the molar mass of a polymer are physical. Methods that depend on the colligative properties of dilute solutions can be used to determine the molar mass of a substance. These include:

1. Lowering of the vapor pressure
2. Elevation of the boiling point
3. Depression of the freezing point
4. Osmotic pressure

A *colligative property* is defined as one that is a function of the number of solute molecules present per unit volume of solution and is unaffected by the chemical nature of the solute. Thus, if Y represents any of the aforementioned colligative properties, then

$$Y = K \frac{\sum N_i}{V} \tag{9.1}$$

where N_i is the number of particles of each solute component i, and K is a proportionality constant. The concentration of a solution per unit volume of solution V is

$$c = \sum w_i / V = \sum N_i M_i / N_A V ,$$

where w_i is the mass of the component and N_A is the Avogadro constant, and so the colligative property can be expressed in the reduced form Y/c as follows:

$$\frac{Y}{c} = K \frac{\sum N_i}{V} \cdot \frac{N_A V}{\sum N_i M_i} = \frac{K N_A}{M_n} \tag{9.2}$$

Hence, any colligative method should yield the number average molar mass M_n of a polydisperse polymer. Polymer solutions do not behave in an ideal manner, and nonideal behavior can be eliminated by extrapolating the experimental (Y/c) data to $c = 0$. For example, in the case of boiling point elevation measurements (ebullioscopy) Equation 9.2 takes the form

$$\lim_{c \to 0} \frac{\Delta T_b}{c} = \frac{RT^2}{\rho \Delta H_v} \left(\frac{1}{M_n} \right) \tag{9.3}$$

where ΔT_b is the elevation of the boiling point temperature, ρ is the density, and ΔH_v is the enthalpy of vaporization.

Measurements of colligative properties such as boiling point elevation and freezing point depression are limited by the sensitivity of the thermometer used to obtain ΔT and, for a precision of 1×10^{-3} K, the limit of accurate measurements of M_n is in the region 25,000 to 30,000 g mol^{-1}.

9.6 OSMOTIC PRESSURE

Measurement of the osmotic pressure π of a polymer solution can be carried out in the type of cell represented schematically in Figure 9.1. The polymer solution is separated from the pure solvent by a membrane permeable only to solvent molecules. Initially, the chemical potential μ_1 of the solvent in the solution is lower than that of the pure solvent μ_1^0, and solvent molecules tend to pass through the membrane into the solution in order to attain equilibrium. This causes a buildup of pressure in

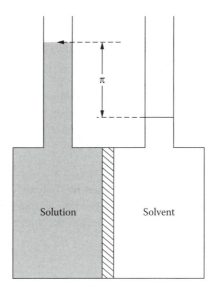

FIGURE 9.1 Osmotic pressure cell.

the solution compartment until, at equilibrium, the pressure exactly counteracts the tendency for further solvent flow. This pressure, P, is the osmotic pressure, π, where

$$\mu_1^o = \mu_1 + \int_P^{P+\pi} \left(\frac{\partial\mu_1}{\partial P}\right)_T dP \tag{9.4}$$

The compressibility of the solvent $(\partial\mu_1/\partial P)_T$ is equal to the molar volume of the solvent in the solution, V_1, and can be assumed to be unchanged over a small range of pressures; thus,

$$\mu_1^o = \mu_1 + V_1 \int_P^{P+\pi} dP = \mu_1 + V_1\,[(P+\pi) - P] \tag{9.5}$$

giving

$$\mu_1 - \mu_1^o = -V_1\,\pi \tag{9.6}$$

Substitution in Equation 8.33 from Chapter 8 gives:

$$\pi V_1 = RT\left[\frac{c_2 V_1}{M_2} + \bar{v}_2^2(\tfrac{1}{2} - \chi_1)c_2^2\right] \tag{9.7}$$

or, by rearranging and including higher-order terms in the virial expansion:

$$\frac{\pi}{c} = \frac{RT}{M_2} + \frac{RT\bar{v}_2^2}{V_1}(\tfrac{1}{2} - \chi_1)c + RT\bar{v}_2^3 c_2^2 \dots \tag{9.8}$$

where the limiting form, valid only at infinite dilution, is

$$\left(\frac{\pi}{c}\right)_{c\to 0} = \frac{RT}{M_n} \tag{9.9}$$

Only under special conditions, i.e., when the polymer is dissolved in a theta solvent, will (π/c) be independent of concentration. Experimentally, a series of concentrations is studied, and the results treated according to one or other of the following virial expansions:

$$\frac{\pi}{c} = RT\left(\frac{1}{M_n} + A_2 c + A_3 c^2 + \dots\right) \tag{9.10}$$

and

$$\frac{\pi}{c} = \left(\frac{\pi}{c}\right)_0 \left(1 + \Gamma_2 c + \Gamma_3 c^2 + \ldots\right) \tag{9.11}$$

The coefficients A_2, Γ_2 and A_3, Γ_3 are the second and third virial coefficients, which describe, respectively, interactions between one polymer molecule and the solvent, and multiple polymer–solvent interactions. When solutions are sufficiently dilute, a plot of (π/c) against c is linear and the third virial coefficients (A_3, Γ_3) can be neglected. The various forms of the second virial coefficient are interrelated by

$$RT\, A_2 = \frac{RT\, \Gamma_2}{M_n} \tag{9.12}$$

Although not normally detected, the third virial coefficient occasionally contributes to the nonideal behavior in dilute solutions, and a curved plot is obtained, as shown in Figure 9.2(a). This increases the uncertainty of the extrapolation, but it can be overcome by recasting Equation 9.11 and introducing a polymer–solvent interaction parameter g

$$\frac{\pi}{c} = \left(\frac{\pi}{c}\right)_0 \left(1 + \Gamma_2 c + g\, \Gamma_2^2 c^2 + \ldots\right) \tag{9.13}$$

It has been found that $g = 0.25$ in good solvents, so Equation 9.13 becomes

$$\frac{\pi}{c} = \left(\frac{\pi}{c}\right)_0 \left(1 + \tfrac{1}{2}\Gamma_2 c\right)^2 \tag{9.14}$$

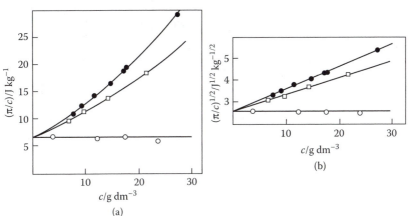

FIGURE 9.2 (a) Plot of (π/c) against c for a sample of poly(methyl methacrylate) ($M_n = 382{,}270$ g mol^{-1}) in three solvents: (●) toluene, (□) acetone, and (○) acetonitrile. (b) Plot $(\pi/c)^{1/2}$ against c for the same data as in (a). (From Fox, T.G. et al., *Polymer*, 3, 71, 1962. With permission.)

A plot of $(\pi/c)^{1/2}$ against c is now linear, and this extrapolation is illustrated in Figure 9.2(b).

Figure 9.2 also shows the differing solubility of poly(methyl methacrylate) in the three solvents at a temperature of 303 K. In a good solvent (toluene), the slope or A_2 is large, but as the solvent becomes poorer (acetone) A_2 decreases, until it is zero in the theta-solvent acetonitrile. Thus, A_2 provides a useful measure of the thermodynamic quality of the solvent and measures the deviation from ideality of the polymer solution.

The value of M_n is calculated from the intercept $(\pi/c)_0 = 6.59$ J kg^{-1} using Equation 9.9.

$$M_n = RT/(\pi/c)_0 = 8.314 \text{ J K}^{-1} \text{ mol}^{-1} \times 303 \text{ K}/6.59 \text{ J kg}^{-1} = 382.27 \text{ kg mol}^{-1}$$

The corresponding values of the second virial coefficient are obtained from the slopes of the plots and are equal to $A_2 = 2.08 \times 10^{-4}$ and 1.63×10^{-4} m^3 mol kg^{-2} for toluene and acetone, respectively.

9.7 LIGHT SCATTERING

Light scattering is one of the most popular methods for determining the weight average molar mass M_w. The phenomenon of light scattering by small particles is familiar to us all; the blue color of the sky, the varied colors of a sunset, and the poor penetration of a car's headlights in a fog are caused by water droplets scattering the light. The obvious presence of dust in a sunbeam or the Tyndall effect in an irradiated colloidal solution are further examples of this effect.

The fundamentals of light scattering were expounded by Lord Rayleigh in 1871 during his studies on gases, where the particle is small compared with the wavelength of the incident radiation. Light is an electromagnetic wave, produced by the interaction of a magnetic and electric field, both oscillating at right angles to one another in the direction of propagation. When a beam of light strikes the atoms or molecules of the medium, the electrons are perturbed or displaced and oscillate about their equilibrium positions with the same frequency as the exciting beam. This induces transient dipoles in the atoms or molecules, which act as secondary scattering centers by reemitting the absorbed energy in all directions, i.e., scattering takes place.

For gases, Rayleigh showed that the reduced intensity of the scattered light R_θ at any angle θ to the incident beam of wavelength λ could be related to the molar mass of the gas M, its concentration c, and the refractive index increment $(d\tilde{n}/dc)$ by

$$R_\theta = \frac{2\pi^2 \tilde{n}^2}{N_A \lambda^4} \left(\frac{d\tilde{n}}{dc} \right)^2 (1 + \cos^2 \theta) M_c. \tag{9.15}$$

The quantity R_θ is often referred to as the Rayleigh ratio and is equal to $(i_\theta r^2/I_0)$, where I_0 is the intensity of the incident beam, i_θ is the quantity of light scattered per unit volume by one center at an angle θ to the incident beam, and r is the distance of the center from the observer. This is valid for a gas, where all the particles are considered to be independent scattering centers, and the addition of more centers, which increases $\tilde{n}$, increases the scattering. The situation changes when dealing with a liquid, as $(d\tilde{n}/dc)$ remains unaffected by the addition of molecules and can be expected to be zero. This conceptual difficulty was overcome in the fluctuation theories of Smoluchowski and Einstein; they postulated that optical discontinuities exist in the liquid arising from the creation and destruction of holes during Brownian motion. Scattering emanates from these centers, created by local density fluctuations, which produce changes in $(d\tilde{n}/dc)$ in any volume element.

When a solute is dissolved in a liquid, scattering from a volume element again arises from liquid inhomogeneities, but now an additional contribution from fluctuations in the solute concentration is present, and for polymer solutions the problem is to isolate and measure these additional effects. This was achieved by Debye in 1944, who showed that for a solute whose molecules are small compared with the wavelength of the light used, the reduced angular scattering intensity of the solute is

$$R_\theta = R_\theta(\text{solution}) - R_\theta(\text{solvent}), \tag{9.16}$$

and that this is related to the change in Gibbs free energy with concentration of the solute. As ΔG is related to the osmotic pressure π, we have

$$R_\theta = \frac{2\pi^2 \tilde{n}_o^2}{\lambda^4} \left(\frac{d\tilde{n}}{dc}\right)^2 (1 + \cos^2 \theta) \frac{N\,M}{N_A} \frac{RT}{\left(d\pi/dc\right)_T} \tag{9.17}$$

Here $\tilde{n}_0$ and $\tilde{n}$ are the refractive indices of the solvent and solution, respectively, and N is the number of polymer molecules. Differentiation of the virial expansion for π with respect to c, followed by substitution in Equation 9.17 and rearrangement leads to

$$\frac{K'\left(1 + \cos^2 \theta\right)c}{R_\theta} = \frac{1}{M_w} + 2A_2 c \tag{9.18}$$

where

$$K' = \frac{2\pi^2 \tilde{n}_0^2 (d\tilde{n}/dc)^2}{\lambda^4 N_A}.$$

These equations are valid for molecules smaller than $(\lambda'/20)$ when the angular scattering is symmetrical, λ' being the wavelength of light in solution, i.e., $\lambda'=(\lambda/\tilde{n}_0)$.

For small particles, M_w can be calculated from Equation 9.18. The important experimental point to remember is that dust will also scatter light and contribute to the

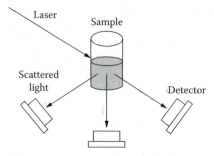

FIGURE 9.3 Schematic diagram of a multiangle light-scattering photometer. Detectors are arranged at set angles with respect to the incident laser beam.

TABLE 9.1
Polystyrene in Benzene at 298 K

C (g dm⁻³)	$10^3 R_{90}$ (m⁻¹)	$(K'c/R_{90})$ (mol g⁻¹)
1.760	5.31	8.56
3.708	8.43	11.36
6.244	11.24	14.35
7.736	12.43	16.07
10.230	13.80	19.15

scattering intensity. Great care must be taken to ensure that solutions are clean and free of extraneous matter, and this is achieved by either centrifugation or membrane filtration.

A number of instruments are available commercially; all of these use a laser as the light source, e.g., a He-Ne laser as light source ($\lambda = 632.8$ nm), offering a highly monochromated wavelength and a light beam of high coherency and stable intensity. The schematic diagram of Figure 9.3 shows a multiangle light-scattering photometer. The sample, which is placed in a cuvette, is illuminated by the light source, and the scattered radiation is detected by a set of high-gain photodiodes (typically 15) arranged at set angles with respect to the incident laser beam in the range 0 to 180°.

For small particles, measurements at a single scattering angle are sufficient to obtain information on the sample's molar mass. For example, M_w can be obtained from measurements of the 90° scattering (R_{90}) as a function of concentration by plotting ($K'c/R_{90}$) against c, followed by linear extrapolation at $c = 0$. Typical results are shown in Table 9.1 for a polystyrene sample dissolved in benzene. The relevant constants are $(d\tilde{n}/dc) = 0.112 \times 10^{-3}$ m³ kg⁻¹, $K' = 2.5888 \times 10^{-5}$ m² mol kg⁻², the intercept $(K'c/R_{90})_{c=0} = 6.9 \times 10^{-3}$ mol kg⁻¹, and $M_w = 148$ kg mol⁻¹.

9.7.1 SCATTERING FROM LARGE PARTICLES

When polymer dimensions are greater than $\lambda'/20$, intraparticle interference causes the scattered light from two or more centers to arrive considerably out of phase at the observation point, and the scattering envelope becomes dependent on the molecular shape. This attenuation, produced by destructive interference, is zero in the direction of the incident beam but increases as θ increases, because the path length difference $\Delta\lambda_f$ in the forward direction is less than $\Delta\lambda_b$ in the backward direction (see Figure 9.4). This difference can be measured from the dissymmetry coefficient Z

$$Z = \frac{R_\theta}{R_{\pi-\theta}} \tag{9.19}$$

which is unity for small particles, but greater than unity for large particles. The scattering envelope reflects the scattering attenuation and is compared with that for small particles in Figure 9.5. The angular attenuation of scattering is measured by the particle scattering factor $P(\theta)$, which is simply the ratio of the scattering intensity to the intensity in the absence of interference, measured at the same angle θ.

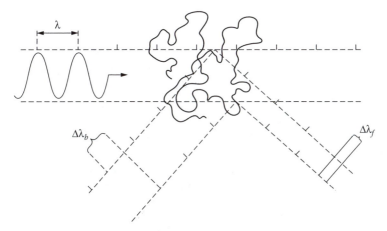

FIGURE 9.4 Destructive interference of light scattered by large particles. Waves arriving at the forward observation point are $\Delta\lambda_f$ out of phase and those arriving at the backward point are $\Delta\lambda_b$ out of phase. For large polymer molecules $\Delta\lambda_f < \Delta\lambda_b$, as shown also in Figure 9.5.

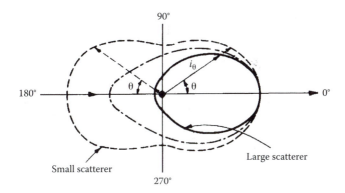

FIGURE 9.5 Intensity distribution of light scattered at various angles. The symmetrical envelope is obtained for small isotropic scatterers in dilute solution, and the two asymmetric envelopes are for much larger scattering particles. The solid line represents the scattering from spheres whose diameters are approximately one half of the wavelength of the incident light.

Guinier showed that a characteristic shape-independent geometric function called the *radius of gyration* $\langle \overline{S}^2 \rangle^{1/2}$ can be measured from large particle scattering. It is defined as an average distance from the center of gravity of a polymer coil to the chain end.

The function $P(\theta)$ is size dependent and can be related to the polymer coil size by the well-known Debye equation

$$P(\theta) = \left(\frac{2}{u^2}\right)\left[e^{-u} - (1-u)\right] \tag{9.20}$$

where $u = \{(4\pi / \lambda)\sin(\theta / 2)\}^2 \langle \overline{S}^2 \rangle$ for monodisperse randomly coiling polymers. In the limit of small θ, the expansion

$$P(\theta)^{-1} = 1 + u/3 - \ldots \tag{9.21}$$

can be used, and the coil size can be estimated from $P(\theta)$ without assuming a particular model. Specific shapes can be related to $P(\theta)$ if desired, as shown in Figure 9.6(a) and Figure 9.6(b).

One of the most common methods used to determine the shape-independent parameter $\langle \overline{S}^2 \rangle^{1/2}$ uses the double extrapolation method proposed by Zimm (1948a). This method is based on the knowledge that, as the scattering at zero angle is independent of size, $P(\theta)$ is unity when $\theta = 0$. Scattering at zero angle is difficult to measure experimentally, and an extrapolation procedure has been devised that makes use of a modified form of Equation 9.18 for large particles,

$$\frac{Kc}{R_\theta} = \frac{1}{M_w P(\theta)} + 2A_2 c + \ldots \tag{9.22}$$

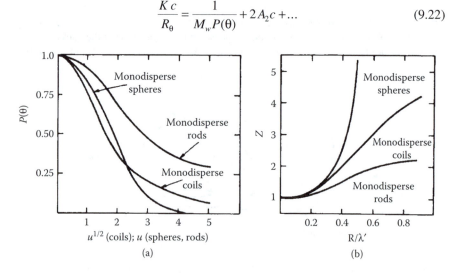

FIGURE 9.6 (a) $P(\theta)$ as a function of u for three model shapes: coils, spheres, and rods. (b) The dissymmetry Z as a function of R/λ', where R is the characteristic linear dimension of either a rod, a sphere, or a coil.

Substituting for $P(\theta)$ and rearranging leads to

$$\frac{Kc}{R_\theta} = \frac{1}{M_w}\left(1 + \frac{Q^2\langle\overline{S}\rangle^2}{3}\right) + 2A_2c + \ldots \tag{9.23}$$

where Q is known as the scattering vector:

$$Q = \frac{4\pi}{\lambda'}\sin\left(\frac{\theta}{2}\right) \tag{9.24}$$

If the scattering intensity for each concentration in a dilution series is measured over an angular range, e.g., 35 to 145°, the data can be plotted as (Kc/R_θ) against $\{\sin^2(\theta/2) + k'c\}$ (or, equivalently, against $\{Q^2 + k'c\}$), where k' is an arbitrary constant chosen to provide a convenient spread of the data in the gridlike graph that is obtained. A double extrapolation is then carried out, as shown in Figure 9.7, by joining all points of equal concentration and extrapolating to zero angle, and then all points measured at equal angles and extrapolating these to zero concentration. For example, on the diagram the points corresponding to concentration c_3 are joined and extrapolated to intersect with an imaginary line corresponding to the value of $k'c_3$ on the abscissa; similarly, all points measured at equal θ are joined and extrapolated until the point corresponding to $\sin^2(\theta/2)$ is reached. This is done for each concentration and each angle, and the extrapolated points are then lines of $\theta = 0$ and $c = 0$. Both lines, on extrapolation to the axis, should intersect at the same point. The intercept is then $(M_w)^{-1}$, the slope of the $\theta = 0$ line yields A_2, whereas $\langle S^2\rangle$ is obtained from the initial slope s_i of the $c = 0$ line, i.e.,

$$\langle S^2\rangle = s_iM_w(3\lambda'^2/16\pi^2). \tag{9.25}$$

Obviously, this reduces to $\langle S^2\rangle = 3s_iM_w$ if (Kc/R_θ) is plotted against $\{Q^2 + k'c\}$. The square of the radius of gyration calculated in this way for a polydisperse sample is a z-average, and it is indicated as $\langle S^2\rangle_z$.

Before M_w can be calculated from light-scattering measurements, the specific refractive index increment $(d\tilde{n}/dc)$ must be known for the particular polymer + solvent system under examination. It is defined as $(\tilde{n} - \tilde{n}_0)/c$, where $\tilde{n}$ and $\tilde{n}_0$ are the refractive indices of the solution and the solvent, and c is the concentration. Measurements of $\Delta\tilde{n} = (\tilde{n} - \tilde{n}_0)$ are made using a differential refractometer employing the same wavelength of light as used in the light scattering.

9.8 DYNAMIC LIGHT SCATTERING

The static light-scattering measurements described in the previous section provide a measure of the time-averaged scattered intensity. On the contrary, dynamic light scattering follows the random motion of molecules in solution called Brownian

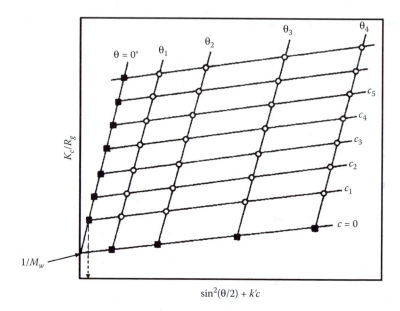

FIGURE 9.7 A typical Zimm plot showing the double extrapolation technique, where $-\bigcirc-$ represents the experimental points and $-\blacksquare-$ represents the extrapolated points.

motion. In solution, particles move randomly, and this motion gives rise to a *Doppler effect*, which manifests itself as a small frequency shift of the scattered light from the frequency of the incident radiation. In practice, the time-dependent fluctuations in the intensity of the scattered light are measured by a fast photon counter, and these are directly related to the rate of diffusion of the solute molecules. Therefore, the fluctuations can be analyzed to determine the translational diffusion coefficient and from this the hydrodynamic radius of the polymer molecule.

9.9 VISCOSITY

When a polymer dissolves in a liquid, the interaction of the two components stimulates an increase in polymer dimensions over that in the unsolvated state. Because of the vast difference in size between solvent and solute, the frictional properties of the solvent in the mixture are drastically altered, and an increase in viscosity occurs that should reflect the size and shape of the dissolved solute, even in dilute solutions. This was first recognized in 1930 by Staudinger, who found that an empirical relation existed between the relative magnitude of the increase in viscosity and the molar mass of the polymer.

One of the simplest methods of examining this effect is by capillary viscometry, although automatic viscometers are commercially available. In a U-tube viscometer such as the Ubbelohde suspended level dilution model shown in Figure 9.8, the flow times of pure solvent t_0 and a polymer solution t are recorded. This is done by pipetting an aliquot of solution of known volume into bulb D. The solution is then pumped into E. The flow time t is the time taken for the solution meniscus to pass from x to y in bulb E.

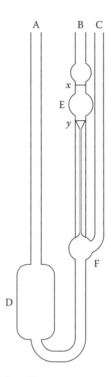

FIGURE 9.8 Suspended level dilution viscometer.

It has been shown that the ratio of the flow time of a polymer solution t to that of the pure solvent t_0 is effectively equal to the ratio of their viscosity (η/η_0) if the densities are equal. This latter approximation is reasonable for dilute solutions and provides a measure of the relative viscosity η_r.

$$\eta_r = (t/t_0) = (\eta/\eta_0) \qquad (9.26)$$

As this has a limiting value of unity, a more useful quantity is the specific viscosity

$$\eta_{sp} = \eta_r - 1 = (t - t_0)/t_0 \qquad (9.27)$$

Even in dilute solutions, molecular interference is likely to occur, and η_{sp} is extrapolated to zero concentration to obtain a measure of the influence of an isolated polymer coil. This is accomplished in either of two ways: η_{sp} can be expressed as a reduced quantity (η_{sp}/c) and extrapolated to $c = 0$ according to the relation

$$(\eta_{sp}/c) = [\eta] + k'[\eta]^2 c, \qquad (9.28)$$

and the intercept is the limiting viscosity number $[\eta]$ that is a characteristic parameter for the polymer in a particular solvent and k' is a shape-dependent factor called the

Huggins constant having values between 0.3 and 0.9 for randomly coiling vinyl polymers. The alternative extrapolation method uses the inherent viscosity as

$$(\log \eta_r)/c = [\eta] + k''[\eta]^2 c, \tag{9.29}$$

where k'' is another shape-dependent factor. The dimensions of $[\eta]$ are the same as the reciprocal of the concentration.

For a given polymer + solvent system at a specified temperature, $[\eta]$ can be related to M through the Mark–Houwink equation

$$[\eta] = K_v M^v \tag{9.30}$$

K_v and v can be established by calibrating with polymer fractions of known molar mass, and once this has been established for a system, $[\eta]$ alone will give M for an unknown fraction. This is normally achieved by plotting log $[\eta]$ against log M, and interpolation is then quite straightforward. Values of v lie between 0.5 for a polymer dissolved in a theta solvent to about 0.8 for linear, randomly coiling vinyl polymers in very good solvents; typical values for systems studied by viscosity are given in Table 9.2. The exponent v is indicative of solvent quality. When the solvent is ideal, i.e., a theta solvent, v is equal to 0.5, but as the solvent becomes thermodynamically better and deviations from ideality become larger, then v increases.

9.9.1 Viscosity-Average Molecular Weight

Polymer samples are normally polydisperse, and it is of interest to examine the type of average molecular weight that might be expected from a measurement of $[\eta]$. As the specific viscosity will depend on the contributions from each of the polymer molecules in the sample, we can write

$$\eta_{sp} = K \sum c_i M_i^v \tag{9.31}$$

If we now divide by the total concentration $c = \Sigma c_i$ and substitute for $c_i = N_i M_i / N_A V$, then

$$\frac{\eta_{sp}}{c} = \frac{K \sum N_i M_i}{N_A V} \frac{N_A V}{\sum N_i M_i} \cdot \sum M_i^v \tag{9.32}$$

or

$$[\eta] = \frac{\eta_{sp}}{c_{c \to 0}} = \frac{K \sum N_i M_i^{1+v}}{\sum N_i M_i} \tag{9.33}$$

TABLE 9.2
A Comparison of Viscosity Constants and Exponents for Several Polymer + Solvent Systems from Equation 9.30

Polymer	Solvent	T (K)	$10^2\,K_v\,(cm^3\,g^{-1})$	v
Polystyrene	Cyclohexene	298	1.63	0.68
Polystyrene	Chloroform	298	0.716	0.76
Polystyrene	Cyclohexane $\theta =$	308	8.6	0.50
Poly(α-methyl styrene)	Cyclohexane $\theta =$	310	7.8	0.50
Poly(α-methyl styrene)	Toluene	310	1.0	0.72
Poly(vinyl acetate)	Butanone	298	4.2	0.62
Cellulose nitrate	Ethyl acetate	303	0.25	1.01
Cellulose	Cadoxen	298	250	0.75

Comparison with Equation 9.30 shows that the viscosity average M_v is then

$$M_v = \left[\frac{\sum N_i M_i^{1+v}}{\sum N_i M_i} \right]^{1/v} \tag{9.34}$$

and that this lies between M_n and M_w in magnitude but will be usually closer to M_w.

9.10 GEL PERMEATION CHROMATOGRAPHY

The molar mass distribution (MMD) of a polymer sample has a significant influence on its properties, and a knowledge of the shape of this distribution is fundamental to the full characterization of a polymer. The determination of the MMD by conventional fractionation techniques is time consuming, and a rapid, efficient, and reliable method that can provide a measure of the MMD in a matter of hours has been developed. This is gel permeation chromatography (GPC). Known alternatively by its more descriptive name *size exclusion chromatography* (SEC), the method depends on the use of mechanically stable, highly cross-linked gels, which have a distribution of different pore sizes and can, by means of a sieving action, effect separation of a polymer sample into fractions dictated by their molecular volume.

The nonionic gel stationary phase is commonly composed of cross-linked polystyrene or macroporous silica particles, which do not swell significantly in the carrier solvents. A range of pore sizes is fundamental to the success of this size fractionation procedure, which depends on two processes. These are (1) separation by size exclusion alone, which is the more important feature, and (2) a dispersion process, controlled by molecular diffusion, which may lead to an artificial broadening of the MMD.

We consider first the mechanism of the separation process; in simple terms, the large molecules, which occupy the greatest effective volume in solution, are excluded from the smaller pore sizes in the gel and pass quickly through the larger channels between the gel particles. This results in their being eluted first from the column.

As the molecular size of the polymer decreases, there is an increasing probability that the molecules can diffuse into the smaller pores and channels in the gel, which slows their time of passage through the column by providing a potentially longer path length before being eluted. By choosing a series of gel columns with an appropriate range of pore sizes, an effective size separation can be obtained.

The efficiency of the separation process is then a function of the dependence of the retention (or elution) volume V_R on the molar mass M, and a reliable relationship between the two parameters must be established. The value of V_R depends on the interstitial void volume V_0 and the accessible part of the pore volume in the gel,

$$V_R = V_0 + K_D V_i \qquad (9.35)$$

where V_i is the total internal pore volume and K_D is the partition coefficient between V_i and the portion accessible to a given solute. Thus, $K_D = 0$ for very large molecules ($V_R = V_0$) and rapid elution takes place, whereas $K_D = 1$ for very small molecules, which can penetrate all the available pore volume. This is shown schematically in Figure 9.9, and clearly the technique cannot discriminate among molecular sizes with $V_R \leq V_0$ or $V_R \geq V_0 + V_i$. For samples that fall within the appropriate range, it has been suggested that a universal calibration curve can be constructed to relate V_R and M, by assuming that the hydrodynamic volume of a macromolecule is related to the product $[\eta] \cdot M$, where $[\eta]$ is the intrinsic viscosity of the polymer in the carrier solvent used at the temperature of measurement. A universal calibration curve is then obtained by plotting $\log [\eta] \cdot M$ against V_R for a given carrier solvent and a fixed temperature. Experimental verification of this is shown in Figure 9.10 for a variety of different polymers and can be utilized in the following way.

To obtain the MMD, the mass of the polymer being eluted must be measured. This can be achieved continuously using refractive index (RI), UV, or IR detectors, which will give a mass distribution as a function of V_R. It is still necessary to estimate the molar mass of each fraction before the MMD curve can be constructed. If the universal calibration curve is valid for the system, then

$$\log [\eta]_s M_s = \log [\eta]_u M_u \qquad (9.36)$$

where the subscripts "s" and "u" denote the standard calibration and the polymer under study, respectively. As the chains of the polymer under examination may swell in the carrier solvent to a different extent compared to an equal molar mass sample of the standard, the hydrodynamic volumes will not necessarily be equivalent. This can be compensated for by applying a correction based on the knowledge of the appropriate Mark–Houwink relations for each, the standard and the unknown, measured in the solvent used for elution (i.e., $[\eta]_s = K_s M_s^{v_s}$ and $[\eta]_u = K_u M_u^{v_u}$). The molar mass M_u can then be obtained from

$$\log M_u = \frac{1}{1+v_u} \cdot \log \left[\frac{K_s}{K_u} \right] + \frac{1+v_s}{1+v_u} \cdot \log M_s \qquad (9.37)$$

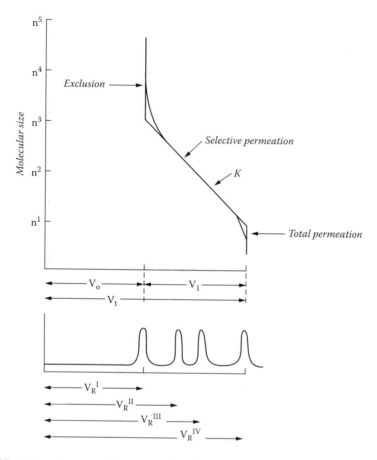

FIGURE 9.9 Elution curve schematically showing the range of elution volumes that are valid for a particular column. In this case, molecules with molecular size $> n^3$ are totally excluded and eluted without discrimination, whereas those $< n^1$ tend to become absorbed or are partitioned if a mixed solvent is used and $K_D = 1$.

Thus, a calibration curve constructed for standard samples of polystyrene can be used to determine M for other polymers if the Mark–Houwink relations are also known. This can be avoided if the polymer's molar mass can be determined *in situ*, by using an online detector sensitive to molar mass. Both the light-scattering and viscosity detectors satisfy this requirement. Instruments are commercially available, where a low-angle laser light scattering (LALLS) or a multiangle laser light-scattering photometer are attached in series with a concentration detector By combining information from these two detectors, absolute molar masses and distributions can be measured without any need for column calibration. The combination of an online viscometer (VIS) and a right-angle light-scattering detector (RALLS) with SEC offers additional advantages, and it is particularly useful for the characterization of molar mass and degree of branching in polymers. SEC instrumentation offering a triple detection system, RI-VIS-RALLS, is now commercially available (Figure 9.11).

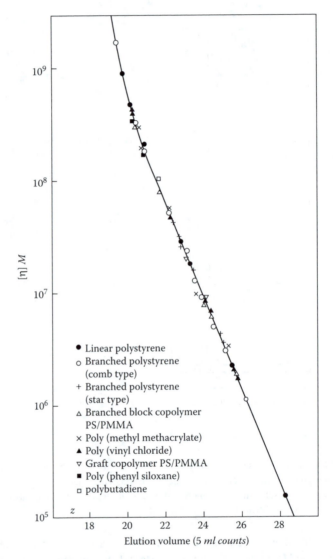

FIGURE 9.10 A universal calibration curve for several polymers in tetrahydrofuran. (Reproduced from data by Grubisic, Z., Rempp, P., and Benoit, H., *J. Polym. Sci.: Polym. Lett.*, 5, 753, 1967. With permission. Copyright John Wiley & Sons, New York.)

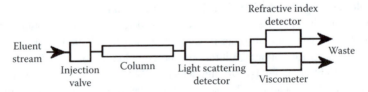

FIGURE 9.11 Schematic diagram of an SEC instrumentation with triple detection system. (Courtesy of Polymer Laboratories.)

9.11 MALDI

Mass spectrometry (MS) is the method of choice for the determination of the molecular weight of small molecules. The technique measures the mass-to-charge ratio (m/z) of analyte ions, and it therefore requires the production of charged species in the gas phase. The difficulty in bringing polymer molecules into the gas phase without degradation would appear to limit the technique to the characterization of low-molecular-weight species. However, during the past decade, new methods such as matrix-assisted-laser-desorption ionization–time-of-flight–mass spectrometry (MALDI-TOF-MS) have been introduced that overcome these limitations.

Among the various mass spectrometry techniques, MALDI is probably the most important as it provides an absolute method for molar mass determination and molar mass distribution, as well as information on end groups and copolymer composition. The MALDI process consists of the ablation of the polymer molecules dispersed in a matrix typically made up of aromatic organic acids. The matrix needs to be able to absorb at the wavelength of a laser (usually 337 nm). This process excites the matrix molecules, which vaporize; at the same time, the polymer molecules desorb into the gas phase, where they are ionized. Thus, the role of the matrix is that of transferring the laser energy to the polymer molecules.

Ions are formed by either removal or addition of an electron, giving a radical cation $M^{+\bullet}$ or radical anion $M^{-\bullet}$, or by the addition of other charged species. Synthetic polymers are not easily charged, unless they contain labile protons. However, as the matrix is usually an organic acid, it can readily furnish a proton or, alternatively, salts can be added to the polymer–matrix mixture for effective ion formation.

Once the polymer molecules have been transferred to the gas phase as ions, they are separated on the basis of their mass-to-charge ratio. Mass spectrometers used for MALDI analysis may differ, but for those that are commercially available, separation is effected by TOF. Other methods such as quadrupole filter and Fourier transform mass spectrometry (FTMS) may also be used.

In a MALDI-TOF spectrometer, a high voltage accelerates the ions, directing them to an evacuated tube before reaching a detector that counts the number of ions. The MALDI intensity represents the total number of ions of any given mass, and it therefore gives a measure of the molar mass distribution. The mass-to-charge ratio can be derived from the flight time of the ions.

Sample preparation is crucial in MALDI analysis. Intimate mixing between matrix and analyte is necessary because only the soluble portion of the analyte can be detected by MALDI. The simplest procedure involves dissolving the analyte and the matrix in a common solvent, and this is best achieved when the polarity of the analyte and matrix are similar.

The main output of a MALDI experiment is the average molar mass of a sample and its distribution. As shown in Figure 9.12, samples with different molar mass distributions can be easily identified. The MALDI-TOF mass spectra shown in Figure 9.12 were obtained by adding $NaBF_4$ to the DHB matrix to promote ion formation. The most intense peaks correspond to ions of the $[M + Na]^+$ series, but weaker $[M + K]^+$ satellites are also evident. The difference between molecular weight distributions of the two samples, both of M_n 2000 g mol^{-1}, is evident on comparison

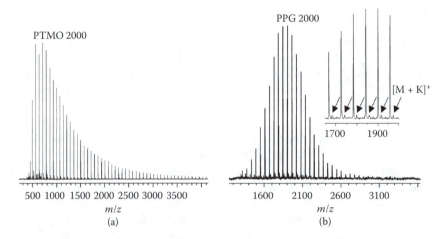

FIGURE 9.12 MALDI-TOF mass spectra (DHB matrix) of (a) poly(tetrahydrofuran) and (b) a low-monol PPG, both with an M_n of 2000 g mol^{-1}. Inset: a minor peak series originates from [M + K]+ adducts. (From Luftmann, H. et al., *Macromolecules*, 36, 6316, 2003. With permission of ACS.)

of the two MALDI-TOF mass spectra. PTMO 2000 displays a broad distribution that tails off toward higher m/z values. By contrast, the spectrum of PPG 2000 shows an almost symmetric distribution centered at m/z 1870.

Generally, comparison between MALDI and GPC data is best carried out by considering the average values, M_n and M_w. The MALDI averages are known to be accurate for samples of low polydispersity, typically below 1.2. Various factors such as sample preparation and mass-dependent desorption/ionization may affect the results for a sample with a relatively large polydispersity above 1.6.

PROBLEMS

1. Predict the change in boiling point of toluene (T_b = 384 K) for a 0.01 g cm^{-3} solution of a polyethylene sample with M_n = 10,000 g mol^{-1} (assume that density of toluene ρ = 0.862 g cm^{-3} and ΔH_v = 38.06 kJ mol^{-1}).

2. A polymer sample was characterized by radioactive end-group analysis and membrane osmometry in acetonitrile at 300 K. The osmotic pressure data are reported below as a function of polymer concentration.

π(N m^{-2})	c (g dm^{-3})
33.2	0.8
66.4	1.6
99.6	2.4
132.8	3.2

a. Demonstrate that acetonitrile is a Θ-solvent for the polymer at 300 K and calculate the polymer molar mass.

    b. End-group analysis showed that 0.1 mmol of radioactive initiator fragments were present in 3 g of polymer. Determine whether, during the polymerization reaction, termination occurred by disproportionation or combination.

3. The following osmotic pressures were measured for solutions of polyisobutylene in benzene at 298 K. Results are reported as centimeters of solution, h, and need to be converted to pressures using the relationship: $\pi = hdg$, where d is the solution density and g the gravitational constant equal to 981 cm s^{-2}.

h (cm solution)	c (g dm^{-3})
0.93	5.0
1.91	10.0
2.94	15.0
4.00	20.0

Calculate the average molar mass of the sample, given that the solution density is 0.88 g cm^{-3}.

4. Light-scattering measurements of three polystyrene standards were carried out using the Zetasizer Nano S by Malvern Instruments (Zetasizer Nano application note, MRK577-01), which measures the intensity of the light scattered at 173°. $Kc/R_{\theta=173}$ values as a function of polymer concentration in toluene solution are reported below. From these, determine M_w and the second virial coefficient A_2.

Sample 1		Sample 2		Sample 3	
c (g dm^{-3})	$Kc/R_\theta \times 10^{-5}$ (mol g^{-1})	c (g dm^{-3})	$Kc/R_\theta \times 10^{-5}$ (mol g^{-1})	c (g dm^{-3})	$Kc/R_\theta \times 10^{-5}$ (mol g^{-1})
0.41	7.33	0.15	11.32	0.22	10.75
0.62	6.29	0.34	11.81	0.41	11.81
0.93	5.22	0.63	11.96	0.62	12.45
		1.28	11.78	0.84	13.20

5. Cyclohexane solutions of a poly(α-methyl styrene) sample prepared by anionic polymerization were examined by light scattering in the temperature range 300 to 320 K (Cowie et al., 1967). Scattered intensity measured at $\theta = 90°$ are listed in the following table as a function of concentration and temperature.

$T = 300.8$ K		$T = 305.6$ K		$T = 311.9$ K		$T = 319.3$ K	
c (g dm^{-3})	$I_{\theta=90}$ (cm^{-1})	c (g dm^{-3})	$I_{\theta=90}$ (cm^{-1})	c (g dm^{-3})	$I_{\theta=90}$ (cm^{-1})	c (g dm^{-3})	$I_{\theta=90}$ (cm^{-1})
0.002186	343	0.002173	338	0.002156	331	0.002136	323
0.004019	643	0.003995	626	0.003965	607	0.003928	587
0.006032	989	0.005996	948	0.005951	912	0.005895	868
0.008060	1337	0.008014	1273	0.007953	1207	0.007880	1138
		0.010175	1628	0.010098	1527	0.010003	1414

From these data determine the second virial coefficient and the theta temperature of poly(α-methyl styrene) in cyclohexane, knowing that $K = K' \tilde{n}_0^2 (d\tilde{n}/dc)^2$, where $K' = 18.17$ mol cm^{-4}, the refractive index increment $(d\tilde{n}/dc)$ is 0.199 ml g^{-1}, and the temperature dependence of the refractive index is expressed by $\tilde{n}_0 = -0.0005327 \times$ T (°C) + 1.446.

6. Static light-scattering measurements were carried out by Zimm (1948b) on polystyrene in butanone at 340 K at two concentrations.

θ	$c = 0.0019$ g cm^{-3}	$c = 0.00038$ g cm^{-3}
	I	I
25.8	—	7.38
36.9	29.8	7.34
53.0	28.5	6.96
66.4	27.9	6.78
90.0	26.2	6.29
113.6	24.7	5.88
143.1	23.5	5.54

Construct a Zimm plot and, using the double extrapolation procedure, determine the polymer molecular weight. Note that relative intensities are reported in the preceding table, and in order to obtain the correct molecular weight, the following relationship needs to be used:

$$\lim_{\substack{c \to 0 \\ \theta \to 0}} \frac{I}{c} = K M_w = 0.0289 M_w$$

7. A poly(2,3-dimethyl-1,3-butadiene) sample was prepared by anionic polymerization, and its molar mass determined via multiangle light-scattering measurements at a wavelength of 516 nm. Kc/R_θ values as a function of scattering angle are tabulated below. From these determine M_w and the mean square radius of gyration.

θ	$Kc/R_\theta \times 10^5$ (mol g^{-1})
45	2.7871
90	2.7988
135	2.8106

8. The following data were obtained for polystyrene dissolved in cyclohexane, when viscosity measurements were made at the Θ temperature of 308 K (solvent flow time, $t_0 = 100$ s).

t (s)	c (g cm^{-3})
109.5	0.001
120.0	0.002
135.0	0.003
144.0	0.004

Calculate the molar mass of the polymer sample from the Mark–Houwink equation:

$$[\eta] = 8.6 \times 10^{-2} M^{0.5}$$

9. Weight-average molecular weights, M_w, for a series of poly-α-methyl styrene samples obtained from light-scattering measurements are tabulated below, together with the corresponding intrinsic viscosities measured in cyclohexane at 310 K (Cowie et al.,1967).

$M_w \times 10^{-6}$ (g mol^{-1})	$[\eta]$
4.00	1.60
1.54	1.01
1.40	0.93
1.25	0.90
0.84	0.69
0.66	0.64
0.48	0.52
0.22	0.36
0.09	0.23

Determine the Mark–Houwink parameters. For the same samples, intrinsic viscosities in toluene are greater than in cyclohexane. Comment on the different solvent quality.

10. A set of polystyrene standards was used to calibrate a GPC column. Elution times obtained for these standards using THF as eluent and a 1.0 ml/min flow rate are:

V (min)	M (g mol^{-1})
10.83	8500000
11.58	3040000
12.45	1030000
13.66	330000
14.45	156000
15.09	66000
15.96	28500
16.75	9200
18.08	3250

Construct a calibration curve and determine the molecular weight of a polymer sample with elution time $V = 13.0$ min.

REFERENCES

Cowie, J.M.G., Bywater, S., and Worsfold, D.J., *Trans. Faraday Soc.*, 105, 1967.
Debye, P., *J. Phys. Coll. Chem.*, 51, 18, 1947.

Fox, T.G., Kinsinger, J.B., Mason, H.F., and Schuele, E.M., *Polymer*, 3, 71, 1962.
Grubisic, Z., Rempp, P., and Benoit, H., *J. Polym. Sci.: Polym. Lett.*, 5, 753, 1967.
Wyatt, P.J., *Anal. Chim. Acta*, 272, 1, 1993.
Zimm, B.H., *J. Chem. Phys.*, 16, 1093, 1948a.
Zimm, B.H., *J. Chem. Phys.*, 16, 1099, 1948b.

BIBLIOGRAPHY

Allen, G. and Bevington, J.C., Eds., *Comprehensive Polymer Science*, Vol. 1, Pergamon Press, 1989.
Billingham, N.C., *Molar Mass Measurements in Polymer Science*, Kogan Page, 1977.
Billmeyer, F.W., *Textbook of Polymer Science*, 3rd ed., John Wiley and Sons, 1979.
Bohdanecky, M. and Kovar, J., *Viscosity of Polymer Solutions*, Elsevier, 1982.
Carroll, B., *Physical Methods in Macromolecular Chemistry*, Vol. 2, Marcel Dekker, 1972.
Huglin, M.B., *Light Scattering from Polymer Solutions*, Academic Press, 1972.
Johnson, J.F. and Porter, R.F., *Analytical Gel Permeation Chromatography*, John Wiley and Sons, 1968.
Kratochvil, P., *Classical Light Scattering from Polymer Solutions*, Elsevier, 1987.
Morawetz, H., *Macromolecules in Solution*, 2nd ed., John Wiley and Sons, 1975.
Pethick, R.A. and Dawkins, J.V., Eds., *Modern Techniques for Polymer Characterisation*, John Wiley and Sons, 1999.
Rabek, J.F., *Experimental Methods in Polymer Chemistry*, John Wiley and Sons, 1980.
Yau, W.W., Kirkland, J.J., and Bly, D.D., *Modern Size-Exclusion Liquid Chromatography. Practice of Gel Permeation and Gel Filtration Chromatography*, John Wiley and Sons, 1979.

10 Polymer Characterization — Chain Dimensions, Structures, and Morphology

As the size and shape of a polymer chain are of considerable interest to the polymer scientist, it is useful to know how these factors can be assessed. Much of the information can be derived from studies of dilute solutions; absolute measurement of polymer chain size can be obtained from light scattering, when the polymer is large compared with the wavelength of the incident light. Sometimes, the absolute measurement cannot be used, but the size can be deduced indirectly from viscosity measurements, which are related to the volume occupied by the chain in solution. Armed with this information, we must now determine how meaningful it is, and for this a clearer understanding of the factors governing the shape of the polymer is required. We will confine ourselves to models of the random coil, as this is usually believed to be most appropriate for synthetic polymers; other models — rods, disks, spheres, and spheroids — are also postulated but need not concern us at this level.

10.1 AVERAGE CHAIN DIMENSIONS

A polymer chain in dilute solution can be pictured as a coil, continuously changing its shape under the action of random thermal motions. This means that at any time, the volume occupied by a chain in solution could differ from that occupied by its neighbors, and these size differences are further accentuated by the fact that each sample will contain a variety of chain lengths. Taking these two points into consideration leads us to the conclusion that meaningful chain dimensions can only be values averaged over the many conformations assumed. Two such averages have been defined: (1) the average root-mean-square distance between the chain ends $\langle \bar{r}^2 \rangle^{1/2}$ and (2) the average root-mean-square radius of gyration $\langle \bar{S}^2 \rangle^{1/2}$, which is a measure of the average distance of a chain element from the center of gravity of the coil. The angular brackets denote averaging due to chain polydispersity in the sample, and the bar indicates averaging of the many conformational sizes available to chains of the same molar mass.

The two quantities are related, in the absence of excluded volume effects, for simple chains by

$$\langle \bar{r}^2 \rangle^{1/2} = \langle 6\bar{S}^2 \rangle^{1/2}, \tag{10.1}$$

but as the actual dimensions obtained can depend on the conditions of the measurement, other factors must also be considered.

10.2 FREELY JOINTED CHAIN MODEL

The initial attempts to arrive at a theoretical representation of the dimensions of a linear chain treated the molecule as a number n of chain elements joined by bonds of length l. By assuming the bonds act like universal joints, complete freedom of rotation about the chain bonds can be postulated. This model allows the chain to be pictured as in Figure 10.1(a), which resembles the path of a diffusing gas molecule, and as random-flight statistics have proved useful in describing gases, a similar approach is used here. In two dimensions, the diagram is more picturesquely called the "drunkard's walk," and r_f is estimated by considering first the simplest case of two links. The end-to-end distance r_f follows from the cosine law that

$$OB^2 = OA^2 + AB^2 - 2(OA)(AB)\cos\theta \tag{10.2}$$

See Figure 10.1(b):

$$r_f^2 = 2l^2 - 2l^2 \cos\theta. \tag{10.3}$$

When the number of bonds, n, is large, the angle θ will vary over all possible values so that the sum of all these terms will be zero, and as $\cos\theta = -\cos(\theta + \pi)$, Equation 10.3 will reduce to

$$r_f^2 = nl^2. \tag{10.4}$$

This shows that the distance between the chain ends for this model is proportional to the square root of the number of bonds, and so is considerably shorter than a fully extended chain.

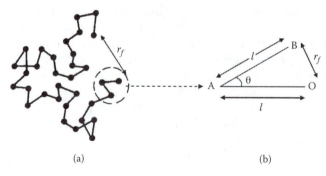

(a)   (b)

FIGURE 10.1 (a) Random walk chain of 32 steps, length l and (b) cosine law for two bonds.

The result is the same if the molecule is thought to occupy three-dimensional space, but if it is centered on a coordinate system, both positive and negative contributions occur with equal probability. To overcome this, the dimension is always expressed as the square, which eliminates negative signs. This model is, however, unrealistic. Polymer chains occupy a volume in space, and the dimensions of any macromolecule are influenced by the bond angles and by interactions between the chain elements. These interactions can be classified into two groups: (1) short-range interactions, which occur between neighboring atoms or groups, and are usually forces of steric repulsion caused by the overlapping of electron clouds; and (2) long-range interactions, which comprise attractive and repulsive forces between segments widely separated in a chain that occasionally approach one another during molecular flexing, and between segments and solvent molecules. These are often termed *excluded volume* effects.

10.3 SHORT-RANGE EFFECTS

The expansion of a covalently bonded polymer chain will be restricted by the valence angles between each chain atom. In general, this angle is θ for a homoatomic chain, and Equation 10.4 can be modified to allow for these short-range interactions.

$$\langle \overline{r}^2 \rangle_{of} = nl^2 (1 - \cos \theta) / (1 + \cos \theta) \tag{10.5}$$

For the simplest case of an all-carbon backbone chain such as polyethylene, $\theta \approx 109°$ and $\cos \theta = -\frac{1}{3}$ so that Equation 10.5 reduces to

$$\langle \overline{r}^2 \rangle_{of} = 2nl^2. \tag{10.6}$$

This indicates that the polyethylene chain is twice as extended as the freely jointed chain model when short-range interactions are considered.

10.4 CHAIN STIFFNESS

As we have already seen in Chapter 1 for butane and polyethylene, steric repulsions impose restrictions on bond rotation. This means that Equation 10.5 has to be modified further, and now becomes

$$\langle \overline{r}^2 \rangle_0 = nl^2 \frac{(1 - \cos \theta)}{(1 + \cos \theta)} \cdot \frac{(1 - \langle \cos \phi \rangle)}{(1 + \langle \cos \phi \rangle)} \tag{10.7}$$

where $\langle \cos \phi \rangle$ is the average cosine of the angle of rotation of the bonds in the backbone chain. The parameter $\langle \overline{r}^2 \rangle_0$ is the average mean square of the unperturbed dimension, which is a characteristic parameter for a given polymer chain.

The freely jointed dimensions are now more realistic when restricted by the skeletal factor ζ, composed of the two terms,

$$\zeta = \sigma^2 (1 - \cos\theta)/(1 + \cos\theta) \tag{10.8}$$

where σ is known as the steric parameter and is $(1 - \langle\cos\phi\rangle)/(1 + \langle\cos\phi\rangle)$ for simple chains.

For more complex chains containing rings or heteroatomic chains, e.g., polydienes, polyethers, polysaccharides, and proteins, an estimate of σ is obtained from

$$\sigma^2 = \langle\bar{r}^2\rangle_0 / \langle\bar{r}^2\rangle_{of}. \tag{10.9}$$

Values of the unperturbed dimension can be obtained experimentally from dilute solution measurements made either directly in a theta solvent (see Chapter 9, Section 9.7) or by using indirect measurements in nonideal solvents and employing an extrapolation procedure. The geometry of each chain allows the calculation of $\langle\bar{r}^2\rangle_{of}$, and results are expressed either as σ or as the characteristic ratio $C_\infty = \{\langle\bar{r}^2\rangle_0 / nl^2\}$. Both provide a measure of chain stiffness in dilute solution. The range of values normally found for σ is from about 1.5 to 2.5, as shown in Table 10.1.

10.5 TREATMENT OF DILUTE SOLUTION DATA

We can now examine some of the ways of calculating the polymer dimensions from experimental data.

10.5.1 THE SECOND VIRIAL COEFFICIENT

An investigation of the dilute solution behavior of a polymer can provide useful information about the size and shape of the coil, the extent of polymer–solvent interaction, and the molar mass. Deviations from ideality, as we have seen in Chapter 9,

TABLE 10.1
Chain Stiffness Parameters and Typical Dimensions

Polymer	T (K)	σ	$[\langle\bar{r}^2\rangle_0/nl^2]$
Polypropylene(isotactic)	408	1.53	4.67
Polypropylene(atactic)	408	1.65	5.44
Natural rubber	293	1.67	4.70
Gutta-percha	333	1.38	7.35
Polystyrene	308	2.23	10.00
Poly(methyl methacrylate)			
(Isotactic)	298	2.28	10.40
(Atactic)	298	2.01	8.10
(Syndiotactic)	308	1.94	7.50

Section 9.6, are conveniently expressed in terms of virial expansions, and when solutions are sufficiently dilute, the results can be adequately described by the terms up to the second virial coefficient A_2 while neglecting higher terms. The value of A_2 is a measure of polymer–solvent compatibility, as the parameter reflects the tendency of a polymer segment to exclude its neighbors from the volume it occupies. Thus, a large positive A_2 indicates a good solvent for the polymer, whereas a low value (sometimes, even negative) shows that the solvent is relatively poor. The virial coefficient can be related to the Flory dilute solution parameters by

$$A_2 = \psi_1 (1 - \Theta / T)\left(\bar{v}_2^2 / V_1 \right) F(x) \tag{10.10}$$

where $\bar{v}_2$ is the partial specific volume of the polymer, V_1 is the molar volume of the solvent, and $F(x)$ is a molar-mass-dependent function of the excluded volume. The exact form of $F(x)$ can be defined explicitly by one of several theories, and though each leads to a slightly different form, all predict that $F(x)$ is unity when theta conditions are attained and the excluded volume effect vanishes. Equation 10.10 can be used to analyze data such as that in Chapter 8, Figure 8.5. Once Θ has been located, the entropy parameter ψ_1 can be calculated by replotting the data as $\psi_1 F(x)$ against T. Extrapolation to $T = \Theta$, where $F(x) = 1$, allows ψ_1 to be estimated for the system under theta conditions. This method of measuring Θ and ψ_1 is only accurate when the solvent is poor, and extrapolations are short.

The dependence of A_2 on M can often be predicted, for good solvents, by a simple equation

$$A_2 = kM^{-\gamma} \tag{10.11}$$

where γ varies from 0.15 to 0.4, depending on the system, and k is a constant.

10.5.2 EXPANSION FACTOR α

The value of A_2 will tell us whether or not the size of the polymer coil, which is dissolved in a particular solvent, will be perturbed or expanded over that of the unperturbed state, but the extent of this expansion is best estimated by calculating the expansion factor α.

If the temperature of a system containing a polymer of finite M drops much below Θ, the number of polymer–polymer contacts increases until precipitation of the polymer occurs. Above this temperature, the chains are expanded, or perturbed, from the equilibrium size attained under pseudoideal conditions, by long-range interactions. The extent of this coil expansion is determined by two long-range effects. The first results from the physical exclusion of one polymer segment by another from a hypothetical lattice site, which reduces the number of possible conformations available to the chain. This serves to lower the probability that tightly coiled conformations will be favored. The second is observed in very good solvents, where the tendency is for polymer–solvent interactions to predominate, and leads to a preference for even more extended conformations. In a given solvent, an equilibrium conformation is eventually achieved when the forces of expansion are balanced by forces of contraction in the molecule. The tendency to contract arises from both polymer–polymer interactions and the resistance to expansion of the chain into overextended and energetically less-favored conformations.

The extent of this coil perturbation by long-range effects is measured by an expansion factor α, introduced by Flory. This relates the perturbed and unperturbed dimensions by

$$\langle \overline{S^2} \rangle^{1/2} = \alpha \langle \overline{S^2} \rangle_0^{1/2} \qquad (10.12)$$

In good solvents (large, positive A_2), the coil is more extended than in poor solvents (low A_2), and α is correspondingly larger. Because α is solvent- and temperature-dependent, a more characteristic dimension to measure for the polymer is $\langle \overline{S^2} \rangle_0^{1/2}$, which can be calculated from light scattering in a theta solvent, or indirectly as described in the next section.

10.5.3 FLORY–FOX THEORY

The molecular dimensions of a polymer chain in any solvent can be calculated directly from light-scattering measurements, using Equation 9.23, if the coil is large enough to scatter light in an asymmetric manner, but when the chain is too short to be measured accurately in this way, an alternative technique has to be used. Because of the smaller wavelength of the associated radiation compared to light, x-ray and neutron scattering provide a means of measuring molecular dimensions of smaller polymer molecules. These techniques will be discussed later in this chapter. Here we are solely concerned with viscosity data.

Flory and Fox suggested that as the viscosity of a polymer solution will depend on the volume occupied by the polymer chain, it should be feasible to relate coil size and $[\eta]$. They assumed that if the unperturbed polymer is approximated by a hydrodynamic sphere, then $[\eta]_\theta$, the limiting viscosity number in a theta solvent, could be related to the square root of the molar mass by

$$[\eta]_\theta = K_\theta \, M^{1/2} \qquad (10.13)$$

where

$$K_\theta = \Phi \left(\frac{\overline{r_0^2}}{M} \right)^{3/2} . \qquad (10.14)$$

Equation 10.13 and Equation 10.14 are actually derived for monodisperse samples, and when measurements are performed with heterodisperse polymers, then appropriate averages to use are M_n and $\langle \overline{r^2} \rangle_0$. The parameter Φ was originally considered to be a universal constant, but experimental work suggests that it is a function of the solvent, molar mass, and heterogeneity. Values can vary from an experimental value of 2.1×10^{23} to a theoretical limit of about 2.84×10^{23}, when $[\eta]$ is expressed in $cm^3 \, g^{-1}$. A most probable value of 2.5×10^{23} has been found to be acceptable for most flexible heterodisperse polymers in good solvents.

For nonideal solvents, Equation 10.13 can be expanded to give

$$[\eta] = K_\theta \, M^{1/2} \alpha_\eta^3 \qquad (10.15)$$

where $\alpha_\eta^3 = [\eta]/[\eta]_\theta$ is the linear expansion factor pertaining to viscosity measurements, and is a measure of long-range interactions. As the derivation is based on an unrealistic Gaussian distribution of segments in good solvents, it has been suggested that α_η is related to the more direct measurement of α in Equation 10.12 by

$$\alpha_\eta^3 = \alpha^{2.43}. \qquad (10.16)$$

Considerable experimental evidence exists to support this conclusion.

10.5.4 INDIRECT ESTIMATES OF UNPERTURBED CHAIN DIMENSIONS

It is not always possible to find a suitable theta solvent for a polymer, and methods have been developed that allow unperturbed dimensions to be estimated in nonideal (good) solvents.

Several methods of extrapolating data for $[\eta]$ have been suggested. The most useful of these was proposed by Stockmayer and Fixman, using the equation:

$$[\eta] \, M^{-1/2} = K_\theta + 0.15 \, \Phi \, B'M^{1/2}, \qquad (10.17)$$

where Φ is assumed to adopt its limiting theoretical value; B' is related to the thermodynamic interaction parameter χ_1 by

$$B' = \bar{v}_2^2 (1 - 2\chi_1)/V_1 N_A, \qquad (10.18)$$

and examination of Equation 10.10 shows that B' is also proportional to A_2. The unperturbed dimension can be estimated by plotting $[\eta]M^{-1/2}$ against $M^{1/2}$; K_θ is obtained from the intercept and $\langle \bar{r}^2 \rangle_0$ is calculated from Equation 10.14.

These extrapolation procedures all depend on the validity of the theoretical treatment, and reliability must be judged in this light. Fortunately, it has been demonstrated that most nonpolar polymers can be treated in this way and results agree well with direct measurements of $\langle \bar{r}^2 \rangle_0^{1/2}$. For more polar polymers, specific solvent effects become more pronounced, and extrapolations have to be regarded with corresponding caution.

10.5.5 INFLUENCE OF TACTICITY ON CHAIN DIMENSIONS

Studies of the dilute solution behavior of polymers with a specific stereostructure have revealed that the unperturbed dimensions may depend on the chain configuration.

This can be seen from the data in Table 10.1, where isotactic, syndiotactic, and atactic poly(methyl methacrylate) (PMMA) have different σ values. If the size of a polymer chain can be affected by its configuration, the microstructure must be well characterized before an accurate assessment of experimental data can be made. This can be achieved using nuclear magnetic resonance (NMR) and infrared techniques.

10.6 NUCLEAR MAGNETIC RESONANCE (NMR)

High-resolution NMR has proved to be a particularly useful tool in the study of the microstructure of polymers in solution, where the extensive molecular motion reduces the effect of long-range interactions and allows the short-range effects to dominate. Interpretation of chain tacticity, based on the work of Bovey and Tiers, can be illustrated using PMMA. The three possible steric configurations are shown in Figure 10.2, where R is the group –COOCH$_3$.

For the purposes of NMR measurements, three or more consecutive monomer units in a chain are considered to define a configuration and are called a *triad*. The term *heterotactic* is used now to define a triad that is neither isotatic nor syndiotactic. In the structures shown, the three equivalent protons of the α-methyl group absorb radiation at a single frequency, but this frequency will be different for each of the three kinds of triad because the environment of the α-methyl groups in each is different. For PMMA samples that were prepared under different conditions to give the three forms, resonances at $\tau = 8.78$, 8.95, and 9.09 were observed, which were assigned to the isotatic, heterotactic, and syndiotactic triads, respectively. Thus, in a sample with a mixture of configurations, a triple peak will be observed, and the area under each of these peaks will correspond to the amount of each triad present in the polymer chain. This is illustrated in Figure 10.3, where one sample is predominantly isotactic but also contains smaller percentages of the heterotactic and syndiotactic configurations.

The analysis can be carried further. The fraction of each configuration, P_i, P_h, and P_s, measured from the respective peak areas, can be related to p_m, which is the probability that a monomer adding on to the end of a growing chain will have the same configuration as the unit it is joining. This leads to the relations

$$P_i = p_m^2, P_s = (1 - p_m)^2 \quad \text{and} \quad P_h = 2 p_m (1 - p_m).$$

Curves plotted according to this simple analysis are shown in Figure 10.4, where they are compared with experimental data obtained for various tactic forms of poly(α-methyl styrene).

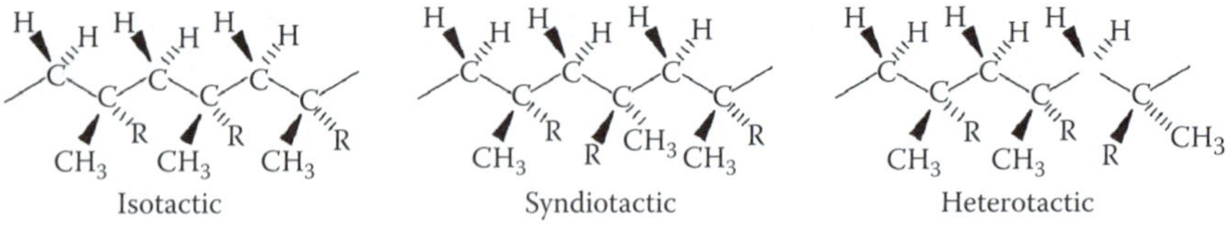

FIGURE 10.2 Stereoregular triads for poly(methyl methacrylate) (PMMA), where R = –COOCH$_3$.

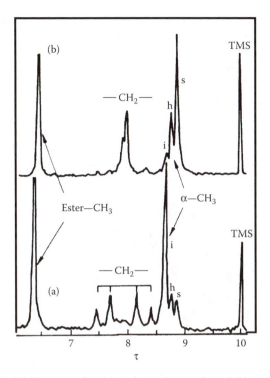

FIGURE 10.3 The NMR spectra for: (a) an isotactic sample and (b) a predominantly syndiotactic sample of poly(methyl methacrylate) (PMMA).

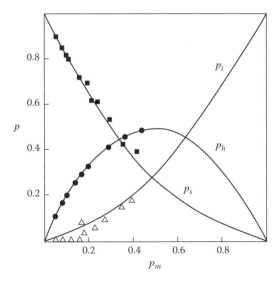

FIGURE 10.4 Theoretical curves for P as a function of p_m for each of the three configurations. Points represent experimental data for poly(α-methyl styrene) and illustrate the validity of the analysis. (From data by Brownstein et al., *Makromol. Chem.*, 58, 127, 1961. With permission.)

Differences in the microstructure of polydienes and copolymers can also be made using NMR. In the polydienes, the difference between 1,2- and 1,4-addition can be distinguished on examination of the resonance peaks corresponding to terminal olefinic protons, found at $\tau = 4.9$ to 5.0, and nonterminal olefinic protons observed at $\tau = 4.6$ to 4.7.

Not only is the local field acting on the nucleus altered by environment, it is also sensitive to molecular motion, and it has been observed that as the molecular motion within a sample increases, the resonance lines become narrower. Determination of the width, or second moment, of an NMR resonance line then provides a sensitive measure of low-frequency internal motions in solid polymers and can be used to study transitions and segmental rotations in the polymer sample. Line widths are also altered by the polymer crystallinity. Partially crystalline polymers present complex spectra as they are multiphase materials, in which the molecular motions are more restricted in the crystalline phase than in the amorphous phase. However, attempts to estimate percentage crystallinity in a sample using NMR have not been particularly successful. The method is illustrated in Figure 10.5 for poly(tetrafluoroethylene), where glass and other transitions are readily detected. Below 200 K the chains are virtually immobile, but above 200 K the lines sharpen as $-CF_2-$ rotation begins. This is associated with the glass transition, but the way the line width increases in this region is governed by sample crystallinity.

10.7 INFRARED SPECTROSCOPY

Infrared spectroscopy can be used to characterize long-chain polymers because the infrared active groups, present along the chain, absorb as if each was a localized group in a simple molecule. Identification of polymer samples can be made by making use of the "figureprint" region, where it is least likely for one polymer to exhibit exactly the same spectrum as another. This region lies within the range 6.67 to 12.50 μm.

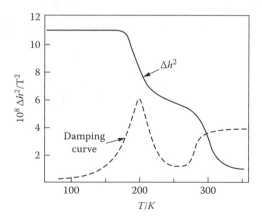

FIGURE 10.5 The NMR line width Δh (the unit is the tesla, T) as a function of temperature for poly(tetrafluoroethylene). The mechanical damping curve (-----) is included for comparison. (Adapted from Sauer, J.A. and Woodward, A.E., *Rev. Mod. Phys.*, 32, 88, 1960.)

In addition to identification, the technique has been used to elucidate certain aspects of polymer microstructure, such as branching, crystallinity, tacticity, and cis–trans isomerism. The relative proportions of *cis*-1,4-, *trans*-1,4-, and 1,2-addition in polybutadienes can be ascertained by making use of the differences in absorption between (CH) out-of-plane bending vibrations, which depend on the type of substitution at the olefinic bond. Terminal and internal groups can also be distinguished, as an absorption band at about 11.0 μm is characteristic of a vinyl group and indicates 1,2-addition. The *cis*-1,4-addition is characterized by an absorption band at about 13.6 μm, whereas the *trans*-1,4 configuration exhibits a band at about 10.4 μm. An estimate of cis–trans isomerism can be made by measuring the absorbance A of each band, where $A = \log_{10}(I_0/I)$, and I_0 and I are the intensities of the incident and transmitted radiation, respectively. If we assume that the 1,2 content is negligible, this is calculated by locating a baseline across the minima on either side of the absorption band, and the vertical height to the top of the band from the baseline is converted into a composition using the equation

$$P_{cis} = 3.65\, A_{cis}/(3.65\, A_{cis} + A_{trans}),$$

where P_{cis} is the fraction of cis configuration, A_{cis} is the absorbance at 13.6 μm, and A_{trans} is the absorbance at 10.4 μm. Polyisoprenes can also be analyzed in this way, only now the bands at 11.0 and 11.25 μm are used to estimate the 1,2- and 3,4-addition, whereas a band at 8.7 μm corresponds to the *trans*-1,4 linkage.

The infrared spectra of highly stereoregular polymers are distinguishable from those of their less regular counterparts, but many of the differences can be attributed to crystallinity rather than tacticity as such. The application of infrared to stereostructure determination in polymers is less reliable than NMR, but has achieved moderate success for PMMA and polypropylene. In PMMA, a methyl deformation at 7.25 μm is unaffected by microstructure, and comparison of this with a band at 9.40 μm, which is present only in atactic or syndiotactic polymers, allows an estimate of the syndiotacticity to be made from the ratio $\{A(9.40\ \mu m)/A(7.25\ \mu m)\}$. Similarly, $\{A(6.75\ \mu m)/A(7.25\ \mu m)\}$ provides a measure of the isotactic content. An alternative method is to calculate the quantity J as an average of the two equations

$$J_1 = 179\{A(9.40\ \mu m)/A(10.10\ \mu m)\} + 27$$

$$J_2 = 81.4\ \{A(6.75\ \mu m)/A(7.25\ \mu m)\} - 43$$

where the absorption band at 10.10 μm is now used. If J lies between 100 and 115, a highly syndiotactic polymer is indicated, and if between 25 and 30, the polymer is highly isotactic. For polypropylene, the characteristic band for the syndiotactic polymer appears at 11.53 μm, and the syndiotactic index I_s is $\{2A(11.53\ \mu m)/\{2.32\ \mu m\} + A(2.35\ \mu m)\}$. Values of I_s about 0.8 indicate highly syndiotactic samples. Spectra can be measured in a number of ways; for soluble polymers a film can be cast, perhaps even on the NaCl plate to be used, and examined directly. Measurements can also be made in solution, if the solvent absorption in any important region is low, or by a differential method.

10.8 THERMAL ANALYSIS

When a substance undergoes a physical or chemical change, a corresponding change in enthalpy is observed. This forms the basis of the technique known as differential thermal analysis (DTA), in which the change is detected by measuring the enthalpy difference between the material under study and an inert standard.

The sample is placed in a heating block and warmed at a uniform rate. The sample temperature is then monitored by means of a thermocouple and compared with the temperature of an inert reference, such as powdered alumina, or simply an empty sample pan, which is subjected to the same linear heating program. As the temperature of the block is raised at a constant rate (5 to 20 K min^{-1}), the sample temperature T_s and that of the reference, T_r, will keep pace until a change in the sample takes place. If the change is exothermic, T_s will exceed T_r for a short period, but if it is endothermic, T_s will temporarily lag behind T_r. This temperature difference ΔT is recorded and transmitted to a chart recorder, where changes such as melting or crystallization are recorded as peaks. A third type of change can be detected. Because the heat capacities of the sample and reference are different, ΔT is never actually zero, and a change in heat capacity, such as that associated with a glass transition, will cause a shift in the baseline. All three possibilities are shown in Figure 10.6 for quenched terylene.

Other changes such as sample decomposition, cross-linking, and the existence of polymorphic forms can also be detected. As ΔT measured in DTA is a function of the thermal conductivity and bulk density of the sample, it is nonquantitative and relatively uninformative. To overcome these drawbacks, an alternative procedure known as differential scanning calorimetry (DSC) is used. This technique retains the constant mean heat input, but instead of measuring the temperature difference during a change, a servo system immediately increases the energy input to either sample or reference to maintain both at the same temperature. The thermograms obtained are similar to DTA, but actually represent the amount of electrical energy supplied to the system, not ΔT, and so the areas under the peaks will be proportional to the change in enthalpy that occurred. An actual reference sample can be dispensed with in practice and an empty sample pan used instead. Calibration of the instrument will allow the heat capacity of a sample to be calculated in a quantitative manner. This information is additional to that gained on crystallization, melting, glass transitions, and decompositions.

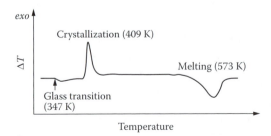

FIGURE 10.6 A DTA curve for quenched terylene showing the glass transition, melting endotherm, and a crystallization exotherm.

10.9 WIDE-ANGLE AND SMALL-ANGLE SCATTERING

Scattering techniques, such as light, x-ray, and neutron scattering, are extensively used to investigate various structural features of polymeric systems, including crystalline order and crystallinity, conformation, local structure, domain sizes, etc.

The basic features of any scattering experiment, whether it uses light, x-rays, or neutrons, are schematically represented in Figure 10.7. They consist of (1) a radiation source, (2) a system of collimation and/or monochromation, (3) the sample, and (4) a detection system. The incident radiation is characterized in terms of its wavelength, λ_0, wave vector k_0 with magnitude $2\pi/\lambda_0$, and energy E_0. In the case of elastic scattering, no energy change occurs during the process, but only a change in the direction of traveling of the wave, defined by the scattering angle, θ. Thus, for elastic scattering, the magnitude of the initial wave vector k_0 coincides with that of the final wave vector k. The wave vector change, $Q = k - k_0$, defines this change, and it is referred to as the *scattering vector* or the *momentum transfer*. For an elastic process, $Q = (4\pi/\lambda_0) \sin(\theta/2)$ (Figure 10.7).

Structural investigations are based on measurements of elastic scattering, and so the scattered intensity needs only be monitored as a function of the scattering angle. The scattering vector Q makes it possible to distinguish between wide- and small-angle regimes, through Bragg's law:

$$n\lambda = 2d \sin\left(\frac{\theta}{2}\right) \tag{10.19}$$

where d is the distance between two crystallographic planes and n is the order of reflection. By combining Equation 10.19 with the definition of Q, a relationship between structure in real space and scattering in Q-space is derived:

$$d = \frac{2\pi}{Q} \tag{10.20}$$

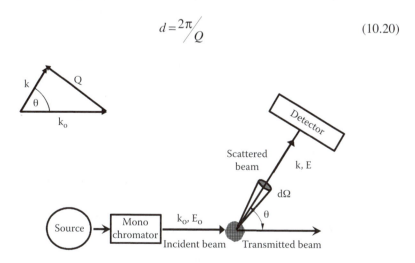

FIGURE 10.7 Schematic diagram of a scattering experiment.

TABLE 10.2
Characteristics of Different Types of Radiation

Parameter	Neutrons	X-Rays	Light
Scattering by	Nuclei	Electrons	Electrons
Availability	Large-scale facility	Laboratory and large-scale facility	Laboratory
Brilliance of source	Low	Medium to high	Medium to high
Wavelength (nm)	0.15–2.5	0.15	400–700
Distance scale	1–2000	0.1–2500	250–25000
Sample volumes (cm³)	0.05–3.5	0.0001–0.5	0.05–5

Equation 10.20 indicates that an inverse relationship exists between d and Q and so low Q values probe large distances characteristic of chain dimensions, lamellar spacing, and domain structures, whereas large Q values give information on the local polymer structure.

X-rays and neutron wavelengths are of the order of interatomic distances (Table 10.2), and small Q values can be accessed by performing measurements at small scattering angles. The Q range typically probed by conventional small-angle scattering instruments ranges between 10^{-3} to 10^{-1} Å^{-1}, in the scattering angle range $0.1°$ to $10°$. This corresponds to length scales of 1 to 2000 Å, giving access to polymer chain dimensions and to polymer aggregates. As shown in Table 10.2, this size range is comparatively smaller than that accessed by light scattering.

The wide-angle regime is defined by $0.6 < Q < 15$ Å^{-1}, and wide-angle scattering, probing length scales of a few angstroms, is used to investigate crystal structures, local conformation, and orientation.

In addition to the size range explored, scattering techniques differ in the way they interact with matter (Table 10.2). X-rays and light are scattered by electrons, whereas neutrons interact with the nuclei. As a result, although the scattering laws modeling the variation of the scattered intensity with scattering vector are common for all three types of radiation, there are important differences among the techniques.

10.9.1 Wide-Angle X-Ray Scattering

The extent of sample crystallinity can influence the behavior of a polymer sample greatly. A particularly effective way of examining partially crystalline polymers is by x-ray diffraction. The crystallites present in a powdered or unoriented polymer sample diffract x-ray beams from parallel planes for incident angles θ, which are determined by the Bragg equation (Equation 10.19). The reinforced waves reflected by all the small crystallites produce diffraction rings, or halos, which are sharply defined for highly crystalline materials and become increasingly diffuse when the amorphous content is high.

If the polymer sample is oriented, by drawing a fiber, or by applying tension to a film, the crystallites tend to become aligned in the direction of the stress, and the x-ray pattern is improved. In some samples of stereoregular or symmetrical polymers, the degree of three-dimensional ordering of the chains may be sufficiently high to allow a structural analysis of the polymer to be accomplished.

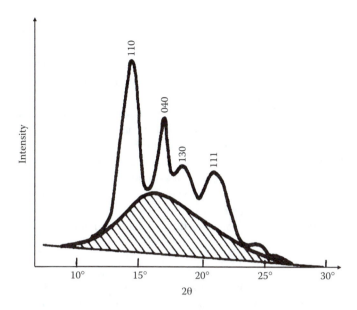

FIGURE 10.8 X-ray diffraction curves; the intensity I as a function of angle for totally amorphous polypropylene (shaded area), and for a sample with a 50% crystalline content.

Sample crystallinity can be estimated from the x-ray patterns by plotting the density of the scattered beam against the angle of incidence. If this can be done for an amorphous sample and a corresponding sample that is highly crystalline, a relative measure of crystallinity for other samples of the same polymer can be obtained. In Figure 10.8 the shaded portion is the amorphous polypropylene, whereas the maxima arise from the crystallites.

10.9.2 SMALL-ANGLE X-RAY SCATTERING (SAXS)

The theoretical outline presented for light-scattering studies in Chapter 9 is valid for electromagnetic radiation of all wavelengths. For x-rays, λ can be as low as 0.154 nm, and as this is much smaller than typical polymer dimensions, structural information over small distances should be available from x-ray scattering. However, as discussed earlier, when operating at small scattering angles of order $1°$, structural features at length scales larger than the separation between crystal planes can be accessed.

SAXS spectrometers are available both as laboratory equipment and as part of large-scale facility instrumentation. The former usually make use of the Cu K_α line with $\lambda = 1.54$ Å, whereas synchrotron sources offer a wavelength range from 0.6 to 3 Å.

The intensity of scattering can be modeled by the same equations developed for light scattering, provided that account is taken of the different matter–radiation interaction. For visible light, scattering is a result of differences in the refractive index of the solute and the solvent, and the optical constant $K^{\bullet}$ (Chapter 9, Equation 9.18) is proportional to $\tilde{n}_0^2\,(\mathrm{d}\tilde{n}/\mathrm{d}c)^2$. For SAXS, the scattered intensity is a function of the electron density, and the molar mass is then related to the excess electron density $\Delta\rho_e$ of solute over solvent. For $\lambda = 1.54$ Å, the Rayleigh ratio at $\theta = 0$, R_0 is:

$$R_0 = (4.8\,\text{cm}^{-1})M_w(\Delta\rho_e)^2 c \qquad (10.21)$$

Similar to light scattering, SAXS provides a measure of M_w, A_2, and the z-average radius of gyration, $\langle S^2 \rangle_z$, the main difference being the smaller size range that can be accessed using SAXS (Table 10.2).

SAXS is extensively used to investigate crystalline polymers and ordered arrangements in block copolymers. The complementarity between SAXS and WAXS is illustrated by the example in Figure 10.9, for measurements on a semicrystalline polymer. Here, the SAXS pattern yields information on the spacing between the regular stacks of lamellae from the peak maximum. Complementary to this, WAXS measurements offer a detailed characterization of the crystal structure and degree of crystallinity. Use of time-resolved SAXS makes it possible to follow the crystallization process through the variation of the lamellar spacing as a function of time.

10.9.3 SMALL-ANGLE NEUTRON SCATTERING (SANS)

Neutrons are produced at either reactor or spallation sources, and spectrometers for neutron-scattering experiments are therefore uniquely available at large-scale facilities. After production, the neutron beam is subjected to a process of moderation during which the neutron energy decreases through collisions with the moderator to

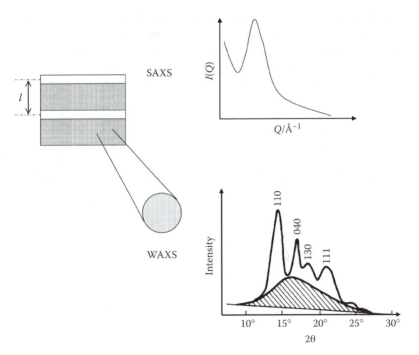

FIGURE 10.9 Schematic diagram of a SAXS–WAXS experiment performed on a semicrystalline polymer: SAXS provides a measure of lamellar spacing, whereas the crystal structure is derived from the WAXS measurements.

values of the order of $k_B T$, where T is the temperature of the moderator (k_B is Boltzmann's constant). For a water moderator at 323 K, the Maxwell–Boltzmann distribution of neutron velocities peaks at about 3000 m s^{-1}, which corresponds to $\lambda = 1.3$ Å, of the order of interatomic spacing. Similar to x-rays, neutrons can be used to probe polymer local structure, conformation, and aggregation.

Solely on the basis of probed length scale, x-rays would be the technique of choice because they are much cheaper and of higher beam intensity compared to the neutron. However, neutrons offer other advantages in that they are efficiently scattered by light elements such as protons, in contrast to x-ray radiation, to which light atoms are virtually transparent (x-rays are scattered by the electron clouds). This feature has played a major role in the development and success of neutron scattering for polymer investigations.

Neutrons interact with the nuclei. Two quantities characterize the nucleus–nuclear interaction: the neutron-scattering length, b, and the scattering cross-section, σ. The former, usually of the order 10^{-12} cm for all nuclei, defines the amplitude of the scattered wave with respect to that of the incident wave. The scattering cross-section represents the probability of a neutron being scattered somewhere in space, and it equals $4\pi b^2$. σ has dimensions of area, and it is usually reported in units of a barn (1 barn = 10^{-24} cm^2).

In contrast to x-rays, for which the scattering increases with increasing atomic number, the scattering length varies randomly across the periodic table. Of even greater significance is the fact that isotopes of the same element have different scattering lengths, as is the case for hydrogen and its heavier isotope, deuterium. Selected b values are reported in Table 10.3 for elements commonly encountered in polymeric structures.

SANS was the first technique to have been used extensively for polymer studies, and perhaps it is still the most widely employed. SANS instruments are available at various large-scale facilities, both reactor-based and pulsed sources. A common feature of these instruments is the presence of a collimated beam that is transmitted through the sample; the scattering is monitored by a large-area detector placed some distance away from the sample. Long wavelengths and large sample-to-detector distances afford small Q values, but being restricted to λ values below 10 Å, it is necessary to operate at distances of at least few meters. This causes a large decrease

---

TABLE 10.3
Coherent Scattering Lengths, b, for Selected Nuclei Commonly Encountered in Polymers

Nucleus	b (10^{-13} cm)
H	− 3.741
D	+ 6.671
C	+ 6.646
N	+ 9.362
O	+ 5.803

of the measured scattered intensity, limiting the smallest Q values that can be experimentally achieved (6×10^{-4} Å^{-1} on the D11 spectrometer at the Institute Laue Langevin, France).

Conventional SANS spectrometers make it possible to probe structures over a length scale from near to atomic (10 Å) dimensions to a few microns. The ultra-small-angle neutron-scattering (USANS) instruments that have recently become available have extended the experimental Q range down to 2×10^{-5} Å^{-1}, reaching distances up to about 30 µm.

The scattered intensity is defined in terms of the normalized differential scattering cross-section per unit volume, $d\Sigma(Q)/d\Omega$ (expressed in units of cm^{-1}), where $d\Sigma(Q)$ (neutrons s^{-1}) represents the number of neutrons scattered per second into a small solid angle, $d\Omega$. In general,

$$\frac{d\Sigma(Q)}{d\Omega} = NV^2 (\Delta\rho)^2 P(Q) S(Q) \tag{10.22}$$

where N is the number density of the scattering centers, V is their volume, and $(\Delta\rho)^2$ is the contrast factor defined as the square of the difference of the scattering length density of the scattering centers, e.g., the particle, and the matrix.

The scattering length density is calculated from

$$\rho = \sum b_i \frac{\delta N_A}{m} \tag{10.23}$$

by adding the coherent scattering lengths b_i of the nuclei building up the repeat unit (here, δ is the density, m is the mass of the repeat unit, and N_A is Avogadro's number).

The particle scattering factor $P(Q)$, or form factor, has already been introduced in Chapter 9 ($P(\theta)$; Section 9.7.1). It describes the average conformation of an individual polymer chain and model functions exist for a variety of particle shapes such as spheres, disks, or rodlike particles. For a random coil, it is expressed in terms of the Debye equation (Chapter 9, Equation 9.20).

The interparticle structure factor, $S(Q)$, is also a dimensionless function describing the modulation of the scattered intensity by interference effects between radiation scattered from different scattering units in the sample. It therefore gives information on the relative positions of the scattering objects.

The large difference between the scattering lengths of hydrogen ($b = -3.74 \times 10^{-13}$ cm) and deuterium ($b = 6.67 \times 10^{-13}$ cm) offers a means of manipulating the contrast, and it forms the basis of all conformational studies in bulk. The pioneering experiments carried out in the 1970s on atactic polystyrene (PS) and PMMA confirmed Flory's hypothesis that polymer dimensions in bulk are unperturbed and follow the expected $M_w^{1/2}$ dependence.

As shown in Figure 10.10 and Figure 10.11 for mixtures of unlabeled and deuterated PMMA, the scattered intensity vs. Q data can be displayed in a Zimm

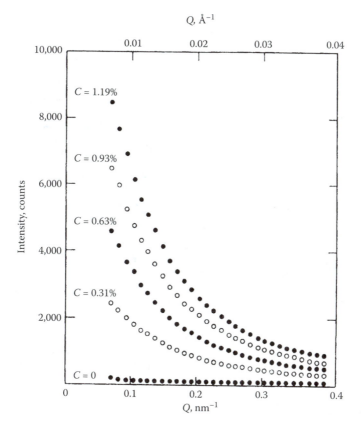

FIGURE 10.10 Intensity vs. Q plot of SANS data for various concentrations c of hydrogenous poly(methyl methacrylate) (PMMA) dissolved in deuterated PMMA. (From Kirste et al., *Polymer* 16, 120, 1975. With permission.)

fashion, similar to that used in light scattering. The double extrapolation in Figure 10.11 yields the radius of gyration from the $c = 0$ line and the second virial coefficient from the slope of the $Q = 0$ extrapolation, in this case equal to zero.

SANS is an expensive tool and, whenever possible, SAXS or light scattering should be used. It is the judicious use of deuterium labeling that usually justifies SANS measurements, giving access to structural features unavailable through other scattering methods or complementary to those.

10.10 MICROSCOPY

The term *microscopy* encompasses a wide range of techniques, which in their different forms make use of a microscope. The discussion in the following sections will include optical microscopy and higher-resolution techniques such as transmission electron microscopy (TEM).

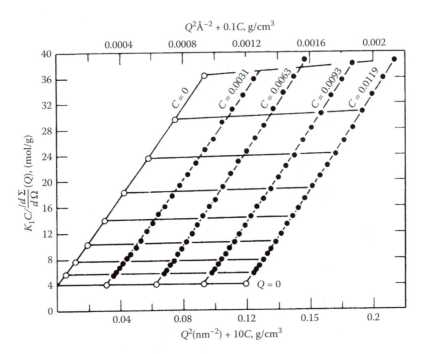

FIGURE 10.11 Zimm plot of SANS data from hydrogenous and deuterated poly(methyl methacrylate) (PMMA) samples. (From Kirste et al., *Polymer* 16, 120, 1975. With permission.)

10.10.1 OPTICAL MICROSCOPY

In its simple form, an optical microscope acts as a magnifying glass because it renders structural features visible that would otherwise be invisible to the naked eye. Optical microscopes may operate in either transmission or reflection mode.

When light passes through a specimen, changes in optical absorption and color may produce amplitude contrast. This technique is referred to as *bright-field illumination*. Transparent samples containing structural features that cause spatial variation in refractive index (i.e., where the phase of the transmitted light varies from place to place) would appear uniformly bright under conditions of bright-field illumination. These samples can be studied using *phase contrast microscopy*.

Optical microscopy is particularly useful for investigating various anisotropic features of polymeric systems. Polymers are intrinsically anisotropic objects and, when arranged in anisotropic structures such as crystalline and liquid crystalline phases, they display macroscopic optical anisotropy. These are investigated using a polarizing microscope, i.e., under crossed polarizers.

A polarizing microscope makes use of polarized light and requires (1) a polarizer to be placed between the light source and the sample and (2) an analyzer placed after the sample. The polarizer generates plane-polarized light when the polarizer and analyzer are at right angles to each other, i.e., when they are crossed, in absence of a sample, or when the sample is isotropic.

Molecules in anisotropic samples will be preferentially oriented along one direction, e.g., the nematic director for crystalline polymers or the fiber axis. The refractive index in the direction parallel to the orientation will differ from that in the perpendicular direction and, therefore, the sample is said to be birefringent. When polarized light passes through an anisotropic sample oriented in different directions, it is split into two components that are not in phase; these will combine to give *elliptically polarized light*. Thus, oriented samples such as crystalline, liquid crystalline polymers and fibers are visible under polarized light.

One of the main disadvantages of optical microscopy is its low resolution, which is of the order of half the wavelength of light.

10.10.2 SCANNING ELECTRON MICROSCOPY

The main advantage of electron microscopy is its much-improved resolution compared to that of light microscopy. In a scanning electron microscope (SEM), a finely focused electron beam is scanned across the specimen, and the scattered electrons emitted at each point are collected by appropriate detectors, forming an image. Figure 10.12 shows the range of events that take place when the incident electron beam strikes the specimen. Backscattered electrons are high-energy electrons that have been elastically scattered by interaction of the incident beam with the nucleus. The energy of these electrons is comparable to that of the incident beam.

SEM images may be formed by using either secondary or backscattered electrons, and both methods offer advantages and disadvantages. For example, images of more uniform illumination are formed when using backscattered electrons; these images do not show the bright areas at corners that are characteristic of the secondary

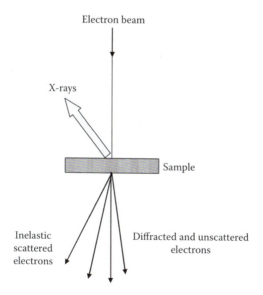

FIGURE 10.12 Illustration of the types of radiation produced following interaction of an electron beam with a specimen.

electron image. However, high-energy electrons such as those in the backscattered beam originate from deeper parts of the sample and have larger interaction volume compared to the secondary beam. As a consequence, images formed from backscattered electrons are of lower resolution compared to those resulting from secondary electrons.

When the incident electrons strike the electrons in the outer shell of the atom, some of these latter electrons are ejected and x-rays are produced. These electrons have lower energy compared to the incident electron beam and are called *secondary electrons*. Due to their low energy, those that reach the detector are generated from the sample's thin top layer. As shown in Figure 10.12, the interaction volume is small and comparable to the irradiated area by the incident beam. This affords higher resolution.

SEM instruments have resolutions better than 5 nm and are useful for the characterization of surfaces and the determination of surface topography. Contrary to transmission electron microscopy (TEM), little sample preparation is required. For polymers, which are poor conductors, problems associated with charge buildup need to be overcome.

Figure 10.13(a) shows an SEM micrograph of a 4-methylphenyl urethane of cellulose (4-MPC) sample photopolymerized from an *N*-vinyl pyrrolidinone solution (43.8 wt%). At polymer concentrations above 35 wt%, 4-MPC forms a cholesteric mesophase (Chapter 11). The SEM image shows period striations with repeat distance equal to 140 ± 30 nm, a value that is comparable to that of half the pitch, P/2, of the cholesteric mesophase.

In addition to detecting backscattered and secondary electrons, SEM instruments offer information on the sample elemental composition when using x-ray detectors. It has been pointed out earlier that the ejection of electrons from an atom is accompanied by emission of x-rays. The x-ray spectrum that is produced is a characteristic feature of any given element, and by measuring the energy or the wavelength of the x-rays that are produced, its identification is possible.

10.10.3 TRANSMISSION ELECTRON MICROSCOPY

Transmission electron microscopy (TEM) is used to analyze very thin samples (less than 100 nm thick), provided that the specimen has structural features that scatter electrons in different amounts. One of the main problems with using TEM for polymers is that these are made up of low-atomic-number elements, which are low scatterers of electrons. Low scattering and the fact that there is little spatial variation in electron density in a polymer sample leads to poor contrast. One way to overcome this problem is to stain the specimen with a heavy metal, such as uranium or osmium, that preferentially attaches itself to certain regions of the sample.

Different techniques can be used to enhance the contrast. For example, dark-field images can be produced by blocking the transmitted electrons and using only the scattered electrons.

10.10.4 ATOMIC FORCE MICROSCOPY AND SCANNING TUNNELING MICROSCOPY

Atomic force microscopy (AFM) and scanning tunneling microscopy (STM) offer a means of obtaining three-dimensional images of polymer surfaces. These techniques

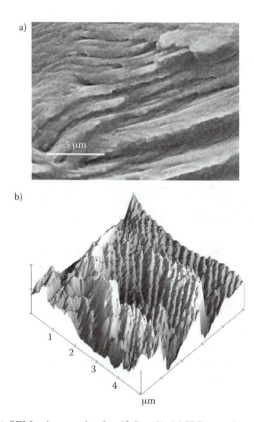

FIGURE 10.13 (a) SEM micrograph of a 43.8 wt% 4-MPC sample polymerized for 12 h and fractured normal to the composite film surface (magnification 4000 ×), (b) AFM image of the same sample fractured normal to the composite film surface (4.5 × 4.5 μm). (From Arrighi, V., Cowie, J.M.G., Vaqueiro, P., and Prior, K.A., Fine structure and optical properties of cholesteric films prepared from cellulose 4-methylphenyl urethane/N-vinyl pyrrolidinone solutions, *Macromolecules*, 35, 7354, 2002. With permission.)

offer advantages compared to electron microscopy because no sample preparation is required such as coating or microtoming.

STM and AFM are scanned-proximity probe microscopy techniques that measure a local property, using a probe or "tip" placed close to the sample. Whereas the STM probe is placed a few angstroms above the surface to measure a tunneling current between the sample and the tip of the probe, the AFM probe is in contact with the sample's surface.

STM is limited to studies of conducting or metal-coating polymers. Because most polymeric materials are insulating, AFM is of more general use in polymer surface studies.

AFM measures attractive or repulsive forces between a probe and the sample. In the repulsive "contact" mode, the tip of the probe is attached to a cantilever. Vertical deflections of the cantilever caused by surface variations are monitored as a raster scan drags the tip over the sample. This imaging mode is used to study "hard" samples and provides images with angstrom resolution. If the sample is "soft," a noncontact mode is preferred.

AFM is widely used in the analysis of polymer surfaces, such as morphology and molecular structure of crystalline and oriented polymers, block copolymers, and polymer blends. The example shown in Figure 10.13(b) is the AFM three-dimensional surface image of the fracture surface of a composite. A lamellar structure is clearly observed, with periodicity of about 200 nm, comparable to values obtained from the SEM micrographs [Figure 10.13(a)].

PROBLEMS

1. Investigations of the hydrodynamic properties of amylose indicate that the chain is sterically hindered. The model proposed by Eliezer and Hayman has the pyranose ring in the chair form and the oxygen valence angle is set at $100°32'$.

   a. If the root mean square end-to-end distance of an amylose chain with 4882 units in DMSO is $\langle \bar{r}^2 \rangle^{1/2} = 1001$ Å and the length of a glucose unit is 5.15 Å, determine the steric parameter σ (the angle θ in Equation 10.7 is the supplement of the oxygen valence angle = $79°28'$).

   b. How much larger is the real chain compared to the one predicted for free rotation?

2. Values of the second virial coefficient derived from osmotic pressure measurements (Krigbaum, 1954) for a polystyrene fraction of molecular weight 125,000 in cyclohexane are:

T/K	$A_2 \times 10^5$
303	−2.45
313	5.78
323	11.63

   a. Determine the Θ-temperature.

   b. Calculate $\Psi_1 F(x)$ and then evaluate the entropy parameter, knowing that the molar volume of the solvent is equal to 109.5, 110.9, and 112.3 cm^3 mol^{-1} and the partial specific volumes of the polymer are 0.930, 0.935, and 0.940 cm^3 g^{-1} at 303, 313, and 323 K, respectively.

3. Styrene-2-vinylpyridine diblock copolymer samples prepared by anionic polymerization were characterized using transmission electron microscopy (TEM) and small-angle x-ray scattering (SAXS) (Matsushita et al., 1990). Using the least-squares method, the following empirical equation was derived for the lamellar domain spacing as a function of molecular weight, in the range 38,000 to 740,000:

$$D = 0.33 \, M_n^{\,0.6 \pm 0.03}$$

Determine the Q value at which a maximum for the first-order reflection is expected to appear in the SAXS pattern of samples with $M_n = 38,000$ and $M_n = 739,000$.

4. Calculate the scattering length and scattering length densities of hydrogenated and perdeuterated polystyrene (density = 1.05 g cm^{-3}) and then evaluate the neutron contrast.
5. In the electron microscope, electrons are generated by a cathode filament and then accelerated through a voltage varying between 40 and 120 kV. For an accelerating voltage of 100 kV, calculate the wavelength of the electrons.

REFERENCES

Arrighi, V., Cowie, J.M.G., Vaqueiro, P., and Prior, K.A., Fine structure and optical properties of cholesteric films prepared from cellulose 4-methylphenyl urethane/N-vinyl pyrrolidinone solutions, *Macromolecules,* 35, 7354, 2002.
Brownstein, S., Bywater, S., and Worsfold, D.J., *Makromol. Chem.,* 58, 127, 1961.
Krigbaum, W.R., *J. Am. Chem. Soc.,* 76, 3758, 1954.
Matsushita, Y., Mori, K., Saguchi, R., Nakao, Y., Noda, I., and Nagasawa, M., *Macromolecules,* 23, 4393, 1990.
Sauer, J.A. and Woodward, A.E., *Rev. Mod. Phys.,* 32, 88, 1960.

BIBLIOGRAPHY

Allen, G. and Bevington, J.C., Eds., *Comprehensive Polymer Science,* Vol. 1, 2, Pergamon Press, 1989.
Balta-Calleja, F.J. and Vonk, C.G., *X-Ray Scattering of Synthetic Polymers,* Elsevier Science, 1989.
Bovey, F.A. and Mirau, P., *NMR of Polymers,* Academic Press, 1996.
Elliott, A., *Infrared Spectra and Structure of Organic Long Chain Polymers,* Edward Arnold, 1969.
Flory, P.J., *Principles of Polymer Chemistry,* Cornell University Press, 1953, chap. 10, 14.
Flory, P.J., *Statistical Mechanics of Chain Molecules,* Interscience Publishers, 1969.
de Gennes, P.G., *Scaling Concepts in Polymer Physics,* Cornell University Press, 1979.
Higgins, J.S. and Benoit, H.C., *Polymers and Neutron Scattering,* Vol. 8, Claredon Press, 1994.
Hunt, B.J. and James, M.I., Eds., *Polymer Characterisation,* Blackie Academic and Professional, 1993.
King, S.M., Small-angle neutron scattering, in *Modern Techniques for Polymer Characterisation,* Pethrick, R.A. and Dawkins, J.V., Eds., John Wiley and Sons, 1999.
Koenig, J.L., *Chemical Microstructure of Polymer Chains,* John Wiley and Sons, 1980.
Mitchell, J., Ed., *Applied Polymer Analysis and Characterization: Recent Developments in Techniques. Instrumentation and Problem Solving,* Hanser Publishers, 1987.
Painter, P.C., Coleman, M.M., and Koenig, J., *The Theory of Vibrational Spectroscopy and its Application to Polymeric Materials,* John Wiley and Sons, 1982.
Rabek, J.C., *Experimental Methods in Polymer Chemistry,* John Wiley and Sons, 1980.
Turi, E.A., *Thermal Characterization of Polymeric Materials,* Academic Press, 1981.

11 The Crystalline State and Partially Ordered Structures

11.1 INTRODUCTION

When polymers are irradiated by a beam of x-rays, scattering produces diffuse halos on the photographic plate for some polymers, whereas for others a series of sharply defined rings superimposed on a diffuse background is recorded. The former are characteristic of amorphous polymers, and illustrate that a limited degree of short-range order exists in most polymeric solids. The latter patterns are indicative of considerable three-dimensional order and are typical of polycrystalline samples containing a large number of unoriented crystallites associated with amorphous regions. The rings are observed to sharpen into arcs or discrete spots if the polymer is drawn or stretched, a process that orients the axes of the crystallites in one direction.

The occurrence of significant crystallinity in a polymer sample is of considerable consequence to a materials scientist. The properties of the sample — the density, optical clarity, modulus, and general mechanical response — all change dramatically when crystallites are present, and the polymer is no longer subject to the rules of linear viscoelasticity, which apply to amorphous polymers as outlined in Chapter 13. However, a polymer sample is rarely completely crystalline, and the properties also depend on the amount of crystalline order.

It is important then to examine crystallinity in polymers and determine the factors that control the extent of crystallinity.

Polymers forming partially ordered phases (polymer liquid crystals) between the solid and the liquid phase have become increasingly important, and these are also reviewed in this chapter, along with the self-ordered structures formed by many block copolymers.

11.2 MECHANISM OF CRYSTALLIZATION

A polymer in very dilute solution can be effectively regarded as an isolated chain whose shape is governed by short- and long-range inter- and intramolecular interactions. In the aggregated state this is no longer true; the behavior of the chain is now influenced largely by the proximity of the neighboring chains and the secondary valence forces that act between them. These factors determine the orientation of chains relative to each other in the undiluted state, and this is essentially an interplay between the entropy and internal energy of the system, which is expressed in the usual thermodynamic form:

$$G = (U + pV) - TS.$$

In the melt, polymers normally attain a state of maximum entropy consistent with a stable state of minimum free energy. Crystallization is a process involving the orderly arrangement of chains and is consequently associated with a large negative entropy of activation. If a favorable free-energy change is to be obtained for crystallite formation, the entropy term has to be offset by a large negative energy contribution.

The alignment of polymer chains at specific distances from one another so as to form crystalline nuclei will be assisted when intermolecular forces are strong. The greater this interaction between chains, the more favorable will be the energy parameter, and this provides some indication of the type of chain that might be expected to crystallize from the melt, i.e.:

1. Symmetrical chains, which allow the regular close packing required for crystallite formation.
2. Chains possessing groups, which encourage strong intermolecular attraction, thereby stabilizing the alignment.

In addition to the thermodynamic requirements, kinetic factors relating to the flexibility and mobility of a chain in the melt must also be considered. Thus, polyisobutylene $+CH_2C(CH_3)_2+_n$ might be expected to crystallize because the chain is symmetrical, but it will only do so if maintained at an optimum temperature for several months. This is presumably a result of the flexibility of the chain, which allows extensive convolution, thereby impeding stabilization of the required long-range alignment.

The creation of a three-dimensional ordered phase from a disordered state is a two-stage process. Just above its melting temperature, a polymer behaves similar to a highly viscous liquid in which the chains are all tangled up with their neighbors. Each chain pervades a given volume in the sample, but as the temperature decreases, the volume available to the molecule also decreases. This in turn restricts the number of disordered conformational states available to the chain because of the constraining influence of intramolecular interactions among chains in juxtaposition. As a result, there is an increasing tendency for the polymer to assume an ordered conformation in which the chain bonds are in the rotational states of lowest energy. However, various other factors will tend to oppose crystallization: chain entanglements will hinder the diffusion of chains into suitable orientations, and if the temperature is above the melting temperature, thermal motions will be sufficient to disrupt the potential nuclei before significant growth can take place. This restricts crystallization to a range of temperatures between T_g and T_m.

The first step in crystallite formation is the creation of a stable nucleus brought about by the ordering of chains in a parallel array, stimulated by intramolecular forces, followed by the stabilization of long-range order by the secondary valence forces, which aid the packing of molecules into a three-dimensional ordered structure.

The second stage is the growth of the crystalline region, the size of which is governed by the rate of addition of other chains to the nucleus. As this growth is counteracted by thermal redispersion of the chains at the crystal–melt interface, the temperature must be low enough to ensure that this disordering process is minimal.

11.3 TEMPERATURE AND GROWTH RATE

Measurable rates of crystallization occur between $(T_m - 10\ K)$ and $(T_g + 30\ K)$, a range in which the thermal motion of the polymer chains is conducive to the formation of stable ordered regions. The growth rate of crystalline areas passes through a maximum in this range, as illustrated in Figure 11.1 for isotactic polystyrene. Close to T_m, the segmental motion is too great to allow many stable nuclei to form, whereas near T_g the melt is so viscous that molecular motion is extremely slow.

As the temperature drops from T_m, the melt viscosity, which is a function of the molar mass, increases and the diffusion rate decreases, thereby giving the chains greater opportunity to rearrange themselves to form a nucleus. This means that there will exist an optimum temperature of crystallization, which not only depends largely on the interval between T_m and T_g, but also on the molar mass of the sample.

The melt usually has to be supercooled by about 5 to 20 K before a significant number of nuclei appear that possess the critical dimensions required for stability and further growth. If a nucleating agent is added to the system, crystallization can be induced at higher temperatures. This is known as *heterogeneous nucleation* and only affects the crystallization rate, not the spherulitic growth rate, at a given temperature.

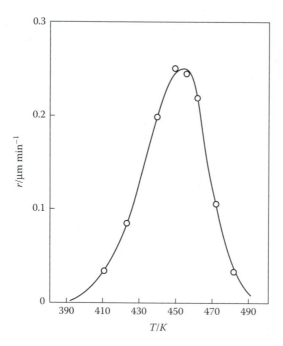

FIGURE 11.1 Radial growth rate r of spherulites of isotactic polystyrene as a function of the crystallization temperature.

11.4 MELTING

The melting of a perfectly crystalline substance is an equilibrium process characterized by a marked volume change and a well-defined melting temperature. Polymers are never perfectly crystalline but contain disordered regions and crystallites of varying size. The process is normally incomplete because crystallization takes place when the polymer is a viscous liquid. In this state, the chains are highly entangled, and as sufficient time must be allowed for the chains to diffuse into the three-dimensional order required for crystallite formation, the crystalline perfection of the sample is affected by the thermal history. Thus, rapid cooling from the melt usually prevents the development of significant crystallinity. The result is that melting takes place over a range of temperatures, and this range is a useful indication of sample crystallinity.

11.4.1 Effect of Crystallite Size on Melting

The range of temperature over which a polymer melts is indicative of the size and perfection of the crystallites in the sample. This is illustrated in a study of the melting of natural rubber samples, which has shown that the melting range is a function of the temperature of crystallization. At low crystallization temperatures, the nucleation density in the rubber melt is high, segmental diffusion rates are low, and small imperfect crystalline regions are formed. Thus, broad melting ranges are measured for samples crystallized at these lower temperatures, and these become narrower as the crystallization temperature increases.

This suggests that careful annealing at the appropriate temperature could produce samples with a high degree of crystallinity. These samples might then exhibit almost perfect first-order phase changes at the melting temperature. A close approximation to these conditions has been attained by Mandelkern (1949), who annealed a linear polyethylene for 40 d. The improvement in the crystalline organization is obvious from examination of the resulting fusion curves in Figure 11.2, where the variation of specific volume with temperature for this sample is compared with that for a branched polyethylene of low crystallinity. The effect of branching is to decrease the percentage crystallinity, broaden the melting range, and reduce the temperatures at which the largest crystallites disappear, which are regarded as the respective melting temperatures T_m for the samples.

The effect of crystal size on T_m is shown more clearly in Figure 11.3. The small crystals melt about 30 K lower than the large ones owing to the greater contribution from the interfacial free energy in the smaller crystallites; i.e., there is an excess of free energy associated with the disordered chains emerging from the ends of ordered crystallites, and this is relatively greater for the small crystallites, resulting in lower melting temperatures.

11.5 THERMODYNAMIC PARAMETERS

Even with carefully annealed specimens, it is thought that the equilibrium melting temperature of the completely crystalline polymer T_m° is never actually attained. The

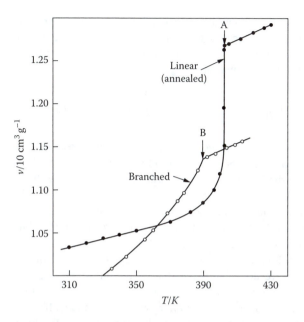

FIGURE 11.2 Specific volume v plotted against temperature T for a sample of linear polyethylene annealed for 40 d and a branched sample. Points A and B are the respective melting temperatures. (From data by Mandelkern, L., *Rubber Chem. Technol.*, 32, 1392, 1949.)

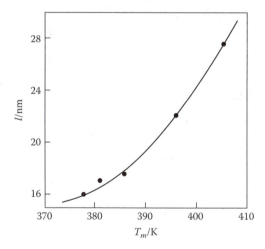

FIGURE 11.3 Dependence of T_m on the length l of the crystallite.

temperature T_m° is related to the change in enthalpy ΔH_u and the entropy change ΔS_u, for the first-order melting transition of pure crystalline polymer to pure amorphous melt, by

$$T_m^\circ = \Delta H_u / \Delta S_u. \tag{11.1}$$

The enthalpy change can be estimated by adding varying quantities of a diluent to the polymer, which serves to depress the observed melting temperature, and measuring T_m for each polymer and diluent mixture. The results are then plotted according to the Flory equation:

$$(1/\phi_1)(1/T_m - 1/T_m^\circ) = (R/\Delta H_u)(V_u/V_1)(1 - BV_1\phi_1/RT_m), \qquad (11.2)$$

where (V_u/V_1) is the ratio of the molar volume of the repeating unit in the chain to that of the diluent, and ϕ_1 is the volume fraction of the diluent. The factor (BV_1/RT_m) is equivalent to the Flory interaction parameter χ_1, indicating that Equation 11.2 is dependent on the polymer–diluent interaction. For practical purposes, T_m° is taken to be the melting temperature of the undiluted polymer irrespective of the crystalline content. Typical values obtained in this manner are shown in Table 11.1.

In many cases, the entropy change is the most important influence on the magnitude of the melting temperature of a polymer. A large part of this entropy is due to the additional freedom that allows the chain conformational changes to occur in the melt; i.e., the restriction of the crystalline lattice no longer applies. In the crystalline phase, the chain bonds are in their lowest energy state. If the energy difference between the rotational states $\Delta\varepsilon$ is low, the population of the higher-energy states will increase in the melt, and considerable flexing of the chain is achieved. The contribution of ΔS_u is then high. When $\Delta\varepsilon$ is large, the tendency to populate the high-energy states is not too great; consequently, the chain is less flexible, and ΔS_u is lower. Two polymers that exist in the all-trans state in the crystal are polyethylene and poly(tetrafluoroethylene). For polyethylene $\Delta\varepsilon$ is about 3.0 kJ mol^{-1}, but it is as

---

TABLE 11.1
Thermodynamic Parameters Derived from Melting

Polymer	T_mK	ΔH_u (J mol^{-1})	ΔS_u (J K^{-1} mol^{-1})	Basic Unit
Polyethylene	410	3,970	9.70	$-(CH_2)-$
Poly(tetrafluoroethylene)	645	2,860	4.76	$-(CF_2)-$
cis-1,4-Polyisoprene	301	4,400	14.60	$\left.\begin{array}{c} \\ \\ \end{array}\right\}$ $-(CH_2-C=CH-CH_2)-$ $\overset{|}{CH_3}$
trans-1,4-Polyisoprene	347	12,700	36.90	
Polypropylene	447	10,880	24.40	$-(CH_2-CH)-$ $\overset{|}{CH_3}$
Poly(decamethylene terephthalate)	411	46,000	114.00	$-\left[(CH_2)_{10}-O-\overset{O}{\overset{\|}{C}}-\bigcirc-\overset{O}{\overset{\|}{C}}-O\right]-$

Note: The quantities refer to unit amount of basic unit shown.

high as 18.0 kJ mol^{-1} for poly(tetrafluoroethylene). Hence, the polyethylene chain is much more flexible in the melt and gains considerably more entropy on melting, so that T_m is correspondingly lower.

11.6 CRYSTALLINE ARRANGEMENT OF POLYMERS

The formation of stable crystalline regions in a polymer requires that (1) an economical closely packed arrangement of the chains be achieved in three dimensions and that (2) a favorable change in internal energy be obtained during this process. This imposes restrictions on the type of chain that can be crystallized with ease and, as mentioned earlier, one would expect symmetrical linear chains such as polyesters, polyamides, and polyethylene to crystallize most readily.

11.6.1 Factors Affecting Crystallinity and T_m

These can be dealt with under the following general headings: symmetry, intermolecular bonding, tacticity, branching, and molar mass.

11.6.1.1 Symmetry

The symmetry of the chain shape influences both T_m and the ability to form crystallites. Polyethylene and poly(tetrafluoroethylene) are both sufficiently symmetrical to be considered smooth stiff cylindrical rods. In the crystal, these rods tend to roll over each other and change position when thermally agitated. This motion within the crystal lattice, called *premelting*, increases the entropy of the crystal and effectively stabilizes it. Consequently, more thermal energy is required before the crystal becomes unstable, and T_m is increased. Flat or irregularly shaped polymers, with bends and bumps in the chain, cannot move in this way without disrupting the crystal lattice, and so have lower T_m values. This is only one aspect.

For crystallite formation in a polymer, easy close-packing of the chains in a regular three-dimensional fashion is required. Again, linear symmetrical molecules are best. Polyethylene, poly(tetrafluoroethylene), and other chains with more complex backbones containing $+O+$, $+COO+$, and $+CONH+$ groups all possess a suitable symmetry for crystallite formation, and usually assume extended zigzag conformations when aligned in the lattice.

Chains containing irregular units, which detract from the linear geometry, reduce the ability of a polymer to crystallize. Thus, cis-double bonds (I), *o*- and *m*-phenylene groups (II), or cis-oriented puckered rings (III) all encourage bending and twisting in the chains and make regular close-packing very difficult. If, however, the phenylene rings are para-oriented, the chains retain their axial symmetry and can crystallize more readily. Similarly, incorporation of a trans-double bond maintains the chain symmetry. This is highlighted when comparing the amorphous elastomeric *cis*-polyisoprene with

I   II   III

the highly crystalline *trans*-polyisoprene, which has no virtue as an elastomer, or *cis*-poly(1,3-butadiene) ($T_m = 262$ K) with *trans*-poly(1,3-butadiene) ($T_m = 421$ K).

11.6.1.2 Intermolecular Bonding

In polyethylene crystallites, the close packing achieved by the chains allows the van der Waals forces to act cooperatively and provide additional stability to the crystallite. Any interaction between chains in the crystal lattice will help to hold the structure together more firmly and raise the melting temperature. Polymers containing polar groups, e.g., Cl, CN, or OH, can be held rigid, and aligned, in a polymer matrix by the strong dipole–dipole interactions between the substituents, but the effect is most obvious in the symmetrical polyamides. These polymers can form intermolecular hydrogen bonds, which greatly enhance crystallite stability. This is illustrated in Figure 11.4 for nylon-6,6, where the extended zigzag conformation is ideally suited to allow regular intermolecular hydrogen bonding. The increased stability is reflected in T_m, which for nylon-6,6 is 540 K compared to 410 K for polyethylene.

The structures of related polyamides do not always lead to this neat arrangement of intermolecular bond formation; for example, the geometry of an extended nylon-7,7 chain allows the formation of only every second possible hydrogen bond when the chains are aligned and fully extended. However, the process is so favorable energetically that sufficient deformation of the chain takes place to enable formation of all possible hydrogen bonds. The added stability that these bonds impart to the crystallite far outweighs the limited loss of energy caused by chain flexing.

Secondary bonds can therefore lead to a stimulation of the crystallization process in the appropriate polymers.

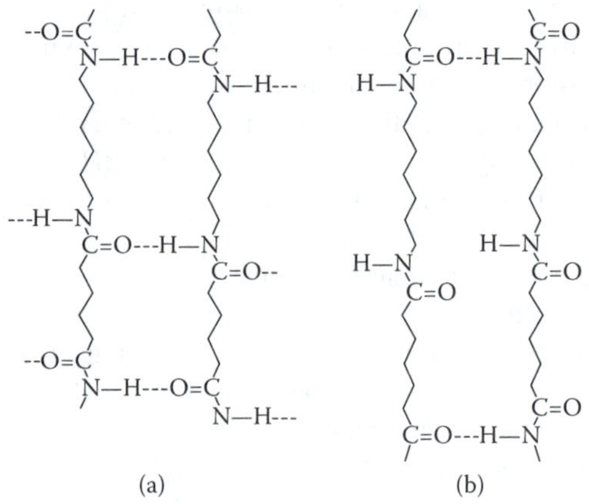

(a)   (b)

FIGURE 11.4 Extended zigzag structures for (a) nylon-6,6 and (b) nylon-7,7 showing the allowed hydrogen bonding.

11.6.1.3 Tacticity

Chain symmetry and flexibility both affect the crystallinity of a polymer sample. If a chain possesses large pendant groups, these will not only increase the rigidity but also increase the difficulty of close packing and, hence, the formation of a crystalline array. This latter problem can be overcome if the groups are arranged in a regular fashion along the chain. Isotactic polymers tend to form helices to accommodate the substituents in the most stable steric positions; these helices are regular forms capable of regular alignment. Thus, atactic polystyrene is amorphous, but isotactic polystyrene is semicrystalline ($T_m = 513$ K).

Syndiotactic polymers are also sufficiently regular to crystallize, not necessarily as a helix but rather as glide planes.

Branching in the side group tends to stiffen the chain and raise T_m, as shown in the series poly(but-1-ene), $T_m = 399$ K; poly(3-methyl but-1-ene), $T_m = 418$ K; and poly(3,3'-dimethyl but-1-ene), $T_m > 593$ K. If the side group is flexible and nonpolar, T_m is lowered.

11.6.1.4 Branching and Molar Mass

If the chain is substantially branched, the packing efficiency deteriorates and the crystalline content is lowered. Polyethylene provides a good example (Figure 11.2), where extensive branching lowers the density and T_m of the polymer.

Molar mass can also alter T_m. Chain ends are relatively free to move, and if the number of chain ends is increased by reducing the molar mass, then T_m is lowered because of the decrease in energy required to stimulate chain motion and melting. For example, polypropylene, with $M = 2000$ g mol^{-1}, has $T_m = 387$ K, whereas a sample with $M = 30,000$ g mol^{-1} has $T_m = 443$ K.

11.7 MORPHOLOGY AND KINETICS

Having once established that certain polymeric materials are capable of crystallizing, fundamental studies are directed along two main channels of interest centered on: (1) the mode and kinetics of crystallization and (2) the morphology of the sample on completion of the process.

Although the morphology depends largely on the crystallizing conditions, we shall consider the macro- and microscopic structure first, before dealing with the kinetics of formation.

11.8 MORPHOLOGY

A number of distinct morphological units have been identified during the crystallization of polymers from the melt, which have helped to clarify the mechanism. We shall now discuss the ordered forms that have been identified.

11.8.1 CRYSTALLITES

In an x-ray pattern produced by a semicrystalline polymer, the discrete maxima observed arise from the scattering by small regions of three-dimensional order, which are called crystallites. They are formed in the melt by diffusion of molecules, or sections of molecules, into close-packed ordered arrays; these then crystallize. The sizes of these crystallites are small relative to the length of a fully extended polymer chain, but they are also found to be independent of the molar mass, and rarely exceed 1 to 100 nm. As a result, various portions of one chain may become incorporated in more than one crystallite during growth, thereby imposing a strain on the polymer, which retards the process of crystallite formation. This will also introduce imperfections in the crystallites that continue to grow until the strains imposed by the surrounding crystallites eventually stop further enlargement. Thus, a matrix of ordered regions with disordered interfacial areas is formed, but unlike materials with small molar masses, the ordered and disordered regions are not discrete entities and cannot be separated by differential solution techniques unless the solvent causes selective degradation of the primary bonds in the amorphous regions.

Crystallites of cellulose have been isolated from wood pulp in this way by treatment with acid to hydrolyze and remove the amorphous regions. Typical dimensions of the remaining crystallites were 46 nm long by 7.3 nm wide, corresponding to bundles of about 100 to 150 chains in each crystallite.

The first attempts to explain the crystalline structure of a polymer sample produced a model called the fringe-micelle structure. The chain was envisaged as meandering throughout the system, entering and leaving several ordered regions along its length. The whole structure was thus made up of crystalline regions embedded randomly in a continuous amorphous matrix. This model has now been virtually discarded in the light of more recent research, which has revealed features incompatible with this picture.

11.8.2 SINGLE CRYSTALS

When a polymer is crystallized from the melt, imperfect polycrystalline aggregations are formed in association with substantial amorphous content. This is a consequence of chain entanglement, the high viscosity of the melt combining to hinder the diffusion of chains into the ordered arrays necessary for crystallite formation.

If these restrictions on free movement are reduced and a polymer is allowed to crystallize from a dilute solution, it is possible to obtain well-defined single crystals. By working with solutions in which the amount of polymer is considerably less than 0.1%, the chance of a chain being incorporated in more than one crystal is greatly reduced, thereby increasing the possibility of isolated single crystals being formed.

These crystals are usually very small, but they have been detected for a range of polymers, including polyesters, polyamides, polyethylene, cellulose acetate, and poly(4-methyl pentene-1). Although small, these single crystals can be studied using an electron microscope. This reveals that they are made up of thin lamellae, often lozenge-shaped, sometimes oval, about 10 to 20 nm thick, depending on the temperature of crystallization. The most surprising feature of these lamellae is that although the molecular chains may be as long as 1000 nm, the direction of the chain axis is across the thickness of the platelet. This means that the chain must be folded many times like a concertina to be accommodated in the crystal.

For a polymer such as polyethylene, the fold in the chain is completed using only three or four monomer units with bonds in the gauche conformation. The extended portions in between have about 40 monomer units all in the trans conformation.

The crystals thus formed have a hollow pyramid shape because of the requirement that the chain folding must involve a staggering of the chains if the most efficient packing is to be achieved. There is also a remarkable constancy of lamellar thickness, but this increases as the temperature increases. Although opinions vary between kinetic and thermodynamic reasons for this constancy of fold distance, it is suggested that the fold structure allows the maximum amount of crystallization of the molecule at a length that produces a free-energy minimum in the crystal. One suggestion is that the folding maintains the appropriate kinetic unit of the chain at any given temperature; as this would be expected to lengthen with increasing temperature, it would account for the observed thickening of the lamellae.

11.8.3 HEDRITES

If the concentration of the polymer solution is increased, a crystalline polyhedral structure emerges, composed of lamellae joined together along a common plane. These have also been detected growing from a melt, which suggests that lamellar growth can take place in the melt and may be a subunit of the spherulite.

11.8.4 CRYSTALLIZATION FROM THE MELT

Whereas crystallization from dilute solutions may result in the formation of single polymer crystals, this perfection is not achieved when dealing with polymers cooled from the melt. The basic characteristic feature is still the lamellar-like crystallite with amorphous surfaces or interfaces, but the way these are formed may be different, based on the careful investigation of melt-crystallized polymers using neutron-scattering techniques. The two models that have been proposed to describe the fine structure of these lamellae and their surface characteristics in semicrystalline polymers differ mainly in the way the chains are thought to enter and leave the ordered lamellae regions. These are:

1. The *regular folded array* with adjacent reentry of the chains, but with some loose folding and emergent chain ends or cilia that contribute to the disordered surface
2. The *switchboard model*, where there is some folding of the chains but reentry is now quite random

Both are represented schematically in Figure 11.5, but the exact nature of the structure has been the subject of considerable controversy. Although the morphology of the single crystals grown from dilute solutions may be more regular and resemble the first model, for polymers that are crystallized from the melt (and this is by far

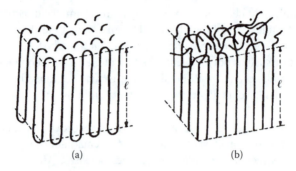

(a)   (b)

FIGURE 11.5 Schematics of possible chain morphology in a single polymer crystal: (a) regular folding with adjacent reentry of chains, (b) "switchboard" model with random reentry of chains.

TABLE 11.2
Comparison between the Radii of Gyration $(\langle S^2 \rangle/M_w)^{1/2}$ Expressed in the Melt and in the Semicrystalline State

Polymer	$M_w \cdot 10^{-3}$	$(\langle S^2 \rangle/M_w)^{1/2} / [\text{Å}/(\text{g mol}^{-1})^{1/2}]$	
		Crystallized	Melt
Polyethylene (quenched)	140	0.46 ± 0.05	0.46 ± 0.05
		0.45	0.45
Polyethylene oxide (crystallized by slow cooling)	150	0.52	0.46
Isotactic polypropylene (isothermally crystallized)	340	0.37	0.34

the more important procedure technically), the mass of evidence tends to favor a form of the switchboard model.

Measurements of the densities of several semicrystalline polymers points to the fact that a significant fraction of the chain units are in a noncrystalline environment. This is not consistent with the regular folded form of the crystallites, where the amorphous part is associated only with loose folding of the chains and cilia. Even more persuasive are small-angle neutron-scattering studies. These have demonstrated that the radii of gyration of several semicrystalline polymers remain essentially unchanged on moving from the melt phase to the semicrystalline phase (Table 11.2).

This means that there is no significant reordering of the chain conformation when crystallization takes place after cooling from the melt, which would be required if a regularly folded chain structure was to be constructed in the lamellae. To explain these observations, Fischer has proposed the *solidification model*, in which crystallization is believed to take place by the straightening of sections of the polymer coil, followed by alignment of these sequences in regular arrays forming the lamellar structure. This precludes the need for the extensive, long-range diffusion of the chain through a highly viscous medium that would be necessary if a regular chain-folded structure was to be constructed. The process is shown schematically in Figure 11.6,

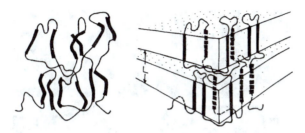

FIGURE 11.6 Solidification model illustrating crystallization from the melt. (Adapted from Dettenmaier, M. et al., *Colloid Polym. Sci.*, 228, 343, 1980. With permission from Dr Dietrich Steinkopff Verlag, Darmstadt.)

and the resulting structure is a variation of the switchboard model. This hypothesis can account for the fact that on cooling, rapid crystal growth is seen to occur, which is inconsistent with the need for long-range diffusion if the regularly folding lamellae were being formed. The solidification model shows that the chains can be incorporated into the basic lamellar form with the minimal amount of movement and that there will be extensive meandering of chains between the lamellae forming the interfacial amorphous regions.

11.8.5 SPHERULITES

Examination of thin sections of semicrystalline polymers reveals that the crystallites themselves are not arranged randomly, but form regular birefringent structures with circular symmetry. These structures, which exhibit a characteristic Maltese cross-optical extinction pattern, are called *spherulites*. Although spherulites are characteristic of crystalline polymers, they have also been observed to form in low-molar-mass compounds that are crystallized from highly viscous media.

Each spherulite grows radially from a nucleus formed either by the density fluctuations that result in the initial chain-ordering process or from an impurity in the system. As the structure is not a single crystal, the sizes found vary from somewhat greater than a crystallite to diameters of a few millimeters. The number, size, and fine structure depend on the temperature of crystallization, which determines the critical size of the nucleating center. This means that large fibrous structures form near T_m, whereas greater numbers of small spherulites grow at lower temperatures. When the nucleation density is high, the spherical symmetry tends to be lost as the spherulite edges impinge on their neighbors to form a mass such as shown in Figure 11.7.

A study of the fine structure of a spherulite shows that it is built up of fibrous subunits; growth takes place by the formation of fibrils, which spread outward from the nucleus in bundles, into the surrounding amorphous phase. As this fibrillar growth advances, branching takes place, and at some intermediate stage in the development, the spherulite often resembles a sheaf of grain. This forms as the fibrils fan out and begin to create the spherical outline. Although the fibrils are arranged radially, the molecular chains lie at right angles to the fibril axis. This has led to the suggestion that the fine structure is created from a series of lamellar crystals winding helically

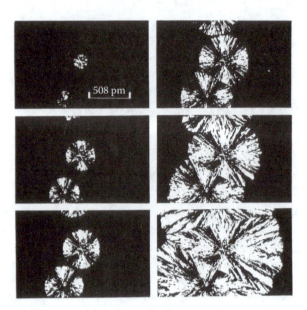

FIGURE 11.7 Sequence of photographs taken, under polarized light, over a period of about 1 min, showing the growth of poly(ethylene oxide) spherulites from the melt. From top left to bottom right: initially, discrete spherulites with spherical symmetry are observed, but the growing fronts eventually impinge on one another to form an irregular matrix. (Photographer: R. B. Stewart. Used with permission.)

along the spherulite radius. Growth proceeds from a small crystal nucleus that develops into a fibril. Low branching and twisting then produces bundles of diverging and spreading fibrils, which eventually fill out into the characteristic spherical structure. In between the branches of the fibrils are amorphous areas and these, along with the amorphous interfacial regions between the lamellae, make up the disordered content of the semicrystalline polymer (Figure 11.8).

Spherulites are classified as positive when the refractive index of the polymer chain is greater across the chain than along the axis, and negative when the greater refractive index is in the axial direction. They also show various other features such as zigzag patterns, concentric rings, and dendritic structures.

11.9 KINETICS OF CRYSTALLIZATION

The crystalline content of a polymer has a profound effect on its properties, and it is important to know how the rate of crystallization will vary with the temperature, especially during the processing and manufacturing of polymeric articles. The chemical structure of the polymer is also an important feature in the crystallization; for example, polyethylene crystallizes readily and cannot be quenched rapidly enough to give a largely amorphous sample, whereas this is readily accomplished for isotactic polystyrene. However, this aspect will be discussed more fully later.

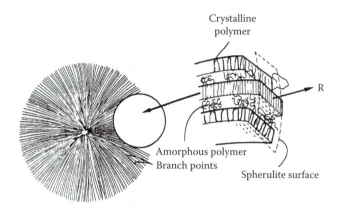

FIGURE 11.8 Fully-developed spherulite grown from the melt, comprising chain-folded lamellae (magnified section) and branching points that help to impart a spherical shape to the structure. Most rapid growth occurs in the direction of the spherulite radius R. (Adapted from McCrum, N.B., Buckley, C.P., and Bucknall, C.B., *Principles of Polymer Engineering,* Oxford University Press, 1988. With permission.)

11.9.1 Isothermal Crystallization

Two main factors influence the rate of crystallization at any given temperature: (1) the rate of nucleation and (2) the subsequent rate of growth of these nuclei to macroscopic dimensions.

The kinetic treatment of crystallization from the melt is based on the radial growth of a front through space and can be likened to someone scattering a handful of gravel onto the surface of a pond. Each stone is a nucleus that, when it strikes the surface, generates expanding circles (similar to spherulites in two dimensions). These grow unimpeded for a while, but the leading edges eventually collide with others and growth rates are altered. When a similar picture is adopted to model the crystallization of a polymer, certain basic assumptions are made first.

The formation of ordered growth centers by the alignment of chains from the melt is called *spontaneous nucleation.* When the temperature of crystallization is close to the melting temperature, nucleation is sporadic and only a few large spherulites will grow. At lower temperatures, nucleation is rapid and a large number of small spherulites are formed. The growth of the spherulites may occur in one, two, or three dimensions, and the rate of radial growth is taken to be linear at any temperature. Finally, the density ρ_c of the crystalline phase is considered to be uniform throughout but different from that of the melt ρ_L. A kinetic treatment has been developed taking these points into account.

11.9.2 The Avrami Equation

The kinetic approach relies on the establishment of a relation between the densities of the crystalline and melt phases and the time. This provides a measure of the overall crystallization rate. It is assumed that the spherulites grow from nuclei whose

TABLE 11.3
Relation between the Avrami Exponent and the
Morphological Unit Formed for Sporadic Nucleation

Growth Unit	Nucleation	Avrami Exponent n
Fibril	Sporadic	2
Disk	Sporadic	3
Spherulite	Sporadic	4
Sheaf	Sporadic	6

relative positions in the melt remain unaltered, and the analysis allows for the eventual impingement of the growing disks on one another. The final relation describing the process is known as the Avrami equation, expressed as

$$w_L / w_0 = \exp(-kt^n), \tag{11.3}$$

where k is the rate constant and w_0 and w_L are the masses of the melt at zero time and that left after time t. The exponent n is the Avrami exponent and is an integer that can provide information on the geometric form of the growth.

Sporadic nucleation is assumed to be a first-order mechanism, and if we consider that a two-dimensional disk is formed, then $n = 2 + 1 = 3$. Rapid nucleation is a zeroth-order process in which the growth centers are formed at the same time, and for each growth unit listed in Table 11.3, the corresponding values of the exponent would be $(n - 1)$. Thus, the Avrami exponent is the sum of the order of the rate process and the number of dimensions the morphological unit possesses.

11.9.3 DEVIATIONS FROM AVRAMI EQUATION

The Avrami equation can describe some, but not all, systems investigated. The crystallization isotherms of poly(ethylene terephthalate) can be fitted by Equation 11.3 using $n = 4$ above 473 K and $n = 2$ at 383 K. The equation should be used with caution, however, as noninteger values have been reported, and the geometric shape of the morphological unit is not always that predicted by the value of n calculated from the experimental data.

Deviations from the Avrami treatment may also be observed toward the end of the crystallization process, as a result of a secondary crystallization process, which is a slower reorganization of the crystalline regions to produce more perfectly formed crystallites.

11.10 BLOCK COPOLYMERS

Block copolymers in which the block components are sparingly soluble in each other have a natural tendency toward phase segregation. However, opposed to this, the covalent bonding between the chemically different portions of the molecule impedes macroscopic phase separation to the extent to which it is observed in polymer–polymer mixtures. As a result, the incompatibility between the covalently bonded blocks results in self-organized nanostructured materials with interesting morphologies.

For the simplest block-copolymer system, i.e., a binary A-B copolymer consisting of only two chemically different blocks, the essential parameters determining the microphase separation are the Flory–Huggins interaction parameter χ, the degree of polymerization, N, and the fraction of A monomers, f_A. Depending on copolymer composition and the product χN, the microphase-separated structure may adopt different morphologies. These are shown in Figure 11.9 with the phase diagrams derived by plotting the product χN, which is inversely proportional to temperature, vs. f_A.

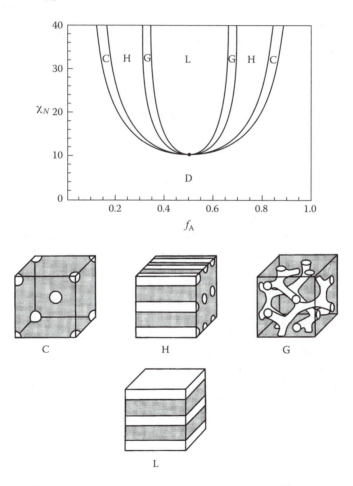

FIGURE 11.9 Phase diagram for conformationally symmetric diblock-copolymer melts showing regions of stability for the disordered (D), lamellar (L), gyroid (G), hexagonal (H), and cubic (C) phases. Dashed lines denote extrapolated phase boundaries and the dot marks the critical point. (Adapted from Matsen, M.W. and Schick, M., *Phys. Rev. Lett.*, 72, 2660, 1994. With permission from the American Physical Society.)

At high temperatures, entropy dominates: the A and B blocks mix homogeneously and form a disordered state. As the temperature is lowered, which corresponds to an increase of the χ parameter in Figure 11.9, the A and B blocks self-organize into one of various possible ordered structures. The classical morphologies are: (1) body-centered cubic (bcc) of spheres (S), (2) hexagonal arrangement of cylinders (C), and (3) lamellae (L) (Figure 11.9). The transition between these structures depends on the relative volume fraction of the two monomers. When the volume fractions of the two components are comparable, a lamellar phase is observed. With decreasing f, the volume fraction of the minority component, the latter will first arrange itself to form a hexagonal packing of cylinders and then spheres in a bcc structure.

Nonclassical phases, more complex morphologies, such as the bicontinuous structure often referred to as gyroid phase (G), were later identified.

The size of the characteristic dimension of the three classical structures, R, was determined by Meier (1969) for spheres (Equation 11.4), cylinders (Equation 11.5), and lamellae (Equation 11.6):

$$R = 1.33\,\alpha\,K\,M^{0.5} \tag{11.4}$$

$$R = 1.0\,\alpha\,K\,M^{0.5} \tag{11.5}$$

$$R = 1.4\,\alpha\,K\,M^{0.5} \tag{11.6}$$

where K is a constant which relates the end-to-end distance to the molecular weight (e.g., $K = 0.670$ Å for polystyrene). The domain structure has additional perturbations that do not exist in the bulk polymer, and these are accounted for by the parameter α. This is related to the interfacial energy and typically has values varying from 1.0 to 1.5.

Experimental techniques such as small-angle x-ray scattering and electron microscopy have been instrumental for the characterization of the ordered structures formed by block copolymers and their dependence on molecular parameters such as molar mass, chemical structure, and composition.

Research in the area of block copolymers has evolved rapidly over the past two decades, stimulated by the development of synthetic routes to novel copolymers with complex architectures, e.g., star- or H-shaped copolymers. This provides us a wealth of nanostructures resulting from self-organization that can be exploited for applications such as preparation of nanoparticles, for templating of nanostructures or nanomotor design.

11.11 HISTORICAL DEVELOPMENT OF POLYMER LIQUID CRYSTALS

The liquid crystalline state was first observed by an Austrian botanist, Friedrich Reintzer, in 1888, when he noted that cholesteryl esters formed opaque liquids on melting which, on heating to higher temperatures, subsequently cleared to form isotropic liquids. This behavior was interpreted by Lehmann as evidence for the existence of a new phase lying between the solid and isotropic liquid state. After further work by Friedel, this new state became known as a *mesophase*, from the Greek *mesos*, meaning in-between or intermediate. These mesophases are not only quite fluid but also show birefringence, and as they appear to have properties associated with both crystals and liquids, they were called *liquid crystals* by Lehmann.

Liquid crystals can be divided into two main classes: those similar to the cholesteryl derivatives, whose liquid crystalline phases are formed when the pure compound is heated, are called *thermotropic*, and those where the liquid crystalline phase forms when the molecules are mixed with a solvent are referred to as *lyotropic*. The thermotropic class also includes enantiotropic types, where the liquid crystalline phases can be seen on both the heating and the cooling cycles, and monotropic types, where the mesophase is stable only on supercooling from the isotropic melt.

Continued investigations led to the identification of three main types of mesophase: a *smectic* state (*smegma* in Greek, meaning soap), a *nematic* state (*nema* in Greek, meaning thread), and a *cholesteric* state observed in systems containing molecules with a chiral center.

Although the early work and many of the recent studies have identified and investigated the liquid crystalline properties of many small molecules, it was suggested that polymeric forms could also exist. In 1956, Flory postulated that concentrated solutions of rigid rodlike polymers should form ordered structures in solution at some critical concentration. This phenomenon was observed initially in 1937 for solutions of tobacco mosaic virus. However, the first systematic experimental verification of this prediction came from work on concentrated solutions of poly(γ-methyl glutamate) and poly(γ-benzyl glutamate), where these polymers exist in extended helical forms that can pack readily into ordered bundles with the long axes generally aligned in one direction. This produces a quasi-parallel distribution of chains in their solutions and anisotropic liquid crystalline properties. Later, it was shown that anisotropic solutions are formed by some aromatic polyamides and cellulose derivatives, where again the molecules are relatively rigid. These are lyotropic systems, but in the 1970s thermotropic liquid crystalline polymers were also synthesized, and this latter group has been developed rapidly since then.

11.12 LIQUID CRYSTALLINE PHASES

Molecules that have a tendency to form liquid crystalline phases usually have either rigid, long lathelike shapes with a high length-to-breadth (aspect) ratio, or disk-shaped molecular structures. Chemically, these may be composed of a central core comprising aromatic or cycloaliphatic units joined by rigid links, and having either polar or flexible alkyl and alkoxy terminal groups. Some typical examples of possible small molecule structures that form liquid crystalline phases are shown in Table 11.4, and these units are called *mesogens*. When polymers exhibiting liquid crystalline properties are formed, they can be constructed from these mesogens in three different ways: (1) incorporation into chainlike structures by linking them together through both terminal units to form main-chain liquid crystalline polymers; (2) attachment through

TABLE 11.4
A Selection of Small Molecule Mesogens and the Associated Liquid Crystalline Behavior

Mesogen	Transition Temperature (°C)
R—⟨⟩—N=CH—⟨⟩—O—CH₃	R = —CN k 106 N 117 i
	—n-C₄H₉ k 20 N 48 i
	—O—C₂H₅ k 83 N 107 i
	—C=O—CH₃ (O) k 79 N 102 i
H₅C₂—O—⟨⟩—N=CH—⟨⟩—CH=N—⟨⟩—O—C₂H₅	k 200 N 320 i
R—⟨⟩—CH=N—⟨⟩—N=CH—⟨⟩—R	R = —H k 239 N 265 i
	—O—CH₃ k 266 N 390 i
	k 110 N 197 i
H₅C₂—O—⟨⟩—N=N—⟨⟩—O—C—CH=CH—CH₃ (O)	
cholesteryl benzoate structure	k 137 N* 155 i
R—⟨⟩—N=N(O)—⟨⟩—R	R = —O—CH₃ k 119 N 135 i
	—O—C₂H₅ k 139 N 169 i
	—O—n-C₄H₉ k 105 N 136 i
	—O—n-C₆H₁₃ k 81 N 128 i
	—C=O—C₂H₅ (O) k 114 N 120 i

Note: Phases are shown as k = crystal phase; N = nematic phase; i = isotropic phase. N* = chiral nematic phase. The temperature of the transition from one to another is represented by the number between the phases on heating. That is, k to N at 106°C and N to i at 117°C.

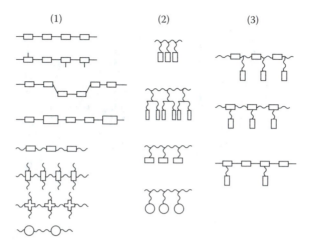

FIGURE 11.10 Schematic representation of various possible arrangements of mesogens in polymer chain structures: (1) main chain, (2) side chain, and (3) combinations of main and side chain. (Adapted from Sek, D., Structural variations of liquid crystalline polymer molecules, *Acta Polymerica*, 39, 599, 1988. With permission from Akademie-Verlag.)

one terminal unit to a polymer backbone to produce a side-chain comb-branch structure; and (3) a combination of both main and side-chain structures. The various possible geometric arrangements are shown schematically in Figure 11.10.

The mesogenic units can then form the ordered structures that are observed in the small molecule systems (though not necessarily the same type of liquid crystalline phase), and are characterized by long-range orientational order, with the long axes of the mesogenic groups arranged in one preferred direction of alignment called the *director*. When this spatial ordering is such that the mesogens are arranged in regular layers with respect to their centers of gravity, they are in one of several possible smectic phases. The lateral forces between the molecules in the smectic phases are stronger than the forces between the layers, and so slippage of one layer over another provides the characteristic fluidity of the system without losing the order within each layer. A number of different smectic phases can be identified in which the ordered packing of the mesogens in the layers differs, and the mesogens are either orthogonal to, or tilted with respect to, the layer structure. These are identified alphabetically. The most ordered is the smectic B (S_B) with a hexagonally, close-packed structure for the mesogens in the layers. The S_B and smectic E (S_E) phases exhibit three-dimensional order and have tilted modifications smectic H (S_H) and smectic G (S_G). A much less ordered structure within the layers produces a phase called the smectic A (S_A) phase, where there is a random lateral distribution of the mesogens in the layers. The tilted modification of this is called the smectic C (S_C) phase, and both S_A and S_C behave like true two-dimensional liquids. Intermediate in order are the smectic F (S_F) and smectic I (S_I), but the most commonly observed phases are S_A, S_B, and S_C.

The nematic phase is much less ordered than the smectic phases. Although the directional ordering of the mesogen long axes is maintained, the centers of gravity are no longer confined in layers but are distributed randomly in the phase. The nematic state is much more fluid than the smectic phases but still exhibits birefringence.

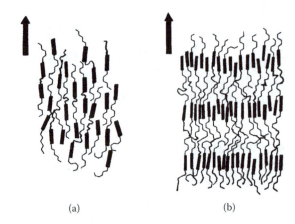

(a)   (b)

FIGURE 11.11 Schematic diagram of (a) the nematic phase and (b) the smectic phase for main-chain liquid crystalline polymers, showing the director as the arrow. The relative ordering is the same for side-chain polymer liquid crystals.

The third important category is a variation of the nematic phase and is called the *chiral nematic* state. It is observed when mesogens that enter a nematic phase also have a chiral center. This imparts a twist to each successive layer in the phase where the orientation of the director changes regularly from layer to layer, forming a helical arrangement of the directors in three-dimensional space. The chiral nematic state was first observed when cholesteryl derivatives were studied, but has now been detected in other chiral mesogens, and can also be induced by adding small chiral molecules to a host nematic liquid crystalline polymer. The main phase types are shown schematically in Figure 11.11.

In some polymer liquid crystals, several mesophases can be identified. In main-chain liquid crystal polymers there is usually a transition from the crystal to a mesophase, whereas in more amorphous systems when a glass transition is present, the mesophase may appear after this transition has occurred. In multiple transition thermotropic systems, the increase in temperature leads to changes from the most-ordered to the least-ordered states, i.e., crystal (k) $\rightarrow$ smectic (S) $\rightarrow$ nematic (N) $\rightarrow$ isotropic (i).

11.13 IDENTIFICATION OF THE MESOPHASES

The liquid crystalline phases are sometimes difficult to identify unequivocally, but several techniques can be used to provide information on the nature of the molecular organization within the phase. If used in a complementary fashion, these techniques can provide reliable information on the state of order of the mesogenic groups.

When an isotropic texture appears on cooling from an isotropic melt, it is formed from the coalescence of droplets that separate from the liquid. This mechanism is indicative of a nematic phase, and one of three possible characteristic textures may result, depending on the way the droplets form larger domains. These are the Schlieren, the threaded, or the marbled nematic textures. An example of Schlieren texture is shown in Figure 11.12.

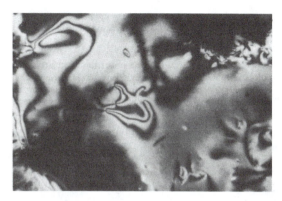

FIGURE 11.12 Photomicrographs of the nematic Schlieren texture that can be observed using a polarizing microscope. (Reproduced from Nöel, C., Synthesis characterization and recent developments of liquid crystalline polymers, *Makromol. Chem. Macromol. Symp.*, 22, 95, 1988. With permission.)

FIGURE 11.13 Photomicrographs of a focal conic and fan texture characteristic of a smectic A phase. Observed using a polarizing microscope. (Reproduced from Nöel, C., Synthesis characterization and recent developments of liquid crystalline polymers, *Makromol. Chem. Macromol. Symp.*, 22, 95, 1988. With permission.)

The Schlieren textures show large dark "brush" patterns, corresponding to the extinction zones where the mesogens are aligned perpendicular to the glass slide. Also noticeable are the points where two or four of these brushes meet; if the texture shows points where only two brushes meet, this is an unambiguous indication of a nematic phase. It should be noted that if a Schlieren texture shows only points where four brushes meet, then this represents a smectic C texture, or its chiral modification.

Smectic phases show a number of characteristic textures including: (1) the focal conic and fan texture, characteristic of smectic A, and often formed from the coalescence of batonnets; (2) the mosaic texture observed when the smectic B phase is formed; and (3) the broken focal conic structure resulting from smectic C phases that can also show the Schlieren pattern described earlier. Examples of focal conic textures are shown in Figure 11.13.

The chiral nematic phases can show a planar Grandjean texture, with oily streaks caused by defects, but they can also show strong reflection colors, depending on the pitch of the helical structure within the phase.

Differential scanning calorimetry (DSC) is widely used as a means of detecting the temperatures of thermotropic transitions, but the technique cannot be used to identify the mesophase. X-ray diffraction measurements on aligned samples obtained on cooling from the isotropic liquid phase under the influence of a strong magnetic field or by drawing a fiber from the mesophase are able to discriminate between different types of mesophases.

11.14 LYOTROPIC MAIN-CHAIN LIQUID CRYSTAL POLYMERS

Certain rodlike polymers, when mixed with small amounts of solvent, form a bi-refringent fluid that is eventually converted into a true isotropic solution when an excess of the solvent is added. These are called *lyotropic liquid crystalline systems* and are formed by the dissolution of amphiphilic molecules in appropriate solvents. The development of liquid crystalline solutions depends on the molar mass of the molecules, the solvent, and the temperature, but most importantly, the structure of the polymer, which should be quite rigid. Polymeric materials that tend to form helical structures, such as the stable α-helical form in polypeptides, are found to be suitable, and esters of poly(L-glutamic acid) are good examples of lyotropic systems. However, one of the most important groups of synthetic polymers is the aromatic polyamides. Those that are lyotropic have a variety of structures and appear to gain their rigidity of structure from the ring systems that are coupled by the amide link. This coupling unit adopts a trans conformation and can conjugate with the phenyl rings, when adjacent, to produce an extended rodlike structure in the polymer chain.

In this respect, the necessary rigidity of the polymers makes them inherently less soluble in common solvents, and dissolution often requires the use of more strongly interacting liquids. Whereas the poly(L-glutamate)s can form lyotropic solutions in dioxane or methylene chloride, the aromatic polyamides require the more agressive, strong, protonating acids (H_2SO_4, CF_3SO_3H, CH_3SO_3H) or aprotic solvents such as dimethylacetamide or *N*-methyl pyrrolidone in conjunction with LiCl or $CaCl_2$ in small (2 to 5) percentages, to effect solution.

In solution, flexible polymers assume a random coil conformation, but the rigid polymers are more rodlike and as their concentration in solution is increased, they tend to cluster together in bundles of quasi-parallel rods. These form domains that are anisotropic and within which there is nematic order of the chains. There is, however, little or no directional correlation between the directors within these domains unless the solutions are sheared. When shearing takes place, the domains tend to become aligned parallel to the direction of flow, thereby reducing the viscosity of the system below what would be expected from a solution of random coils. Lyotropic liquid crystalline polymers exhibit characteristic viscosity behavior as the concentration of the solutions is changed. Typically, the viscosity follows the trend shown in Figure 11.14 for partially chlorinated poly(1,4-phenylene-2,6-naphthalamide) dissolved in a

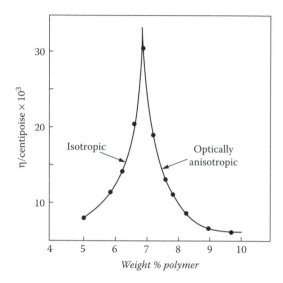

FIGURE 11.14 Variation of the viscosity of solutions of partially chlorinated poly(1,4-phenylene-2, 6-naphthalamide) dissolved in a solvent mixture of hexamethylene phosphoramide and *N*-methylpyrrolidone containing 2.9% LiCl as a function of the solution concentration, showing the transition from isotropic to anisotropic solutions. (From Morgan, P.W., Aromatic polyamides, *CHEMTECH*, 316, 1979. With permission of American Chemical Society, Washington, D.C.)

(1:1) mixture of hexamethylene phosphoramide and *N*-methyl pyrrolidone containing 2.9% LiCl. As the polymer is added to the solvent, the viscosity increases, but the solution remains isotropic and clear. At a critical concentration (which depends on the system), the solution becomes opaque and anisotropic, and there is a sharp decrease in the viscosity with further increase in the polymer concentration. This reflects the formation of the oriented nematic domains in which the chains are now aligned parallel to the direction of flow, thereby reducing the frictional drag on the molecules. The critical concentration that must be reached before the nematic phase is achieved is a function of the solvent, and tends to decrease with the increase in molar mass of the polymer.

A number of the polyamides have achieved commercial importance because of the very high tensile strengths of the fibers that can be spun from the nematic solutions. The additional chain orientation in the direction of the fiber long axis, obtained from the nematic self-ordering in the system, leads to a dramatic enhancement of the properties and makes them attractive alternatives to metal or carbon fibers for use in composites as reinforcing material.

The most significant of these aramid fibers are:

1. Poly(*m*-phenylene isophthalamide), trade name Nomex:

2. Poly (*p*-benzamide), or Fiber B:

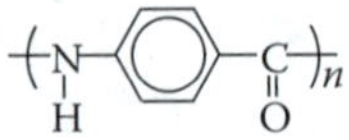

$$\left[N-\bigcirc-\underset{O}{\overset{\|}{C}}\right]_n$$

3. Poly(*p*-phenylene terephthalamide), trade name Kevlar:

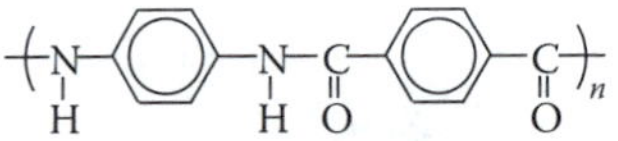

$$\left[N-\bigcirc-N-\underset{O}{\overset{\|}{C}}-\bigcirc-\underset{O}{\overset{\|}{C}}\right]_n$$

A more extensive description of aramid fibers is given in Chapter 15, Section 15.11.

Lyotropic polymers incorporating heterocyclic structures are also known, e.g.,

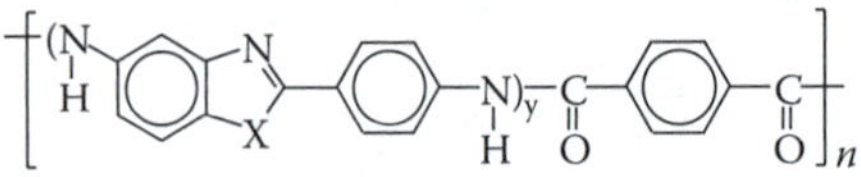

$$\left[\underset{H}{\overset{(N}{\underset{X}{\bigcirc}}}\bigcirc-\bigcirc-N)_y-\underset{O}{\overset{\|}{C}}-\bigcirc-\underset{O}{\overset{\|}{C}}\right]_n$$

where X = O, S, or –NH. These can form nematic phases in strongly protonating acids and will also spin into high-tensile-strength fibers.

Other structures capable of forming lyotropic solutions are:

1. Polyisocyanates:

$$\left[\underset{O}{\overset{\|}{C}}-\underset{R}{\overset{|}{N}}\right]_n$$

2. Poly(alkyl isonitrile)s:

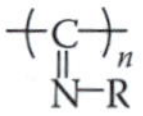

$$\left[\underset{\underset{R}{\overset{\|}{N}}}{C}\right]_n$$

3. Poly(organophosphazine)s:

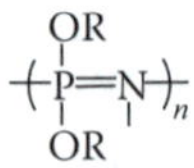

$$\left[\overset{OR}{\underset{OR}{\overset{|}{P}}}=N\right]_n$$

where *R* = alkyl or aryl.

11.15 THERMOTROPIC MAIN-CHAIN LIQUID CRYSTAL POLYMERS

The first examples of thermotropic main-chain liquid crystal polymers were produced by Roviello and Sirigu in 1975 when they reacted alkyl acylchlorides with *p*, *p*′-dihydroxy-α,-α′-dimethylbenzalazine, giving the structure (I), which exhibited anisotropic fluid phases after melting.

$$\left(\!\!+\!O\!\!-\!\!\bigcirc\!\!-\!\!\underset{CH_3}{\overset{|}{C}}\!\!=\!\!N\!\!-\!\!N\!\!=\!\!\underset{CH_3}{\overset{|}{C}}\!\!-\!\!\bigcirc\!\!-\!\!O\!\!-\!\!\underset{O}{\overset{||}{C}}\!\!-\!(CH_2)_n\!\!-\!\!\underset{O}{\overset{||}{C}}\!\!+\!\!\right)$$

(I)   n= 6, 8, 10

Subsequently, it was found that by linking mesogens together to form polymeric chains, these chains often showed the presence of a liquid crystal phase appearing at temperatures just above the melting point of the polymer. Many of these materials are polyesters that are synthesized by condensation reactions, including interfacial polymerizations, or by high-temperature-solution polymerizations using diols and diacid chlorides. However, the preferred method is often an ester interchange reaction in the melt.

Among the commonly used monomer units are hydroxybenzoic acid, p-terephthalic acid, 2-6-naphthalene dicarboxylic acid, 2-hydroxy-6-naphthoic acid, and 4,4'-biphenol. Thus, polymers can be prepared from simply one unit, as with poly(p-hydroxybenzoic acid):

$$(n+1)\ HO\!\!-\!\!\bigcirc\!\!-\!\!\underset{O}{\overset{||}{C}}\!\!-\!\!OH \longrightarrow HO\!\!-\!\!\bigcirc\!\!\left[\!\!\underset{O}{\overset{||}{C}}\!\!-\!\!O\!\!-\!\!\bigcirc\!\!\right]_n\!\!\underset{O}{\overset{||}{C}}\!\!-\!\!OH$$

or from more than one, i.e., structure II:

$$\left[\!\!+\!O\!\!-\!\!\bigcirc\!\!-\!\!CO\!\!+\!\!O\!\!-\!\!\bigcirc\!\!\bigcirc\!\!-\!\!CO\!\!+\right]$$

(II)

where the ratios of the components may be varied to alter the properties of the product.

The materials prepared in this way tend to be very insoluble polymers with high melting points and mesophase ranges, e.g., poly(p-hydroxybenzoic acid) melts at ~883 K. This makes them difficult to process, and alternative structures with much lower melting points are more useful. The melting points of the main-chain liquid crystal polymers can be reduced in a number of different ways, i.e.:

1. Incorporation of flexible space units
2. Copolymerization of several mesogenic monomers of different sizes to give a random and more irregular structure
3. Introduction of lateral substituents to disrupt the chain symmetry
4. Synthesis of chains with kinks, such as unsymmetrically linked aromatic units

The use of flexible spacers is a popular approach and consists of two cyclic units, normally joined by a short rigid bridging unit to form the mesogenic moiety. These are then linked through functional groups to flexible units of varying length

TABLE 11.5
Group Arrangements Typical of Thermotropic Main-Chain Polymer Liquid Crystals

Cyclic Unit	Linking Group	Functional Group	Spacer

$$\left[\begin{array}{c} \bigcirc \end{array}\right]_x$$

x = 1–3

| | $$\begin{array}{c} O \\ \| \\ -C-O- \end{array}$$ | $$\begin{array}{c} O \\ \| \\ -C-O- \end{array}$$ | $$-(CH_2)_n-$$ |

1,4
1,5
2,6

| | $$\begin{array}{c} -C=N-N=C- \\ \quad R \qquad R \end{array}$$ | $$\begin{array}{c} O \\ \| \\ -C-O- \end{array}$$ | $$-(CH_2CHO)_n- \\ \qquad R$$ |

| | $$\begin{array}{c} -CH=C- \\ \quad R \end{array}$$ | $$-O-$$ | $$-S-R-S-$$ |

| | $$\begin{array}{c} -CH=N \\ -N=N- \\ -N=N- \\ \quad O \end{array}$$ | $$-(CH_2)_n-$$ | $$\begin{array}{c} R \\ -Si-O- \\ R \end{array}$$ |

| Cyclic Unit | — | Linking Unit | — | Cyclic Unit | — | Functional Group | — | Spacer |

Mesogenic Group

that space the mesogens along the chain and reduce the overall rigidity. The schematic chemical constitutions of these chains, together with examples of the various types of groups that have been used, are shown in Table 11.5. The bridging groups must be rigid to maintain the overall stiffness of the mesogens, and they are usually multiple bond units. Ester groups also serve this purpose, particularly in conjunction with aryl rings, where the conjugation leads to a stiffening of the overall structure.

The majority of these main-chain liquid crystal polymers show a nematic phase after melting, but in some cases small variations in structure can lead to formation of a smectic mesophase. Thus, for polyesters with structure

$$\left[O-\bigcirc-\bigcirc-\underset{O}{\overset{O}{\underset{\|}{C}}}-(CH_2)_n-\underset{O}{\overset{O}{\underset{\|}{C}}} \right]_x$$

when the number of methylene units (n) in the spacer is odd, a nematic phase is observed; however, when n is even, a smectic mesophase results. These observations agree with de Gennes' predictions. Similarly, for polyesters with multiple rings but different orientation of the ester units, the phases can also be changed, i.e.,

(a)

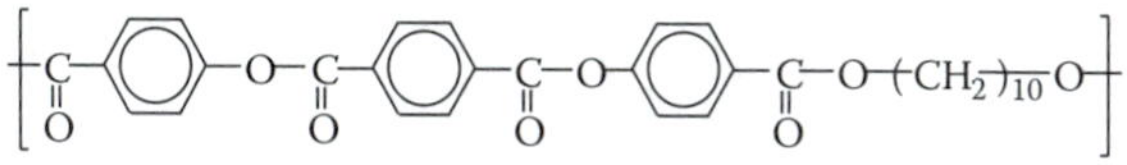

$$T_m = 493K; T_i = 540K; \text{SMECTIC}$$

(b)

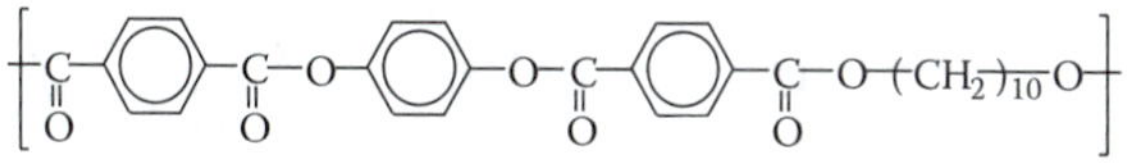

$$T_m = 503K; T_i = 538K; \text{NEMATIC}$$

The introduction of a flexible spacer can lower the melting point and increase the temperature range in which the mesophase is stable. The magnitude of the effect will depend on the length of the spacer unit. Thus, in systems such as the poly(α-cyanostilbene alkanoate)s, structure IV,

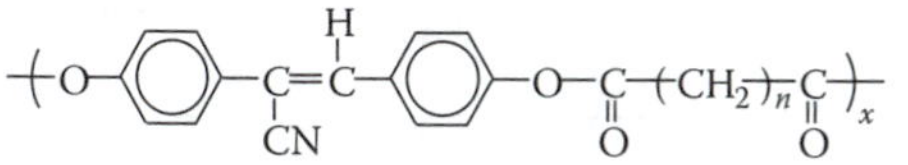

(IV)

in which a nematic phase is detected, both T_m and the nematic-to-isotropic transition temperature T_i decrease as the length of the methylene sequence (n) increases. This is clearly seen in Figure 11.15, where an odd–even alternation is also evident. The polymers with spacers having an even number of (CH$_2$) units usually have higher melting and clearing temperatures T_i than those with an odd number, and this suggests that the spacer length influences the ordering in the liquid crystal phase. The long-range ordering will tend to try to maintain the orientation of the mesogen parallel to the director axis, and this may be easier for even-numbered methylene unit spacers if they are in the all-trans zigzag conformation, as shown in Figure 11.16.

Copolymerization reactions can use many other combinations, but one interesting reaction involves the modification of poly(ethylene terephthalate) by reacting the preformed polymer with p-acetoxybenzoic acid. This has the effect of introducing a mesogenic unit to the structure at the points where the two units combine, producing a thermotropic liquid crystal polymer with a flexible spacer.

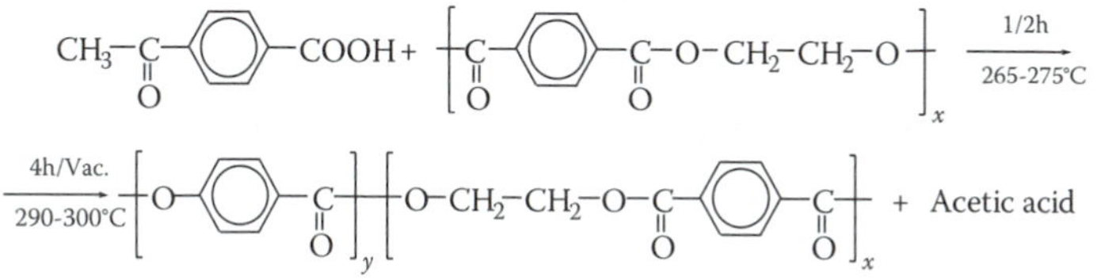

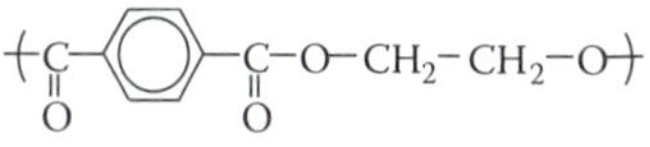

normal PET unit 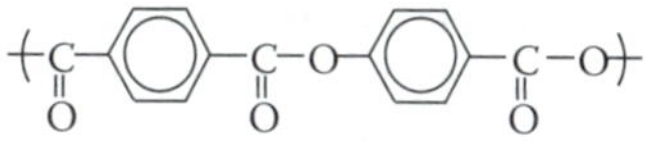   mesogenic unit

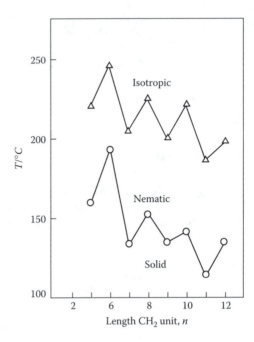

FIGURE 11.15 The transition temperatures $T_i \to T_N$ $(-\triangle-)$ and $T_N \to T_K$ $(-O-)$ for structure IV as a function of the length of the methylene sequence. (From Imura, K. et al., Synthesis and characterization of some thermotropic liquid crystalline polymers, *Curr. Top. Polym. Sci.*, 1, 1987. With permission. © Carl Hansser Verlag, Munich.)

The liquid crystalline phase can then be controlled by introducing specific amounts of the oxybenzoate units, and at about 30 mol% incorporation, a nematic phase appears in the melt. The optimum mechanical properties are obtained when 60 to 70 mol% of the oxybenzoate is present in the chain, as reflected by a large increase in the tensile strength of the material, which is accompanied by a corresponding decrease in the melt viscosity.

Introduction of a lateral substituent into the mesogen will also lower T_m and T_i, as the bulky side groups will tend to force the chains apart, thereby reducing the intermolecular forces of attraction. Ring substitution is the easiest way to achieve this, and Lenz has shown that the structure VI,

$$\left[O-\bigcirc-\underset{O}{\overset{\parallel}{C}}-O-\underset{X}{\bigcirc}-O-\underset{O}{\overset{\parallel}{C}}-\bigcirc-O-(CH_2)_{10}\right]_n$$

(VI)

which forms a nematic phase, can be altered by varying the group X to given lower T_m and T_i values, as shown in Table 11.6.

The introduction of kinks by using meta-substituted monomers or a crankshaft monomer such as 6-hydroxy-2-naphthoic acid (HNA) can be equally effective.

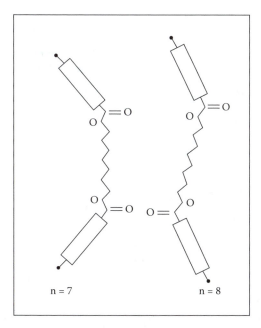

FIGURE 11.16 Schematic diagram showing the effect of odd and even numbers of spacer units on the relative orientation of the mesogenic units in a main-chain liquid crystal polymer. (Adapted from Krigbaum, W.R. et al., *Macromolecules*, 16, 1271, 1983. With permission from the American Chemical Society.)

TABLE 11.6
Effect of a Lateral Substituent on the Transition Temperatures of Structure VI

X	T_g/K	T_m/K	T_i/K	ΔT/K
H	340	509	540	31
$-CH_3$	317	427	463	36
$-C_2H_5$	308	344	400	56

Hence, when HNA is incorporated by copolymerization in the poly(hydroquinone terephthalate) structure, T_m is lowered.

11.16 SIDE-CHAIN LIQUID CRYSTAL POLYMERS

It has been demonstrated that polymers with mesogens attached as side chains can exhibit liquid crystalline properties, and much of the basic knowledge we have concerning side-chain liquid crystal polymers is due to the work of Ringsdorf, Finkelmann, Shibaev, and Platé. The extent to which the mesophases can develop in these systems is influenced by the flexibility of the backbone chain and whether

---

TABLE 11.7
Effect of Chain Flexibility on the Transition Temperatures of
Side-Chain Liquid Crystal Polymers Having a Common Mesogen

Polymer	Transitions (K)	ΔT
$\begin{array}{c} CH_3 \\ \mid \\ +CH_2\text{-}C+_n \\ \mid \\ COOR \end{array}$	$g \xrightarrow{369} N \xrightarrow{394} i$	25
$\begin{array}{c} +CH_2\text{-}CH+_n \\ \mid \\ COOR \end{array}$	$g \xrightarrow{320} N \xrightarrow{350} i$	30
$\begin{array}{c} CH_3 \\ \mid \\ +O\text{-}Si+_n \\ \mid \\ CH_2R \end{array}$	$g \xrightarrow{288} N \xrightarrow{334} i$	46

---

the mesogen is attached directly to the chain or is pushed further away by the insertion of a flexible spacing unit.

The polymer chain to which the mesogens are bonded can have different degrees of flexibility, and this can affect both the T_g and the temperature of the liquid crystal to isotropic phase transition T_i. This is illustrated in Table 11.7 for a series of polymers with a constant mesogen unit but a chain flexibility that decreases in the order methacrylate > acrylate > siloxane. The temperatures of the transitions are seen to decrease in the same order.

The conclusions drawn from these trends are that the thermal range of the mesophase (ΔT) is greatest when the chain is most flexible, and its conformational changes largely do not interfere with, or disrupt, the anisotropic alignment of the mesogens in the liquid crystalline phase. The influence of the backbone can be minimized by decoupling the motions of the chain from those of the side-chain mesogens. This can be accomplished by introducing long flexible spacing units between the backbone and the mesogen.

When longer spacer units are introduced, the observed effects are that the T_g of the polymer is usually lowered by internal plasticization, and the tendency for the more ordered smectic phases to develop is increased. As the spacer length is increased, there is an enhancement of the ordering, and the nematic phase gives way to a smectic phase. This ordering effect can also be encouraged by lengthening the alkyl tail unit. Both these alterations reflect the tendency of long alkyl side chains to order, and eventually to crystallize when long enough, and show that this is also imposed on the liquid crystal state.

The orientational order in nematic polymers can be described by a parameter S, defined by

$$S = 3/2(\overline{\cos^2 \theta - 1/3}) \tag{11.7}$$

where θ is the angle of mean deviation of the molecular axes of the mesogens with respect to the director; i.e., $S = 1$ means perfect parallel orientation. The value of S for side-chain liquid crystal polymers is approximately 75% of that obtained for the corresponding low-molar-mass mesogens and decreases with rising temperature as thermal agitation disturbs the orientation of the mesogenic groups. Although extension of the flexible spacer length does not appear to change S significantly when the system is in the nematic states, there is a marked change at a smectic–nematic transition, as might be expected.

The ordered state of the mesophase in these side-chain liquid crystalline polymers is readily frozen into the glassy state if the temperature is dropped rapidly below T_g and the value of S remains unchanged. This means that the liquid crystalline phase can be locked into the glassy polymer and remain stable until disturbed by heating above T_g again. This phenomenon offers several interesting applications in optoelectronics and information storage (see Chapter 16, Section 16.19 and Section 16.22). These applications often depend on the ability of mesogenic groups to align under the influence of an external magnetic or electric field.

It was stated earlier that liquid crystal polymers can be oriented by interaction with the surfaces of the measuring cell, which may be as simple as two glass slides with the polymer layer sandwiched between. This can give two extreme cases of either (1) homogeneous or (2) homeotropic alignment, where the long axes of the mesogens are respectively parallel to or perpendicular to the cell surface, and, when viewed vertically to the glass plates through crossed polarizers, the systems are then opaque or transparent, respectively. Because of the high viscosity experienced in polymer systems, this alignment may take some time or be incomplete. As the dielectric constant and diamagnetic susceptibility of many mesogens are anisotropic, side-chain liquid crystal polymers in the nematic state can also be oriented quite rapidly by the application of a magnetic or electric field. Now, the parameter of interest is the magnitude of the critical field that is required to effect the Fredericks transition. This is the transition from the homogeneous to the homeotropic aligned state. Whereas the relaxation time for this transition to take place in low-molar-mass mesogens is of the order of seconds, viscosity effects in polymer systems can push this up several orders of magnitude larger. This may make the use of polymeric liquid crystals less attractive in rapid response display devices. However, the additional stability that can be gained in polymeric systems can be advantageous in other ways, as will be illustrated in Chapter 16.

11.17 CHIRAL NEMATIC LIQUID CRYSTAL POLYMERS

A special form of the nematic phase, first observed in low-molar-mass esters of cholesterol, and originally called the "cholesteric mesophase," can often be detected in mesogenic systems containing a chiral center. The structure is a helically disturbed nematic phase, shown schematically in Figure 11.17, where the nematic order is preserved in each alternate layer, but where the director of each layer is displaced regularly by an angle θ relative to its immediate neighbors. This imparts a helical twist with a pitch p to the phase. This type of ordering results in a system with a

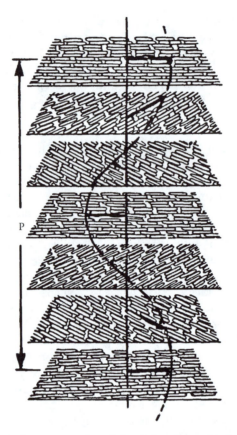

FIGURE 11.17 Schematic representation of the cholesteric or chiral nematic phase, where p is the helix pitch length.

very high optical activity and the ability to selectively reflect circularly polarized light of a specific wavelength λ_R when irradiated by normal light. The wavelength of this reflected light is related to the pitch of the helical structure by

$$\lambda_R = \tilde{n}p \qquad (11.8)$$

where $\tilde{n}$ is the average refractive index of the liquid crystalline phase.

The synthesis of polymers capable of entering into a chiral nematic phase initially proved difficult as many of the acrylate and methacrylate comb-branch polymers to which a cholesterol unit was attached as a side chain tended to give a smectic phase. This was overcome by either copolymerizing the cholesterol-containing monomers with another potential mesogenic monomer, or by synthesizing mesogens with a chiral unit in the tail moiety. Examples of both types are shown as structures VIII and IX.

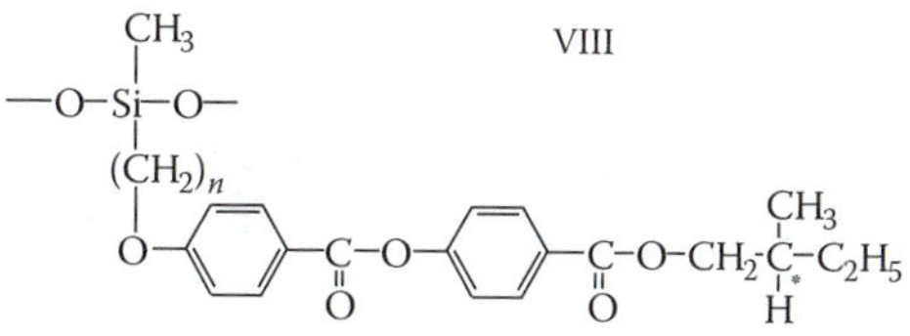

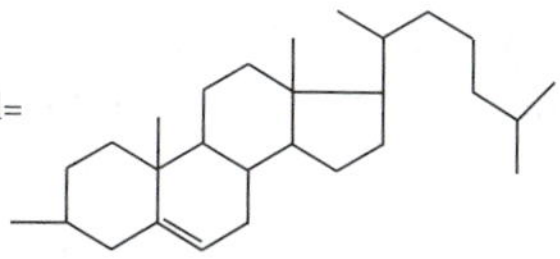

The data in Table 11.8 indicate that as the content of the chiral monomer (cholesterol) is varied in the copolymer, the wavelength of the reflected light changes, for an average temperature region between T_g and T_i. This suggests that the composition changes alter the pitch of the helical structure in the chiral nematic phase, and so the wavelength of reflected light will also change. For structure IX, the increase in % cholesterol causes the helix twist to become tighter, i.e., p becomes smaller, so λ_g also moves to shorter wavelengths. The pitch is also sensitive to temperature, and when the temperature is raised, the helix tends to unwind, with a consequential increase in λ_R.

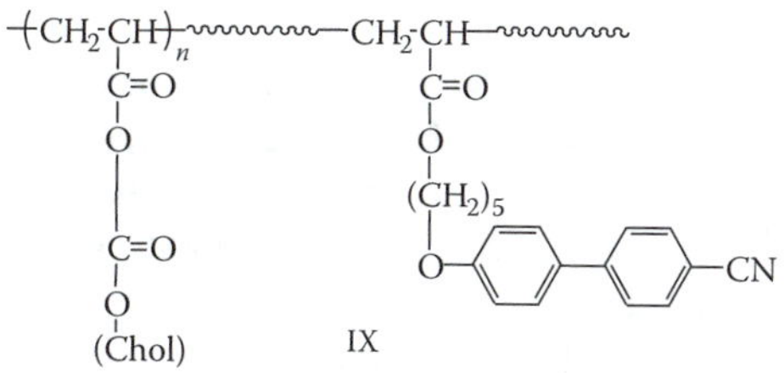

Similar to the other liquid crystal polymers described in Section 11.16, these materials offer the possibility of locking the chiral nematic phase into the glassy state by rapid supercooling to temperatures below T_g. This leads to a preservation of the structure and, of course, the reflected color, thereby leading to the formation of stable, lightfast, monochromatic films, when suitable systems are used.

TABLE 11.8
Variation of λ_R with Copolymer Composition for Structure IX

% Cholesterol	T_g (K)	T_i (K)	λ_R (nm)
34	50	90	850
40	50	102	660
55	55	105	555
65	55	150	500

PROBLEMS

1. Explain why the ratio of the density of the crystalline regions, ρ_c, to the density of the amorphous regions, ρ_a, is larger for nylon-6,6 than for isotactic polystyrene.

2. Poly(vinylidene fluoride) forms an orthorhombic unit cell in which the angles between the b and c, a and c, and a and b axes (α, β, and γ, respectively) are $\alpha = \beta = \gamma = 90°$. The dimensions of the orthorhombic unit cell are: $|a| = 8.47$ Å, $|b| = 4.70$ Å, and $|c| = 2.56$ Å, and there are two monomer units per unit cell. Calculate the theoretical density of the crystal.

3. The heat of fusion of a (polyethylene oxide) (PEO) sample was determined by DSC and found to be 6.70 kJ mol⁻¹. By considering that the heat of fusion of 100% crystalline PEO is 8.29 kJ mol⁻¹, estimate the degree of crystallinity of the sample.

4. The degree of crystallinity can be determined from density measurements and knowledge of the density of a 100% amorphous polymer, ρ_a, and 100% crystalline polymer, ρ_b. If constant mass is assumed, ρ_a and ρ_b are linked to the density of the specimen by the relationship:

$$\rho V = \rho_c V_c + \rho_a V_a$$

where V_c and V_a are the volumes of crystals and amorphous material, respectively, and V is the total sample volume. Using the preceding equation, evaluate the volume fraction of crystals for a high-density polyethylene (HDPE) sample ($\rho = 0.960$ g cm⁻³) and a linear low-density polyethylene (LLDPE) sample ($\rho = 0.930$ g cm⁻³). ($\rho_a = 0.851$ g cm⁻³ and $\rho_b = 0.990$ g cm⁻³.)

5. Fuller and Erickson (1937) prepared a series of oriented polyesters and examined these by means of x-rays. Which of the two structures (a) and (b) shown below would you expect to have the lower T_m value?

	Structure	Fiber Period /Å
(a)	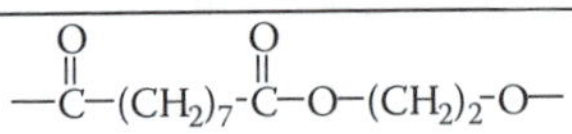 $-\overset{O}{\overset{\|}{C}}-(CH_2)_7-\overset{O}{\overset{\|}{C}}-O-(CH_2)_2-O-$ polyethylene azelate	31.2
(b)	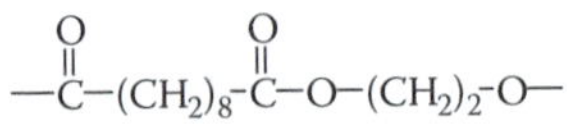 $-\overset{O}{\overset{\|}{C}}-(CH_2)_8-\overset{O}{\overset{\|}{C}}-O-(CH_2)_2-O-$ polyethylene sebacate	16.67

The highly oriented fiber diagrams showed uniaxial orientation and the fiber period determined from the x-ray measurements are reported in the preceding table. How do these values compare with the length of the polymer repeat units (C-C bond distance = 1.54 Å and C-O bond distance = 1.43 Å)?

6. The effect of side-chain structure on the crystalline melting point of isotactic poly(α-olefins) has been investigated. Consider the structures listed below as substituents into a methylene chain and discuss trends among melting temperatures. List structures in order of increasing T_m.

	Structure
(a)	$-CH_3$
(b)	$\begin{matrix} CH_3 \\ \mid \\ -CH_2-C-CH_2CH_3 \\ \mid \\ CH_3 \end{matrix}$
(c)	$-CH_2CH_2CH_3$
(d)	$-CH_2CH_2CH_2CH_3$
(e)	$-CH_2CH_3$
(f)	$\begin{matrix} H \\ \mid \\ -CH_2-C-CH_2CH_3 \\ \mid \\ CH_3 \end{matrix}$

7. Gechele and Crescentini (1963) measured the melting temperature, T_m, of polycaprolactam (nylon-6) as a function of m-cresol volume fraction, ϕ_1:

ϕ_1	T_m (K)
0.057	495
0.113	489
0.167	481
0.220	473
0.272	468
0.324	462
0.370	453
0.424	446
0.522	433
0.618	413

The absolute melting temperature T_m° for the polymer ($M_n = 12,200$) was reported to be 501 K. From these data, determine the interaction parameter and the heat of fusion (density of m-cresol = 1.03 g cm^{-3} and density of nylon 6 = 1.12 g cm^{-3}).

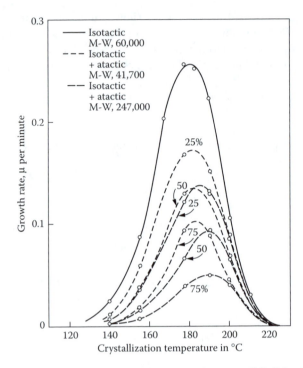

FIGURE 11.18 Radial growth rates for pure isotactic polystyrene (PS) (M_w= 60,000 g mol^{-1}) and two samples blended with atactic PS of molecular weight 41,700 and 247,000. (From Keith, H.D. and Padden, F.J., *J. Appl. Phys.*, 35, 1286, 1964. With permission.)

8. Radial growth rates for a pure isotactic polystyrene (PS) sample (M_w= 60,000 g mol^{-1}) and two other samples blended with atactic PS of molecular weight 41,700 and 247,000 are plotted in Figure 11.18. Comment on the changes observed due to addition of the diluents (atactic PS) (Keith and Paden, 1964).

9. The rate of crystallization of polyethylene terephthalate was measured using the density-gradient method (Keller et al., 1954) by following the changes in density as a function of time for a given crystallization temperature. Data for T = 383 K are tabulated as follows:

t (min)	Density (g cm⁻³)
0.00	1.340
1.78	1.340
6.11	1.341
10.19	1.343
15.29	1.345
20.13	1.348
25.48	1.350
30.32	1.356
35.67	1.359
40.76	1.362
45.61	1.364
50.45	1.367
60.64	1.369
70.32	1.370
80.00	1.371

By plotting $\ln(-\ln(\theta))$ vs. $\ln t$, determine the Avrami exponent n. In this case,

$$\theta = \frac{\rho_f - \rho_t}{\rho_f - \rho_i}$$

where ρ_t is the density at time t, and ρ_f and ρ_i are the final and initial densities for the crystallization.

10. Electron microscopy measurements on a very thin film of a styrene-isoprene block copolymer of 15,000 and 75,000 block molecular weights indicated that the polystyrene blocks form a spherical morphology with domains of 260 Å diameter (Meier, 1969). Calculate the predicted domain size for this morphology ($K = 0.670$ Å and assume $\alpha = 1.2$).

11. For the two liquid crystalline polymers (i) and (ii) shown below, sketch the expected DSC trace, consistent with the transition temperatures.

(i)

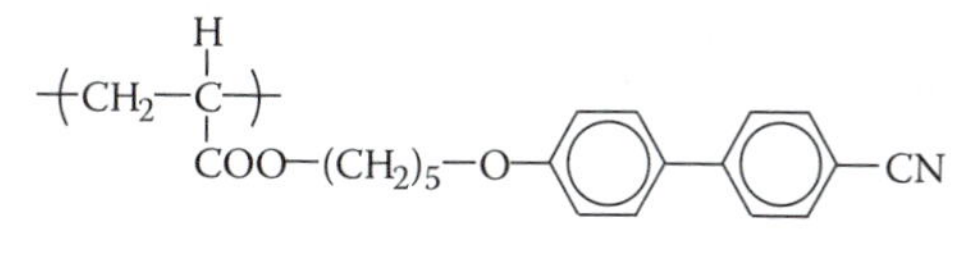

$$\text{glass (g)} \xrightarrow{308 \text{ K}} S_A \xrightarrow{393 \text{ K}} N \xrightarrow{397 \text{ K}} I$$

(ii)

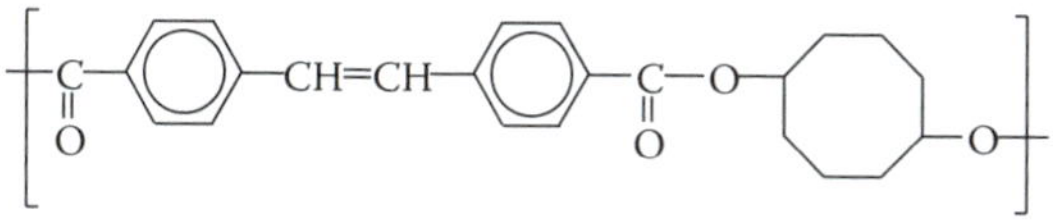

$$k \xrightarrow{495 \text{ K}} S_A \xrightarrow{540 \text{ K}} N \xrightarrow{563 \text{ K}} I$$

12. The polyether poly-*n*-nonyl-4,4′-biphenyl-2-chloro-ethane showed mono-
    tropic liquid crystalline behavior. The DSC trace indicated a sharp high-
    temperature exotherm (about 70°C) due to liquid-to-liquid crystal trans-
    formation and a broader exotherm at lower temperature (about 38°C) due
    to liquid crystal-to-crystal phase transition, on cooling. Subsequent heat-
    ing gave rise to only one endotherm corresponding to the crystal-to-liquid
    transition (about 90°C). Construct a graph of heat flow vs. temperature
    for both the heating and cooling runs.

REFERENCES

Benthack-Thomas, H. and Finkelmann, H., *Makromol. Chem.*, 186, 1895, 1985.
Decobert, G., Dubois, J.C., Esselin, S., and Nöel, C., *Liq. Cryst.*, 1, 307, 1986.
Dettenmaier, M., Fischer, E.W., and Stamm, M., *Colloid Polym. Sci.*, 228, 343, 1980.
Fuller, C.S. and Erickson, C.L., *J. Am. Chem. Soc.*, 59, 344, 1937.
Gechele, G.B. and Crescentini, L., *J. Appl. Polym. Sci.*, 7, 1349, 1963.
Imura, K., Koide, N., and Takeda, M., Synthesis and characterization of some thermotropic
       liquid crystalline polymers, *Curr. Top. Polym. Sci.*, 1, 1987.
Keith, H.D. and Padden, F.J., *J. Appl. Phys.*, 35, 1286, 1964.
Keller, A., Lester, G.R., and Morgan, L.B., *Phylos. Trans. R. Soc. London*, 247A, 13, 1954.
Krigbaum, W.R., Watanabe, J., and Ishikawa, T., *Macromolecules*, 16, 1271, 1983.
Leibler, L., *Macromolecules*, 13, 1602, 1980.
Lenz, R.W. and Jin, J.I., Liquid crystal polymers: a new state of matter, *Polymer News*, 11,
       200, 1986.
Mandelkern, L., *Rubber Chem. Technol.*, 32, 1392, 1949.
Marker, L., Early, R., and Aggarwal, S.L., *J. Polym. Sci.*, 38, 369, 1959.
Matsen, M.W. and Schick, M., *Phys. Rev. Lett.*, 72, 2660, 1994.
McCrum, N.B., Buckley, C.P., and Bucknall, C.B., *Principles of Polymer Engineering*, Oxford
       University Press, 1988.
Meier, D.J., *J. Polym. Sci.: Part C*, 26, 81, 1969.
Morgan, P.W., Aromatic polyamides, *CHEMTECH*, 316, 1979.
Nöel, C., Synthesis characterization and recent developments of liquid crystalline polymers,
       Makromol. Chem. Macromol. Symp., 22, 95, 1988.
Sek, D., Structural variations of liquid crystalline polymer molecules, *Acta Polymerica*, 39,
       599, 1988.

BIBLIOGRAPHY

Allen, G. and Bevington, J.C., Eds., *Comprehensive Polymer Science*, Vol. 2, Pergamon Press,
       1989.
Bassett, D.C., *Principles of Polymer Morphology*, Cambridge University Press, 1981.
Blumstein, A., Ed., *Polymeric Liquid Crystals*, Plenum Press, 1985.
Ciferri, A., Krigbaum, W., and Meyer, R., Eds., *Polymer Liquid Crystals*, Academic Press,
       1982.
Gordon, M. and Platé, N.A., *Liquid Crystal Polymers (Advances in Polymer Science*, Vol.
       59/60/61), Springer-Verlag, 1984.
Gray, G., Ed., *Thermotropic Liquid Crystals*, CRC Series, Vol. 22, John Wiley and Sons, 1987.

Gray, G. and Goodby, J., *Smectic Liquid Crystals: Textures and Structures,* Leonard Hill (Blackie), 1984.

Hall, I.H., Ed., *Structure of Crystalline Polymers*, Elsevier, 1984.

Mandelkern, L., *Crystallization of Polymers*, McGraw-Hill, 1964.

Mark, J., Ngai, K., Graessley, W., Mandelkern, L., Samulski, E., Koenig, J., and Wignall, G., *Physical Properties of Polymers*, 3rd ed., Cambridge University Press, 2004.

McArdle, C.B., Ed., *Side Chain Liquid Crystals Polymers,* Blackie and Sons, 1989.

McCrum, N.B., Buckley, C.P., and Bucknall, C.B., *Principles of Polymer Engineering*, 2nd ed., Oxford University Press, 1997.

McCrum, N.B., Buckley, C.P., and Bucknall, C.B., *Principles of Polymer Engineering,* Oxford University Press, 1988.

Platé, N.A. and Shibaev, V.P., *Comb-Shaped Polymers and Liquid Crystals,* Plenum Press, 1987.

Sadler, D.M., *Static and Dynamic Properties of the Polymeric Solid State*, Pethrick, R.A. and Richards, R.W., Eds., D. Reidel Publishing Co., 1982.

Sharples, A., *Introduction to Polymer Crystallization,* Edward Arnold, 1966.

Sperling, L.H., *Introduction to Physical Polymer Science*, 4th ed., Wiley-Interscience, 2005.

Tobolsky, A.V. and Mark, H., *Polymer Science and Materials*, Wiley-Interscience, 1971, chap 8.

12 The Glassy State and Glass Transition

12.1 THE AMORPHOUS STATE

In the amorphous state, the distribution of polymer chains in the matrix is completely random, with none of the strictures imposed by the ordering encountered in the crystallites of partially crystalline polymers. This allows the onset of molecular motion in amorphous polymers to take place at temperatures below the melting point of such crystallites. Consequently, as the molecular motion in an amorphous polymer increases, the sample passes from a glass through a rubberlike state until finally it becomes molten. These transitions lead to changes in the physical properties and material application of a polymer, and it is important to examine physical changes wrought in an amorphous polymer as a result of variations in the molecular motion.

12.2 THE GLASSY STATE

When a linear amorphous polymer is in the glassy state, the material is rigid and brittle because the flow units of the chain are cooperatively immobile and effectively frozen in position. The polymer sample is also optically transparent, as the chains are distributed in a random fashion and present no definite boundaries or discontinuities from which light can be reflected. An amorphous polymer in this state has been likened to a plate of frozen spaghetti. If a small stress is applied to a polymer glass, it exhibits a rapid elastic response resulting from purely local, bond angle deformation. Consequently, although the modulus is high, specimen deformation is limited to about 1%, due to the lack of glide planes in the disordered mass. This means that the sample has no way of dissipating a large applied stress other than by bond rupture, and so a polymer glass is prone to brittle fracture.

12.3 RELAXATION PROCESSES IN THE GLASSY STATE

Polymers do not form perfectly elastic solids, as a limited amount of bond rotation can occur in the glass, which allows slight plastic deformation; this makes them somewhat tougher than inorganic glass.

There is now ample evidence to support the suggestion that relaxation processes can be active in polymer glasses at temperatures well below T_g. Whereas the cooperative, long-range, chain motion that is released on passing from the glass to the rubberlike state is not possible at $T < T_g$, other relaxations can take place. Many of these processes can be identified as secondary loss peaks in dynamic mechanical or dielectric measurements, as will be seen in Chapter 13. They often find their origin in the movement of groups that are attached pendant to the main chain, but relaxation of limited sections of the main chain can also be identified.

The molecular mechanisms for a number of these subglass transition relaxations have now been established, often by relating the temperature dependence of the frequency of the observed mechanical or dielectric relaxations to specific molecular motions, identified using either experiments (e.g., NMR and neutron scattering) or, more recently, computer simulations. By way of illustration, some examples of group motions that have been found to be active in a series of poly(alkyl methacrylate)s will be described in the following text.

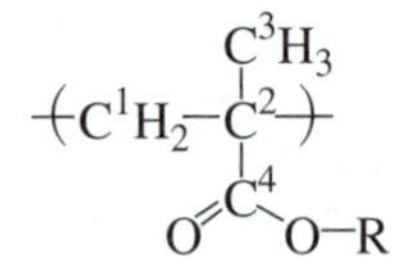

$$\begin{array}{c} C^3H_3 \\ | \\ -(C^1H_2-C^2)- \\ | \\ C^4 \\ O^{\diagup}\diagdown O-R \end{array}$$

For the methyl, ethyl, and propyl derivatives, a broad mechanically active damping peak is observed at 280 K (1 Hz), which is below the T_g of each polymer; rotation of the oxycarbonyl group about (C^2-C^4) has been identified as the cause. However, if the group R is an alkyl or cycloalkyl unit, then these can relax at even lower temperatures. Thus, when R is a methyl unit, rotation is possible in the glass at temperatures below 100 K, and the α-methyl unit will also be capable of rotation at low temperatures. When R is larger, -$(CH_2)_n$-CH_3, another relaxation process is seen at around 120 K, which is common to all of these polymers with $n = 3$ to 11. This is believed to involve the relaxation of a four-atom unit (-O-C-C-C-) or (-C-C-C-C-), which has been variously described by mechanisms including, and similar to, the Schatzki or Boyer crankshaft motions, as shown schematically in Figure 13.6 and Figure 12.1. The latter mechanisms have also been used to account for limited segmental relaxations in the backbone of all carbon chain, single strand polymers. Even larger units can relax, and Heijboer has demonstrated that if R is a cyclohexyl ring, a relaxation at 180 K (1 Hz) can be located in the glass. This can be attributed to an intramolecular chair–chair transition in the ring.

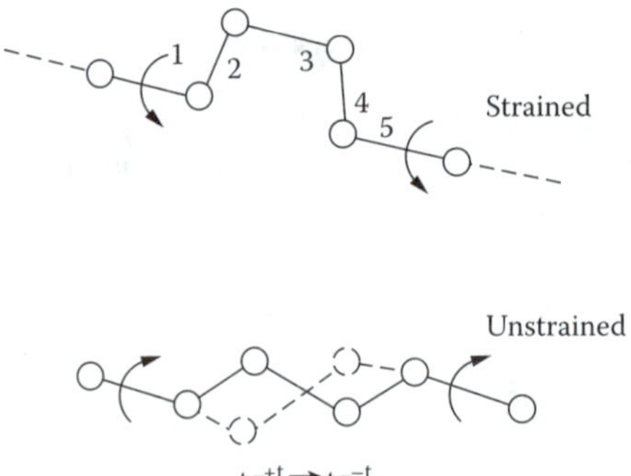

FIGURE 12.1 Schematic representation of the Boyer crankshaft motion.

As these relaxations require energy and are associated with a characteristic activation energy, it has been suggested that they may improve the impact resistance of some materials. This point still requires confirmation as a general phenomenon, but there is little doubt that polymer molecules are not totally frozen or immobile when in the glassy state, and that small subunits in the chain can remain mechanically and dielectrically active below T_g.

12.4 GLASS TRANSITION REGION

When the polymer is at a temperature below its glass temperature, chain motion is frozen. The polymer then behaves like a stiff spring, storing all the available energy in stretching as potential energy when work is performed on it. If sufficient thermal energy is supplied to the system to allow the chain segments to move cooperatively, a transition from the glass to the rubberlike state begins to take place. Motion is still restricted at this stage, but as the temperature increases further, a larger number of chains begin to move with greater freedom. In mechanical terms, the transition can be likened to the transformation of a stiff spring to a weak spring. As weak springs can only store a fraction of the potential energy that a strong spring can hold, the remainder is lost as heat, and if the change from a strong to a weak spring takes place over a period of time, equivalent to the observation time, then the energy loss is detected as mechanical damping. Finally, when molecular motion increases to a sufficiently high level, all the chains behave like weak springs the whole time. This means that the modulus is much lower but so, too, is the damping, which passes through a maximum in the vicinity of T_g. The maximum appears because the polymer is passing from the low-damping glassy state, through the high-damping transition region, to the lower-damping rubberlike state.

Treloar has described a very apt demonstration of the transition. A thin rubber rod is wound around a cylinder to create the shape of a spring, and then frozen in this shape using liquid nitrogen. The cylinder (possibly of paper) is then removed, leaving behind only the rubber spring. The rubber is now in the glassy state, and it acts like a stiff metal spring by regaining its shape rapidly after an extension. As the temperature is raised, a gradual loss in the elastic recovery is observed after each applied stress, until a stage is reached when there is no recovery and the rubber remains in the deformed shape. With a further increase in temperature, the rod straightens under its own weight and eventually regains its rubberlike elasticity at slightly higher temperatures.

12.4.1 THE GLASS TRANSITION TEMPERATURE, T_g

The transition from the glass to the rubberlike state is an important feature of polymer behavior, marking as it does a region where dramatic changes in the physical properties, such as hardness and elasticity, are observed. The changes are completely reversible, however, and the transition from a glass to a rubber is a function of molecular motion, not polymer structure. In the rubberlike state or in the melt, the chains are in relatively rapid motion, but as the temperature is lowered, the movement becomes progressively slower, until eventually the available thermal energy is insufficient to overcome the rotational energy barriers in the chain. At this temperature, which is known as the glass transition temperature T_g, the chains become locked in whichever conformation they possessed when T_g was reached. Below T_g, the polymer is in the glassy state and is, in effect, a frozen liquid with a completely random structure. This is reflected in the lack of ordering detected by means of x-rays, where only diffuse halos are observed for amorphous polymers, contrary to the sharp rings for the crystalline materials.

Although the glass–rubber transition itself does not depend on polymer structure, the temperature at which T_g is observed depends largely on the chemical nature of the polymer chain and, for most common synthetic polymers, lies between 170 and 500 K. It is quite obvious that T_g is an important characteristic property of any polymer, as it has an important bearing on the potential application of a polymer. Thus, for a polymer with a flexible chain, such as polyisoprene, the thermal energy available at about 300 K is sufficient to cause the chain to change shape many thousands of times in a second. This polymer has $T_g = 200$ K. On the other hand, virtually no motion can be detected in atactic poly(methyl methacrylate) at 300 K, but at 450 K, the chains are in rapid motion; in this case $T_g = 378$ K. This means that at 300 K, polyisoprene is likely to exhibit rubberlike behavior and be useful as an elastomer, whereas poly(methyl methacrylate) will be a glassy material. If the operating temperature was lowered to 100 K, both polymers would be glasses.

12.4.2 EXPERIMENTAL DEMONSTRATION OF T_g

The glass transition is not specific to long chain polymers. Any substance that can be cooled to a sufficient degree below its melting temperature without crystallizing will form a glass. The phenomenon can be conveniently demonstrated using glucose penta-acetate (GPA). A crystalline sample of GPA is melted, then chilled rapidly in ice water to form a brittle amorphous mass. By working the hard material between one's fingers, the transition from glass to rubber will be felt as the sample warms up. A little perseverance, with further rubbing and pulling, will eventually result in the recrystallization of the rubbery phase, which then crumbles to a powder.

The transition from a glass to a rubberlike state is accompanied by marked changes in the specific volume, the modulus, the heat capacity, the refractive index, and other physical properties of the polymer. The glass transition is not a first-order transition, in the thermodynamic sense, as no discontinuities are observed when the entropy or volume of the polymer is measured as a function of temperature (Figure 12.2). If the first derivative of the property–temperature curve is measured, a change in the vicinity of T_g is found; for this reason, it is sometimes called a second-order transition (Figure 12.2). Thus, whereas the change in a physical property can be used to locate T_g, the transition bears many of the characteristics of a relaxation process, and the precise value of T_g can depend on the method used and the rate of the measurement.

Techniques for locating T_g can be divided into two categories: dynamic and static. In the static methods, changes in the temperature dependence of an intensive property, such as density or heat capacity, are followed and measurements carried

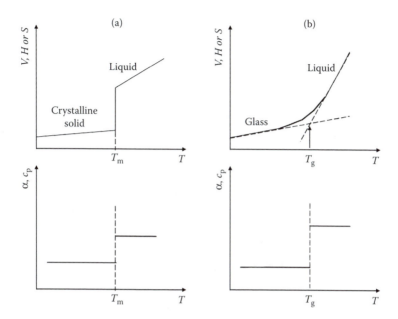

FIGURE 12.2 Schematic representation of the changes with temperature of volume (V), enthalpy (H), and entropy (S), and the first derivative of V and H for a (a) first-order and (b) second-order transition.

out slowly to allow the sample to equilibrate and relax at each observation temperature. In dynamic mechanical methods, a rapid change in modulus is indicative of the glass transition, but now the transition region is dependent on the frequency of the applied force. If we assume that, in the transition region, the restrictions to motion still present in the sample allow only a few segments to move in some time interval, say 10 s, then considerably fewer will have moved if the observation time is less than 10 s. This means that the location of the transition region and T_g will depend on the experimental approach used, and T_g is found to increase from 5 to 7 K for every tenfold increase in the frequency of the measuring techniques. This time dependence of segmental motion corresponds to the strong-weak transformation of a hypothetical spring and results in the high damping that imparts the lifeless leathery consistency to the polymer in this region. The temperature of maximum damping is usually associated with T_g, and at low frequencies, the value assigned to T_g is within a few kelvins of that obtained from the static methods. As the static methods lead to more consistent values, some of these can be described.

12.4.2.1 Measurement of T_g from V–T Curves

This was one of the most frequently used methods of locating T_g, and it consists in following the change in the volume of the polymer as a function of temperature. The polymer sample is placed in the bulb of a dilatometer, degassed, and a confining liquid such as mercury is added. If the bulb is attached to a capillary, the change in polymer volume can be traced by noting the overall change in volume registered by

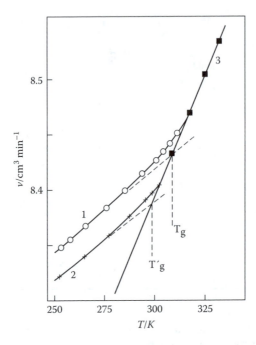

FIGURE 12.3 Specific volume v plotted against temperature for poly(vinyl acetate) measured after rapid cooling from above the T_g: measured 0.02 h after cooling and measured 100 h after cooling. T_g and T'_g are the glass transition temperatures measured for the different equilibration times. (From Kovacs, A.J., *J. Polym. Sci.*, 30, 131, 1958. With permission.)

the movement of the mercury level in the capillary. A variation of this method makes use of a density gradient column.

Typical specific volume–temperature curves are shown in Figure 12.3 for poly(vinyl acetate). These consist of two linear portions whose slopes differ and a closer inspection reveals that over a narrow range of temperature, i.e., between 2 and 5 K, the slope changes continuously. To locate T_g, the linear portions are extrapolated and intersect at the point that is taken to be the characteristic transition temperature of the material. Each point on the curve is normally recorded after allowing the polymer time to equilibrate at the chosen temperature, and as the rate of measurement affects the magnitude of T_g quite noticeably, the equilibration time should be several hours at least. The effect of the measuring rate on T_g was demonstrated by Kovacs (1958), who recorded the volume of a polymer at each temperature over a range including the transition, using two rates of cooling. If the sample was cooled rapidly (0.02 h), at each temperature the value of T_g derived from the resulting curve was some 8 K higher than that measured from results obtained using a slow cooling rate (100 h).

12.4.2.2 Thermal Methods

Phase transitions, including the glass transition, can be detected by following the changes in thermal properties. It has been shown in Figure 12.2 that the glass

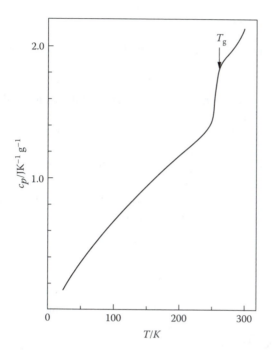

FIGURE 12.4 Specific heat capacity c_p plotted against temperature for atactic polypropylene, showing the glass transition in the region of 260 K. (From O'Reilly, J.M. and Karasz, F.E., *J. Polym. Sci.*, C (14), 49, 1966. With permission.)

transition is accompanied by a change in the heat capacity. It therefore can be detected calorimetrically by following the change in heat capacity with change in temperature. The curve for atactic polypropylene is shown in Figure 12.4, where the abrupt increase in c_p, at about 260 K, corresponds to the T_g.

Perhaps the most useful techniques now in use to locate the T_g are differential thermal analysis (DTA) and differential scanning calorimetry (DSC) (see Chapter 10). Both methods provide a means of characterizing thermal transitions as either exothermic or endothermic peaks (melting and crystallization), or as changes in heat capacity (glass transition).

Among other methods, dielectric loss measurements (see Chapter 13) offer additional information on polymer dynamics across the glass transition and have been used extensively.

12.4.3 FACTORS AFFECTING T_g

We have seen that the magnitude of T_g varies over a wide temperature range for different polymers. As T_g depends largely on the amount of thermal energy required to keep the polymer chains moving, a number of factors that affect rotation about chain links will also influence T_g. These include (1) chain flexibility, (2) molecular structure (steric effects), (3) molar mass (see Section 12.6), and (4) branching and cross-linking.

12.4.3.1 Chain Flexibility

The flexibility of the chain is undoubtedly the most important factor influencing T_g. It is a measure of the ability of a chain to rotate about the constituent chain bonds; hence, a flexible chain has a low T_g, whereas a rigid chain has a high T_g.

For symmetrical polymers, the chemical nature of the chain backbone is all important. Flexibility is obtained when the chains are made up of bond sequences that are able to rotate easily, and polymers containing $-(CH_2-CH_2)-$, $-(CH_2-O-CH_2)-$, or $-(Si-O-Si)-$ links will have correspondingly low values of T_g. The value of T_g is raised markedly by inserting groups that stiffen the chain by impeding rotation, so that more thermal energy is required to set the chain in motion. The p-phenylene ring is particularly effective in this respect, but when carried to extremes, produces a highly intractable, rigid structure poly(p-phenylene)

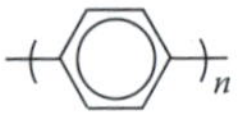

with no softening point. The basic structure can be modified by introducing flexible groups in the chain, and some examples are given in Table 12.1.

12.4.3.2 Steric Effects

When the polymer chains are unsymmetrical with repeat units of the type $-(CH_2CHX)-$, an additional restriction to rotation is imposed by steric effects. These arise when bulky pendant groups hinder the rotation about the backbone and cause

TABLE 12.1
Influence of Bond Flexibility on T_g

Polymers	Repeat Unit	T_g (K)
Poly(dimethylsiloxane)		150
Polyethylene	$-(CH_2-CH_2)-$	180
cis-Polybutadiene	$-(CH_2-CH=CH-CH_2)-$	188
Poly(oxyethylene)	$-(CH_2-CH_2-O)-$	206
Poly(phenylene oxide)	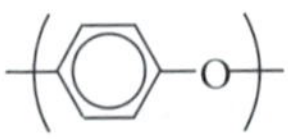	356
Poly(arylene sulfone)	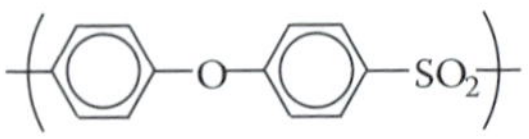	523
Poly(p-xylylene)	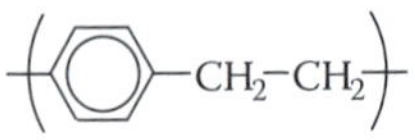	About 553

T_g to increase. The effect is accentuated by increasing the size of the side group, and there is some evidence of a correlation between T_g and the molar volume V_X of the pendant group. It can be seen in Table 12.2 that T_g increases with increasing V_X in the progressive series: polyethylene, polypropylene, polystyrene, and poly(vinyl naphthalene). Superimposed on this group size factor are the effects of polarity and the intrinsic flexibility of the pendant group itself. An increase in the lateral forces in the bulk state will hinder molecular motion and increase T_g. Thus, polar groups tend to encourage a higher T_g than nonpolar groups of similar size, as seen when

TABLE 12.2
Glass Transition Temperatures for Atactic Polymers of the General Type
$-\!\!\left(CH_2 \cdot CXY\right)_{\overline{n}}$

Polymer Type $-\!\!\left(CH_2CHX\right)_{\overline{n}}$	T_g (K)	V_X (cm^3 mol^{-1})[a]	Group X
Polyethylene	188	3.7	$-H$
Polypropylene	253	25.9	$-CH_3$
Poly(but-1-ene)	249	48.1	$-C_2H_5$
Poly(pent-1-ene)	233	70.3	$-C_3H_7$
Poly(hex-1-ene)	223	92.5	$-C_4H_9$
Poly(4-methyl pent-1-ene)	302	92.5	$-CH_2-CH(CH_3)_2$
Poly(vinyl alcohol)	358	11.1	$-OH$
Poly(vinyl chloride)	354	22.1	$-Cl$
Polyacrylonitrile	378	30.0	$-CN$
Poly(vinyl acetate)	301	60.1	$-O\cdot\overset{\displaystyle \parallel}{\underset{\displaystyle O}{C}}-CH_3$
Poly(methyl acrylate)	279	60.1	$-\overset{\displaystyle \parallel}{\underset{\displaystyle O}{C}}-O-CH_3$
Poly(ethyl acrylate)	249	82.3	$-COOC_2H_5$
Poly(propyl acrylate)	225	104.5	$-COOC_3H_7$
Poly(butyl acrylate)	218	126.7	$-COOC_4H_9$
Polystyrene	373	92.3	⌬
Poly(α-vinylnaphthalene)	408	143.9	⌬⌬

Polymer Type $-\!\!\left(CH_2C(CH_3)X\right)_{\overline{n}}$	T_g (K)	$V(X+Y)$ (cm^3 mol^{-1})[a]
Poly(methyl methacrylate)	378	86.0
Ply(ethyl methacrylate)	338	108.2
Poly(propyl methacrylate)	308	130.4
Polymethacrylonitrile	393	55.9
Poly(α-methylstyrene)	445	118.2

[a] Calculated using LeBas volume equivalents. (See Glasstone *Textbook of Physical Chemistry,* Macmillan, 1951, chap. 8.)

comparing polypropylene, poly(vinyl chloride), and polyacrylonitrile. The influence of side-chain flexibility is evident on examination of the polyacrylate series from methyl through butyl, and also in the polypropylene to poly(hex-1-ene) series.

A further increase in steric hindrance is imposed by substituting an α-methyl group that restricts rotation even further and leads to higher T_g. For the pair poly-styrene-poly(α-methyl styrene), the increase in T_g is 70 K, whereas the difference between poly(methyl methacrylate) and poly(methyl acrylate) is 100 K.

These steric factors all affect the chain flexibility and are simply additional contributions to the main chain effects.

12.4.3.3 Configurational Effects

Cis-trans isomerism in polydienes and tacticity variations in certain α-methyl-sub-stituted polymers alter chain flexibility and affect T_g. Some examples are shown in Table 12.3. It is interesting to note that when no α-methyl group is present in a polymer, tacticity has little influence on T_g.

12.4.3.4 Effect of Cross-Links on T_g

When cross-links are introduced into a polymer, the density of the sample is increased proportionally. As the density increases, the molecular motion in the sample is restricted and T_g rises. For a high cross-link density, the transition is broad and ill defined, but at lower values, T_g is found to increase linearly with the number of cross-links.

12.5 THEORETICAL TREATMENTS

Before embarking on a rather brief description of the theoretical interpretations of the glass transition, a word of caution should be given. In the foregoing sections, several features of the results point to the fact that, in the vicinity of T_g, rate effects are closely associated with changes in certain thermodynamic properties. This has engendered two schools of thought on the origins of this phenomenon, together with variations on each theme. The elementary level of this text precludes detailed critical

TABLE 12.3
The Effect of Microstructure on T_g

Polymer	Stereostructure	T_g (K)
Poly(methyl methacrylate)	Isotactic	318
	Atactic	378
	Syndiotactic	388
Polybutadiene	Cis	165
	Trans	255
Polyisoprene	Cis	200
	Trans	220

discussion of the relative merits of any particular treatment, and to avoid prejudicing the issue with personal comments, the main ideas of each are outlined together with a more recent and possibly unifying approach to complete the picture.

12.5.1 THE FREE-VOLUME THEORY

The free-volume concept has been mentioned in previous sections, but it is instructive now to consider this idea more closely and to draw together the various points alluded to earlier. *Free volume*, V_f, is defined as the unoccupied space in a sample, arising from the inefficient packing of disordered chains in the amorphous regions of a polymer sample. The presence of these empty spaces can be inferred from the fact that when a polystyrene glass is dissolved in benzene, there is a contraction in the total volume. This and similar other observations indicate that the polymer can occupy less volume when surrounded by benzene molecules, and that there must have been unused space in the glassy matrix to allow this increase in packing efficiency to occur.

On that basis, the observed specific volume of a sample, V, will be composed of the volume actually occupied by the polymer molecules, V_0, and the free volume in the system, i.e.,

$$V = V_0 + V_f \tag{12.1}$$

Each term will, of course, be temperature dependent. The free volume is a measure of the space available for the polymer to undergo rotation and translation, and when the polymer is in the liquid or rubberlike state the amount of free volume will increase with temperature as the molecular motion increases. If the temperature is decreased, this free volume will contract and eventually reach a critical value where there is insufficient free space to allow large-scale segmental motion to take place. The temperature at which this critical value is reached is the glass transition temperature. Below T_g, the free volume will remain essentially constant as the temperature decreases further because the chains have now been immobilized and frozen in position. In contrast, the occupied volume will alter because of the changing amplitude of thermal vibrations in the chains and, to the first approximation, will be a linear function of temperature, irrespective of whether the polymer is in the liquid or glassy state.

The glass transition can then be visualized as the onset of coordinated segmental motion made possible by an increase of the holes in the polymer matrix to a size sufficient to allow this type of motion to occur. This is manifest as a change in the specific volume due solely to an increase in the free volume and is shown schematically as the crosshatch area in Figure 12.5, where the broken line indicates the temperature dependence of V_0.

The precise definition of the average amount of free volume present in a totally amorphous polymer remains unclear, but it must also depend to some extent on the thermal history of the sample. A number of suggestions have been made.

Simha and Boyer (S-B) observed that a general empirical relationship exists between the T_g and the difference in expansion coefficients of the liquid and glass states. From the examination of a wide range of polymers, they concluded that

$$(\alpha_l - \alpha_g) T_g = K_1 \tag{12.2}$$

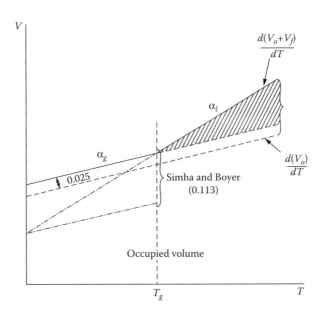

FIGURE 12.5 Schematic representation of the free volume as defined by Flory and Fox (0.025) and by Simha and Boyer (0.113).

where K_1 is a constant with a value of 0.113. This implies that the free-volume fraction is the same for all polymers, i.e., 11.3% of the total volume in the glassy state. The definition of the S-B free volume can be seen, from Figure 12.5, to be

$$V_f = V - V_{0l}(1 + \alpha_g T) \tag{12.3}$$

where V_{0l} is the hypothetical liquid volume at absolute zero. This definition is perhaps too rigid and discounts differing chain flexibilities, so a more accurate representation is thought to be given by

$$\Delta\alpha \cdot T_g = 0.07 + 10^{-4} T_g \tag{12.4}$$

The values obtained are still much higher than the estimates from the Williams–Landel–Ferry (WLF) equation. This is an empirical equation, but it can also be derived from free-volume considerations by starting with a description of the viscosity of the system, based on an empirical equation proposed by Doolittle to describe the temperature dependence of viscous flow

$$\ln \eta = \ln A + B \left\{ \frac{V - V_f}{V_f} \right\} \tag{12.5}$$

where A and B are constants and $(V - V_f) = V_0$. On a molecular level, the ratio (V_0/V_f) is then a measure of the average volume of the polymer relative to that of the holes.

Thus, when $V_0 > V_f$, i.e., the polymer chain is larger than the average "hole" size, the viscosity will be correspondingly high, whereas when $V_0 < V_f$, the viscosity will be low.

We can now introduce a free-volume fraction f

$$f = \left[\frac{V_f}{V_0 + V_f}\right] \approx \left[\frac{V_f}{V_0}\right] \tag{12.6}$$

and substitute in Equation 12.5

$$\ln \eta = \ln A + B / f \tag{12.7}$$

Next, a comparison can be made between the viscosity, η, of a polymer melt at a temperature T (η_T), and that at a reference temperature such as T_g (η_g) and so

$$\ln\left(\frac{\eta_T}{\eta_g}\right) = B\left[\frac{1}{f_T} - \frac{1}{f_g}\right] \tag{12.8}$$

Here f_T and f_g are the fractional free volumes at T and T_g, respectively. From Figure 12.5, it can be seen that V_f is assumed to remain constant during expansion of the polymer in the glassy state, but that above T_g there is a steady increase with rising temperature. If α_f is the expansion coefficient of the free volume above T_g, then the temperature dependence of f_T can be written

$$f_T = f_g + \alpha_f (T - T_g) \tag{12.9}$$

Substitution of Equation 12.9 in Equation 12.8, after rearranging and dividing by α_f, gives

$$\ln\left(\frac{\eta_T}{\eta_g}\right) = -\frac{(B/f_g)(T - T_g)}{(f_g/\alpha_f) + (T + T_g)} \tag{12.10}$$

Equation 12.10 is one form of the WLF equation, but as viscosity is a time-dependent quantity and is proportional to the flow time, t, and density, ρ, then

$$\left(\frac{\eta_T}{\eta_g}\right) = -\left(\frac{\rho_T}{\rho_g} \cdot \frac{t_T}{t_g}\right) \tag{12.11}$$

and

$$\log_{10}\left(\frac{t_T}{t_g}\right) \approx \frac{-(B/2.303 f_g)(T - T_g)}{(f_g \alpha_f) + (T - T_g)} \tag{12.12}$$

where the small differences in density have been neglected. This can be compared with the form of the WLF equation

$$\log_{10} a_T = \frac{-C_1(T - T_g)}{C_2 + (T - T_g)} \tag{12.13}$$

where a_T is the reduced variables shift factor, and C_1 and C_2 are constants that can be evaluated from experimental data and are found to be $C_1 = 17.44$ and $C_2 = 51.6$ when T_g is the reference temperature. A more general description can be used, such as

$$\log_{10} a_T = \frac{-8.86(T - T_s)}{101.6 + (T - T_s)} \tag{12.14}$$

where T_s is an arbitrary reference temperature usually located ≈ 50 K above T_g. C_1 and C_2 now have different values, and the shift factor is expressed as a ratio of relaxation times, τ, at T and T_s

$$a_T = \tau(T)/\tau(T_s) \tag{12.15}$$

As we shall see in Chapter 13, the relaxation time is a function of the viscosity and modulus (G) of the polymer and, according to the Maxwell model, $\tau = (\eta/G)$. The modulus will be much less temperature dependent than the viscosity, so we can write $a_T = (\eta_T/\eta_s)$, which demonstrates the equivalence of the empirical Equation 12.13 with that derived from the free-volume theory, shown in Equation 12.10 and Equation 12.12.

The WLF equation can be used to describe the temperature dependence of dynamic mechanical and dielectric relaxation behavior of polymers near the glass transition where the response is no longer described by an Arrhenius relation. This will be dealt with in Chapter 13.

Equation 12.10 and Equation 12.13 can be used to evaluate f_g as we see that $(B/2.303\,f_g) = 17.44$ and $(f_g/\alpha_f) = 51.6$. On the basis of viscosity data, B can be assigned a value of unity leading to $f_g = 0.025$ and $\alpha_f = 4.8 \times 10^{-4}$ K^{-1}. If α_f is assumed to be equivalent to $(\alpha_l - \alpha_g)$, this value compares well with the average value of $\Delta\alpha = 3.6 \times 10^{-4}$ K^{-1} determined for 18 polymers covering a wide range of T_g. The free-volume fraction of 2.5% is low compared with the S-B estimation but is comparable to that derived from the Gibbs–Di Marzio theory. Other values of 8% from the hole theory of Hirai and Eyring and 12% calculated by Miller from heats of vaporization and liquid compressibilities illustrate the uncertainty surrounding the magnitude of this free-volume parameter. The free-volume theory deals with the need for space to be available before cooperative motion, characteristic of the glass transition, can be initiated, but it tells us little about the molecular motion itself. Other approaches have chosen to base their description of the glass transition on a thermodynamic analysis.

12.5.2 GIBBS–DI MARZIO THERMODYNAMIC THEORY

Comments on the thermodynamic theories will be restricted to the proposals of Gibbs and Di Marzio (1958) (G-D) who, although acknowledging that kinetic effects are inevitably encountered when measuring T_g, consider the fundamental transition to be a true equilibrium. The data reported by Kovacs (1958) in Section 12.3 imply that the observed T_g would decrease further if a sufficiently long time for measurement was allowed. This aspect is considered in the G-D theory by defining a new transition temperature T_2 at which the configurational entropy of the system is zero. This temperature can be considered, in effect, to be the limiting value T_g would reach in a hypothetical experiment taking an infinitely long time. On this basis, the experimentally detectable T_g is a time-dependent relaxation process, and the observed value is a function of the timescale of the measuring technique. The theoretical derivation is based on a lattice treatment. The configurational entropy is found by calculating the number of ways that n_x linear chains, each x segments long, can be placed on a diamond lattice for which the coordination number $z = 4$, together with n_0 holes. The restrictions imposed on the placing of a chain on the lattice are embodied in the hindered rotation, which is expressed as the *flex energy* $\Delta\varepsilon$, and ε_h, which is the energy of formation of a hole. The flex energy is the energy difference between the potential energy minimum of the located bond and the potential minima of the remaining $(z - 2)$ possible orientations that may be used on the lattice. Thus, for polyethylene the trans position is considered most stable, and the gauche positions are the flexed ones with $\Delta\varepsilon$ as the energy difference between the ground and flexed states. This of course varies with the nature of the polymer. The quantity ε_h is a measure of the cohesive energy. The configurational entropy S_{conf} is derived from the partition function describing the location of holes and polymer molecules.

As the temperature drops toward T_2, the number of available configurational states in the system decreases, until at the temperature T_2, the system possesses only one degree of freedom. This leads to

$$\frac{S_{conf}(T_2)}{n_x k T_2} = 0 = \phi\left(\frac{\varepsilon_h}{kT_2}\right) + \lambda\left(\frac{\Delta\varepsilon}{kT_2}\right) + \frac{1}{x}\ln\left[\{(z-2)x+2\}\frac{(z-1)}{2}\right] \quad (12.16)$$

where $\phi(\varepsilon_h/kT) = \ln(\varepsilon_h/S_0)^{z/2-1} + f_0/f_x \ln(f_0/S_0)$ and

$$\lambda\left(\frac{\Delta\varepsilon}{kT}\right) = \frac{x-3}{x}\ln\left\{1+(z-2)\exp(-\Delta\varepsilon/kT)+(\Delta\varepsilon/kT)\right.$$

$$\left. \times\left[\frac{(z-2)\exp(-\Delta\varepsilon/kT)}{1+(z-2)\exp(-\Delta\varepsilon/kT)}\right]\right\} \quad (12.17)$$

The fractions of unoccupied and occupied sites are f_0 and f_x, respectively, whereas S_0 is a function of f_0, f_x, and z. The main weaknesses of this theory are (1) that a chain of zero stiffness would have a T_g of 0 K and (2) that the T_g would be essentially independent of any intermolecular interaction. In spite of these limitations, various aspects of the behavior of copolymers, plasticized polymers, and the chain length dependence of T_g can be predicted in a reasonably satisfactory manner. The temperature T_2 is not, of course, an experimentally measurable quantity but is calculated to lie approximately 50 K below the experimental T_g, and it can be related to T_g on this basis.

12.5.3 ADAM–GIBBS THEORY

Whereas the kinetic approach embodied in the WLF equation and the equilibrium treatment of the G-D theory have both been successful in their way, the one-sided aspect of each probably masks the fact that they are not entirely incompatible with one another. An attempt to reunite both channels of thought has been made by Adam and Gibbs (1965), who have outlined a molecular kinetic theory.

In this, they relate the temperature dependence of the relaxation process to the temperature dependence of the size of a region, which is defined as a volume large enough to allow cooperative rearrangement to take place without affecting a neighboring region. This "cooperatively rearranging region" is large enough to allow a transition to a new conformation, and hence is determined by the chain conformation and by definition will equal the sample size at T_2, where only one conformation is available to each molecule. Evaluation of the temperature dependence of the size of such a region leads to an expression for the cooperative transition probability, $W(T)$, which is simply the reciprocal of the relaxation time.

The WLF was derived as:

$$-\log_{10} a_T = C_1(T - T_s)/C_2 + (T - T_s) \tag{12.18}$$

and expressed in terms of the activation energy for a cooperative rearrangement per monomer segment, $\Delta\mu$, a critical configurational entropy S_c^* that, from the definition, can be approximated by $k \ln 2$, a macroscopic configurational entropy for the ensemble, S_c, and the change in the heat capacity at T_g, i.e.,

$$C_1 = \left\{ \frac{2.303 \Delta\mu S_c^*}{k \Delta C_p T_s \ln(T_s/T_2)} \right\} \tag{12.19a}$$

$$C_2 = \frac{T_s \ln(T_s/T_2)}{1 + \ln(T_s/T_2)} \tag{12.19b}$$

Results plotted according to the WLF equation could be predicted also from the molecular kinetic equation and show that the two approaches are compatible. The Adam–Gibbs equations also lead to a value of $(T_g - T_2) = 55$ K, so the theory appears to resolve most of the differences between the kinetic and thermodynamic interpretations of the glass transition.

These theories point to the fundamental importance of T_2 as a true second-order transition temperature and to the experimental T_g as the temperature governed by the time scale of the measuring technique. The latter value has great practical significance, however, and is a parameter that is essential to the understanding of the physical behavior of a polymer.

12.6 DEPENDENCE OF T_g ON MOLAR MASS

The value of T_g depends on the way in which it is measured, but it is also found to be a function of the polymer chain length. At high molar masses the glass temperature is essentially constant when measured by any given method, but decreases as the molar mass of the sample is lowered. In terms of the simple free-volume concept, each chain end requires more free volume in which to move about than a segment in the chain interior. With increasing thermal energy, the chain ends will be able to rotate more readily than the rest of the chain, and the more chain ends a sample has the greater the contribution to the free volume when these begin moving. Consequently, the glass transition temperature is lowered. Bueche (1962) expressed this as

$$T_g(\infty) = T_g + K/M = T_g + (2\rho N_A \theta/\alpha_f M_n), \tag{12.20}$$

where $T_g(\infty)$ is the glass temperature of a polymer with a very large molar mass, θ is the free-volume contribution of one chain end and is 2θ for a linear polymer, ρ is the polymer density, N_A is Avogadro's constant, and α_f is the free-volume expansivity defined as

$$\alpha_f = (\alpha_L - \alpha_G). \tag{12.21}$$

The linear expression in Equation 12.20 has been widely used and describes the behavior of many polymer systems over a reasonable range of molar mass (> 5000). For short chains, the relationship is no longer valid, and it has been shown that if T_g is plotted against $\log x$, where x is the number of atoms or bonds in the polymer backbone, then three distinct regions can be identified for a number of common amorphous polymers (Figure 12.6). Region I denotes the range of chain lengths at which T_g reaches its asymptotic value $T_g(\infty)$, and the critical value x_c at which this occurs increases as the chain becomes more rigid. Thus, x_c is approximately 90 for a flexible polymer poly(dimethyl siloxane) but nearer 600 for the more rigid poly(α-methyl styrene).

The relationship between $T_g(\infty)$ and x_c is then

$$T_g(\infty) = 372.6 \log x_c - 595 \tag{12.22}$$

In region II, T_g is dependent on the molar mass and can be described by Equation 12.22, but, on entering region III where the decrease in T_g accelerates, this is no longer true. The latter region incorporates the material that is olilgomeric, and the

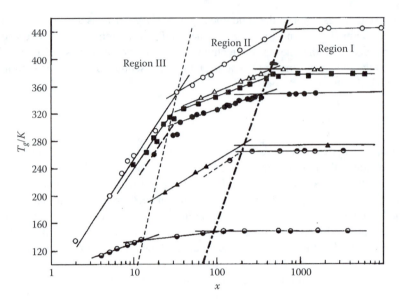

FIGURE 12.6 Plot of the glass transition temperature as a function of log x, where x is the number of chain atoms or bonds in the backbone. Data for $(-\bigcirc-)$ poly(α-methyl styrene), $(-\triangle-)$ poly(methyl methacrylate) $(-\blacksquare-)$ polystyrene, $(-\bullet-)$ poly(vinylchloride), $(-\blacktriangle-)$ isotactic poly-propylene, $(-\ominus-)$ atactic polypropylene, and $(-\ominus-)$ poly(dimethylsiloxane). (From Cowie, J.M.G., *Eur. Polym. J.*, 11, 297, 1975. With permission of Pergamon Press.)

line separating II and III represents the oligomer-polymer transition where the chains begin to become long enough to be considered capable of adopting a Gaussian coil conformation.

12.7 STRUCTURAL RELAXATION AND PHYSICAL AGING

Rapid cooling of amorphous polymers from the equilibrium liquid state to below the glass transition produces glassy materials that are in a nonequilibrium thermodynamic state. This is schematically illustrated in Figure 12.7 where the change in enthalpy, H, which a glass-forming liquid undergoes during cooling through the transition region, is reported. The starting point, A, is the equilibrium liquid state. Quenching below T_g involves a gradual slowing down of molecular motion and generates a glass that is not in conformational equilibrium with its temperature (point B). Annealing at the aging temperature T_a causes the system to change its physical properties, such as specific volume and enthalpy, toward the thermodynamic equilibrium state corresponding to that temperature. With reference to Figure 12.7, the result of annealing at temperature T_a for annealing time t_a is a change from enthalpy state B to C. Point E on the equilibrium enthalpy line would be reached if the sample were to be annealed at T_a for an infinitely long time. The rate at which the system approaches equilibrium becomes longer with increasing separation from the T_g and is particularly pronounced for temperatures of more than 15°C below the glass

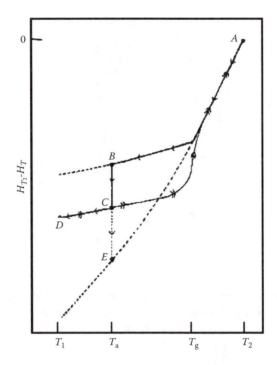

FIGURE 12.7 Illustration of thermal history used in enthalpic relaxation experiments via an enthalpy vs. temperature plot. (Adapted from Cowie, J.M.G. and Ferguson, R., *Macromolecules*, 22, 2307, 1989.)

transition. Upon reheating, starting from point *D* below T_a, the structure initially follows the *D-C* path until molecular motion is high enough, after which *H* converges to the equilibrium value. In this temperature range, the slope is greater than that of the equilibrium curve, and the heat capacity displays a characteristic peak.

The phenomenon described here is termed *physical aging* or *structural relaxation*. It can be detected through the time evolution of not only thermodynamic properties such as specific volume or enthalpy but also mechanical or dielectric properties.

It is believed that physical aging involves some change in the average structure of the liquid, and it should not be confused with nonreversible processes due to chemical aging, such as thermal degradation and photooxidation. Nevertheless, physical aging effects have practical implications, and there have been many attempts to develop models that quantitatively describe the aging processes. In practice, aging tends to render the material more brittle, and this may impair any long-term applications.

PROBLEMS

1. Specific volume against temperature data for poly(vinyl acetate) are tabulated in the following (Meares, 1957). Locate the glass transition and estimate its width, α_s and α_l (the specific volume at the glass transition, V_g, is equal to 0.839 cm³ mol⁻¹).

T (°C)	V (cm³ mol⁻¹)
10.1	0.8354
12.0	0.8359
14.0	0.8362
16.0	0.8366
17.9	0.8370
20.0	0.8374
22.1	0.8379
24.2	0.8384
26.2	0.8393
27.8	0.8403
30.2	0.8413
32.0	0.8426
33.9	0.8435
36.1	0.8449
37.9	0.8460
39.0	0.8471
42.0	0.8484

2. Consider the following polymer pairs and explain, giving reasons, which of the two polymers should display a glass transition at higher temperature:
    a. Poly(p-chlorostyrene) or poly(p-methyl styrene)
    b. Poly(1-vinyl naphthalene) ($V_m = 143.9$ cm³ mol⁻¹) or poly(1-vinyl biphenyl) ($V_m = 184.0$ cm³ mol⁻¹)
    c. Poly(tetramethylene sebacate) or poly(hexamethylene adipamide) (nylon-6,6)
    d. Poly(ethylene adipate) or poly(ethylene terephthalate)
    e. Poly(1,4-cyclohexane dimethylene terephthalate) or poly(hexamethylene terephthalate)
    f. Cellulose triacetate or cellulose (2,3) acetate.

3. Figure 12.8 shows a plot of specific volume vs. temperature for the poly(vinyl acetate)/acetonyl acetone system at a solvent volume fraction of 0.14 (Kokes et al., 1952). By comparing these data with your results for Problem 1, discuss the effect of solvent addition on the specific volume and T_g.

4. Polypropylene (PP) is available in three different tactic forms: (a) atactic ($T_g = 260$ K), isotactic ($T_g = 260$ K and $T_m = 440$ K), and syndiotactic ($T_g = 260$ K and $T_m = 400$ K).
    a. Sketch on the same graph specific volume–temperature curves for (i) atactic PP, (ii) a highly crystalline isotactic PP sample, and (iii) a syndiotactic polymer with lower degree of crystallinity.
    b. Draw DSC traces for these three samples.

5. Which of the following polymers will have the highest and which the lowest glass transition: (a) poly(methyl acrylate) ($R_1 = H$ and $R_2 = CH_3$), (b) poly(methyl methacrylate) ($R_1 = CH_3$ and $R_2 = CH_3$), and (c) poly(pentyl acrylate)? Explain.

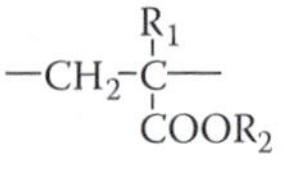

$$-CH_2-\overset{\overset{R_1}{|}}{\underset{\underset{COOR_2}{|}}{C}}-$$

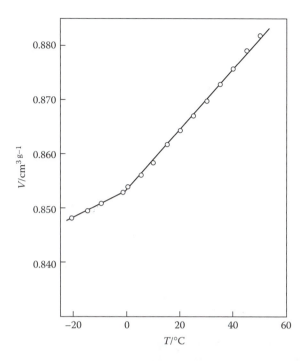

FIGURE 12.8 Volume–temperature plot for poly(vinyl acetate)/acetonyl acetone system (solvent volume fraction is 0.14).

6. Poly(di-n-alkyl itaconate)s have the general structure:

$$-CH_2-C-\begin{matrix}CH_2-CO\,(CH_2)_nCH_3\\|\\CO\,(CH_2)_nCH_3\end{matrix}$$

If the glass transitions of the dimethyl ($n = 0$), dipropyl ($n = 2$), and dihexyl ($n = 5$) polymers are 368, 307, and 255 K, respectively, estimate the glass transition temperatures of the diethyl and dibutyl itaconates.

7. Poly(ethylene oxide) (PEO) is able to complex with cations such as Li$^+$ and Na$^+$, and such systems have potential interest as battery electrolytes. Would you expect the glass transition of the PEO complex to be higher or lower than the pure polymer T_g ($= -60°C$)?

8. The glass transition temperature of 15 poly(dimethyl siloxane) (PDMS) samples and related oligomers were measured by means of differential scanning calorimetry (Cowie and McEwen, 1973):

M_n (g mol^{-1})	x	T_g (K)
162	4	112
236	6	123
310	8	128
384	10	133
540	14	136
1200	32	141
2400	64	146
4200	114	147
7700	208	148
12200	330	148
15500	418	148
19000	514	148
27000	730	148
57000	1540	148
136000	3680	149

Demonstrate that T_g varies with M_n according to Equation 12.20 and determine $T_g(\infty)$ and K. Then considering the expression for K given by Bueche (1962), and $\alpha_L = 9.5 \times 10^{-4}$ K^{-1}, $\alpha_L = 4.5 \times 10^{-4}$ K^{-1}, and $\rho = 1.105$ g cm^{-3}, evaluate θ, the free-volume contribution of a chain end.

9. Copolymers of ethylene and vinyl acetate are commercial materials prepared by free-radical polymerization. Explain how copolymerization with vinyl acetate affects the degree of crystallinity of polyethylene and the polymer chain flexibility at room temperature. How would you expect T_g to vary with increasing vinyl acetate content?

REFERENCES

Adam, G. and Gibbs, J.H., *J. Chem. Phys.*, 43, 139, 1965.
Beevers, R.B. and White, E.F.T., *Trans. Faraday Soc.*, 56, 744, 1960.
Boyer, R.F., *Rubber Chem. Technol.*, 36, 1303, 1963.
Cowie, J.M.G., *Eur. Polym. J.*, 11, 297, 1975.
Cowie, J.M.G. and Ferguson, R., *Macromolecules*, 22, 2307, 1989.
Cowie, J.M.G. and McEwen, I.J., *Polymer*, 14, 423, 1973.
de Gennes, P.G., *Physics Today*, 33, 1983.
Fox, T.G. and Flory, P.J., *J. Phys. Chem.*, 55, 221, 1951.
Gibbs, J.H. and DiMarzio, E.A., *J. Chem. Phys.*, 28, 373, 1958.
Kokes, R.J., Long, F.A., and Hoard, J.L., *J. Chem Phys.*, 20, 1711, 1952.
Kovacs, A.J., *J. Polym. Sci.*, 30, 131, 1958.
Meares, P., *Trans. Faraday Soc.*, 53, 31, 1957.
O'Reilly, J.M. and Karasz, F.E., *J. Polym. Sci.*, C (14), 49, 1966.
Petrie, S.E.B., *J. Macromol. Sci.: Phys.*, B12(2), 225, 1976.

BIBLIOGRAPHY

Allen, G. and Bevington, J.C., Eds., *Comprehensive Polymer Science,* Vol. 2, Pergamon Press, 1989.

Bueche, F., *Physical Properties of Polymers,* Interscience Publishers, 1962.

Doi, M. and Edwards, S.F., *The Theory of Polymer Dynamics,* Oxford University Press, 1986.

Ferry, J.D., *Viscoelastic Properties of Polymers,* 3rd ed., John Wiley and Sons, 1979.

de Gennes, P.G., *Scaling Concepts in Polymer Physics,* Cornell University Press, 1979.

Mark, J.E., Ngai, K., Graessley, W., Mandelkern, L., Samulski, E., Koenig, J.L., and Wignall, G., *Physical Properties of Polymers,* 3rd ed., Cambridge University Press, 2004.

Meares, P., *Polymers: Structures and Bulk Properties,* Van Nostrand 1965, chap. 10.

Pethrick, L.H. and Richards, R.W., Eds., *Static and Dynamic Properties of the Polymeric Solid State,* D. Reidel Publishing Co., 1982.

Sperling, L.H., *Introduction to Physical Polymer Science,* 4th ed., John Wiley and Sons, 2005.

Tobolsky, A.V. and Mark, H., *Polymer Science and Materials,* Wiley-Interscience, 1971, chap. 6.

Ward, I.M., *Mechanical Properties of Solid Polymers,* 2nd ed., John Wiley and Sons, 1983.

13 Rheology and Mechanical Properties

13.1 INTRODUCTION TO RHEOLOGY

The fabrication of an article from a polymeric material in the bulk state, whether it be the molding of a thermosetting plastic or the spinning of a fiber from the melt, involves deformation of the material by applied forces. Afterward, the finished article is inevitably subjected to stresses; hence, it is important to be aware of the mechanical and rheological properties of each material, as well as understand the basic principles underlying their response to such forces.

In a broad sense, the term *rheology* (*rheo* meaning "to flow") refers to both studies of deformation as well as flow of materials under the influence of applied forces. Thus, the rheological behavior of polymers encompasses a wide range of macroscopic phenomena including

1. Flow of viscous liquids, i.e., irreversible deformation
2. Mechanical properties of elastic solids
3. Viscoelasticity, i.e., time dependent properties of polymers

In classical terms, the mechanical properties of elastic solids can be described by Hooke's law, which states that an applied stress is proportional to the resultant strain but is independent of the rate of strain. For liquids, the corresponding statement is known as Newton's law, with the stress now independent of the strain but proportional to the rate of strain. Both are limiting laws, valid only for small strains or rates of strain, and although it is essential that conditions involving large stresses, leading to eventual mechanical failure, be studied, it is also important to examine the response to small mechanical stresses. Both laws can prove useful under these circumstances.

In many cases, a material may exhibit the characteristics of both a liquid and a solid, and neither of the limiting laws will adequately describe its behavior. The system is then said to be in a *viscoelastic state*. A particularly good illustration of a viscoelastic material is provided by a silicone polymer known as "bouncing putty." If a sample is rolled into the shape of a sphere, it can be bounced like a rubber ball, i.e., the rapid application and removal of a stress causes the material to behave like an elastic body. If, on the other hand, a stress is applied slowly over a longer period the material flows like a viscous liquid, and the spherical shape is soon lost if left to stand for some time. Pitch behaves in a similar, if less spectacular, manner.

13.2 THE FIVE REGIONS OF VISCOELASTIC BEHAVIOR

The physical nature of an amorphous polymer is related to the extent of the molecular motion in the sample, which in turn is governed by the chain flexibility and the temperature of the system. Examination of the mechanical behavior shows that there are five distinguishable states in which a linear amorphous polymer can exist, and these are readily displayed if a parameter such as the elastic modulus is measured over a range of temperatures.

The general behavior of a polymer can be typified by results obtained for an amorphous atactic polystyrene sample. The relaxation modulus E_r was measured at a standard time interval of 10 s and $\log_{10}E_r$ is shown as a function of temperature in Figure 13.1. Five distinct regions can be identified on this curve as follows:

1. *The glassy state:* This is section A to B lying below 363 K, and it is characterized by a modulus between $10^{9.5}$ and 10^9 Nm^{-2}. Here, cooperative molecular motion along the chain is frozen, causing the material to respond like an elastic solid to stress, and the strain-time curve is of the form shown in Figure 13.1(a).

2. *Leathery or retarded highly elastic state:* This is the transition region B to C where the modulus drops sharply from about 10^9 to about $10^{5.7}$ Nm^{-2} over the temperature range 363 to 393 K. The glass transition temperature T_g is located in this area, and the rapid change in modulus reflects the constant increase in molecular motion as the temperature rises from T_g to about $(T_g + 30$ K). Just above T_g the movement of the chain segments is still rather slow, imparting what can best be described as leathery properties to the material. The strain–time curve is shown in Figure 13.1(b).

3. *The rubbery state:* At approximately 30 K above the glass transition the modulus curve begins to flatten out into the plateau region C to D in the modulus interval $10^{5.7}$ to $10^{5.4}$ Nm^{-2} and extends up to about 420 K.

4. *Rubbery flow:* After the rubbery plateau the modulus again decreases from $10^{5.4}$ to $10^{4.5}$ Nm^{-2} in the section D to E. The effect of applied stress to a polymer in states (3) and (5) is shown in Figure 13.1(c), where there is instantaneous elastic response followed by a region of flow.

5. *Viscous state:* Above a temperature of 450 K, in the section E to F, there is little evidence of any elastic recovery in the polymer, and all the characteristics of a viscous liquid become evident [Figure 13.1(d)]. Here, there is a steady decrease of the modulus from $10^{4.5}$ Nm^{-2} as the temperature increases.

The overall shape of the curve shown in Figure 13.1 is typical for linear amorphous polymers in general, although the temperatures quoted are specific to polystyrene and will differ for other polymers. Variations in shape are found for different molar masses when the sample is cross-linked or partly crystalline. The value of the modulus provides a good indication of the state of the polymer and can be obtained from the curve.

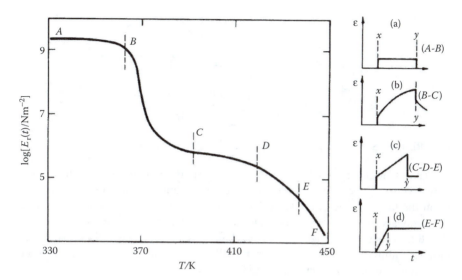

FIGURE 13.1 Five regions of viscoelasticity, illustrated using a polystyrene sample. Also shown are the strain–time curves for stress applied at x and removed at y: (a) glassy state, (b) leathery state, (c) rubbery state, and (d) viscous state.

Before examining the viscoelastic behavior of amorphous polymeric substances in more detail, some of the fundamental properties of polymer melts and elastic solids will be reviewed.

13.3 THE VISCOUS REGION

The application of a force to a simple liquid of low molar mass is relieved by the flow of molecules past one another into new positions in the system. This situation is schematically represented in Figure 13.2, where a volume element undergoing deformation by simple shear due to a force applied parallel to the upper plate is shown. The bottom plate is stationary, as is the layer of fluid immediately adjacent to it.

The resistance to deformation, i.e., the fluid's viscosity, is defined in terms of the *shear stress*, τ:

$$\tau = \frac{F}{A} \qquad (13.1)$$

given by the ratio between the force, F, and the cross-sectional area of the upper plate, A, and the *shear strain*, γ, which, for small deformations, is equal to

$$\gamma = \tan\theta = \frac{X}{Y} \qquad (13.2)$$

where θ is the angle that characterizes the deformation, and X and Y are defined in Figure 13.2.

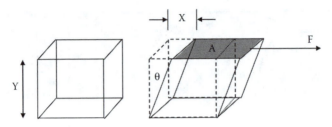

FIGURE 13.2 Schematic illustration of a simple shear experiment.

The situation described in Figure 13.2 corresponds to one where the upper plate moves with a velocity v_0 and the bottom plate is stationary; a velocity profile sets up in the fluid between the plates. Molecules that are close to the bottom surface are stationary, whereas those in contact with the upper plate move with velocity v_0. The fluid can be subdivided into ideal layers, with molecules within each layer moving with the same velocity, v_x; there will be a velocity gradient from the top to the bottom surface, varying from v_0 to 0.

An ideal fluid is one that obeys Newton's law

$$\tau = \eta \, \dot{\gamma} \tag{13.3}$$

according to which a linear relationship between shear stress and shear rate (or velocity gradient) exists:

$$\dot{\gamma} = \frac{d\gamma}{dt} \tag{13.4}$$

The proportionality constant η is the coefficient of viscosity of the fluid, and for a Newtonian fluid η is independent of $\dot{\gamma}$, but substances that show deviations from this flow pattern, with either decreasing or increasing $(\tau/\dot{\gamma})$ ratios, are termed non-Newtonian (Figure 13.3). Most polymers fall into this latter category, with η decreasing as the shear rate increases.

Examples of non-Newtonian behavior are shown in Figure 13.3. Shear-thickening (*dilatant*) is observed when the resistance to deformation increases with the shear rate, whereas the opposite is true for a system undergoing shear-thinning (*pseudoplastic*). In some cases, a minimum critical value of stress, τ_c, is needed for flow to occur (*Bingham* flow). This situation is characteristic of fluids where a structured arrangement of the molecules exists, and therefore a critical τ_c is required to break down the structure.

Units of viscosity are:

$$\eta = \frac{\tau}{\dot{\gamma}} = \left(\frac{\text{Force}}{\text{Area}} \right) \cdot \text{time} = \text{Pa} \cdot \text{s} \tag{13.5}$$

but it is usually expressed in 1 P (poise) = 1 dyne s/cm².

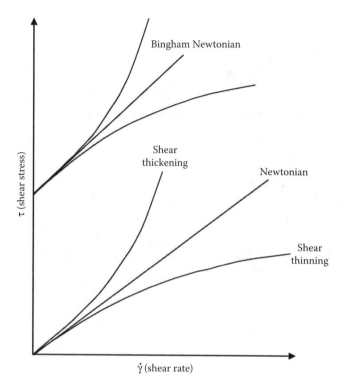

FIGURE 13.3 Newtonian and non-Newtonian flow curves.

Various instruments are available to measure the viscosity of polymer melts and solutions, and more generally their rheological behavior, which include capillary and rotational viscometers. The former can be used to measure parameters such as shear viscosity, melt fracture, and extensional viscosity, which are important for many polymer processes. The latter type of device can be used in either steady or oscillatory mode, thus providing a measure of the viscosity as well as viscoelasticity (G'', G', and tan δ) as a function of frequency and temperature.

Figure 13.4 gives an example of a rheometer used to investigate polymer fluids. Whereas the instrument operates on the basic principle that a fluid is interspersed between two components, moving one past the other, different measuring devices are available, of which the most common are concentric cylinders, cone-and-plate, and parallel plate (Figure 13.4).

These different geometries present advantages and disadvantages. For example, the cone-and-plate arrangement offers a constant shear rate across the sample $\dot{\gamma}$, whereas in the parallel plate geometry, though it is easy to set and vary the gap, the shear rate is zero at the center and a maximum at the rim.

13.3.1 SHEAR DEPENDENCE OF VISCOSITY

Polymer fluids, including polymer solutions and melts, show deviations from ordinary Newtonian behavior, usually of the shear-thinning type. If the flow properties

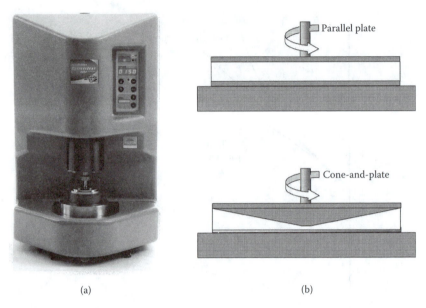

FIGURE 13.4 (a) Bohlin Gemini HRnano Rheometer by Malvern Instruments, and (b) parallel plate and cone-and-plate geometry.

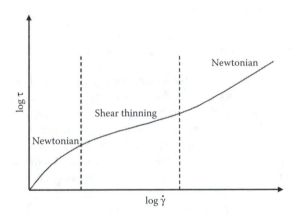

FIGURE 13.5 Idealized flow behavior of polymer melts and solutions indicating Newtonian and non-Newtonian regions.

of a polymer solution or melt were to be measured over a wide range of shear rates, one would expect to observe a region of Newtonian behavior at very low $\dot{\gamma}$, followed by a shear dependent viscosity, and then, at very high shear rates, a further Newtonian region where η is independent of the shear rate (Figure 13.5).

This behavior can be rationalized by considering that, at low shear rates, the molecules are in a highly entangled state, and the rate at which entanglements are disrupted by the flow field is counteracted by the formation of new entanglements. However, with increasing shear rate, the entanglement density decreases and therefore the resistance to flow, i.e., the viscosity. In an ideal case, when reentanglement can no longer take place, a further region of Newtonian behavior at high shear rate is reached.

Whereas a variety of *constitutive equations* have been proposed to describe non-Newtonian flow, one of the most useful relationships is the so-called power law

$$\tau = K \cdot (\dot{\gamma})^n \tag{13.6}$$

where K is a constant called the *consistency* and n is the *power-law index.*

One approach that is consistent with the generalized behavior in Figure 13.5 was developed by Cross (1965), and it gives a relationship between the viscosity and shear rate of the form:

$$\eta = \eta_\infty + \frac{\eta_0 - \eta_\infty}{1 + (a\dot{\gamma})^p} \tag{13.7}$$

where a and p are constants for any given system, η_0 is the zero-shear rate viscosity (the maximum value of viscosity at low shear rates) and η_∞ is the minimum viscosity obtained in the Newtonian region at high shear rates.

13.3.2 Kinetic Units in Polymer Chains

Resistance to flow in polymer systems is greater than in low molar mass fluids, because now the molecules are covalently bonded into long chains, which are coiled and entangled, and translational motion must, of necessity, be a cooperative process. It would be unreasonable to expect easy cooperative motion along the entire polymer chain, but as there is normally some degree of flexibility in the chain, local segmental motion can take place more readily. The polymer can then be considered as a series of kinetic units; each of these moves in an independent manner and involves the cooperative movement of a number of consecutive chain atoms.

Crankshaft motion. If we now consider an arbitrary kinetic unit that involves the movement of six atoms by rotation about two chain bonds, the movement can be visualized as shown diagrammatically in Figure 13.6. The amorphous or molten polymer is a conglomeration of badly packed interlacing chains, and the extra empty space caused by this random molecular arrangement is called the *free volume*, which

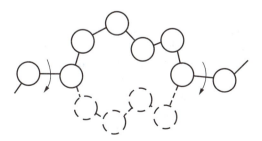

FIGURE 13.6 Crankshaft motion in a polymer chain.

essentially consists of all the holes in the matrix. When sufficient thermal energy is present in the system the vibrations can cause a segment to jump into a hole by cooperative bond rotation, and a series of such jumps will enable the complete polymer chain eventually to change its position. Heating will cause a polymer sample to expand, thereby creating more room for movement of each kinetic unit, and the application of a stress in a particular direction will encourage flow by segmental motion in the direction of the stress. The segmental transposition involving six carbon atoms is called *crankshaft motion* and is believed to require an activation energy of about 25 kJ mol^{-1}.

13.3.3 Effect of Chain Length

Although it is thought that translation of a polymer chain proceeds by means of a series of segmental jumps involving short kinetic units, which may each consist of 15 to 30 chain atoms, the complete movement of a chain cannot remain unaffected by the surrounding chains. As stated previously, considerable entanglement exists in the melt, and any motion will be retarded by other chains.

According to Bueche (1962), the polymer molecule may drag along several others during flow, and the energy dissipation is then a combination of the friction between the chain plus those that are entangled and the neighboring chains as they slip past each other. It would seem reasonable to assume from this that the length of the chains in the sample must play a significant role in determining the resistance to flow and the effect of chain length on log η_0, the logarithm of the so-called zero shear rate viscosity, i.e., the viscosity measured at low shear rates to ensure Newtonian flow, is illustrated in Figure 13.7. The plot comprises two linear portions meeting at a critical chain length $\mathcal{Z}_c$. Above $\mathcal{Z}_c$ the relation describing the flow behavior is

$$\log \eta = 3.4 \log \mathcal{Z} + \log K_2, \tag{13.8}$$

and η is proportional to the 3.4 power of $\mathcal{Z}$. Below $\mathcal{Z}_c$, η is directly proportional to $\mathcal{Z}$, and the expression becomes

$$\log \eta = \log \mathcal{Z} + \log K_1, \tag{13.9}$$

where K_1 and K_2 are temperature dependent constants.

The critical chain length $\mathcal{Z}_c$ is interpreted as representing the dividing point between chains, which are too short to provide a significant contribution to η from entanglement effects and those large enough to cause retardation of flow by intertwining with their neighbors. If $\mathcal{Z}$ is defined as the number of atoms in the backbone chain of a polymer, then typical values for $\mathcal{Z}_c$ are 610 for polyisobutylene, 730 for polystyrene, and 208 for poly(methyl methacrylate). In general, $\mathcal{Z}_c$ is lower for polar polymers than for nonpolar polymers.

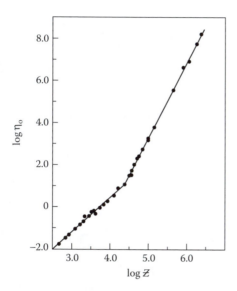

FIGURE 13.7 Dependence of melt viscosity on chain length $\mathcal{Z}$, for polyisobutylene fractions measured at low shear rates and at 490 K. (Data by Fox, T.G. and Flory, P.J., *J. Phys. Chem.* 55, 221, 1951.)

13.3.4 TEMPERATURE DEPENDENCE OF η

The temperature dependence of η can normally be expressed in the form

$$\eta = A \exp (\Delta H/RT) \tag{13.10}$$

where A is a constant and, according to the Eyring rate theory, ΔH is the activation enthalpy of viscous flow, a more representative parameter than the energy. Values of ΔH vary slowly over a range of from 20 to 120 kJ mol^{-1}.

Equation 13.10 describes the temperature dependence of the viscosity of a polymer melt at temperatures in excess of 100 K above T_g, provided the material is stable. When the temperature is lowered toward T_g, ΔH changes dramatically and a simple equation such as Equation 13.10 is no longer valid. The increase in ΔH, observed with temperature lowering, can be equated with a rapid loss of free volume as T_g is approached. Hence, ΔH becomes dependent on the availability of a suitable hole for a segment to move into, rather than being representative of the potential energy barrier to rotation. This approach suggests that the jump frequency decreases when there is an increasing cooperative motion among the chains needed to produce holes. At these temperatures the WLF equation (Equation 13.44), of which Equation 13.10 is an approximated form, provides a more adequate representation of the viscosity changes.

13.3.5 CONCENTRATION DEPENDENCE OF VISCOSITY

The rheological properties of polymer solutions are remarkably similar to those of polymers in the undiluted state: parameters such as molar mass, temperature, and

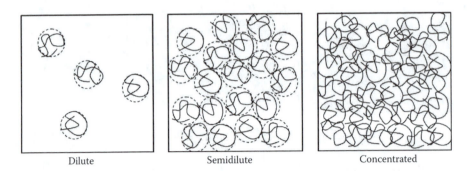

| | Dilute | Semidilute | Concentrated |

FIGURE 13.8 Schematic representation of dilute, semidilute, and concentrated polymer systems.

shear rate affect the viscosity of polymer solutions in a way similar to polymer melts. There are two main differences: (1) the presence of small solvent molecules makes the viscosity of polymer solutions decrease compared to that of a polymer in the undiluted state, and (2) concentration is an additional variable.

Polymer solutions can be classified according to one of three concentration regimes and the diagram in Figure 13.8 gives a schematic representation of the dilute, semidilute, and concentrated states. In the dilute regime, the polymer coils are well separated, and the viscosity behavior, which is determined by the individual contributions of the molecules, has been described in Chapter 9.

In semidilute solutions the polymer coils start to overlap, and this situation is realized when the spatial extent of one molecule is close to the volume per polymer molecule of the solution. If the rms end-to-end distance, $\langle r^2 \rangle^{1/2}$, is taken as a measure of spatial extent, then overlap will take place when the solution volume per molecule $V^*/N = (2 \langle r^2 \rangle^{1/2})^3$ and the overlap concentration, c^*:

$$c^* \propto M/(N_A \langle r^2 \rangle^{3/2}) \tag{13.11}$$

By considering that $\langle r^2 \rangle^{1/2}$ scales as M^ν, it follows that the critical concentration c^* will consequently scale as $M^{1-3\nu}$.

At concentrations above the overlap concentration c^*, the viscosity varies with molecular weight in a way similar to that described in Section 13.6. Thus, at a critical molecular weight, M_c, the behavior of the viscosity varies from $\eta \propto M$ to $\eta \propto M^{3.4}$. As the change is a result of the entanglements between polymer coils, then as the polymer concentration decreases so does the tendency of the chains to become entangled, and therefore M_c will increase.

13.3.6 TIME-DEPENDENT BEHAVIOR

One further variable that greatly influences flow behavior is time. It is often found that, at constant shear rate, a fluid's viscosity varies as a function of time. The two types of time dependent flow behavior are shown in Figure 13.9. Thixotropy is encountered when the viscosity decreases with time, returning to its original value

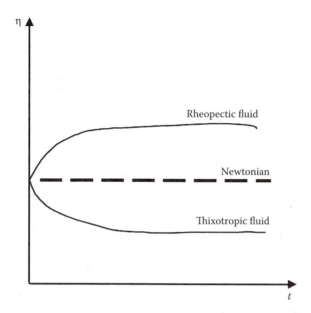

FIGURE 13.9 Schematic representation of time-dependent behavior.

after a period of rest. This situation is encountered when an internal network structure exists that is broken down by shearing.

The opposite situation is encountered when the viscosity increases as a function of time and shear rate. This behavior, albeit less common, is termed *antithixotropic* or *rheopectic* behavior.

Having reviewed some of these aspects of viscous flow, we now turn to the low-temperature behavior of elastic solids.

13.4 MECHANICAL PROPERTIES

Homogeneous, isotropic, elastic materials possess the simplest mechanical properties, and three elementary types of elastic deformation can be observed when such a body is subjected to (1) simple tension, (2) simple shear, and (3) uniform compression.

Simple tension. Consider a parallelepiped of length x_0 and cross-sectional area $A_0 = y_0 z_0$ [Figure 13.10(a)]. If this is subjected to a balanced pair of tensile forces F, its length changes by an increment dx so that $x_0 + dx = x$. When dx is small, Hooke's law is obeyed, and the tensile *stress* σ is proportional to the tensile strain, ε. The constant of proportionality is known as the *Young's modulus*, E, and for elastic solids

$$\sigma = E\varepsilon \qquad (13.12)$$

The stress σ is a measure of the force per unit area (F/A), and the strain or elongation is defined as the extension per unit length, i.e., $\varepsilon = (dx/x_0)$. Of course,

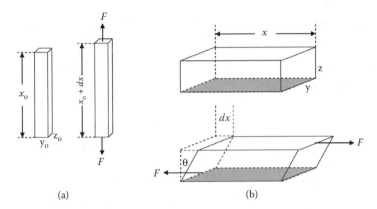

FIGURE 13.10 (a) Tensile stressing of a bar. (b) Shearing of a rectangular block, with a balanced pair of forces F.

the extension dx will be accompanied by lateral contractions dy and dz, but although normally negative and equal, they can usually be assumed to be zero.

For an isotropic body, the change in length per unit length is related to the change in width per unit of length, such that

$$v_P = (dy/y_0)/(dx/x_0) \tag{13.13}$$

where v_P is known as *Poisson's ratio*, and varies from 0.5, when no volume change occurs, to about 0.2.

Simple shear. In simple shear the shape change is not accompanied by any change in volume. If the base of the body, shown shaded in Figure 13.10(b) is firmly fixed, a transverse force F applied to the opposite face is sufficient to cause a deformation dx through an angle θ. The shear modulus G is then given by the quotient of the shearing force per unit area and the shear per unit distance between shearing surfaces, and so

$$G = \sigma_s/\varepsilon_s = (F/yz)/(dx/y) = F/A \tan\theta. \tag{13.14}$$

For very small shearing strains $\tan\theta \approx \theta$ and

$$G = F/A\theta. \tag{13.15}$$

Both E and G depend on the shape of the specimen, and it is usually necessary to define the shape carefully for any measurement.

Uniform compression. When a hydrostatic pressure, p, is applied to a body of volume V_0, causing a change in volume ΔV, a bulk modulus B can be defined as

$$B = -p/(\Delta V/V_0). \tag{13.16}$$

The quantity B is often expressed in terms of the compressibility, which is the reciprocal of the bulk modulus. Similarly, E^{-1} and G^{-1} are known as the *tensile* and *shear* compliances, and given the symbols D and J, respectively.

13.4.1 INTERRELATION OF MODULI

The relations given previously pertain to isotropic bodies, and for nonisotropic bodies the equations are considerably more complex. Polymeric materials are normally either amorphous or partially crystalline with randomly oriented crystallites embedded in a disordered matrix. However, any symmetry possessed by an individual crystallite can be disregarded, and the body as a whole is treated as being isotropic.

The various moduli can be related to each other in a simple manner because an isotropic body is considered to possess only two independent elastic constants; therefore,

$$E = 3B(1 - 2v_P) = 2(1 + v_P)G. \tag{13.17}$$

This indicates that for an incompressible elastic solid, i.e., one having a Poisson ratio of 0.5, Young's modulus is three times larger than the shear modulus. These moduli have dimensions of pressure, and typical values for several polymeric and nonpolymeric materials can be compared at ambient temperatures in Table 13.1 $(GN = 10^9 N)$.

The response of polymers to mechanical stresses can vary widely and depends on the particular state the polymer is in at any given temperature.

13.5 MECHANICAL MODELS DESCRIBING VISCOELASTICITY

A perfectly elastic material obeying Hooke's law behaves like a perfect spring. The stress-strain diagram is shown in Figure 13.11(a), and can be represented in mechanical terms by the model of a weightless *spring* whose modulus of extension represents the modulus of the material.

TABLE 13.1
Comparison of Various Moduli for Some Common Materials

Material	E (GN m^{-2})	v_P	G (GN m^{-2})
Steel	220	0.28	85.9
Copper	120	0.35	44.4
Glass	60	0.23	24.4
Granite	30	0.30	15.5
Polystyrene	34	0.33	1.28
Nylon-6,6	20	—	—
Polyethylene	24	0.38	0.087
Natural rubber	0.02	0.49	0.00067

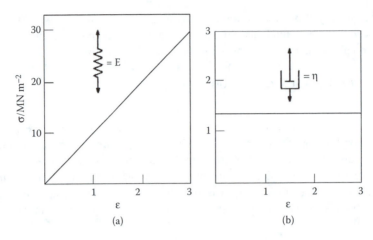

FIGURE 13.11 Stress–strain (σ–ε) behavior of (a) spring of modulus E and (b) a dashpot of viscosity η.

The application of a shear stress to a viscous liquid on the other hand, is relieved by viscous flow, and for small values of σ_s can be described by Newton's law

$$\sigma_s = \eta \, d\varepsilon_s / dt, \tag{13.18}$$

where η is the coefficient of viscosity and $(d\varepsilon_s/dt)$ is the rate of shear sometimes denoted by $\dot{\gamma}$. As stress is now independent of the strain, the form of the diagram changes and can be represented by a *dashpot*, which is a loose fitting piston in a cylinder containing a liquid of viscosity η [Figure 13.11(b)].

Comparison of the two models shows that the spring represents a system storing energy that is recoverable, whereas the dashpot represents the dissipation of energy in the form of heat by a viscous material subjected to a deforming force. The dashpot is used to denote the retarded nature of the response of a material to any applied stress.

Because of their chain-like structure, polymers are not perfectly elastic bodies, and deformation is accompanied by a complex series of long and short range cooperative molecular rearrangements. Consequently, the mechanical behavior is dominated by viscoelastic phenomena, in contrast to materials such as metal and glass where atomic adjustments under stress are more localized and limited.

The Maxwell model. One of the first attempts to explain the mechanical behavior of materials such as pitch and tar was made by James Clark Maxwell. He argued that when a material can undergo viscous flow and also respond elastically to a stress it should be described by a combination of both the Newton and Hooke laws. This assumes that both contributions to the strain are additive so that $\varepsilon = \varepsilon_{elast} + \varepsilon_{visc}$. Expressing this as the differential equation leads to the equation of motion of a Maxwell unit

$$d\varepsilon/dt = (1/G)(d\sigma/dt) + \sigma/\eta \tag{13.19}$$

Under conditions of constant shear strain ($d\sigma/dt = 0$) the relation becomes

$$d\sigma/dt + G\sigma/\eta = 0, \tag{13.20}$$

and if the boundary condition is assumed that $\sigma = \sigma_0$ at zero time, the solution to this equation is

$$\sigma = \sigma_0 \exp(-tG/\eta), \tag{13.21}$$

where σ_0 is the initial stress immediately after stretching the polymer. This shows that when a Maxwell element is held at a fixed shear strain, the shearing stress will relax exponentially with time. At a time $t = (\eta/G)$ the stress is reduced to $1/e$ times the original value, and this characteristic time is known as the *relaxation time* τ.

The equations can be generalized for both shear and tension, and G can be replaced by E. The mechanical analogue for the Maxwell unit can be represented by a combination of a spring and a dashpot arranged in series so that the stress is the same on both elements. This means that the total strain is the sum of the strains on each element as expressed by Equation 13.19. A typical stress–strain curve predicted by the Maxwell model is shown in Figure 13.12(a). Under conditions of constant stress, a Maxwell body shows instantaneous elastic deformation first, followed by a viscous flow.

Voigt–Kelvin model. A second simple mechanical model can be constructed from the ideal elements by placing a spring and dashpot in parallel. This is known as a *Voigt–Kelvin model.* Any applied stress is now shared between the elements, and each is subjected to the same deformation. The corresponding expression for strain is

$$\varepsilon(t) = \sigma_0 J\{1 - \exp(t/\tau_R)\}. \tag{13.22}$$

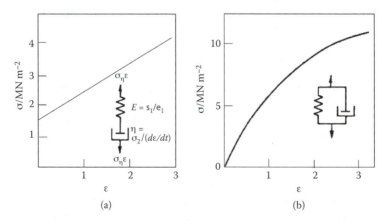

FIGURE 13.12 Stress–strain (σ–ε) behavior of two simple mechanical models: (a) the Maxwell model and (b) the Voigt–Kelvin model.

Here $\tau_R = (\eta/G)$ is known as the *retardation time* and is a measure of the time delay in the strain after imposition of the stress. For high values of the viscosity, the retardation time is long, and this represents the length of time the model takes to attain $(1 - 1/e)$ or 0.632 of the equilibrium elongation.

Such models are much too simple to describe the complex viscoelastic behavior of a polymer, nor do they provide any real insight into the molecular mechanism of the process, but in certain instances they can prove useful in assisting the understanding of the viscoelastic process.

13.6 LINEAR VISCOELASTIC BEHAVIOR OF AMORPHOUS POLYMERS

A polymer can possess a wide range of material properties, and of these the hardness, deformability, toughness, and ultimate strength are among the most significant. Certain features such as high rigidity (modulus) and impact strength combined with low creep characteristics are desirable in a polymer if eventually it is to be subjected to loading. Unfortunately, these are conflicting properties, as a polymer with a high modulus and low creep response does not absorb energy by deforming easily and hence has poor impact strength. This means a compromise must be sought depending on the use of the polymer, and this requires knowledge of the mechanical response in detail.

The early work on viscoelasticity was performed on silk, rubber, and glass, and it was concluded that these materials exhibited a *delayed elasticity* manifest in the observation that the imposition of a stress resulted in an instantaneous strain, which continued to increase more slowly with time. It is this delay between cause and effect that is fundamental to the observed viscoelastic response, and the three major examples of this hysteresis effect are (1) *creep*, where there is a delayed strain response after the rapid application of a stress, (2) *stress-relaxation* (Section 13.15), in which the material is quickly subjected to a strain and a subsequent decay of stress is observed, and (3) *dynamic response* (Section 13.17) of a body to the imposition of a steady sinusoidal stress. This produces a strain oscillating with the same frequency as, but out of phase with, the stress. For maximum usefulness, these measurements must be carried out over a wide range of temperature.

13.6.1 Creep

To be of any practical use, an object made from a polymeric material must be able to retain its shape when subjected to even small tensions or compressions over long periods of time. This dimensional stability is an important consideration in choosing a polymer for use in the manufacture of an item. No one wants a plastic telephone receiver that sags after sitting in its cradle for several weeks, or a car tire that develops a flat spot if parked in one position for too long, or clothes made from synthetic fibers that become baggy and deformed after short periods of wear. Creep tests provide a measure of this tendency to deform and are relatively easy to carry out.

Creep can be defined as a progressive increase in strain, observed over an extended time period, in a polymer subjected to a constant stress. Measurements are carried out on a sample clamped in a thermostat. A constant load is firmly fixed to one end, and the elongation is followed by measuring the relative movement of two fiducial marks, made initially on the polymer, as a function of time. To avoid excessive changes in the sample cross-section, elongations are limited to a few percent and are followed over approximately three decades of time.

The initial, almost instantaneous, elongation produced by the application of the tensile stress is inversely proportional to the rigidity or modulus of the material, i.e., an elastomer with a low modulus stretches considerably more than a material in the glassy state with a high modulus. The initial deformation corresponds to portion OA of the curve (Figure 13.13), increment a. The rapid response is followed by a region of creep, A to B, initially fast, but eventually slowing down to a constant rate represented by the section B to C. When the stress is removed the instantaneous elastic response OA is completely recovered, and the curve drops from C to D, i.e., the distance $a' = a$. There follows a slower recovery in the region D to E which is never complete, falling short of the initial state by an increment $c' = c$. This is a measure of the viscous flow experienced by the sample and is a completely non-recoverable response. If the tensile load is enlarged, both the elongation and the creep rate increase, so results are usually reported in terms of the creep compliance $J(t)$, defined as the ratio of the relative elongation y at time t to the stress so that

$$J(t) = yE/\sigma \tag{13.23}$$

At low loads, $J(t)$ is independent of the load.

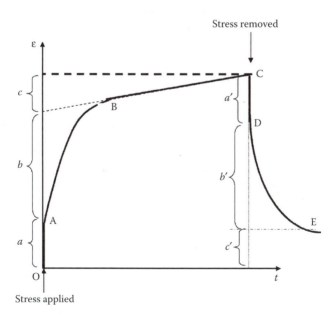

FIGURE 13.13 Schematic representation of a creep curve: a, initial elastic response; b, region of creep; c, irrecoverable viscous flow. This curve can be represented by the four-element model shown in Figure 13.14.

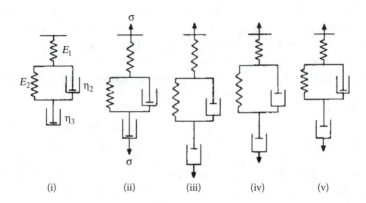

FIGURE 13.14 Use of mechanical models to describe the creep behavior of a polymeric material.

This idealized picture of creep behavior in a polymer has its mechanical equivalent constructed from the springs and dashpots described earlier. The changes a and a' correspond to the elastic response of the polymer, and so we can begin with a Hookean spring. The Voigt-Kelvin model is embodied in Equation 13.22, and this reproduces the changes b and b'. The final changes c and c' represent viscous flow, and can be shown with a dashpot so that the whole model is a four-element model (Figure 13.14).

The behavior can be explained in the following series of steps. In diagram (i) the system is at rest. The stress σ is applied to spring E_1 and dashpot η_3; it is also shared by E_2 and η_2, but in a manner that varies with time. In diagram (ii), representing zero time, the spring E_1 extends by an amount $\sigma/E_1 (= a)$. This is followed by a decreasing rate of creep with a progressively increasing amount of stress being carried by E_2 until eventually none is carried by η_2, and E_2 is fully extended [diagram (iii)]. Such behavior is described by

$$\varepsilon(t) = (\sigma_0/E_2)\{1 - \exp(-t/\tau_R)\} \tag{13.24}$$

where the retardation time τ_R provides a measure of the time required for E_2 and η_2 to reach 0.632 of their total deformation. A considerably longer time is required for complete deformation to occur. When spring E_2 is fully extended, the creep attains a constant rate corresponding to movement in the dashpot η_3. Viscous flow continues and the dashpot η_3 is deformed until the stress is removed. At that time, E_1 retracts quickly along section a' and a period of recovery ensues (b'). During this time spring E_2 forces the dashpot plunger in η_2 back to its original position. As no force acts on η_3, it remains in the extended state and corresponds to the nonrecoverable viscous flow; region $c' = \sigma t/\eta_3$. The system is then as shown in diagram (v). In practice, a substance possesses a large number of retardation times, which can be expressed as a distribution function $L_1(\tau)$ where

$$L_1(\tau) = d\{J(t) - (t/\eta)\}/d \ln t. \tag{13.25}$$

To the first approximation, this is estimated from a plot of creep compliance against $\ln t$, and (t/η) is the contribution from viscous flow.

13.6.2 STRESS–STRAIN MEASUREMENTS

The data derived from stress-strain measurements on thermoplastics are important from a practical viewpoint, providing as they do, information on the modulus, the brittleness, and the ultimate and yield strengths of the polymer. By subjecting the specimen to a tensile force applied at a uniform rate and measuring the resulting deformation, a curve of the type shown in Figure 13.15 can be constructed.

The shape of such a curve is dependent on the rate of testing; consequently, this must be specified if a meaningful comparison of data is to be made. The initial portion of the curve is linear, and the tensile modulus E is obtained from its slope. The point L represents the stress beyond which a brittle material will fracture, and the area under the curve to this point is proportional to the energy required for brittle fracture. If the material is tough, no fracture occurs, and the curve then passes through a maximum or inflection point Y, known as the *yield point*. Beyond this, the ultimate elongation is eventually reached, and the polymer breaks at B. The area under this part of the curve is the energy required for tough fracture to take place.

13.6.3 EFFECT OF TEMPERATURE ON STRESS–STRAIN RESPONSE

Polymers such as polystyrene and poly(methyl methacrylate) with a high E at ambient temperatures fall into the category of hard brittle materials which break before point Y is reached. Hard tough polymers can be typified by cellulose acetate and several curves measured at different temperatures are shown in Figure 13.16(a). Stress–strain curves for poly(methyl methacrylate) are also shown for comparison [Figure 13.16(b)].

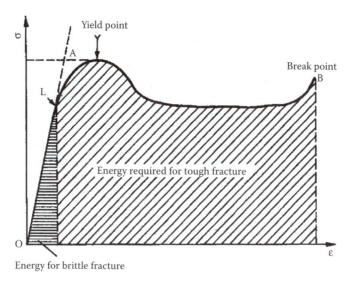

FIGURE 13.15 Idealized stress-strain curve. The slope of line OA is a measure of the true modulus.

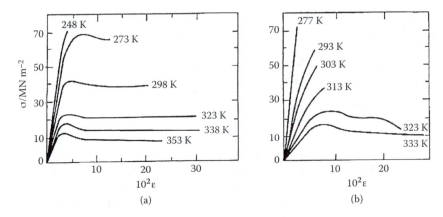

FIGURE 13.16 Influence of temperature on the stress-strain response of (a) cellulose acetate and (b) poly(methyl methacrylate). (From data by Carswell and Nason.)

It can be seen that the effect of temperature on the characteristic shape of the curve is significant. As the temperature increases, both the rigidity and the yield strength decrease, whereas the elongation generally increases. For cellulose acetate there is a transformation from a hard brittle state below 273 K to a softer but tougher type of polymer at temperatures above 273 K. For poly(methyl methacrylate) the hard brittle characteristics are retained to a much higher temperature, but it eventually reaches a soft tough state at about 320 K. Thus, if the requirements of high rigidity and toughness are to be met, the temperature is important. Cellulose acetate meets these requirements if used at 298 K more satisfactorily than when used at 350 K where the modulus is smaller, and the ability to absorb energy (represented by the area under the curve) is also lower.

13.6.4 BOLTZMANN SUPERPOSITION PRINCIPLE

If a Hookean spring is subjected to a series of incremental stresses at various times, the resulting extensions will be independent of the loading or past history of the spring. A Newtonian dashpot also behaves in a predictable manner. For viscoelastic materials the response to mechanical testing is time dependent, but the behavior at any time can be predicted by applying a superposition principle proposed by Boltzmann. This can be illustrated by a creep test using a simple Voigt–Kelvin model with a single retardation time τ_R, placed initially under a stress σ_0 at time t_0. If after times $t_1, t_2, t_3, \ldots$ the system is subjected to additional stresses $\sigma_1, \sigma_2, \sigma_3, \ldots$, then the principle states that the creep response of the system can be predicted simply by summing the individual responses from each stress increment. Thus, if the stress alters continually, the summation can be replaced by an integral, and σ_n by a continually varying function, so that at time t^* when the stress $\sigma(t^*)$ existed, the strain is given by

$$\varepsilon(t^*) = \int_0^{t^*} \frac{d\sigma(t^*)}{dt^*} \phi(t^* - t_n)\, dt. \qquad (13.26)$$

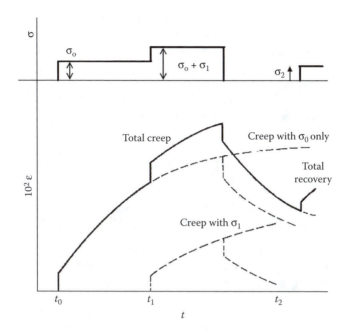

FIGURE 13.17 Application of the Boltzmann superposition principle to a creep experiment.

The principle has been applied successfully to the tensile creep of amorphous and rubber-like polymers, but it is not too successful if appreciable crystallinity exists in the sample. Graphical representation of the principle is shown in Figure 13.17.

13.6.5 STRESS RELAXATION

Stress relaxation experiments involve the measurement of the force required to maintain the deformation produced initially by an applied stress as a function of time. Stress relaxation tests are not performed as often as creep tests because many investigators believe they are less readily understood. The latter point is debatable, and it may only be that the practical aspects of creep measurements are simpler. As will be shown later, all the mechanical parameters are in theory interchangeable, and so all such measurements will contribute to the understanding of viscoelastic theory. Whereas stress relaxation measurements are useful in a general study of polymeric behavior, they are particularly useful in the evaluation of antioxidants in polymers, especially elastomers, because measurements on such systems are relatively easy to perform and are sensitive to bond rupture in the network.

In a stress-relaxation experiment, the sample under study is deformed by a rapidly applied stress. As the stress is normally observed to reach a maximum as soon as the material deforms and then decreases thereafter, it is necessary to alter this continually in order to maintain a constant deformation or measure the stress that would be required to accomplish this operation. The apparatus used varies in complexity with the physical nature of the sample, being simplest for an elastomer and becoming more sophisticated when the polymer is more rigid.

The results are expressed as a relaxation modulus $E_r(t)$, which is a function of the time of observation. Typical data for polyisobutylene are shown in Figure 13.25, where the logarithm of the relaxation modulus $\log E_r(t)$ is plotted against $\log t$. From the curves it can be seen that there is a rapid change in $\log E_r(t)$ over a narrow range of temperature corresponding to the glass transition.

Again, a simple model with a single relaxation time is too crude, and the stress-relaxation modulus $E_r(t)$ is better represented by

$$E_r(t) = \int_0^\infty H(\tau) \exp(-t/\tau) \mathrm{d}(\ln \tau), \qquad (13.27)$$

where $H(\tau)$ is the distribution function of relaxation times. This is suitable for a linear polymer, but requires the additional term E_∞ if the material is cross-linked.

13.7 DYNAMIC MECHANICAL AND DIELECTRIC THERMAL ANALYSIS

Nondestructive testing methods are particularly useful for assessing the physical properties of polymeric materials when an understanding of the performance at a molecular level is important. The foregoing techniques for measuring mechanical properties are transient or nonperiodic methods and typically cover time intervals of up to 10^6 s. For information relating to short times, two approaches that have been widely used are dynamic mechanical thermal analysis (DMTA) and dielectric thermal analysis (DETA). These are both particular kinds of relaxation spectroscopy in which the sample is perturbed by a sinusoidal force (either mechanical or electrical), and the response of the material is measured over a range of temperatures and at different frequencies of the applied force. From an analysis of the material response it is possible to derive information about the molecular motions in the sample, and how these can affect the modulus, damping characteristics, and structural transitions. Both techniques can be used to probe molecular motions in liquid or solid polymers, but when dielectric spectroscopy is used, the relaxation or transition must involve movement of a dipole or a charge displacement if it is to be detected. Thus, whereas both DMTA and DETA can provide similar information about a sample, they can also be used in a complementary fashion, particularly when trying to identify the molecular mechanism of a particular process and in ascertaining whether or not the group is polar.

13.7.1 DYNAMIC MECHANICAL THERMAL ANALYSIS (DMTA)

In DMTA a small sinusoidal stress is imparted to the sample in the form of a torque, push-pull, or a flexing mode of angular frequency ω. If the polymer is treated as a classical damped harmonic oscillator, both the elastic modulus and the damping characteristics can be obtained. Elastic materials convert mechanical work into potential energy, which is recoverable; for example, an ideal spring, if deformed by a stress, stores the energy and uses it to recover its original shape after removal of the stress. No energy is converted into heat during the cycle, and so no damping is experienced.

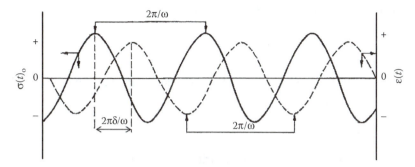

FIGURE 13.18 Harmonic oscillation of a Maxwell model, with the solid line representing the stress and the broken line representing the strain curve. This shows the lag in the response to an applied stress.

Liquids, on the other hand, flow if subjected to a stress; they do not store the energy but dissipate it almost entirely as heat and thus possess high damping characteristics. Viscoelastic polymers exhibit both elastic and damping behavior. Hence, if a sinusoidal stress is applied to a linear viscoelastic material, the resulting strain will also be sinusoidal, but will be out of phase when there is energy dissipation or damping in the polymer.

Harmonic motion of a Maxwell element. The application of a sinusoidal stress to a Maxwell element produces a strain with the same frequency as, but out of phase with, the stress. This can be represented schematically in Figure 13.18 where δ is the phase angle between the stress and the strain. The resulting strain can be described in terms of its angular frequency ω and the maximum amplitude ε_0 using complex notation, by

$$\varepsilon^* = \varepsilon_0 \exp(i\omega t), \tag{13.28}$$

where $\omega = 2\pi v$, the frequency is v and $i = -1^{1/2}$. The relation between the alternating stress and strain is written as

$$\sigma^* = \varepsilon^* E^*(\omega), \tag{13.29}$$

where $E^*(\omega)$ is the frequency-dependent complex dynamic modulus defined by

$$E^*(\omega) = E'(\omega) + iE''(\omega). \tag{13.30}$$

This shows that $E^*(\omega)$ is composed of two frequency-dependent components: $E'(\omega)$ is the real part in phase with the strain called the *storage modulus*, and $E''(\omega)$ is the *loss modulus* defined as the ratio of the component 90° out of phase with the stress to the stress itself. Hence, $E'(\omega)$ measures the amount of stored energy and $E''(\omega)$, sometimes called the *imaginary part*, is actually a real quantity measuring the amount of energy dissipated by the material.

The response is often expressed as a complex dynamic compliance

$$J*(\omega) = J'(\omega) - iJ''(\omega), \tag{13.31}$$

especially if a generalized Voigt model is used. For a Maxwell model

$$\sigma*/\varepsilon* = E'\omega^2\tau^2/(1+\omega^2\tau^2) + iE''\omega\tau/(1+\omega^2\tau^2). \tag{13.32}$$

In more realistic terms, there is a distribution of relaxation times, and a continuous distribution function can be derived, if required.

The damping in the system or the energy loss per cycle can be measured from the "loss tangent" tan δ. This is a measure of the internal friction and is related to the complex moduli by

$$\tan\delta = 1/\omega\tau = E''(\omega)/E'(\omega) = J''(\omega)/J'(\omega). \tag{13.33}$$

In general, one can select a dissipation factor or loss tangent derived from the ratio (G''/G') or (E''/E') to represent the energy conversion per cycle. To a first approximation it is also possible to write

$$(E''/E') = (G''/G') \tag{13.34}$$

thereby allowing use of the data from either type of measurement to characterize the sample. It should also be noted that if complex moduli are used, the corresponding complex compliances are given by $(G''/G') = (J''/J')$. Moduli can also be related to the viscosity, $G' = \omega\eta''$ and $G'' = \omega\eta'$, where η is known as the dynamic viscosity.

The onset of molecular motion in a polymer sample is reflected in the behavior of E' and E''. A schematic diagram (Figure 13.19) of the variation of E' and E'' as a function of ω, assuming only a single value for τ in the model, shows that a maximum in the loss angle is observed where $\omega = 1/\tau$. This represents a transition point, such as T_g, T_m or some other region where significant molecular motion occurs in the sample. The maximum is characteristic of the dynamic method as the creep and relaxation techniques merely show a change in the modulus level.

The approximations $E \approx E'$ and $G \approx G'$ can be made when damping is low, and the absolute value for the modulus $|G|$ or $|E|$ can be related to the complex components by $|E| = \{(E')^2 + (E'')^2\}^{1/2}$. A similar expression holds for $|G|$.

Several types of instrument can be used for this type of test, and these are usually limited to measurements on rigid polymers or rubbers. Commercial instruments offer several damping arrangements for the sample so that measurements may be made in the bending, shear, or tensile modes.

In the bending mode, the sample, in the form of a small bar, is clamped firmly at both ends and the central point is vibrated by means of a ceramic drive shaft. This can be driven at frequencies selected from the range of 0.01 to 200 Hz. The applied

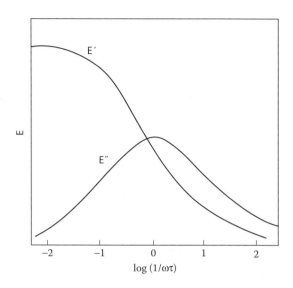

FIGURE 13.19 Behavior of E' and E'' as a function of the angular frequency for a system with a single relaxation time.

stress is proportional to the AC current fed to the drive shaft, and the strain is detected using a transducer that measures the displacement of the drive clamp. Temperature can be controlled over the range of 120 to 770 K, either isothermally or more normally by ramping up and down at various fixed rates.

13.7.2 DIELECTRIC THERMAL ANALYSIS (DETA)

Dry polymers are very poor conductors of electricity and can be regarded as insulators. Application of an electric field to a polymer can lead to polarization of the sample, which is a surface effect, but if the polymer contains groups that can act as permanent dipoles, then the applied field will cause them to align along its direction. When the electric field is released, the dipoles can relax back into a random orientation, but due to the frictional resistance experienced by the groups in the bulk polymer, this will not be instantaneous. The process of disordering can be characterized by a relaxation time but may not be easily measured. It is more convenient to apply a sinusoidally varying voltage to the sample and to study the dipole polarization under steady-state conditions.

In DETA, a small alternating electric field is applied to the sample, and the electric charge displacement Q is measured by following the current i ($= dQ/dt$). The complex dielectric permittivity ε^* can be measured from the change in amplitude and, if the phase lag between the applied voltage and the outcoming current is determined (see Figure 13.20), then ε^* can be resolved into the two components: ε', the storage (dielectric constant), and ε'', the loss (dielectric loss). The frequencies used in the measurements must now be in the range where orientational polarization of the dipoles in the polymer is active. These frequencies are much higher than used normally for DMTA and typically lie in the range 20 Hz to 100 kHz.

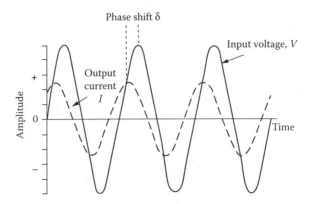

FIGURE 13.20 A schematic diagram of the behavior of input voltage and output current in DETA. (Adapted from Grentzer, T. and Leckenby, J., *Intl. Lab.*, 19(6), 34–38, July–August, 1989. Copyright by International Scientific Communications, Inc. With permission.)

Although the main variable is temperature, the factors ε' and ε'' can be studied as a function of the angular frequency ω, and in the frequency region where there is a relaxation, ε' decreases as shown in Figure 13.21. The magnitude of this decrease $(\varepsilon_0 - \varepsilon_\infty)$ is a measure of the strength of the molecular dipole involved in the relaxation, where ε_0 is the static dielectric constant related to the actual dipole moment of the polymer, and ε_∞ is the dielectric constant measured at high frequencies. When the dielectric loss factor is measured at a characteristic frequency ω_{max}, and a given temperature, it passes through a maximum when a relaxation occurs, and the dipole relaxation time $\tau = 1/\omega_{max}$ can be obtained. At frequencies above ω_{max}, the dipoles cannot move fast enough to follow the alternating field so both ε' and ε'' are low. When the frequency is lower than ω_{max}, the permanent dipoles can follow the field quite closely and so ε' is high because the dipoles align easily with each change in polarity; ε'' on the other hand is low again because now the voltage and the current are approximately 90° out of phase.

The dielectric relaxation process can be described formally by the following relations:

$$\varepsilon' = \varepsilon_0 + \frac{(\varepsilon_0 - \varepsilon_\infty)}{(1 + \omega^2 \tau^2)} \tag{13.35}$$

and

$$\varepsilon'' = \omega\tau + \frac{(\varepsilon_0 - \varepsilon_\infty)}{(1 + \omega^2 \tau^2)} \tag{13.36}$$

A useful way of examining the data is to measure the ratio of the two factors, which gives the dielectric loss tangent

$$\tan \delta_D = (\varepsilon'' / \varepsilon') \tag{13.37}$$

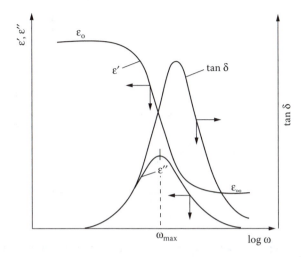

FIGURE 13.21 A schematic representation of the behavior of the storage and loss factors ε' and ε'' as a function of the log of angular frequency ω. Also shown is the dielectric loss tangent as a function of ω.

Dipolar groups in a polymer coil may not be able to relax at the same speed because of the variable steric restrictions they may experience, imposed by their environment. This can be caused by the disordered packing of chains in the amorphous glassy phase and a random distribution of the available free volume, or perhaps even by the random coil structure of the chain itself, causing local environmental changes. The result is that a distribution of relaxation times is to be expected for a given process, and this results in a broadening of the dielectric loss peak. Thus, the more mobile a dipolar group, the easier it is for it to follow the electric field up to higher frequencies, whereas the less mobile groups can only orient at lower frequencies.

13.7.3 COMPARISON BETWEEN DMTA AND DETA

Data from mechanical and dielectric measurements can be related, certainly in a qualitative, if not always in a quantitative, way. Formally, the dielectric constant (ε') can be regarded as the equivalent of the mechanical compliance (J'), rather than the modulus, and this highlights the fact that mechanical techniques measure the ability of the system to resist movement, whereas the dielectric approach is a measurement of the ability of the system to move, given that the groups involved must also be dipolar. Interestingly, the dielectric loss (ε'') appears to match the loss modulus $(E''$ or $G'')$ more closely than the loss compliance when data are compared for the same system.

Both techniques respond in a similar fashion to a change in the frequency of the measurement. When the frequency is increased, the transitions and relaxations that are observed in a sample appear at higher temperatures. This is illustrated from work on poly(ethylene terephthalate) where the loss peak representing the glass transition has been measured by both DMTA and DETA at several frequencies between 0.01 Hz and 100 kHz (Figure 13.22). The maximum of this loss peak (T_{max}) is seen to

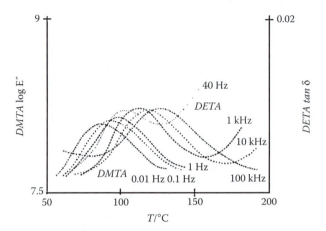

FIGURE 13.22 Comparison of the loss peaks for the glass transition in poly(ethylene terephthalate) measured at different frequencies from both dynamic mechanical and dielectric techniques. T_{max} shifts to higher temperatures as the frequency of the measurement increases. (From Wetton, R.E. et al., *Intl. Lab.*, March, 1986. Figure 7, p. 80. With permission from International Laboratory.)

move from a temperature of about 360 K (0.01 Hz) to about 400 K (100 kHz), which is an increase of 40 K over a frequency change of seven orders of magnitude. This is close to the rule of thumb that the temperature for the maximum of a loss peak (T_{max}) (or a relaxation process) will change by approximately 7 K for each decade of change in frequency.

This type of measurement can be used to estimate the activation energy (ΔE^*) for a transition or relaxation process, if the frequency v, at T_{max} is expressed as a function of reciprocal temperature according to the relation

$$\Delta E^* = d(\log v)/d(1/T_{max}) \qquad (13.38)$$

Data plotted using Equation 13.38 for the β-relaxation process in a series of poly(alkylmethacrylate)s are shown in Figure 13.23. Both techniques have been used and separately give good straight lines with the same slope, but the fact that the lines do not overlap precisely indicates that the measurements may not be exactly equivalent.

The results from DMTA and DETA can be used in a complementary manner to distinguish between relaxations involving polar and nonpolar units relaxing in the polymeric system. Thus, if the response of poly(ethylene terephthalate) to both DMTA and DETA is examined, two major loss peaks can be identified in each, as seen in Figure 13.24. The high-temperature loss peak (α-peak) can be assigned to the glass transition, and this can be confirmed by DSC measurements. There is a second loss (β-relaxation) that appears at lower temperatures and suggests that there is a relaxation process active in the glassy state. It is not immediately obvious which group is responsible for this process, but it is active both mechanically and dielectrically. Examination of the polymer structures suggests that the relaxation in the

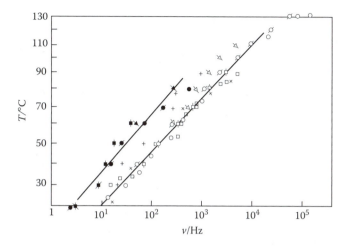

FIGURE 13.23 Plot of $1/T_{max}$ against frequency for the β-relaxation in poly(methyl methacrylate) PMMA (○, ●); poly(ethyl methacrylate) PEMA (⊟, ▪, □); poly(n-propyl methacrylate) PnPMA (×); poly(iso-propyl methacrylate) PiPMA (+); and poly(t-butyl methacrylate) PtBMA (◁,◀). Open symbols represent dielectric data whereas filled symbols refer to dynamic mechanical data. (From McCrum, N.G. et al., *Anelastic and Dielectric Effects in Polymeric Solids*, John Wiley and Sons, New York, 1976. With permission.)

glass could involve liberation of the phenyl ring, motion of the oxycarbonyl unit, or rearrangement of the (-O-C-C-O-) unit. From the spectra it can be seen that the intensity of the β-peak relative to the α-peak is much stronger in the dielectric response compared to the mechanical measurements. This suggests that the group undergoing relaxation is associated with a dipole moment, and thus the phenyl ring libration is ruled out as a likely process. This does not give irrefutable evidence of the participation of the oxycarbonyl unit, but it does point in this direction.

13.8 TIME–TEMPERATURE SUPERPOSITION PRINCIPLE

A curve of the logarithm of the modulus against time and temperature is shown in Figure 13.25. This provides a particularly useful description of the behavior of a polymer and allows one to estimate, among other things, either the relaxation or retardation spectrum.

The practical timescale for most stress relaxation measurements ranges from 10^1 to 10^6 s but a wider range of temperature is desirable. Such a range can be covered relatively easily by making use of the observation, first made by Leaderman, that for viscoelastic materials time is equivalent to temperature. A composite isothermal curve covering the required extensive time scale can then be constructed from data collected at different temperatures.

This is accomplished by translation of the small curves along the log t axis until they are all superimposed to form a large composite curve. The technique can be illustrated using data for polyisobutylene at several temperatures. An arbitrary temperature T_0 is first chosen to serve as a reference, which in the present case is 298 K.

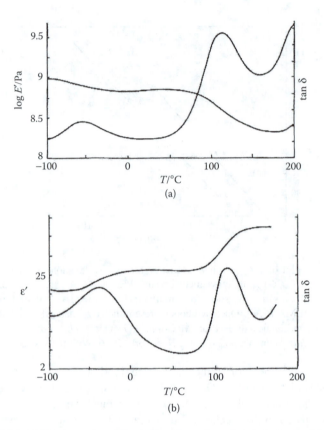

FIGURE 13.24 (a) The dynamic mechanical spectrum for poly(ethylene terephthalate) (PET) showing the storage modulus and tan δ as a function of temperature. (b) The dielectric (storage and tan δ) behavior for PET in comparison with the mechanical response. (From Wetton, R.E. et al., *Intl. Lab.*, March, 1986. Figure 13a and Figure 13b, p. 60. With permission from International Laboratory.)

As values of the relaxation modulus $E_r(t)$ have been measured at widely differing temperatures, they must be corrected for changes in the sample density with temperature to give a reduced modulus, where ρ and $ρ_0$ are the polymer densities at T and T_0, respectively. This correction is small and can often be neglected.

$$[E_r(t)]_{red} = (T_0 ρ_0 / T_ρ) E_r(t).$$ (13.39)

Each curve of reduced modulus is shifted with respect to the curve at T_0 until all fit together forming one master curve. The curve obtained at each temperature is shifted by an amount

$$(\log t - \log t_0) = \log (t/t_0) = \log a_T$$ (13.40)

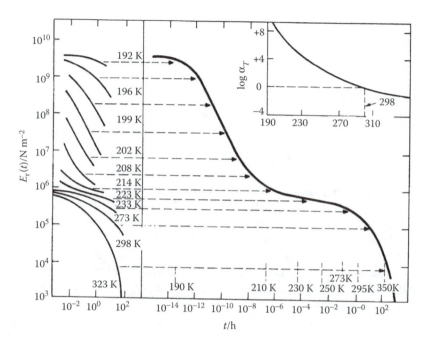

FIGURE 13.25 Illustration of the time–temperature superposition principle using stress-relaxation data for polyisobutylene. Curves are shifted along the axis by an amount represented by α_T as shown in the inset. The reference temperature in this instance is 298 K. (Adapted from Castiff, E. and Tobolsky, A.V., *J. Colloid Sci.*, 10, 375, 1955.)

The parameter $\log a_T$ is the shift factor and is positive if the movement of the curve is to the left of the reference and negative for a move to the right. The shift factor is a function of temperature only and decreases with increasing temperature; it is, of course, unity at T_0. The superposition principle can also be applied to creep data. Curves exhibiting the creep behavior of polymers at different temperatures can be compared by plotting $J(t)T$ against $\log t$. This reduces all the curves at various temperatures to the same shape but displaced along the $\log t$ axis. Superposition to form a master curve is readily achieved by movement along the $\log t$ axis, where the shift factor a_T has the same characteristics as for the relaxation data. This shift factor has also been defined as the ratio of relaxation or retardation times at temperatures T and T_0, i.e.,

$$a_T = \tau/\tau_0 = (\eta/\eta_0)(T_0\rho_0/T_\rho), \tag{13.41}$$

and is related to the viscosities. If the viscosities obey the Arrhenius equation, then by neglecting the correction factor, we can express a_T in an exponential form as

$$a_T = \exp b(1/T - 1/T_0) \tag{13.42}$$

or

$$\log_{10} a_T = -b(T - T_0)/2.303\ TT_0, \tag{13.43}$$

where b is a constant.

This equation is very similar in form to the WLF equation,

$$\log_{10} a_T = -a_1(T - T_0)/(a_2 + T - T_0) \tag{13.44}$$

For polyisobutylene, the shift factor a_T can be predicted if $T_0 = (T_g + 45\ K)$ is used with $a_1 = 8.86$ and $a_2 = 101.6\ K$. As outlined in Chapter 12, the reference temperature is often chosen to be T_g with $a_1 = 17.44$ and $a_2 = 51.6\ K$, from which a_T can be calculated for various amorphous polymers.

The superposition principle can be used to predict the creep and relaxation behavior at any temperature if some results are already available, with the proviso that the most reliable predictions can be made for interpolated temperatures rather than long extrapolations.

The principle can also be applied to dielectric data, which can be shifted either along the temperature or the frequency axis. An example of the latter type of shift is shown in Figure 13.26, where instead of time-dependence measurements the frequency dependence of the β-relaxation in poly(vinyl acetate) has been studied at fixed temperature in the range 212 to 266 K. A master curve can be constructed for this relaxation region by plotting $(\varepsilon''/\varepsilon''_{max})$ against $\log_{10}(\omega/\omega_{max})$, where the "max" subscript refers to the peak maximum at each experimental temperature.

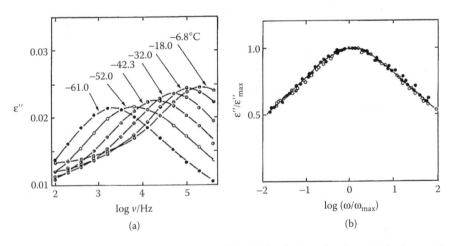

FIGURE 13.26 (a) Frequency dependence of ε'' for the β-relaxation in poly(vinyl acetate) measured at different temperatures. (b) Master dielectric loss curve for the data in (a) (○) compared with similar data for the β-relaxation of poly(vinyl benzoate) (●). (From Ishida, Y. et al., *Koll. Z.* 180, 108, 1962. With permission from Dr. Dietrich Steinkopff Verlag, Darmstadt.)

13.9 DYNAMIC VISCOSITY

Rheological experiments in the oscillatory mode provide a means of characterizing the viscoelastic properties of polymers and plastics in solution or molten state. These are crucial for determining molecular architecture, e.g., molecular weight, molecular weight distribution, and degree of branching, and to understand how these parameters affect polymer processability and end product performance.

Similar to the dynamic mechanical measurements described in Section 13.7, the fluid here is subjected to an oscillatory stress, and the oscillatory response is analyzed. The complex dynamic modulus, $G^*(\omega)$, is then resolved into two components:

$$G^*(\omega) = G'(\omega) + iG''(\omega). \tag{13.45}$$

the storage and loss moduli, $G'(\omega)$ and $G''(\omega)$, respectively. The complex modulus is then related to the dynamic viscosity:

$$\eta^* = \eta' - i\eta'' \tag{13.46}$$

with $\eta' = G''/\omega$ and $\eta'' = G'/\omega$ and so the in-phase, real component describes the viscous dissipation, whereas the out-of-phase component represents the stored energy.

The frequency dependences of $G'(\omega)$ and $G''(\omega)$ for a typical polymer melt are reported in Figure 13.27. At low frequency $G'(\omega) < G''(\omega)$ and viscous behavior is dominant. This is the long-time or *terminal* region from which a measure of the zero shear rate viscosity can be derived. At intermediate frequencies, $G'(\omega) > G''(\omega)$, and

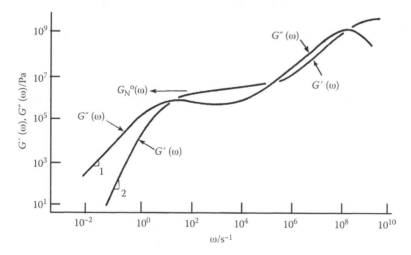

FIGURE 13.27 Frequency dependences of $G'(\omega)$ and $G''(\omega)$ for a typical polymer melt. (From Graessley, W.W., Viscoelasticity and flow in polymer melts and concentrated solutions in *Physical Properties of Polymers*, 3rd ed., Cambridge University Press, 2004, Chap. 3. With permission.)

the polymer melt's behavior is elastic. The plateau modulus G_N^o can be estimated from the constant $G'(\omega)$ value but other more rigorous methods are available, as shown in Figure 13.27, whereas in the rubbery region $G'(\omega)$ shows a plateau, and $G'(\omega)$ displays a maximum.

G_N^o is related to the entanglement molecular weight, M_e:

$$G_N^o = \frac{\rho RT}{M_e} \tag{13.47}$$

where ρ is the density of the polymer. For broad distribution polymers, the rubbery plateau is no longer clearly evident, and other methods need to be used to determine G_N^o such as integration of the $G''(\omega)$ vs. ln ω curve.

13.10 A MOLECULAR THEORY FOR VISCOELASTICITY

So far the interpretation of viscoelastic behavior has been largely phenomenological, relying on the application of mechanical models to aid the elucidation of the observed phenomena. These are, at best, no more than useful physical aids to illustrate the mechanical response and suffer from the disadvantage that a given process may be described in this way using more than one arrangement of springs and dashpots. In an attempt to gain a deeper understanding on a molecular level, Rouse, Zimm, Bueche, and others have attempted to formulate a theory of polymer viscoelasticity based on a chain model consisting of a series of subunits. Each subunit is assumed to behave like an entropy spring and is expected to be large enough to realize a Gaussian distribution of segments (i.e., > 50 carbon atoms). This approach, although still somewhat restrictive, has led to reasonable predictions of relaxation and retardation spectra.

One starts with a single isolated chain and the assumption that it exhibits both viscous and elastic behavior. If the chain is left undisturbed, it will also adopt the most probable conformation or segmental distribution so that, with the exception of high frequencies, the observed elasticity is predominantly entropic. Thus, the application of a stress to the molecule will cause distortion by altering the equilibrium conformation to a less probable one, resulting in a decrease in the entropy and a corresponding increase in the free energy of the system. When the stress is removed, the chain segments will diffuse back to their unstressed positions even though the whole molecule may have changed its spatial position in the meantime. If, on the other hand, the stress is maintained, strain relief is sought by converting the excess free energy into heat, thereby stimulating the thermal motion of the segments back to their original positions. Stress-relaxation is then said to have occurred. For a chain molecule composed of a large number of segments, movement of the complete molecule depends on the cooperative movement of all the segments, and as stress-relaxation depends on the number of ways the molecule can regain its most probable conformation, each possible coordinated movement is treated as a mode of motion with a characteristic relaxation time. For simplicity, we can represent the polymer as in Figure 13.28.

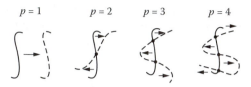

$p = 1$ $p = 2$ $p = 3$ $p = 4$

FIGURE 13.28 First four normal modes of movement of a flexible polymer molecule.

The first mode $p = 1$ represents translation of the molecule as a whole and has the longest relaxation time τ_1 because the maximum number of coordinated segmental movements are involved. The second mode $p = 2$ corresponds to the movement of the chain ends in opposite directions; for $p = 3$, both chain ends move in the same direction, but the center moves in the opposite direction. Higher modes 4, 5, ... m follow involving a progressively decreasing degree of cooperation for each succeeding mode and correspondingly lower relaxation times τ_p. This means that a single polymer chain possesses a wide distribution of relaxation times. Using this concept, Rouse considered a molecule in dilute solution under sinusoidal shear and derived the relations

$$\eta' = (G''/\omega) = \eta_s + (nkT/\omega) \sum_{p=1}^{m} \omega \tau_p /(1+\omega^2\tau_p^2)$$ (13.48)

$$\eta'' = (G'/\omega) = (nkT/\omega) \sum_{p=1}^{m} \omega^2 \tau_p /(1+\omega^2\tau_p^2)$$ (13.49)

$$\tau_p = 6(\eta - \eta_s)/(\pi^2 p^2 nkT)$$ (13.50)

where η and η_s are the viscosities of the solution and the solvent respectively, n is the number of molecules per unit volume, k is the Boltzmann constant, and ω is the angular frequency of the applied stress, which is zero for steady flow.

These equations are strictly applicable only to dilute solutions of nondraining monodisperse coils, but can be extended to undiluted polymers above their glass temperature if suitably modified. This becomes necessary when chain entanglements begin to have a significant effect on the relaxation times. The undiluted system is represented as a collection of polymer segments dissolved in a liquid matrix composed of other polymer segments, and η_s can be replaced by a monomeric frictional coefficient ζ_0. This provides a measure of the viscous resistance experienced by a chain and is characteristic of a given polymer at a particular temperature. Thus, the longest relaxation time, τ_R, for a Rouse chain is given by:

$$\tau_R = \frac{2\zeta_o N \langle r_o^2 \rangle}{3\pi^2 kT} \propto M^2$$ (13.51)

where r_0^2 is the unperturbed mean square end-to-end distance of a chain of molar mass M and N monomer units. By recalling that $\langle r_o^2 \rangle = Nb^2$, it follows that $\tau_R \propto M^2$. Similarly, the diffusion coefficient of a Rouse chain is given by:

$$D_R = \frac{kT}{\zeta_o N} \propto M^{-1} \qquad (13.52)$$

The continuous relaxation and retardation spectra calculated from the Rouse theory are

$$H(\tau) = (\rho N_A / 2\pi M)(r_0^2 NkT\zeta_0 / 6\tau)^{1/2} \qquad (13.53)$$

and

$$L(\tau) = (2M/\pi\rho N_A)(6\tau_R / r_0^2 NkT\zeta_0)^{1/2} \qquad (13.54)$$

where M is the molar mass and ρ is the density. The equations predict linearity in the plots $\log H(\tau)$ and $\log L(\tau_R)$ against $\log \tau$ with slopes of $-\frac{1}{2}$ and $+\frac{1}{2}$, respectively. Comparison with experimental results for poly(methyl acrylate) shows validity only for longer values of the relaxation and retardation times.

The Rouse model only pertains to the region covering intermediate τ values. The reason for this lies in the response of a polymer to an alternating stress. At low frequencies, Brownian motion can relieve the deformation caused by the stress before the next cycle takes place, but as the frequency increases the conformational change begins to lag behind the stress, and energy is not only dissipated but also stored as well. Finally, at very high frequencies only enough time exists for bond deformation to occur. As it was stipulated that each segment be long enough to obey Gaussian statistics, short relaxation times may not allow a segment sufficient time to rearrange and regain this distribution. Thus, the contribution from short segments to the distribution functions tends to be lost and deviations from the theoretical represent departure from ideal Gaussian behavior.

This approach to viscoelastic theory is reasonably successful in the low modulus regions, but it requires considerable modification if the high modulus and rubbery plateau regions are to be described.

13.11 THE REPTATION MODEL

The theory proposed by Bueche (1962) tends to suggest a very clear-cut distinction between the movement of chains of length less than Z_c and the relative immobility of the entangled chains with lengths greater than Z_c. As independent chain mobility cannot be discounted for these longer chains after the onset of entanglement, a modified model is required to account for the ability of long chains to translate and diffuse through the polymer matrix, i.e., the entanglement network must be considered as being transient. Such a concept is embodied in the "reptation" model proposed by de Gennes. In this approach, the chain is assumed to be contained in a hypothetical tube, which is placed initially in a three-dimensional network formed from the other entangled chains. Although for simplicity, these network "knots" are regarded as a set of fixed obstacles around which the isolated chain under consideration must wriggle during translation, in practice the network "knots" would also be in motion. The contours of the tube are then defined by the position of the entanglement points in the network.

Two types of chain motion can be envisaged: a conformational change taking place within the confines of the tube, and more importantly, reptation. The latter is imagined to be a snake-like movement that translates the chain through the tube and allows it to escape at the tube ends. Mechanistically, it can be regarded as the movement of a kink in the chain along its length (see Figure 13.29) until this reaches the end of the chain and leaves it. Motion of this kind translates the chain through the tube, like a snake moving through grass, and successive defects moving the chain in this way will eventually carry it completely out of the hypothetical tube.

The motion can be characterized by a reptation time, or more accurately, by a relaxation time, τ, that is, a measure of the time required for a chain to escape completely from its tube. If the tube is defined as having the same length as the unperturbed chain, nl_0, where l_0 is the bond length under θ conditions (corrected for short-range interactions), then the time required for the chain to reptate out of the tube is proportional to the square of the distance traveled, i.e.,

$$\tau = \frac{(nl_0)^2}{2D_t}$$

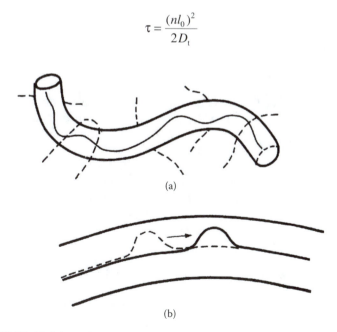

(a)

(b)

FIGURE 13.29 (a) Schematic of polymer chain restrained in a hypothetical tube. (b) Movement of a "kink" along the chain.

Here D_t is the diffusion constant within the tube, and is distinguished from translation outside the tube, which will be slower and more difficult. This can be expressed in terms of the frictional coefficient for the chain, f_t, again within the tube confines ($D_t = kT/f_t$). However, because the reptation is assumed to occur by migration of a segmental kink along the chain, the force needed to do this is applied one segment at a time, and so it is more appropriate to use the frictional factor per segment ζ ($f_t = \zeta n$). Thus,

$$\tau = (nl_0)^2 (n\zeta/2kT) = \left(\frac{l_0^2 \zeta}{2kT} \right) n^3 = \tau_0 n^3 \tag{13.55}$$

This is the fundamental result of the reptation model: the relaxation time is proportional to the cube of the chain length. The cube dependence is not a precise match with the 3.4 exponent obtained from viscosity measurements of long chains, but it is acceptable, particularly as the model gives a satisfactory picture of how a polymer chain can overcome the restraining influence of entanglements and move within the matrix.

Typically, τ_0 is of the order of 10^{-10} s for $n = 1$, and so the relaxation time τ for a polymer chain with $n = 10^4$ would be about 100 s.

Reptation theory has been developed further by Doi and Edwards (1986) and is being applied to both viscoelastic and solution behavior. It has been shown that for a chain moving in the melt, over time-scales that greatly exceed the lifetime of the tube τ, a reptation self-diffusion coefficient D_{rept} can be measured, which is inversely proportional to n^2, i.e., the diffusion law is

$$D_{rept} \propto 1/n^2 \tag{13.56}$$

This law holds for the "welding" of polymers at an interface which can be explained by reptation. When two blocks of the same polymer are brought together and held at a temperature just above the T_g for a time t, interdiffusion of the chains takes place from each block across the interface (see Figure 13.30) thereby joining the blocks together. The strength of the junction formed will depend on t, which should be smaller than the reptation time τ, i.e., the mixed layer ought to be smaller than the size of the coil if an interfacial link is to be formed.

PROBLEMS

1. Using the *power law* $\tau = K \cdot (\dot{\gamma})^n$ derive a relationship for the apparent viscosity, and discuss how differences between Newtonian and non-Newtonian behavior could be detected, as a function of shear rate.

2. Give a relationship between shear stress τ and shear rate $\dot{\gamma}$ that would describe the flow properties of a fluid following Bingham Newtonian behavior.

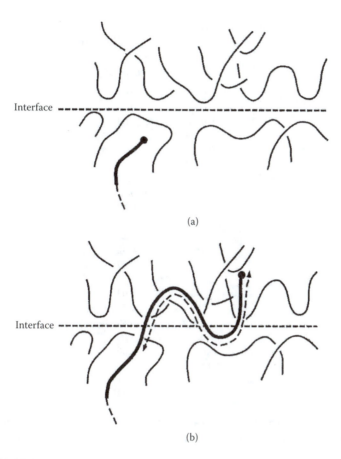

(a)

(b)

FIGURE 13.30 Schematic of chain movement across an interface: (a) first contact and (b) after having been in contact for some time, with a chain from one surface having reptated across into the interface of the adjoining polymer block.

3. The flow behavior of commercial polysaccharide solutions above the overlap concentration can be described by a master curve. Scaling on the x and y axes is achieved by plotting log (η/η_0) vs. log $(\dot{\gamma}/\dot{\gamma}_f)$, where η_0 is the zero-shear rate viscosity and $\dot{\gamma}_f$ is the shear rate at which the viscosity has decreased to an arbitrarily chosen fraction f of the initial value. The numerical master curve reported by Morris (1990) is given in the table that follows for $f = 0.5$.

   a. Identify the Newtonian plateau region.

   b. Starting from Equation 13.7, show that when η_∞ (as is the case here) and $f = 0.5$:

$$\eta = \frac{\eta_0}{(\dot{\gamma}/\dot{\gamma}_{0.5})^p}$$

   and that $p = 0.76$ at high shear rates.

c. Classify the shear dependent behavior.
d. How would you expect this to vary for dilute solutions at $c < c^*$?

$\dot{\gamma}/\dot{\gamma}_{0.5}$	$\log(\eta/\eta_o)$
0.0018	0.00
0.0057	0.00
0.0180	0.00
0.0286	0.00
0.0453	-0.02
0.0717	-0.04
0.114	-0.07
0.180	-0.10
0.286	-0.14
0.453	-0.18
0.717	-0.24
1.14	-0.32
1.80	-0.40
2.86	-0.49
4.53	-0.60
7.17	-0.72
11.4	-0.85
18.0	-1.00
57.0	-1.36
180	-1.73
570	-2.11
1800	-2.48

4. Rheological measurements were carried out by Dvornic et al. (1993) on poly(dimethyl siloxane) (PDMS) samples as a function of molecular weight M_w.

η_o (P)	M_w (g mol^{-1})
0.178	1948
0.638	4787
2.290	12186
6.240	26547
11.750	38298
53.200	56902
141.800	71699
208.000	84211
602.000	108691
2154.000	156672

a. Determine Z_w, the weight-average number of chain atoms from the relation:

$$Z_w = 1 + 2\frac{(M_w - 88)}{74}$$

which accounts for the methyl end groups.

b. Characterize the molecular weight dependence of the zero shear rate viscosity, η_o, and find the critical Z_w and η_o values for the onset of entangled behavior.

5. The molecular weight dependence of the viscosity for a series of n-alkanes and polyethylene samples was reported by Pearson et al. (1987). At 448 K and for low-molecular-weight samples, the viscosity follows a power law $\eta \propto M^{1.8}$, whereas at high molecular weight $\eta \propto M^{3.64}$. The critical values at the intersection between low- and high-molecular-weight behavior are $M_c = 5200$ and $\eta = 1.26$ P. From these determine the viscosities of $C_{36}H_{74}$ and a sample with molecular weight 119,600.

6. For unentangled chains the monomer friction coefficient, ζ, can be calculated from the self-diffusion coefficient. For $C_{44}H_{90}$ the self-diffusion coefficient is reported to be 2.4×10^{-6} cm^2 s^{-1} and 4.9×10^{-6} cm^2 s^{-1} at 400 and 450 K, respectively. Calculate ζ and comment on its temperature dependence.

7. The rheological behavior of an anionically polymerized styrene-butadiene-styrene (SBS) block copolymer sample was investigated by Chung and Gale (1976). Highly non-Newtonian behavior, high viscosity, and high elasticity were observed in the temperature range 398 to 423 K. At 448 K, the sample displayed Newtonian behavior and deviations from the master curve obtained from lower temperature data were observed. Suggest reasons why:
   a. The non-Newtonian, lower-temperature behavior may be considered as characteristic of tri-block copolymers.
   b. A change is observed above 423 K.

8. Master curves for narrow M_W polystyrene samples of different molecular weight obtained by applying time-temperature superposition are reported in Figure 13.31 for a reference temperature of 433 K. Comment on the effect of molecular weight on the terminal and plateau regions. The value of G_N^o is 2.0×10^5 Pa. Calculate M_e and the average chain length Z_e (consider the density of polystyrene equal to 1 g cm^{-3}).

9. The load/elongation curve of a spider drag line silk of length 25.3 mm and diameter 4.20 µm is shown in Figure 13.32. From the graph determine:
   a. The breaking load and breaking stress.
   b. Breaking elongation and breaking strain.
   c. At what load and elongation does the fiber cease to obey Hooke's law?

10. For materials that obey the time–temperature superposition principle, the dynamic moduli can be rescaled using appropriate shift factors to obtain a master curve. The temperature dependence of the horizontal shift factor is usually described by the WLF equation. Explain how the WLF parameters a_1, a_2, and T_o can be determined from a set of log a_T vs. temperature data.

11. Williams et al. (1955) have shown that the temperature dependence of the viscosity and relaxation is similar for all polymers, and it can be described by the WLF equation, at least in the temperature range T_g to $T_g + 100°C$. As shown in Figure 13.33 for polystyrene (PS) and polyisobutylene (PIB),

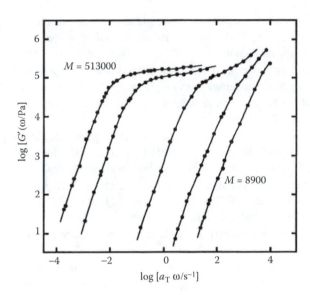

FIGURE 13.31 Selected master curves (reference temperature = 433 K) of G' for a series of narrowly distributed PS samples. (From Graessley, W.W., Viscoelasticity and flow in polymer melts and concentrated solutions, in *Physical Properties of Polymers*, 3rd ed., Cambridge University Press, 2004, chap. 3. With permission.)

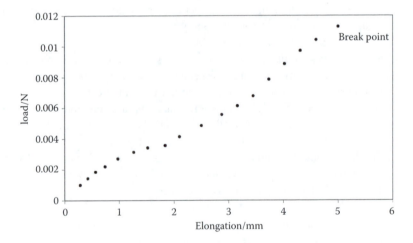

FIGURE 13.32 Load vs. elongation curve of a spider drag line silk.

divergence from a common behavior is already evident at T_g + 50°C. Explain.

12. The viscosity of a polyisobutylene sample (M_w = 133,000) is reported to be 1.39×10^7 P at 298 K. The temperature dependence is described by the WLF parameters, a_1 = 6.92, a_2 = 180 K, and T_o = 298 K; evaluate the sample's viscosity at 373 K.

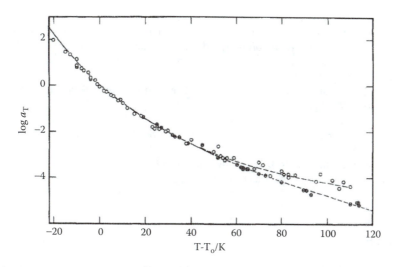

FIGURE 13.33 Log a_T vs. $T\text{-}T_o$ for polystyrene (open circles) and polyisobutylene (slotted circles). (From Williams, M.L. et al., *J. Am. Chem. Soc.*, 77, 3701, 1955. With permission.)

13. The storage modulus data of amorphous poly(ethylene terephthalate) (PET) at 1.2 Hz display a secondary relaxation at 253 K, and glass transition at 353 K (Thompson and Woods, 1956).
    a. Sketch the modulus vs. temperature curve for this sample and predict the tan δ behavior.
    b. How do you expect these transitions and the storage modulus vs. temperature curve to alter for a crystalline PET sample?
14. The tan δ curve at 1 Hz for a poly(vinyl chloride) sample shows two distinct relaxations: (1) one at 210 K, attributed to main chain motion, and (2) a main tan δ peak at 343 K, associated with the glass–rubber transition.
    a. Sketch the corresponding storage modulus vs. temperature curve.
    b. Are these processes likely to be dielectrically active?
15. Estimate the longest Rouse relaxation time, τ_R, for a polyethylene chain with 154 carbon atoms, C_{154} at 448 K, making use of the following parameters:
    a. The mean square radius of gyration divided by M, $\langle S^2 \rangle^{1/2}/M = 0.170$ Å^2 mol g^{-1}.
    b. The self-diffusion coefficient, $D = 0.35 \times 10^{-6}$ cm^2 s^{-1}.

REFERENCES

Castiff, E. and Tobolsky, A.V., *J. Colloid Sci.*, 10, 375, 1955.
Chung, C.I. and Gale, J.C., *J. Polym. Sci.: Polym. Phys.* 14, 1149, 1976.
Cross, M.M., *J. Colloid Sci.*, 20, 417, 1965.
Dvornic, P.R., Javanovic, J.D., and Govedarica, M.N., *J. Appl. Polym. Sci.*, 49, 1497, 1993.

Fox, T.G. and Flory, P.J., *J. Phys. Chem.*, 55, 221, 1951.

Graessley, W.W., Viscoelasticity and flow in polymer melts and concentrated solutions in *Physical Properties of Polymers*, 3rd ed., Cambridge University Press, 2004.

Grentzer, T. and Leckenby, J., *Intl. Lab.*, 19(6), 34–38, July–August, 1989.

Ishida, Y. et al., *Koll. Z.* 180, 108, 1962.

McCrum, N.G. et al., *Anelastic and Dielectric Effects in Polymeric Solids,* John Wiley and Sons, NY, 1976.

Morris, E.R., *Carbohydr. Polym.*, 13, 85, 1990.

Pearson, D.S., ver Strate, G., von Meerwall, E., and Schilling, F.C., *Macromolecules*, 20, 1133, 1987.

Thompson, A.B. and Woods, D.W., *Trans. Faraday Soc.*, 52, 1383, 1956.

Wetton, R.E. et al., *Intl. Lab.*, March, 1986.

Williams, M.L., Landel, R.F., and Ferry, J.D., *J. Am. Chem. Soc.*, 77, 3701, 1955.

BIBLIOGRAPHY

Aklonis, J.J. and MacKnight, W.J., *Introduction to Polymer Viscoelasticity*, John Wiley and Sons, 1983.

Allen, G. and Bevington, J.C., Eds., *Comprehensive Polymer Science*, Vol. 2, Pergamon Press, 1989.

Bailey, R.T., North, A.M., and Pethrick, R.A., *Molecular Motion in High Polymers*, Oxford University Press, 1981.

Bueche, F., *Physical Properties of Polymers,* Interscience Publishers, 1962.

Doi, M. and Edwards, S.F., *The Theory of Polymer Dynamics*, Oxford University Press, 1986.

Ferry, J.D., *Viscoelastic Properties of Polymers*, 3rd ed., John Wiley and Sons, 1980.

Hedvig, P., *Dielectric Spectroscopy of Polymers*, Adam Hilger, 1977.

Mark, J.E., Ngai, K., Mandelkern, L., Samulski, E.T., Koenig, J.L., and Wignall, G.D., *Physical Properties of Polymers*, 3rd ed., Cambridge University Press, 2004.

McCrum, N.G., Read, B.E., and Williams, G., *Anelastic and Dielectric Effects in Polymeric Solids,* John Wiley and Sons, 1967.

Meares, P., *Polymers: Structure and Bulk Properties*, Van Nostrand, 1965, chap. 9, 11.

Nielsen, L.E., *Mechanical Properties of Polymers*, Reinhold Publishing, 1962.

Nielsen, L.E., *Mechanical Properties of Polymers and Composites*, Vol. 1, 2, Marcel Dekker, 1974.

Sperling, L.H., *Introduction to Physical Polymer Science,* 4th ed., John Wiley and Sons, 2005.

Tobolsky, A.V., *Properties and Structure of Polymers*, Interscience Publishers, 1960.

14 The Elastomeric State

14.1 GENERAL INTRODUCTION

Most materials when stressed exhibit a limited elastic region where the material regains its original dimensions if the stress is removed. As the resulting strain is related to the extent of movement of atoms from their equilibrium conditions, substances such as metals and glass have elastic limits rarely exceeding 1% because atomic adjustments are localized. For long-chain polymers, under certain conditions, the situation is different; the extensive covalent bonding between the atoms to form chains allows considerable deformation, which is accompanied by long- and short-range cooperative molecular rearrangement arising from the rotation about chain bonds.

One of the first materials found to exhibit a sizeable elastic region was a natural substance obtained from the tree *Hevea brasiliensis*, now known to us as *cis*-polyisoprene, or more commonly referred to as *rubber*. A large number of polymers with rubberlike characteristics at ambient temperatures are now available, and it is preferable to call this general group of polymers *elastomers*. Elastomers possess several significant characteristics:

1. The materials are above their glass temperature.
2. They possess the ability to stretch and retract rapidly.
3. They have high modulus and strength when stretched.
4. The polymers have a low or negligible crystalline content.
5. The molar mass is large enough for network formation, or they must be readily cross-linked.

The most important factor is, of course, the T_g, as this determines the range of temperatures where elastomeric behavior is important and defines its lower limiting temperature. Hence, polymers with T_g below ambient may be useful elastomers if they are essentially amorphous, whereas this is unlikely if T_g is in excess of 400 K. Environmental temperature is, of course, an important factor; on a different planet, much colder than our own, latex rubber could prove to be a good glassy material, whereas on a hot planet, Perspex could well be adapted for use as an elastomer.

One of the many features of an elastomer is its ability to deform elastically by elongating up to several hundred percent. Chain slippage occurs under prolonged tension and the sample deforms. This flow under stress can be greatly reduced by introducing cross-links between the chains. In effect, these cross-links act as anchors or permanent entanglements and prevent the chains slipping past one another. The process of cross-linking is known generally as *vulcanization*, and the resultant polymer is a network of interlinked molecules now capable of maintaining an equilibrium tension. This is most important, as it changes the properties of an elastomer to a marked degree and extends the usefulness of the polymer as a material.

14.1.1 NATURAL RUBBER

Naturally occurring rubber is a linear polymer of isoprene units linked 1,4, and because the chain is unsaturated, two forms are found. Natural rubber is the cis form; it has low crystallinity, $T_g = 200$ K and $T_m = 301$ K, whereas the trans form, called *gutta-percha*, or *balata*, has medium crystallinity with $T_g = 200$ K and $T_m = 347$ K.

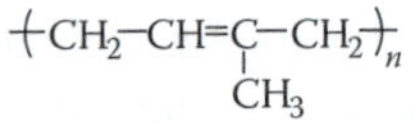

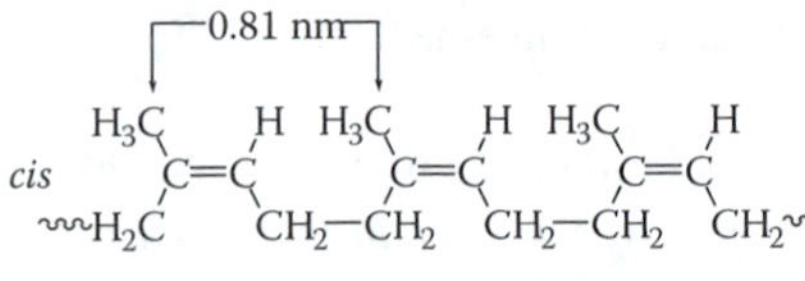

$$\begin{array}{c} +CH_2-CH=C-CH_2+_n \\ | \\ CH_3 \end{array}$$

The remarkable effect of cis and trans isomerism on the properties is well illustrated in the case of the polyisoprenes. The more extended all-trans form of balata allows the polymer to develop a greater degree of crystallinity and order. This is reflected in its hard, tough consistency, which made it suitable, in the vulcanized state, for golf ball covers. X-ray diffraction reveals two forms of gutta-percha: an α-form with a molecular repeat distance of 0.88 nm, slightly larger than a cis unit

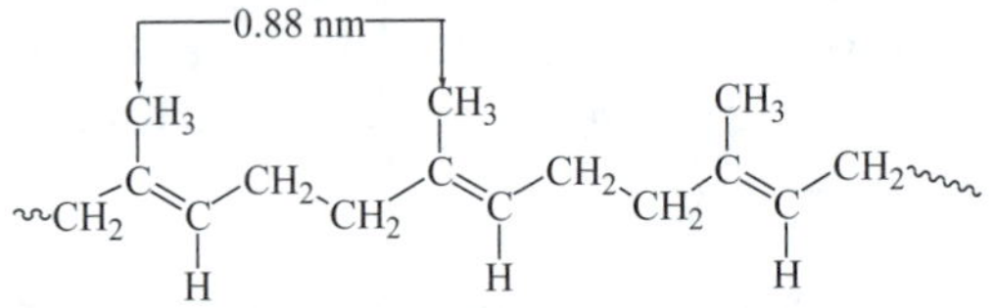

and a more compact β-form with a repeat distance of 0.47 nm.

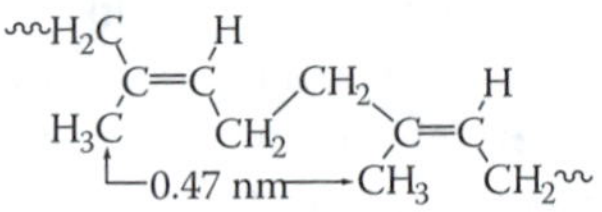

The chains of the *cis*-polyisoprene are more easily rotated than their trans counterparts, and as a result the molecules prefer to coil up into a compact conformation. The viscoelastic behavior and long-range elasticity arise from this random arrangement of long, freely moving chains, and recoverable deformations of up to 1000% can be observed. In the raw state, natural rubber is a tacky substance that is rather difficult to handle, has poor abrasion resistance, and is sensitive to oxidative degradation. It can be used for making crêpe soles and adhesives, but it is immensely improved if vulcanized, a process that cross-links the rubber. Vulcanization enhances resistance to degradation and increases both the tensile strength and elasticity.

14.2 EXPERIMENTAL VULCANIZATION

The process of vulcanization was discovered independently by Goodyear (in 1839) in the U.S. and Hancock (in 1843) in the U.K. Both found that when natural rubber was heated with sulfur, the undesirable properties of surface tackiness and creep under stress could be eliminated. The chemical reaction involves the formation of interchain links, composed of two, three, or four sulfur atoms, between sites of unsaturation on adjacent chains. It has been found that about three parts sulfur per hundred parts rubber produces a useful elastomer, capable of reversible extensions of up to 700%. Increasing the sulfur, up to 30 parts per hundred, alters the material drastically and produces a hard, highly cross-linked substance called *ebonite*. The actual mechanism of the cross-linking reaction is still unclear, but it is thought to proceed via an ionic route.

The quantitative introduction of cross-links using sulfur is difficult to achieve. An alternative method involves heating the polymer with either dicumyl peroxide or ditertiary butyl peroxide, and this is applicable to both polydienes and elastomers with no unsaturated sites in the chain, e.g., the ethylene-propylene copolymers or the polysiloxanes. The peroxide radical abstracts hydrogen from the polymer chain and creates a radical site on the interior of the chain. Two such sites interact to form the cross-link. It is claimed that one peroxide molecule produces one cross-link, but side reactions may impair the attainment of such a precise ratio. The major disadvantage of this technique is its commercial inefficiency; better results can be obtained by synthesizing precursors that are more suitable for cross-linking, such as in ethylene-propylene terpolymers. Room-temperature vulcanization techniques have been developed for silicone elastomers. These are based on linear polydimethylsiloxane chains terminated by hydroxyl groups. Curing can be achieved either by adding a cross-linking agent and a metallic salt catalyst such as tri- or tetra-alkoxysilane with stannous octoate, or by incorporating in the mixture a cross-linking agent sensitive to atmospheric water, which initiates vulcanization.

Cross-linking can also be achieved by compounding a polymer with a peroxide such as dicumyl peroxide or di-*t*-butyl peroxide, followed by heating. This technique, although economically disadvantageous compared to sulfur vulcanization, is particularly useful for saturated polymers that cannot be easily cross-linked, particularly polyethylene and other polyolefins. Similar results can be obtained through radiation cross-linking, and this method is employed commercially in the production of electrical wires and cable insulation made of polyolefins and poly(vinyl chloride).

As discussed in previous chapters, copolymerization may generate cross-linked structures. For example, dissolution of low-molecular-weight unsaturated polyesters, such as prepolymers with a radical initiator, in monomers such as methyl methacrylate or diallyl phthalate leads, upon heating, to a cross-linked structure.

14.3 PROPERTIES OF ELASTOMERS

Elastomers exhibit several other unusual properties that can be attributed to their chainlike structure. It has been found that (1) as the temperature of an elastomer increases, so too does the elastic modulus; (2) an elastomer becomes warm when stretched; and (3) the expansivity is positive for an unstretched sample but negative for a sample under tension. As these properties are so different from those observed for other materials, they are worth examining in detail.

In simple mechanistic terms, the elastic modulus is simply a measure of the resistance to the uncoiling of randomly oriented chains in an elastomer sample under stress. Application of a stress eventually tends to untangle the chains and align them in the direction of the stress, but an increase in temperature will increase the thermal motion of the chains and make it harder to induce orientation. This leads to a higher elastic modulus. Under a constant force, some chain orientation will take place, but an increase in temperature will stimulate a reversion to a randomly coiled conformation, and the elastomer will contract.

Although this is a satisfactory picture to describe properties (1) and (3), it is possible to derive a more rigorous thermodynamic explanation, which follows.

14.4 THERMODYNAMIC ASPECTS OF RUBBERLIKE ELASTICITY

As early as 1806, John Gough made two interesting discoveries when studying natural rubber. He found that (1) the temperature of rubber changed when a rapid change in sample length was induced, and (2) a rubber sample under constant tension changed length as the temperature changed, i.e., observations (2) and (3) in the previous section.

The first point is readily demonstrated with a rubber band. If the center of the band is placed lightly touching the lips and then extended rapidly by pulling on both ends, a sensation of warmth can be felt as the temperature of the rubber rises.

In thermodynamic terms, process (1) is analogous to the change in temperature undergone by a gas subjected to a rapid volume change and can be treated formally in a like manner. An ideal gas can only store energy in the form of kinetic energy, and when work is performed on a gas during compression, the energy appears as kinetic energy or heat, causing the temperature to rise. Extension of an elastomer results in the evolution of heat for similar reasons.

This effect can be examined further by studying the *reversible adiabatic extension* of an elastomer. Although this experiment is more easily carried out under conditions of constant pressure rather than constant volume, it is best to derive the relevant equations for constant volume. For this reason, we will consider first the Helmholtz function for the system, i.e.,

$$A = U - TS. \tag{14.1}$$

If the applied force is f, and l° and l are the lengths of the sample in the unextended and extended states, then differentiation of Equation 14.1 with respect to l at constant temperature gives

$$(\partial A/\partial l)_T = (\partial U/\partial l)_T - T(\partial S/\partial l)_T. \tag{14.2}$$

The work done by the system during a reversible extension of the sample by an amount dl against the restoring force f is given by $dA = -f\, dl$ and so

$$f = (\partial U/\partial l)_T - T(\partial S/\partial l)_T = f_U + f_S. \tag{14.3}$$

The force f is seen to be composed of two contributions: the energy f_U and the entropy f_S. For an ideal elastomer, the contribution of f_U to the total force is negligible because there is no energy change during extension, and

$$f = -T(\partial S/\partial l)_T. \tag{14.4}$$

This is the expression for an entropy spring, and it shows that the strain in a stretched elastomer is caused by a reduction in conformational entropy of the chains under stress. If the interdependence of length and temperature is now examined, the quantity of interest is $(\partial T/\partial l)$ and

$$(\partial T/\partial l)_{S,p} = -(\partial T/\partial S)_{l,p}(\partial S/\partial l)_{T,p}. \tag{14.5}$$

Each factor can now be evaluated separately as follows:

$$(\partial T/\partial S)_{l,p} = (\partial T/\partial H)_{l,p}(\partial H/\partial S)_{l,p} = T/C_{p,l}, \tag{14.6}$$

whereas starting from the Maxwell equation $(\partial A/\partial T)_V = -S$ and differentiating both sides with respect to l leads to

$$(\partial/\partial l)(\partial A/\partial T) = (\partial/\partial T)(\partial A/\partial l) = (\partial f/\partial T)_l = -(\partial S/\partial l)_T. \tag{14.7}$$

It follows then from Equation 14.5, Equation 14.6, and Equation 14.7 that

$$(\partial T/\partial l)_{S,p} = (T/C_{p,l})(\partial f/\partial T)_{l,p}. \tag{14.8}$$

This shows that for a rapid reversible adiabatic extension when $(df/dT)_{l,p}$ is positive, the temperature of the elastomer increases. The equation also tells us that the elastomer will contract, not expand, on heating.

Results showing the temperature-dependent behavior of an elastomer held under constant tension are plotted in Figure 14.1. Between 210 and 330 K, the tensile force increases as T increases, indicating a contraction of the elastomer, but the behavior is reversed below 210 K as the material passes through the glass transition, and the polymer reacts normally like a glassy solid.

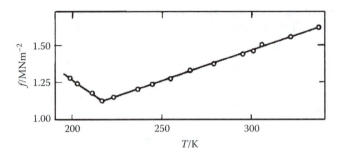

FIGURE 14.1 Tensile force f–temperature plot for rubber. The minimum occurs in the region of the glass transition temperature. (From data by Meyer, K.H. and Ferri, C., *Helv. Chim. Acta,* 18, 570, 1935. With permission.)

14.5 NONIDEAL ELASTOMERS

The behavior of most elastomers under stress is far from ideal, and a significant contribution from f_U is found. This can be expressed in several ways and following on from Equation 14.3 and Equation 14.7:

$$f_s = T(\partial f/\partial T)_{V,l}, \tag{14.9}$$

so that

$$(f_U/f) = (1 - f_s/f) = 1 - (T/f)(\partial f/\partial T)_{V,l}, \tag{14.10}$$

or

$$(f_U/f) = -\{\partial \ln(f/T)/\partial \ln T\}_{V,l}. \tag{14.11}$$

The experimental determination of the quantity of $(\partial f/\partial T)_{V,l}$ at constant volume is extremely difficult to perform, and it is much more convenient to work at constant pressure. Approximate relations between the quantities have been suggested by Flory (1946), who proposed that Equation 14.11 be modified by an additional term

$$f_U/f = -\{\partial \ln(f/T)/\partial \ln T\}_{p,l} - \alpha T/\lambda^3 - 1) \tag{14.12}$$

where λ is the extension ratio (l/l°) and α is the expansivity. The modifying factor is the difference between the values of $(\partial f/\partial T)$ measured at constant volume and constant pressure.

Thermoelastic data showing the relative contributions of f_U and f_S appear as the curves in Figure 14.2, and it can be seen that at low extension f_U remains small but begins to increase as the elongation rises.

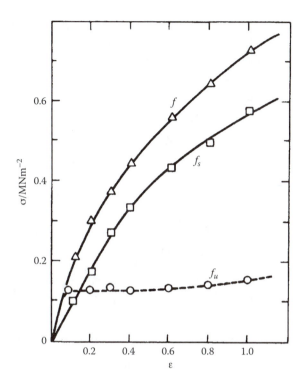

FIGURE 14.2 Stress–strain (σ–ε) curves derived from data in Figure 14.3, showing the magnitude of the contributions from f_U and f_S. (Adapted from Beevers, R.B., *Experiments in Fibre Physics*, Butterworths, 1970.)

The curves in Figure 14.2 can be derived by measuring the force required to maintain a constant elongation in a rubber sample at various temperatures. If a number of extensions are examined, a family of curves (Figure 14.3) can be constructed, starting with the highest temperature at each strain. As the temperature is reduced incrementally, the load must be adjusted accordingly. Here the stress is taken as the ratio of the load to the unstrained cross-sectional area of the elastomer, whereas the strain is just the extension of the sample. Values of f_S and f_U at each strain are calculated from the slope and the intercept at 0 K, respectively. In the diagram $f_S = -(T\partial S/\partial l)$ has been calculated for $T = 298$ K.

14.6 DISTRIBUTION FUNCTION FOR POLYMER CONFORMATION

The retractive force in a rubbery material is a direct result of the chain in the extended form trying to regain its most probable, highly coiled conformation. Thus, it is of considerable interest to calculate, in addition to the average dimensions of the polymer chain, the distribution of all the possible shapes available to the molecules experiencing thermal vibrations.

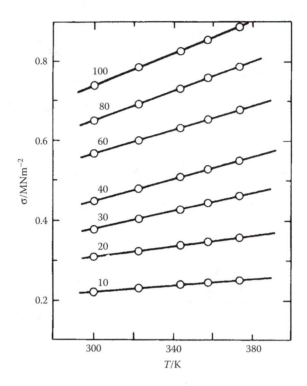

FIGURE 14.3 Thermoelastic behavior of a rubber sample. Stress–temperature (σ–T) curves for a series of extension values. The percentage strain is shown against each curve. (Adapted from Beevers, R.B., *Experiments in Fibre Physics,* Butterworths, 1970.)

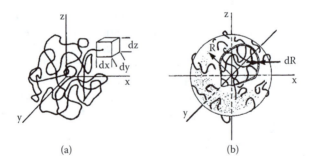

FIGURE 14.4 (a) Schematic diagram of a flexible polymer coil with one end contained in a small volume element (dx, dy, dz) and the other located at the origin of the coordinates. (b) Hypothetical sphere containing the coil.

This can be accomplished by first considering a chain in three-dimensional space with one end located at the origin of a set of Cartesian coordinates (see Figure 14.4). The probability that the other end will be found in a volume element (dx, dy, dz) at the point (x, y, z) is given by $p(x, y, z)dx, dy, dz$, where $p(x, y, z)$ measures the number of possible conformations the chain can adopt in the range ($x + dx$), ($y + dy$), and

$(z + dz)$. This is known as the *probability density* and can be expressed in terms of the parameter $\beta = (3/2nl^2)^{1/2}$ as

$$(dx, dy, dz)p(x, y, z) = (\beta^2 /\pi^{3/2}) \exp\{-\beta^2(x^2 + y^2 + z^2)\} \, dx.dy.dz \quad (14.13)$$

It now remains to calculate the probability $p(r)$ that the chain is located in a spherical shell of thickness dr and distance r from the origin. This is given by

$$p(r)dr = 4\pi r^2 dr\{(\beta^3 /\pi^{3/2}) \exp(-\beta^2 r^2)\}, \quad (14.14)$$

where $4\pi r^2 \, dr$ is the volume of the shell. This function has the form shown in Figure 14.5, where the maximum corresponds to the most probable distance between chain ends and is given by

$$r^2 = \int_0^\infty r^2 p(r)dr = 3/2\beta^2 = l^2 n. \quad (14.15)$$

It has already been stated that although a highly coiled conformation is the most probable for an elastomer, the energy of an ideal elastomer in the extended form is the same as that in the coiled state. The elastic retractive force is entropic and not energetic in origin, and depends on the fact that the number of possible ways a polymer coil can exist in a highly compact form is overwhelmingly greater than the number of available arrangements of the chain segments in an extended ordered form. This means that the probability of finding a chain in a coiled state is high, and as probability and entropy are related by the Boltzmann equation

$$S = k \ln p, \quad (14.16)$$

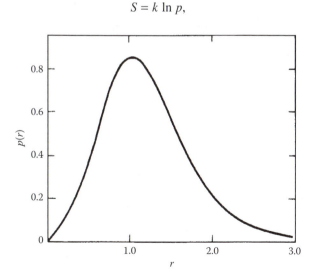

FIGURE 14.5 Distribution function $p(r)$ for end-to-end distances r calculated from Equation 14.14. The maximum occurs at $r = 1/\beta$.

then the most likely state for the chain is one of maximum entropy. Substitution in Equation 14.14 leads to

$$S = C - k\beta^2 r^2, \tag{14.17}$$

where C is a constant. This relation provides a measure of the entropy of an ideal flexible chain whose ends are held a distance r apart.

Although an elastomer is considered to be solely an entropy spring, it is never perfect and small changes in the internal energy are observed when it is under stress.

14.7 STATISTICAL APPROACH

Having examined the thermodynamic approach, we can now outline briefly the stress–strain behavior of an elastomer in terms of the chain conformations.

Consider a lightly cross-linked network with the junction points sufficiently well spaced to ensure that the freedom of movement of each chain section is unrestricted. If the length of a chain between two cross-linking points is assumed to be r, the probability distribution calculated earlier can be applied to the network structure. The entropy of a single chain, as described by Equation 14.17, can be used to calculate S for a chain in the network, but if the stress–strain relations are required, then S for chains in both the deformed and undeformed states must be calculated, and for the complete network an integration over all the chains in the sample should be performed.

When a unit cube of elastomer is stretched, the resulting entropy change is

$$\Delta S = -\tfrac{1}{2} Nk \left(\lambda_1^2 + \lambda_2^2 + \lambda_3^2 - 3 \right), \tag{14.18}$$

where N is the number of individual chain segments between successive cross-links per unit volume, and λ_1, λ_2, and λ_3 are the principal extension ratios. For an ideal elastomer, there is no change in the internal energy, and the work of deformation w is derived entirely from $w = -T\Delta S$, so that

$$w = \tfrac{1}{2} NkT \left(\lambda_1^2 + \lambda_2^2 + \lambda_3^2 - 3 \right). \tag{14.19}$$

14.7.1 Experimental Stress–Strain Results

The treatment of mechanical deformation in elastomers is simplified when it is realized that the Poisson ratio is almost 0.5. This means that the volume of an elastomer remains constant when deformed, and if one also assumes that it is essentially incompressible ($\lambda_1 \lambda_2 \lambda_3 = 1$), the stress–strain relations can be derived for simple extension and compression using the stored energy function w.

14.7.1.1 Simple Extension

The required conditions are $\lambda = \lambda_1$ and $\lambda_2 = \lambda_3 = \lambda^{-1/2}$. Elimination of λ_3 and substitution of λ leads to

$$w = \tfrac{1}{2} NkT \left[\lambda^2 + (2/\lambda) - 3 \right]. \tag{14.20}$$

As the force f is simply $= (dw/d\lambda)$, differentiation provides a relation between f and λ

$$f = G(\lambda - \lambda^{-2}), \tag{14.21}$$

where the modulus factor is $G = NkT$ and shows that G will depend on the number of cross-links in the sample. If we neglect the fact that the chains in a network possess free ends and assume that all network chains end at two cross-linking points, then

$$G = RT \, \rho \, M_s^{-1}, \tag{14.22}$$

where R is the gas constant, ρ the polymer density, and M_s the number average molar mass of a chain section between two junction points in the network. The product (ρM_s^{-1}) is then a measure of the cross-link density of a sample. This also shows that the modulus will increase with temperature. The statistically derived equation (Equation 14.21) was first tested by Treloar for both extension and compression of a rubber vulcanizate. A value of $G = 0.392$ MN m^{-2} was chosen to fit the data at low extensions, where the application of Gaussian statistics to the chain might be expected to be valid. The experimental data for extension (see Figure 14.6) drops below the theoretical curve at $\lambda \approx 2$, then sweeps up sharply when λ exceeds 6. This rapid increase was originally attributed to the crystallization of the polymer

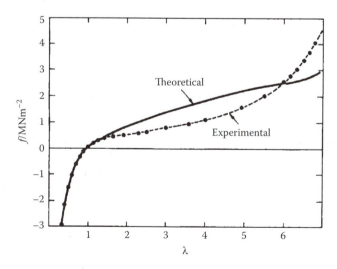

FIGURE 14.6 Extension or compression ratio λ as a function of the tensile or compressive force f for a rubber vulcanizate. Theoretical curve is derived from Equation 14.21 using $G = 0.392$ MN m^{-2}. (From data by Treloar, L.R.G., *Trans. Faraday Soc.*, 40, 59, 1944. With permission.)

under tension, but it is now believed to reflect the departure of the network from the assumed Gaussian distribution.

14.7.1.2 Simple Compression

The agreement between experiment and theory for the equivalent of simple compression is somewhat better. The experimental determination of compression is rather difficult to carry out, but a two-dimensional extension with $\lambda_2 = \lambda_3$ serves to provide the same information. This is achieved by clamping a circular sheet of rubber around the circumference and then inflating it to provide a stress. Correlation between the results and a curve derived from Equation 14.21, using the same G factor as before, is good.

The statistical approach, using a Gaussian distribution, thus appears to predict the stress–strain response except at moderately high elongations.

14.7.1.3 Pure Shear

The corresponding equation to describe the behavior of an elastomer under shear is $f = G(\lambda - \lambda^{-3})$ but, again, agreement is reasonable only at low λ.

14.7.1.4 Large Elastic Deformation

At high extensions, departure from the Gaussian chain approximation becomes significant and has led to the development of a more general but semiempirical theory based on experimental observations. This is expressed in the Mooney, Rivlin, and Saunders (MRS) equation

$$\tfrac{1}{2} f(\lambda - \lambda^{-2})^{-1} = C_1 + C_2 \lambda^{-1}, \tag{14.23}$$

where C_1 and C_2 are constants. Unfortunately, this simple form is still only capable of predicting data over a range of low-to-moderate extension, but not for samples under compression.

14.8 SWELLING OF ELASTOMERIC NETWORKS

A cross-linked elastomer cannot dissolve in a solvent. Dispersion is resisted because the cross-links restrict the movement and complete separation of the chains, but the elastomer does swell when the solvent molecules diffuse into the network and cause the chains to expand. This expansion is counteracted by the tendency of the chains to coil up and, eventually, an equilibrium degree of swelling is established that depends on the solvent and the cross-link density; i.e., the higher the cross-link density, the lower the swelling.

The behavior is predicted in the Flory–Huggins treatment of swelling, which leads to a relation between the degree of swelling Q for a particular solvent and the shear modulus G of the unswollen rubber

$$G = RT\, A/V_1 Q^{5/3}, \tag{14.24}$$

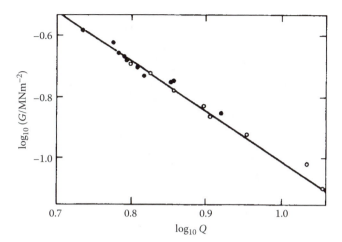

FIGURE 14.7 The swelling of butyl vulcanizates by cyclohexane plotted as a function of the elastic modulus using Equation 14.24. (From Flory, P.J., *Ind. Eng. Chem.*, 38, 417, 1946. With permission.)

where A is a constant. The validity of this expression was tested by Flory (1946) as shown in Figure 14.7, where the expected slope of –5/3 was obtained. The theory also predicts that the equilibrium swelling of an elastomer increases when under a tensile stress.

The statistical theory will also describe the response under stress of elastomers swollen by solvents and, in general, it is found that the greater the degree of swelling, the better the agreement between theory and experiment. The modifications necessary for the treatment of swollen networks are relatively straightforward, and the stored energy function becomes

$$w = \tfrac{1}{2} NkT \phi_{re}^{1/3} \left(\lambda_1'^2 + \lambda_2'^2 + \lambda_3'^2 - 3 \right), \qquad (14.25)$$

where ϕ_{re} is the volume fraction of the elastomer and λ_x represents the swollen unstrained state of the network. Similarly, for simple extension, $\phi_{re}^{1/3}$ is inserted in the right-hand side of Equation 14.21.

14.9 NETWORK DEFECTS

The number average molar mass M_s of a chain section between two junction points in the network is an important factor controlling elastomeric behavior; when M_s is small, the network is rigid and exhibits limited swelling, but when M_s is large, the network is more elastic and swells rapidly when in contact with a compatible liquid. Values of M_s can be estimated from the extent of swelling of a network, which is considered to be ideal but rarely is, and interpretation of the data is complicated by the presence of network imperfections. A real elastomer is never composed of chains linked solely at tetrafunctional junction points but will inevitably contain defects such as (1) loose chain ends, (2) intramolecular chain loops, and (3) entangled chain loops.

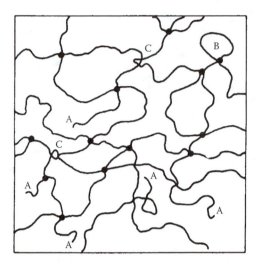

FIGURE 14.8 Diagram showing defects in an elastomeric network. (A) Loose chain ends, (B) intramolecular chain loops, and (C) entangled chain loops.

Both M_s and the interaction parameter χ_1 can be calculated from measurements of network swelling when the elastomer is brought in contact with a solvent. We can assume that the total free-energy change will then be composed of

$$\Delta G = \Delta G^{el} + \Delta G^{M} \tag{14.26}$$

where ΔG^M, the free-energy change on mixing the elastomer chains with solvent, is given by the Flory–Huggins equation (see Chapter 8, Equation 8.30), in which $N_2 = 1$ for a cross-linked network, and ΔG^{el} is the change in the elastic free energy. The latter can be estimated from simple statistical theory, but we now have a choice. The Flory *affine network model* assumes that the junction points are embedded in the network; hence, affine deformation means that any movement of the network chains is in proportion to the change in the macroscopic dimensions of the sample. On the other hand, the James and Guth *phantom network model* assumes that the junction points can fluctuate independently of the macroscopic deformation.

If N is replaced by a front factor F whose value depends on the model used, the general form for ΔG^{el} derived is essentially the same for both theories and is given by Equation 14.19. Experimental work has shown that the behavior of a swollen network is best described by the phantom network model and further equations are derived on that basis.

For isotropic swelling $\lambda_1 = \lambda_2 = \lambda_3 = (\phi_r/\phi_{re})^{1/3}$, where ϕ_r is the volume fraction of the dry elastomer and usually has a value of unity in the unswollen state, and ϕ_{re} is the volume fraction of the elastomer at the maximum degree of swelling (equilibrium) when the network is in contact with excess solvent. Also, for a phantom network,

$$\Delta G^{el} = \tfrac{3}{2} kT \gamma \left[\frac{\phi_r^{2/3}}{\phi_{re}} - 1 \right] \tag{14.27}$$

where

$$\gamma = \frac{V_0 \rho N_A}{M_s}(1 - 2/f) \qquad (14.28)$$

Here, f is the functionality of a junction in the network, ρ is the network density, and V_0 is the volume of solvent plus polymer. Equation 14.27 can be differentiated with respect to N_1 and combined with Equation 8.31 in Chapter 8 (the corresponding mixing term) to give at equilibrium

$$\ln(1 - \phi_{re}) + \chi_1 \phi_{re}^2 + \phi_{re} + B(\phi_{re}/\phi_r)^{1/3} = 0 \qquad (14.29)$$

where $B = (V_1/RT)(\gamma k t V_0)$ and V_1 is the molar volume of the solvent. From this, it follows that

$$M_s = \frac{-\rho(1 - 2/f)V_1 \phi_r^{2/3} \phi_{re}^{1/3}}{\ln(1 - \phi_{re}) + \chi_1 \phi_{re}^2 + \phi_{re}} \qquad (14.30)$$

or alternatively,

$$\chi_1 = \frac{-\ln(1 - \phi_{re}) + \phi_{re} - B\phi_r^{2/3}\phi_{re}^{1/3}}{\phi_{re}^2} \qquad (14.31)$$

14.10 RESILIENCE OF ELASTOMERS

When an elastomer, in the form of a ball, is dropped from a given height onto a hard surface, the extent of the rebound provides an indication of the resilience of the elastomer. A set of elastomer balls, manufactured by Polysar Corp. in Canada, provides an excellent demonstration of this phenomenon and has been used to measure the rebound height of several elastomers, recorded schematically in Figure 14.9. If h_0 is the original height and h is the recovery height, then the rebound resilience is defined as (h/h_0) and the relative energy loss per half cycle is $(1 - h/h_0)$.

It should be remembered that an elastomer exhibiting good elastic properties under slow deformation may not possess good resilience; the relative recoveries of natural and butyl rubber provide a good example of this fact. As resilience is the ability of an elastomer to store and return energy when subjected to a rapid deformation, it can be shown that temperature also plays an important part in determining resilience. If the butyl and natural rubber balls are now heated to about 373 K, both will rebound to about the same extent. The importance of the two variables, time and temperature, is illustrated in a plot of the rebound resilience against temperature for three elastomers and a recognized plastic (see Figure 14.10). The sharply defined minima are characteristic of such curves, and the broad butyl curve is anomalous.

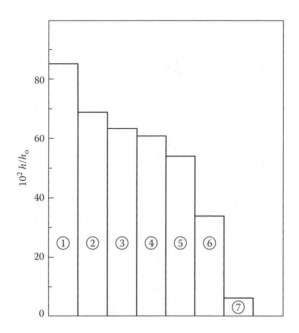

FIGURE 14.9 Percentage rebound recovery measured at 298 K, with balls made from: (1) *cis*-polybutadiene, (2) synthetic *cis*-polyisoprene, (3) natural *cis*-polyisoprene, (4) ethylene-propylene copolymer, (5) styrene–butadiene (SBR), (6) *trans*-polyisoprene, and (7) butyl rubber.

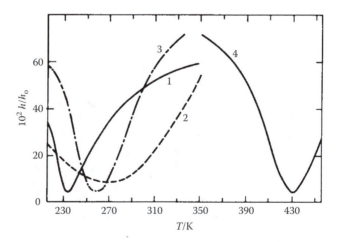

FIGURE 14.10 Change of rebound resilience (h/h_0) with temperature T for: (1) natural rubber, (2) butyl rubber, (3) neoprene, and (4) poly(methyl methacrylate). (After Mullins, L., *I. R. I. Trans.*, 22, 235, 1947.)

The minimum for curve 1 is closely related to the loss of long-range elasticity at the glass temperature $T_g = 218$ K, although it actually occurs at the higher temperature of 238 K. A similar situation is found for neoprene and poly(methyl methacrylate).

One can conclude from this that resilience is closely related to the molecular structure and the intermolecular forces affecting the ability of the chain to rotate.

When the chains are deformed during a bounce, a stress is applied and then rapidly removed. The time required for the chains to regain their original positions is measured by the relaxation time τ, as defined in Chapter 13, Section 13.4. Thus, relaxation times are a measure of the ability of the chains to rotate; at room temperature the butyl rubber with the bulky methyl groups will not rotate as readily as the *cis*-polyisoprene, so that when the deformation of chains in the sample of butyl rubber occurs, the chains do not return to their equilibrium positions as rapidly as the natural rubber; i.e., τ is longer.

The elastomer showing superior rebound potential at room temperature is *cis*-polybutadiene. This sample is noncrystalline and has no pendant groups to impede free segmental rotation, so the relaxation time is correspondingly shorter than other elastomers. The response of the butyl sample improves as the temperature increases, because additional thermal energy is available to enhance chain rotation and decrease the relaxation time correspondingly. This leads to an improved resilience, and the rebound potential now matches the natural rubber whose τ is not so sensitive to temperature change in this range.

PROBLEMS

1. Schematically describe, using equations, the following reactions leading to cross-linked structures:
   a. Reaction between styrene and divinylbenzene
   b. Polyesterification reaction to form unsaturated polyester from maleic anhydride and ethylene glycol, followed by cross-linking via radical polymerization with methyl methacrylate
   c. Radiation cross-linking of polyethylene
2. Cross-linking is essential in the production of elastomeric materials. Peroxide cross-linking has also been used for some plastic applications of polyethylene. What are the advantages?
3. X-ray diffraction photographs for unstrained, unvulcanized rubber and a sample stretched up to 650% extension are shown in Figure 14.11 (Treloar, 1970). Comment on those sample features that are responsible for the differences between the two photographs.
4. The stress–temperature behavior of natural rubber at various extension ratios has been measured by Shen et al. (1967) and shown in Figure 14.12. Compare trends between data at varying extension ratios with those shown in Figure 14.3 and provide an explanation for the changes at low elongations (a phenomenon that is termed *thermoelastic inversion*).
5. Equation 14.14 defines the mean square end-to-end distance of a polymer chain. Prove that this quantity is equal to $r^2 = l^2 n$. (*Note:* To solve the integral, use the substitution $x = \beta^2 r^2$. The gamma function is given by the integral:

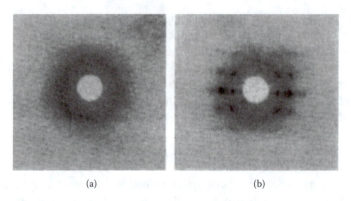

FIGURE 14.11 X-ray diffraction photographs for (a) unstrained unvulcanized rubber and (b) a sample stretched up to 650% extension. (From Treloar, L.R.G., *Introduction to Polymer Science*, Wykeham Publications, 1970. With permission.)

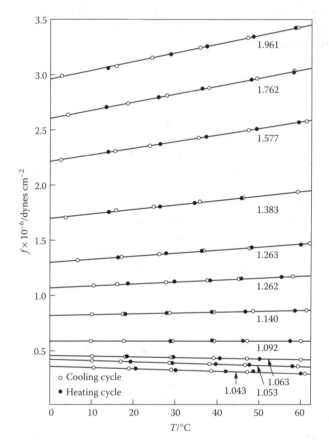

FIGURE 14.12 Stress–temperature curves of natural rubber at various extension ratios (as indicated). (Adapted from Shen et al., *J. Appl. Phys.* 38, 791, 1967. With permission of the American Institute of Physics.)

$$\int_0^\infty x^{n-1}e^{-x}dx$$

and it can be evaluated by considering that:

$$\Gamma\left(\frac{1}{2}\right) = \sqrt{\pi} \qquad \Gamma(1) = 1 \qquad \Gamma(n+1) = n\Gamma(n)$$

6. Using Equation 14.14 as your starting point, demonstrate that the most probable end-to end distance is $r = 1/\beta$.
7. Calculate the root-mean-square end-to-end distance of a polypropylene chain with number average molecular weight equal to 100,000 g mol^{-1}. Consider the bond length to be 2.52 Å and that the chain is freely jointed.
8. A vulcanized polyisoprene sample has a shear modulus of 0.392 MN m^{-2} at room temperature. Determine the average number of monomers between cross-links ($\rho = 1.00$ g cm^{-3}).
9. Give an estimate of the entropy change at 298 K following a 100% elongation for a cross-linked elastomer with density of 1.0 g cm^{-3} and molecular weight between cross-links equal to 10,000.
10. Stress measurements for a natural rubber vulcanizate as a function of extension ratio, λ, were carried out by Grumbell et al. (1953) and the resulting data are listed below in Table 14.1. Confirm that these data follow the Mooey-Rivlin-Saunders equation and determine the constants c_1 and c_2.

TABLE 14.1
Stress as a Function of Extension Ratio, λ, for a Natural Rubber Vulcanizate

λ	f/MN m^{-2}
1.92	0.544
1.69	0.451
1.62	0.436
1.48	0.366
1.39	0.322
1.28	0.248
1.28	0.255
1.26	0.239
1.15	0.155
1.06	0.074
1.08	0.091

Source: S.M. Grumbell, L. Mullins, and R.S. Rivlin, *Trans. Faraday Soc.*, 49, 1495, 1953.

REFERENCES

Beevers, R.B., *Experiments in Fibre Physics,* Butterworths, 1970.

Flory, P.J., *Ind. Eng. Chem.,* 38, 417, 1946.

Grumbell, S.M., Mullins, L., and Rivlin, R.S., *Trans. Faraday Soc.,* 49, 1495, 1953.

Mark, J.E., *J. Chem. Ed.,* 58, 898, 1981.

Meyer, K.H. and Ferri, C., *Helv. Chim. Acta,* 18, 570, 1935.

Mullins, L., *I. R. I. Trans.,* 22, 235, 1947.

Shen, M., McQuarrie, D.A., and Jackson, J.L., *J. Appl. Phys.,* 38, 791, 1967.

Treloar, L.R.G., *Introduction to Polymer Science*, Wykeham Publications, 1970.

Treloar, L.R.G., *Trans. Faraday Soc.,* 40, 59, 1944.

BIBLIOGRAPHY

Allen, G. and Bevington, J.C., Eds., *Comprehensive Polymer Science,* Vol. 2, Pergamon Press, 1989.

Beevers, R.B., *Experiments in Fibre Physics,* Butterworths, 1970.

Blackley, D.C., *Synthetic Rubbers: Their Chemistry and Technology,* Elsevier, 1983.

Bueche, F., *Physical Properties of Polymers,* Interscience Publishers, 1962, chap. 1.

Flory, P.J., *Principle of Polymer Chemistry,* Cornell University Press, 1953, chap. 11.

Koenig, L., *Physical Properties of Polymers,* American Chemical Society, 1984.

Mark, J.E. and Erman, B., *Rubberlike Elasticity — A Molecular Primer,* John Wiley and Sons, 1988.

Mark, J.E., Ngai, K., Graessley, W., Mandelkern, L., Samulski, E., Koenig, J.L., and Wignall, G., *Physical Properties of Polymers,* 3rd ed., Cambridge University Press, 2004.

Odian, G., *Principles of Polymerization*, 4th ed., Wiley, 2004.

Sperling, L.H., *Introduction to Physical Polymer Science,* 4th ed., Wiley-Interscience, 2005.

Tobolsky, A.V. and Mark, H., *Polymer Science and Materials,* Wiley Interscience, 1971, chap. 9.

Treloar, L.R.G., *Introduction to Polymer Science*, Wykeham Publications, 1970.

Treloar, L.R.G., *Physics of Rubber Elasticity,* 3rd ed., Clarendon Press, 1975.

Ward, I.M. and Sweeney, J., *An Introduction to the Mechanical Properties of Solid Polymers*, John Wiley and Sons, 2004.

15 Structure–Property Relations

15.1 GENERAL CONSIDERATIONS

The increasing use of synthetic polymers by industrialists and engineers to replace or supplement more traditional materials, such as wood, metals, ceramics, and natural fibers, has stimulated the search for even more versatile polymeric structures covering a wide range of properties. For such a quest to be efficient, a fundamental knowledge of structure–property relations is required.

The problem can be examined initially on two broad planes:

1. *The chemical level:* This deals with information on the fine structure, namely, what type of monomer constitutes the chain and whether more than one type of monomer (copolymer) is used, i.e., the parameters that relate ultimately to the three-dimensional aggregated structure and influence the extent of sample crystallinity and the physical properties.
2. *The architectural aspects:* These are concerned with the chain as a whole; we are now required to ask such questions as the following: is the polymer linear, branched, or cross-linked? What distribution of chain lengths exists? What is the chain conformation and rigidity?

Having considered these general points, one must then establish the suitability of a polymer for a particular purpose. This depends on whether it is glass-like, rubber-like, or fiber forming, and whether the characteristics depend primarily on chain flexibility, chain symmetry, intermolecular attractions and, of course, environmental conditions. Excluding the environment, these parameters in turn are reflected in the more tangible factors, T_m, T_g, modulus, and crystallinity, which, being easier to assess, are commonly used to characterize the polymer and ascertain its potential use.

As the relative values of both T_m and T_g play such an important role in determining the ultimate behavior of a polymer, we can begin an examination of their structures and properties by finding out how a polymer scientist might control these parameters.

15.2 CONTROL OF T_m AND T_g

We have already seen, in earlier chapters, how chain symmetry, flexibility, and tacticity can influence the individual values of both T_m and T_g. Thus, a highly flexible chain has a low T_g, which increases as the rigidity of the chain becomes greater. Similarly, strong intermolecular forces tend to raise T_g and also increase crystallinity.

Steric factors play an important role. A high T_g is obtained when large pendant groups attached to the chain restrict its internal rotation, and bulky pendant groups tend to impede crystallization, except when arranged regularly in isotactic or syndiotactic chains.

Chain flexibility is undoubtedly the controlling factor in determining T_g, but it also has a strong influence on T_m. Hence, we must consider these parameters together from now on and determine how to effect control of both.

15.2.1 CHAIN STIFFNESS

It is important to be able to regulate the degree of chain stiffness, as rigid chains are preferred for fiber formation whereas flexible chains make better elastomers. The flexibility of a polymer depends on the ease with which the backbone chain bonds can rotate. Highly flexible chains will be able to rotate easily into the various available conformations, whereas the internal rotations of bonds in a stiff chain are hindered and impeded.

Variations in chain stiffness can be brought about by incorporating different groups in linear chains, and the results can be appraised by following the changes in T_m and T_g in a series of different polymers. The effects can be assessed more easily if an arbitrary reference is chosen; the simplest synthetic organic polymer, polyethylene with $T_m \approx 400$ K and $T_g \approx 188$ K, is suitable for this purpose.

One can begin by considering a general structure $+(CH_2)_m X +$, where m and X vary. The effect on T_m of incorporating different links in the carbon chain is illustrated in Table 15.1. The chain flexibility is increased by groups such as $+O+$, $+COO+$, and $+OCOO+$, and also increased with the length of the $+CH_2+$ section. This is shown by a lowering of T_m relative to polyethylene. Insertion of the polar $+SO_2+$ and $+CONH+$ groups raises T_m because the intermolecular bonding now assists in stabilizing the extended forms in the crystallites.

Chain stiffness is also greatly increased when a ring is incorporated in the chain, as this restricts the rotation in the backbone and reduces the number of conformations

TABLE 15.1
Influence of Various Links on T_m when Incorporated in an All-Carbon Chain

Polymer Group	Repeat Unit	$T_m(K)$					
		m	2	3	4	5	6
Polyethylene	$+(CH_2)_m+$	400	—	—	—	—	
Polyester	$+(CH_2)_m COO+$	395	335	329	335	325	
Polycarbonate	$+(CH_2)_m OCOO+$	312	320	330	318	320	
Polyether	$+(CH_2)_m CH_2-O+$	308	333	—	—	—	
Polyamide	$+(CH_2)_m CO NH+$	598	538	532	496	506	
Polysulphone	$+(CH_2)_m CH_2 SO_2+$	573	544	516	493	—	

TABLE 15.2
Effect of Aromatic Rings on Chain Stiffness, as Shown by the Values of T_m and T_g

	Structure	$T_g(K)$	$T_m(K)$
1.	$+CH_2-CH_2\frac{}{n}$	188	400
2.	$+CH_2-CH_2-O\frac{}{n}$	206	339
3.	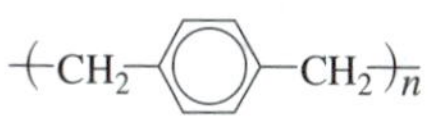 $+CH_2-\bigcirc-CH_2\frac{}{n}$	—	About 653
4.	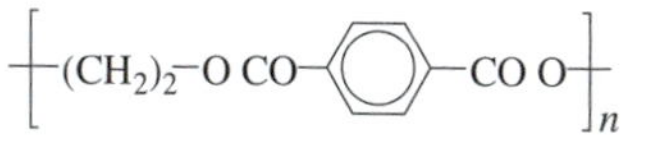 $+(CH_2)_2-O\,CO-\bigcirc-CO\,O+_n$	342	538
5.	$+NH(CH_2)_6NHCO(CH_2)_4CO+_n$	320	538
6.	 $-NH-\bigcirc-NHCO(CH_2)_4-CO-$	—	613
7.	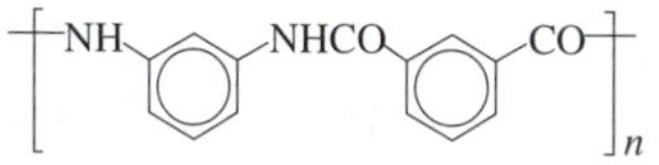 $+NH\diagdown NHCO\diagdown CO+_n$	546	About 635 (decomposition)
8.	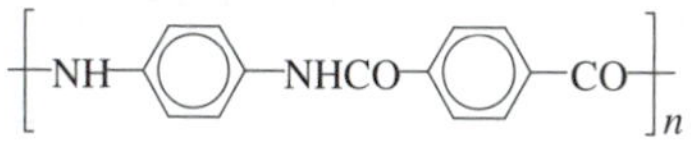 $+NH-\bigcirc-NHCO-\bigcirc-CO+_n$	—	About 773

a polymer can adopt. This is an important aspect, as fiber properties are enhanced by stiffening the chain. The effect of aromatic rings on T_g and T_m is shown in Table 15.2.

The *p*-phenylene group in structure 3 causes a big increase in T_m, and this can be modified by introducing a flexible group as in terylene, structure 4. This shows that a judicious combination of units can lead to a wide variety of chain flexibilities and physical properties. The effect of the aromatic ring is again obvious in structures 5 through 8. The influence of chain symmetry is also noticeable when comparing the decomposition temperatures of the symmetrical chain 8 with the unsymmetrical chain 7. The latter is also an inefficient close-packing structure, as there is sufficient disorder for the glass transition to appear.

15.2.2 INTERMOLECULAR BONDING

An increase in the lattice energy of a crystallite is obtained when the three-dimensional order is stabilized by intermolecular bonding. In the polyamide and polyurethane series, the additional cohesive energy of the hydrogen bond (about 24 kJ mol^{-1}) strengthens the crystalline regions and increases T_m. The effect is strongest when regular, evenly spaced groups exist in the chain, as with nylon-6,6. The importance of secondary bonding in the polyamide series is illustrated quite dramatically when the crucial hydrogen atom of the amide group is replaced by a methylol group. The loss of the hydrogen-bonding capability impairs the tendency toward regular chain

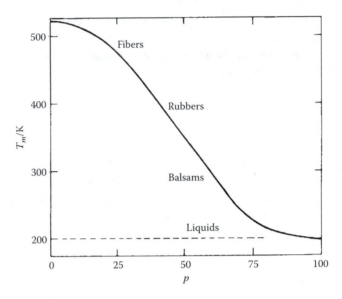

FIGURE 15.1 Change in the properties and melting temperature T_m of nylon-6,6 as the hydrogen-bonding capacity is reduced by changing the percentage p of amide substitution. (Adapted from Hill, R., *Fibres from Synthetic Polymers,* Elsevier, 1953.)

TABLE 15.3
Melting Temperatures T_m of Linear Aliphatic Polyamides

Monadic Nylon	$T_m(K)$	Dyadic Nylon	$T_m(K)$
4	533	4,6	581
6	496	5,6	496
7	506	6,6	538
8	473	4,10	509
9	482	5,10	459
10	461	6,10	495
11	463	6,12	482
12	452		

alignment, and the character of the polyamide changes dramatically. With little substitution, they are suitable fibers, but as the hydrogen is replaced, they change and become more elastomeric, then like balsams and, finally, liquids.

An alternative method of reducing the hydrogen-bonding potential (and so, T_m) in the polyamides is to increase the length of the $+CH_2+_n$ sequence between each bonding site. This leads to nylons with a variety of properties; for example, nylon-12 has properties intermediate between those of nylon-6 and polyethylene. The effect is shown in Table 15.3.

15.3 RELATION BETWEEN T_m AND T_g

Most of the factors discussed so far influence T_g and T_m in much the same way, but in spite of this, the fact that T_m is a first-order thermodynamic transition, whereas T_g is not, precludes the possibility of a simple relation between them.

There is, however, a crude correlation, represented in Figure 15.2, where a broad band covers most of the results for linear homopolymers, and the ratio (T_g/T_m) lies between 0.5 and 0.8 for about 80% of these.

Obviously then, a synthetic chemist attempting to control the T_m and T_g of a simple chain structure by varying flexibility, symmetry, tacticity, etc., is limited to structures with either a high T_m and T_g or a low T_m and T_g. In effect, neither T_m nor T_g can be controlled separately to any great degree.

To exercise this additional control, another method of chain modification must be sought, and this leads to the use of copolymers.

15.4 RANDOM COPOLYMERS

Axial symmetry in a chain is a major factor in determining its ability to form crystallites, and one method of altering the crystalline content is to incorporate some structural irregularity in the chain. The controlled inclusion of linear symmetrical

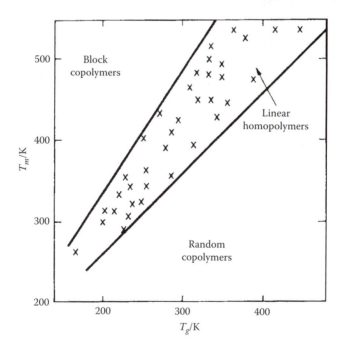

FIGURE 15.2 Plot of melting temperature T_m against glass transition temperature T_g for linear homopolymers with (T_g/T_m) lying in the range 0.5 to 0.8.

homopolymer chains $\vdash(A)_n$ in a crystal lattice can be achieved by copolymerizing A with varying quantities of monomer B, whose purpose is to destroy the regularity of the structure. This leads to a gradual decrease in T_m, as shown schematically in Figure 15.3. The broken line represents the possibility that, in the middle composition range, the decrease in regularity is so great that the material is amorphous. This situation is sometimes obtained when a terpolymer is prepared.

A practical application is found in the polyamides. An improvement in the elastic qualities of the polyamide fiber is obtained if the modulus is reduced; as the factors that affect the melting temperature also affect the modulus, this can be achieved by starting from nylon-6,6 or nylon-6,10 and forming (66/610) copolymers. The random inclusion of the two types of unit in the chain disturbs both the symmetry and the regular spacing of the hydrogen-bonding sites, resulting in a drop in T_m.

The glass transition T_g is not affected in the same way as T_m because T_g is more a function of the difference in chain flexibility than the packing efficiency. This means that the response of T_g to a change in copolymer composition is quite different, and this affords a means of controlling the magnitudes of T_m and T_g independently, a most important feature not readily achieved by other methods.

15.5 DEPENDENCE OF T_m AND T_g ON COPOLYMER COMPOSITION

A quantitative expression for the depression of the melting temperature can be derived thermodynamically in terms of the composition and enthalpy of fusion ΔH_u of polymer A by

$$1/T_m^{AB} - 1/T_m^A = -(R/\Delta H_u)\ln x_A, \tag{15.1}$$

where T_m^A and T_m^{AB} are the melting temperatures of pure polymer A and the copolymer AB, respectively, and x_A is the mole fraction of A in the copolymer.

The simple linear relation between T_g and x shown in Figure 15.3 is found only for a few copolymers composed of compatible monomer pairs, such as styrene copolymerized with either methyl acrylate or butadiene. A simple ideal mixing rule can be applied to these systems, but when the comonomer properties differ markedly, the linear dependence is lost and a nonlinear equation has to be developed.

One simple relation, the Fox equation, which usefully describes the behavior of many vinyl monomer pairs is

$$1/T_g^{AB} = w_A/T_g^A + w_B/T_g^B, \tag{15.2}$$

where w_A and w_B are the mass fractions of monomers A and B, respectively. For a system where the conditions $T_g^B < T_g^{AB} < T_g^B$ hold, the free-volume concept can be used to formulate a relation between T_g and w. Gordon and Taylor have proposed

$$(T_g^{AB} - T_g^A)w_A + K(T_g^{AB} - T_g^B)w_B = 0. \tag{15.3}$$

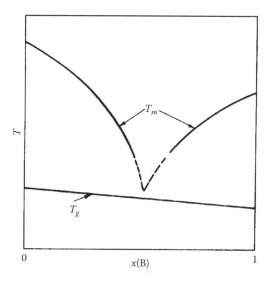

FIGURE 15.3 Schematic representation of T_m and T_g plotted as functions of copolymer composition, shown as mole fraction $x(B)$ of B. The broken lines represent the possibility that structural irregularities are so great that no crystallization of the copolymer can occur.

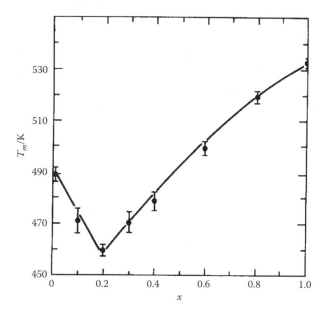

FIGURE 15.4 Melting temperatures of random copolymers of nylon-6,6 and nylon-6,10 as a function of the mole fraction of 6,6 (adipamide) units in the copolymer. (From data by Cowie and Mudie.)

The evaluation of the constant K depends on the model preferred when describing the glass transition. When the expression assumes that the free-volume contribution from a monomer is the same for both homo- and copolymers, then for a given pair of monomers the constant K is calculated from the corresponding expansivities of the homopolymers.

$$K = (\alpha_1^B - \alpha_g^B)/(\alpha_1^A - \alpha_g^A) = \Delta\alpha^B/\Delta\alpha^A \tag{15.4}$$

If a thermodynamic approach is used, as suggested by Couchman and Karasz, then

$$K = \Delta C_p^B/\Delta C_p^A \tag{15.5}$$

A similar relation has been proposed by Gibbs and Di Marzio:

$$(T_g^{AB} - T_g^A)n_o^A + (T_g^{AB} - T_g^B)n_o^B = 0, \tag{15.6}$$

where now the fraction of rotatable bonds n_o is introduced in place of the composition term.

In principle, a simple approach to obtaining a range of properties is to mix two homopolymers in various proportions. It would then be expected that the mixtures would provide a spectrum of properties reflecting those of the individual components. However, as illustrated in Chapter 9, most polymers are inherently immiscible when mixed and produce heterogeneous blends. Although there are a number of common polymers and copolymers that form homogeneous mixtures, it has been found that mixing is enhanced when secondary intermolecular interactions (e.g., hydrogen bonds, dipole–dipole interactions, etc.) are present.

When the T_gs of blends, where secondary intermolecular interactions are present, are measured, it is often observed that these deviate in a positive manner from the simple mixing linear expression:

$$T_g^{AB} = w_A T_g^A + w_B T_g^B \tag{15.7}$$

This is illustrated schematically in Figure 15.5 as curve 3, where Equation 15.7 is curve 1.

An equation has been proposed by Kwei that attempts to address this:

$$T_g^{AB} = \frac{w_A T_g^A + k w_B T_g^B}{w_A + k w_B} + q w_A w_B \tag{15.8}$$

where k and q are adjustable parameters used to describe the positive deviations from Equation 15.7. The parameter q can be regarded as a qualitative estimate of the strength of the specific intermolecular interactions (SII) that serve to stabilize the blends, whereas k is usually close to unity.

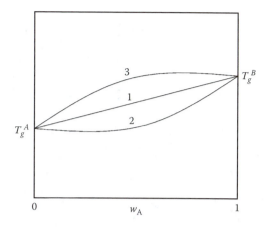

FIGURE 15.5 Schematic representation of T_g variation with copolymer or blend composition.

Equation 15.8 can also be used to describe the behavior of statistical copolymers in which SIIs are present, and monomer units tend to attract rather than repel one another, leading to stiffening of the chain. As the groups A and B in the copolymer tend to repel one another, the free volume is increased and the T_g^{AB} is usually lower than that predicted by Equation 15.7, i.e., curve 2 in Figure 15.5.

15.6 BLOCK COPOLYMERS

Random or statistical copolymers can be prepared if one wishes to narrow the gap between T_m and T_g in a sample, and so cover one property region not readily satisfied by homopolymers.

If a wider interval between T_m and T_g is required, a different class of copolymer — the block copolymer — must be investigated. These are usually {AB} or {ABA} block sequences. By synthesizing sequences that are long enough to crystallize independently, the combination of a high-melting block A with a low-melting block B will provide a material with a high T_m from A and a low T_g from B. A slight depression of T_m, arising from the presence of block B, is sometimes encountered, but this is rarely large. Combinations of this type allow the scientist to cover the remaining area of property combinations shown in Figure 15.2.

The change of T_g in block copolymers is rather variable, and certain pairs of monomers will form a block copolymer possessing two glass transitions. Interesting changes in the mechanical properties can be obtained when SBR block copolymers are synthesized using a lithium catalyst. A material is produced that behaves as though cross-linked at ambient temperatures. This is due to the presence of the two glass transitions associated with each block; the butadiene block has one at 210 K, and the styrene block has one at 373 K. Above 373 K plastic flow is observed, but between 210 and 373 K the glassy polystyrene blocks act as cross-links for the elastomeric polybutadiene, and the copolymer exhibits high resilience and low creep characteristics.

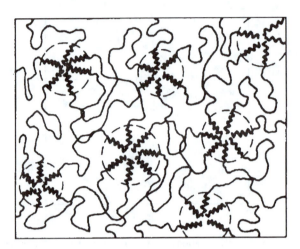

FIGURE 15.6 Schematic representation of elastoplastic sandwich block copolymers, showing areas of aggregation of the glassy {A} blocks, joined by the amorphous rubberlike chains of {B}.

The arrangement of the blocks is important; high-tensile-strength materials, with elastomeric properties similar to a filler-reinforced vulcanizate, are obtained only when the copolymer contains two or more polystyrene (S) blocks per molecule. Thus, copolymers with the structure {S.B.} or {B.S.B.} (where B is a polybutadiene block) are as brittle as polystyrene, but {S.B.S} and {S.B.S.B} copolymers are much tougher. At ambient temperatures these behave like conventional cross-linked rubbers, but they have the additional advantage that their thermal behavior is reproducible.

The property enhancement of these block copolymers is usually explained in terms of the *domain concept*. The glassy polystyrene blocks tend to aggregate in domains (see Figure 15.6) that act as both cross-linking points and filler particles. The glassy regions serve to anchor the central elastomeric polydiene blocks securely at both ends and act as effective cross-linking points, thereby precluding the necessity of vulcanizing the material.

One unexpected application arises from the observation that the presence of more than 10% block copolymer in natural rubber prevents bacterial growth on the polymer surface; thus, incorporation of the copolymer in butchers' chopping blocks can lead to more hygienic conditions in meat handling.

The synthesis of {ABA} blocks from a glassy thermoplastic A and an elastomeric B produces other "elastoplastics" with attractive properties. Polyester chains can be extended with di-isocyanate, which is then treated with cumene hydroperoxide to leave a peroxide group at both ends of the chain. By heating this in the presence of styrene, a vinyl polymerization is initiated and an {ABA} block created. The modulus–temperature curves show how the mechanical properties can be modified in this way (Figure 15.7). These block copolymers are known as *thermoplastic elastomers*.

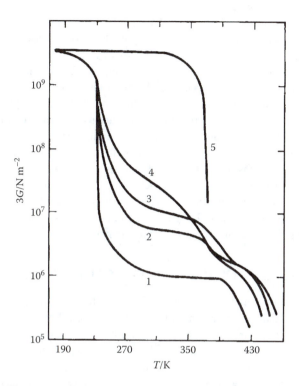

FIGURE 15.7 Modulus–temperature behavior of polyester–polystyrene block copolymers: 1, polyester; 2, polyester containing 20% of polystyrene; 3, containing 45% polystyrene; 4, containing 60% polystyrene; 5, pure polystyrene.

15.7 PLASTICIZERS

A polymer sample can be made more pliable by lowering its T_g, and this can be achieved by incorporating quantities of high-boiling, low-molar-mass compounds in the material. These are called *plasticizers* and must be compatible with the polymer. The extent to which T_g is depressed depends on the amount of plasticizer present and can be predicted from the relation

$$1/T_g^M = w/T_g + w_1/T_g^1, \tag{15.9}$$

where T_g^M and T_g^1 correspond to the mixture and the liquid and w and w_1 are the mass fractions of the polymer and plasticizer in the system, respectively.

The action of the plasticizer is one of a lubricant, where the small molecules ease the movement of the polymer chains by pushing them further apart. As this lowers both T_g and the modulus, their main use is to increase the flexibility of a polymer for use in tubing and in films.

Poly(vinyl chloride), whose T_g is 354 K, usually contains 30 to 40 mass percent of plasticizers, such as dioctyl or dinonyl phthalate, to increase its toughness and flexibility at ambient temperatures. This depresses T_g to about 270 K and makes the polymer suitable for plastic raincoats, curtains, and "leather-cloth." The low volatility of the plasticizer ensures that it is not lost by evaporation, a mistake made in the early post-war years, which eventually led to a brittle product and considerable customer disaffection. In the rubber industry, plasticizers are usually called *oil extenders*.

In fiber technology, water absorption is an important factor governing the mechanical response because the water tends to act as a plasticizer. Thus, as the moisture content increases, the modulus drops, but there is a corresponding improvement in the impact strength. In fibers, such as nylon-6,6, water acts as a plasticizer to depress T_g below room temperature. Thus, when nylon shirts are washed and hung up to drip-dry, the polymer is above the T_g, and this helps creases to straighten out, thereby giving the clothing an "ironed" appearance.

15.8   CRYSTALLINITY AND MECHANICAL RESPONSE

The mechanical properties are dependent on both the chemical and physical nature of the polymer and the environment in which it is used. For amorphous polymers, the principles of linear viscoelasticity apply, but these are no longer valid for a semicrystalline polymer. The mechanical response of a polymer is profoundly influenced by the degree of crystallinity in the sample.

The importance of both crystallinity and molar mass is illustrated by the range of properties displayed by polyethylene. This is shown schematically in Figure 15.8 and provides some indication of the effect of these variables.

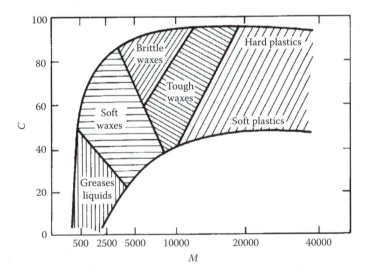

FIGURE 15.8 Influence of crystallinity and chain length on the physical properties of polyethylene. (After Richards, R.B., *J. Appl. Chem.*, 1, 370, 1951.) The percentage crystallinity c is plotted against molar mass M.

The interpretation of the mechanical behavior is further complicated by the presence of glide planes and dislocations, which lead to plastic deformation, but these also serve to provide the materials scientist with a wider variety of property combinations, and can prove useful.

The major effect of the crystallite in a sample is to act as a cross-link in the polymer matrix. This makes the polymer behave as though it was a cross-linked network, but, as the crystallite anchoring points are thermally labile, they disintegrate as the temperature approaches the melting point, and the material undergoes a progressive change in structure until, beyond T_m, it is molten. Thus, crystallinity has been aptly described by Bawn (1969) as a form of "thermoreversible cross-linking."

The restraining influence of the crystallite alters the mechanical behavior by raising the relaxation time τ and changing the distribution of relaxation and retardation times in the sample. Consequently, there is an effective loss of short τ, causing both the modulus and yield point to increase. The creep behavior is also curtailed and stress relaxation takes place over much longer periods. Semicrystalline polymers are also observed to maintain a relatively higher modulus over a wider temperature range than an amorphous sample.

These points can be illustrated by comparing the elastic relaxation modulus $E_r(t)$ for crystalline (isotactic), amorphous, and chemically cross-linked (atactic) polystyrene samples, as shown in Figure 15.9. Crystallinity has little effect below T_g, but as the molecular motion increases above T_g, the modulus of the amorphous polymer drops more sharply. The value of $E_r(t)$ remains high for the crystalline polymer throughout this range until the rapid decrease at the melting temperature is recorded. The cross-linked sample maintains its modulus level at this temperature as the cross-links are not thermally labile and do not melt.

Rapid quenching of the isotactic polymer destroys the crystallinity and produces behavior identical to the atactic material. The spherulite size also affects the response;

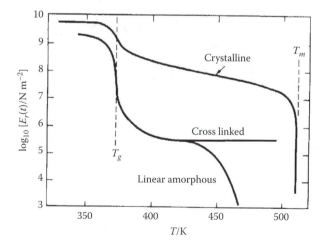

FIGURE 15.9 Illustration of the variation in the modulus–temperature curves for three types of polystyrene.

slow cooling from the melt promotes the formation of large spherulites and produces a polymer with a lower impact strength than one cooled rapidly from the melt, whose spherulites are much smaller and more numerous. This effect can be seen as a shift in the damping maxima.

In practical terms, the use of poly(vinyl chloride) in the manufacture of plastic raincoats provides a good illustration of the effect of crystallite cross-linking. The polymer is plasticized, until T_g is below ambient, to make the material flexible, and one might expect that if the coat was hung on a hook (i.e., subjected to a tensile load), it would eventually flow onto the floor after prolonged tension. This is not so; the material behaves as though it was a chemically cross-linked elastomer because it contains a sufficient number of crystallites to act as restraining points and prevent flow.

Similarly, the glass transition of polyethylene is well below ambient temperature, and if the polymer was amorphous, it is likely that it would be a viscous liquid at room temperature. It is in fact, a tough, leathery, or semirigid plastic because it is highly crystalline and the crystallite cross-links impart a high modulus and increased strength to the polymer between 188 and 409 K, a very useful temperature range.

The main points can now be restated briefly:

1. Crystallinity only affects the mechanical response in the temperature range T_g to T_m, and below T_g the effect on the modulus is small.
2. The modulus of a semicrystalline polymer is directly proportional to the degree of crystallinity and remains independent of temperature if the amount of crystalline order remains unchanged.

15.9 APPLICATION TO FIBERS, ELASTOMERS, AND PLASTICS

We have seen how various parameters can be altered and combined to produce a material with given responses, but the lines of demarcation dividing polymers into the three major areas of application — fibers, elastomers, and plastics — are by no means well defined. It is most important then to establish criteria that determine whether a polymer is a superior fiber, an excellent elastomer, or a particularly suitable plastic before overindulging in the interesting stages of molecular design and engineering.

15.10 FIBERS

Superficially, a fiber is a polymer with a very high length-to-diameter ratio (at least 100:1), but most polymers capable of being melted or dissolved can be drawn into filaments. They may, however, have no technical advantages if they cannot meet the requirements of a good fiber, i.e., high tensile strength, pliability, and resistance to abrasion. In addition, to be useful for clothing, it is preferable that the polymer has $T_m > 470$ K to allow ironing without damage, but lower than 570 K to enable spinning from the melt. Also, T_g must not be so high that ironing is ineffective. Some typical fibers with useful temperature ranges are shown in Table 15.4. These all have T_g lower than 380 K but above room temperature, so that in a cloth the fibers will soften

TABLE 15.4
Values of T_m and T_g of Some Typical Fibers

Polymer	Structure	T_g(K)	T_m(K)
Poly(ethylene terephthalate)	$-(CH_2)_2O\ OC-\bigcirc-CO\ O-$	343	538
Nylon-6,6	$-NH(CH_2)_6NHCO(CH_2)_4CO-$	333	538
Polyacrylonitrile	$-CH_2-\underset{CN}{CH}-$	378	590
Polypropylene (isotactic)	$-CH_2-\underset{CH_3}{CH}-$	268	435

when ironed at about 420 K. This will remove creases or allow pleats to be made, which will be retained on cooling. Subsequent washing is normally carried out at temperatures too low to resoften the polymer significantly and destroy the pleats. This "permanent" crease is a desirable feature of some clothing.

The main distinguishing feature of a fiber is that it is an oriented polymer and, as such, is anisotropic, being much stronger along the fiber axis than across it. Thus, the most important technical requirement for fiber formation is the ability to draw or orient the chains in the direction of the fiber axis, and retain this after removal of the drawing force. Clearly then, factors that aid this retention of orientation are prime requirements for a good fiber, and these include all structural features contributing to intermolecular binding.

This means that a polymer should be symmetrical and unbranched to encourage a high degree of crystallinity; it should preferably have a high cohesive energy; and it should have an average length of about 100 nm when fully extended. These properties can be conveniently examined under two main headings, the chemical requirements and the mechanical response. The important factors to consider are (1) melting and glass transition temperatures; (2) modulus; (3) elasticity; (4) tensile strength; and (5) moisture absorption and dyeability.

15.10.1 Chemical Requirements

If the polymer chains are quite short, they are not entangled to any great extent in the solid state and are relatively free to move; hence, they cannot add to the fiber strength. As the chain length increases (and so, the intertwining), the fiber strength improves. The optimum range of molar mass for a good fiber is 10,000 to 50,000 g mol^{-1}. It has been found that fiber properties deteriorate outside these limits. However, as we shall see later, chain entanglement can detract from the tensile strength and modulus; therefore, alignment of long chains is an important and desirable feature.

We have already mentioned the importance of T_g and T_m and know that these can be affected by chain symmetry, stiffness, and intermolecular bonding. The tensile strength of a fiber is observed to increase with crystallinity; consequently, this is a desirable quality and linear chains will be preferred for fiber formation. As the shape and symmetry of a linear chain governs its ability to crystallize, chains containing irregular units, which detract from the linear geometry, should be avoided in fiber-forming polymers. This is obvious when comparing terylene I, which is an excellent fiber, with its isomer II, prepared using o-phthalic acid.

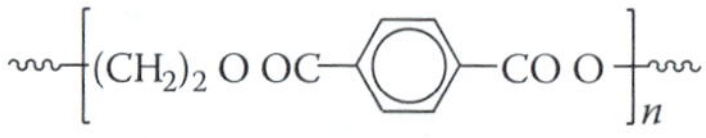

$$\text{~~}\!\!\left[\!(CH_2)_2\,O\,OC\!-\!\!\bigcirc\!\!-\!CO\,O\right]_n\text{~~}$$

I

$$\text{~~}\!\!\left[\!(CH_2)_2\,O\,OC\!\diagdown\!\!\bigcirc\!\!\diagup\!CO\,O\right]_n\text{~~}$$

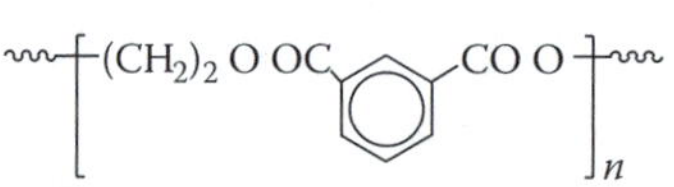

II

Structure II has lost its regularity, is less crystalline, has a lower T_g, and makes a much poorer fiber. This is also true in the polyamide series, where the regular polymer III has $T_m = 643$ K and $T_g = 453$ K, but the irregular form IV has $T_m = 516$ K and $T_g = 363$ K.

$$\left[\!NH(CH_2)_6NH\,CO\!-\!\!\bigcirc\!\!-\!CO\right]_n$$

III

$$\left[\!NH\,CH_2\diagdown\!\!\bigcirc\!\!\diagup\!CH_2NH\,CO\,(CH_2)_4CO\right]_n$$

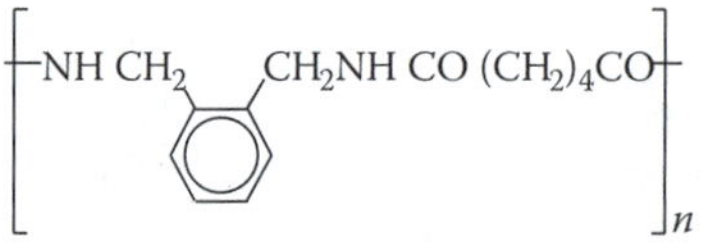

IV

Stereoregular polymers also have symmetrical structures, and the helices of isotactic polymers can be close-packed to produce highly crystalline material. Isotactic polypropylene is crystalline and an important fiber-forming polymer, whereas the atactic form has virtually no crystalline content and has little value as a fiber; indeed, it is considerably more elastomeric in nature.

Although crystallinity and stereoregularity are important factors in fiber formation, atactic amorphous polymers can also prove useful if intermolecular forces are present. Dipolar interactions between side groups such as $-$(CN) (energy of interaction about 36 kJ mol^{-1}) are significantly stronger than hydrogen bonds or van der Waals forces and serve to greatly improve the molecular alignment. This interaction stabilizes orientation during fiber manufacture and enhances the fiber-forming potential of polymers such as polyacrylonitrile and poly(vinyl chloride), both essentially amorphous and atactic. This point highlights the fact that molecular alignment is the most important factor in fiber formation, and not crystallinity, which is only one method of obtaining a stable orientation of chains.

The importance of hydrogen bonding has already been described and will not be dealt with further.

15.10.1.1 Linear Polyesters

Many of the general points discussed can be illustrated conveniently by referring to the numerous linear polyesters that have been prepared. These are grouped together in Table 15.5.

A comparison of structure 1 with structures 2(i) and 2(ii) indicates a drop in T_m caused by the increase in chain flexibility arising from the ethylene and ethylene dioxy groups inserted between the phenylene rings. The change is even more dramatic on comparing structure 1 with structure 6, when two phenylene rings are used instead of the $-(CH_2)_4$ sequence and the difference in T_m is 205 K.

TABLE 15.5
Values of T_m and T_g for Linear Polyesters

Structure	Group R	T_m(K)	T_g(K)
1. $\left[OC-\langle\bigcirc\rangle-\langle\bigcirc\rangle-CO\,O(CH_2)_2O\right]_n$		528	—
2. $\left[OC-\langle\bigcirc\rangle-R-\langle\bigcirc\rangle-CO\,O(CH_2)_2O\right]_n$	(i) $-(CH_2)_4-$	443	—
	(ii) $-O-(CH_2)_2-O-$	513	—
	(iii) $-NH-(CH_2)_2-NH-$	546	—
3. $\left[OC-\langle\bigcirc\rangle-CO\,O\,R\,O\right]_n$	(i) $-(CH_2)_2-$	538	342
	(ii) $-(CH_2)_4-$	503	353
	(iii) $-CH_2-\underset{CH_3}{\overset{CH_3}{C}}-CH_2-$	413	—
	(iv) $-CH_2-\underset{CH_3}{CH}-$	non-crystalline	341
4. $\left[OC-\langle\bigcirc\rangle-CO\,O\,R\,O\right]_n$	(i) $-(CH_2)_2-$	513	324
	(ii) $-(CH_2)_4-$	426	—
5. $\left[OC-\langle\underset{CH_3}{\bigcirc}\rangle-CO\,O\,(CH_2)_2\,O\right]_n$		343	—
6. $\left[OC\,(CH_2)_4CO\,O\,(CH_2)_2\,O\right]_n$		323	—

The influence of symmetry is seen in the terephthalic 3 and isophthalic 4 series. The unsymmetrical ring placement in 4(i) and 4(ii) lowers T_m by 25 K and 77 K compared with their counterparts 3(i) and 3(ii), respectively.

Bulky side groups interfere with the close-packing capabilities of a chain, as evidenced by the effect on T_m of the methyl groups in 3(iii), 3(iv), and 5 compared with 3(i) and 3(ii). The additional asymmetry in 3(iv) actually prevents crystallization.

The added stability of secondary bonding in the crystallite is reflected in the increase in T_m in the series 2(i) to 2(iii), as we move from simply van der Waals forces to dipolar and hydrogen-bond interactions. The hydrogen bonding is also sufficient to raise the T_m of 2(iii) above that of 3(i) in spite of the extra flexible sequence present in the 2(iii) chain.

These points cover most of the chemical requirements, and we can now look at the mechanical properties.

15.10.2 MECHANICAL REQUIREMENTS FOR FIBERS

Fibers are subject to a multitude of mechanical deformations such as stretching, abrasion, bending, twisting, and shearing. Now the properties of interest are (1) *tenacity*, which is the stress at the breaking point of the material; (2) *toughness*, defined as the total energy input to the breaking point; (3) *initial modulus*, the measure of resistance to stretching (portion A–B of the stress–strain curve, Figure 15.10); and (4) the extent of *permanent set*.

In technological terminology, the textile industry recognizes the following qualities as suitable: (1) tenacity: 1 to 10 g denier^{-1} (about 5 g denier^{-1} optimum for clothing), (2) modulus of elasticity: 20 to 200 g denier^{-1}, and (3) extensibility: 2 to 50%. (The denier is the mass in grams of 9000 m of yarn).

As the mechanical response of a fiber can be controlled to some extent in the spinning process, this will be discussed briefly.

15.10.2.1 Spinning Techniques

The process of converting a bulk polymer sample into a thread or yarn is known as *spinning*, and several methods can be used, depending on the nature of the sample.

15.10.2.1.1 Melt Spinning

This is used when polymers are readily melted without degradation and the molten polymer is forced through a spinneret comprising 50 to 1000 fine holes. On emerging from the holes, the threads solidify, often in an amorphous glassy state, and are wound into a yarn. Orientation and crystallinity are important requirements in fibers, and the yarn is subjected to a drawing procedure that orients the chains and strengthens the fiber. This technique is applied to polyesters, polyamides, and polyolefins.

15.10.2.1.2 Wet and Dry Spinning

Acrylic polymers cannot be melt-spun because they are thermally labile, and spinning is carried out using concentrated solutions of the polymer. The solvent is removed by evaporation after extrusion, leaving an amorphous filament, which is then said to have been *dry*-spun. When the solution filaments are extruded into a vat of a nonsolvent, the polymer precipitates in the form of a thread and is then a *wet*-spun fiber.

15.10.2.2 Drawing, Orientation, and Crystallinity

A fiber, in its amorphous state, can be strengthened by *drawing*, a process that extends its length by several times the original, and in doing so aligns the chains in the sample. The process is irreversible and corresponds to the section C–D of the stress–strain curve in Figure 15.10, where deformation up to the yield point C is elastic, but beyond this irreversible plastic deformation occurs. At C the polymer suddenly thins down, or "necks," at one point, and subsequent drawing increases the length of the reduced region at the expense of the undrawn region, until the process is complete. Further extension causes rupture at D, the breaking point.

The effect of molecular order is far more important in fiber production than in any other area of polymer application, and drawing ability is a fundamental require-ment in good fiber-forming materials. It is worth pointing out again that crystallinity and orientation are not necessarily synonymous terms and that there is a difference between orientation of crystallites and orientation of chains in the amorphous regions of a polymer. It is the amorphous part of a fiber that will distort and elongate under stress, and these are the areas that must be oriented to improve the intermolecular attraction if the fiber modulus is to be enhanced. Drawing only improves a highly crystalline fiber to a small extent by orienting the crystallites, but the amorphous fiber is improved immensely. Drawing affects the mechanical properties of a fiber in several ways. It makes the fiber tough and tenacious, increases the modulus and the density, and can alter T_g by orienting the chains $+O-(CH_2)_2-O+$ in the amor-phous regions. For example, the group in poly(ethylene terephthalate), which has a gauche conformation in the amorphous phase, is "drawn" into the trans conformation. This improves the sample's crystallinity, and T_g also rises by 10 to 15 K.

The tenacity and physical characteristics of the fiber can also be controlled by the extent of the draw. Limited orientation, produced at low draw stresses, leads to a medium-tenacity nylon yarn with low tensile strength, low modulus, and high extensibility, which are all properties associated with a flexible soft material suitable for clothing. Higher draw rates yield high-tenacity, high-strength yarns, which are

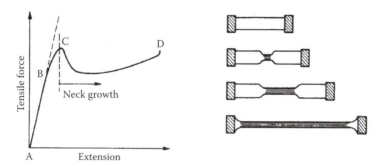

FIGURE 15.10 Successive stages in the drawing of polymer, showing the necking down and subsequent neck growth resulting in increased chain alignment.

more suited for tire cord production. Thus, some fiber properties are subject to the art of the spinner.

Both tenacity and modulus can be controlled by crystallinity in a fiber. Low-pressure polyethylene is highly crystalline and has a fiber tenacity of about 6 g denier^{-1}, but the high-pressure, highly branched, and consequently less-crystalline polyethylene has a fiber tenacity of only 1.2 g denier^{-1}. We have already seen that conversion of the amide group to a non-hydrogen-bonding group, such as a methylol group $-CON(CH_2OH)-$ with formaldehyde, curtails the intermolecular bonding in polyamides. This also increases hydrophilicity and makes the polymer more water soluble, but the hydrophobic characteristics can be restored by methylating the group to $-CON(CH_2OCH_3)-$. At low degrees of substitution, the modulus is reduced, and a more elastic fiber is obtained. As the substitution increases, the crystallinity is completely destroyed, and the fiber-forming capacity disappears.

15.10.2.3 Modulus and Chain Stiffness

Fiber modulus can be regulated by orientation and crystallinity, but a third parameter, chain stiffness, is available for modification if additional control is required.

The effect of chain stiffness on the initial modulus is seen in Figure 15.11. The increase in chain rigidity on $+NH(CH_2)_6NHCO(CH_2)_4CO+_n$ moving from nylon-6,6 to terylene and to poly(m-phenylene isophthalimide), Nomex, is manifest in an increase in the initial modulus, with the inclusion of aromatic rings in the chain. This is discussed more fully in Section 15.11.

15.10.2.4 Other Factors

The moisture regain of a fiber is important when comfort is being considered. In hot weather, the ability to absorb perspiration makes clothing more comfortable, and polar polymers are best adapted for this purpose. High moisture retention also decreases resistivity of the fiber and reduces its tendency to build up static charges, which attract dirt and increase discomfort. Most synthetic fibers have poor moisture-regain characteristics and have to be modified in some way to improve moisture uptake. Grafting of poly(ethylene oxide) or acrylic acid on to nylons greatly improves moisture uptake without affecting the mechanical properties.

Dyeing is also a problem, and chemical modification is often necessary. Sites are provided for dyeing by substituting a number of $-SO_3H$ groups in the phenylene rings of the terylene chain or by copolymerizing acrylonitrile with small quantities of vinyl sulfonic acid. These modifications also improve moisture uptake.

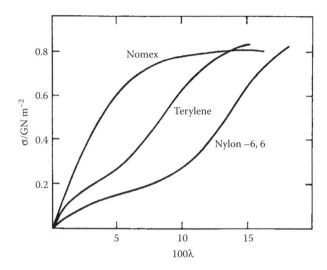

FIGURE 15.11 Stress–strain curves for three fibers, showing the changes wrought by the introduction of one (terylene), then two (Nomex) phenylene groups into the repeat unit of the chain; the stress σ is plotted against 100λ, the percentage elongation.

When selecting a fiber for clothing, one should avoid material with a high value of *permanent set*. This is a measure of the amount of irreversible flow (C–D in Figure 15.10) left in the polymer and is reflected in an increase in fiber length after being subjected to a stress. Obviously, in clothing, there is considerable knee or elbow bending, and a high permanent set value will result in gross fiber deformation, such as "baggy trousers" or "kneed" stockings. This is partially or totally offset by drawing, but high draw ratios may make the fiber hard. Hence, whereas poor creep recovery in an article results in loss of shape, its capacity to absorb energy may deteriorate, and if overcompensated for, may result in actual material failure.

It is not uncommon to be forced into a compromise when faced with a choice between two incompatible properties.

15.11 AROMATIC POLYAMIDES

Synthetic fiber-forming polymers with the qualities of stiffness and heat resistance to make them competitive substitutes for steel wire or glass fibers have been much sought after for use as reinforcing materials in composites, or in the production of ropes, cables, hoses, and coated fabrics. One of the most successful groups developed for this purpose is the aromatic polyamides or aramids, defined as fiber-forming substances comprising long-chain polyamides with >85% of the amide groups attached directly to aromatic rings.

The structures of some of the commercially significant aramids are shown in Table 15.6. Of these, the most important is poly(*p*-phenylene terephthalamide) PPD-T, which exhibits superior properties and can serve as a model for the others. The polymer is prepared by the condensation polymerization of *p*-phenylene diamine

TABLE 15.6
Commercially Important Aramid of Fiber-Forming Polymers

Polymer		Trade Name
I	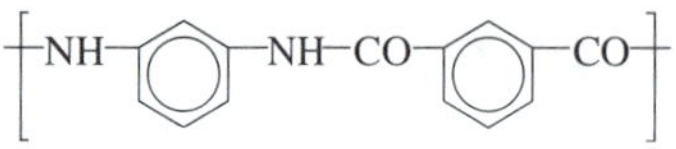 MPD-1	Nomex
II	 PPB	Fiber B / Terlon
III	 PPD-T	Kevlar / Twaron
IV	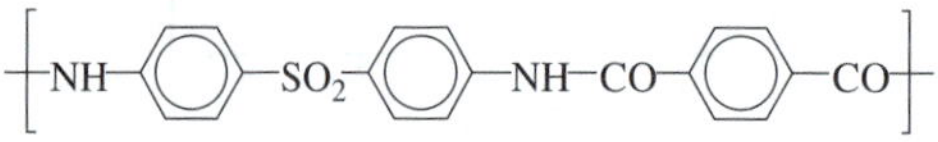	Sulfon T
		SVM
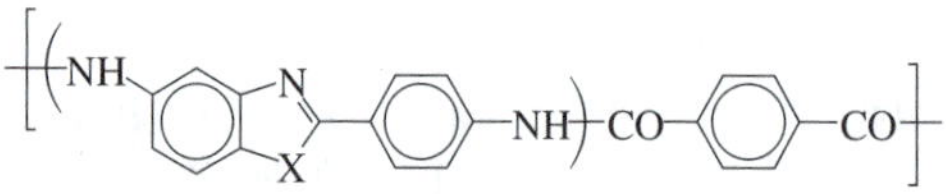 X = —O—, —S—, —NH—		
		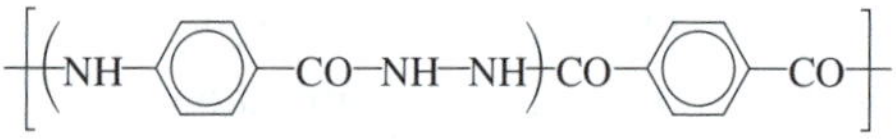 X-500-Series
non-ordered structure		
 ordered structure		Flexten

and terephthaloyl chloride in a solvent comprising *N*-methyl pyrrolidone (NMP) and CaCl$_2$. This solvent combination keeps the growing polymer in solution longer, thereby increasing the molar mass of the product to levels that make it suitable for fiber formation. This is achieved by reducing the intermolecular hydrogen bonding (the salt is a competitive hydrogen-bonding agent), whereas NMP acts as an acid acceptor and good solvator of the polymer chain. High-speed stirring is also necessary. The final product is relatively insoluble but can be redissolved in 98% sulfuric acid, and solutions of PPD-T with greater than 6 to 7% solids form anisotropic solutions characteristic of lyotropic liquid crystalline polymers.

When a flexible, nonaromatic polyamide such as nylon-6,6 is dissolved in a solvent, the chains behave like random coils. As the polymer concentration increases, these become entangled. Subsequent spinning and drawing produces a fiber in which the chains retain this entangled structure and are only partially extended, thereby reducing their potential modulus and tenacity. In solutions of the rigid, rodlike aramids, random coil structures do not form and, instead, the rigid chains pack in quasi-parallel bundles when the polymer concentration in the solution is increased. When spinning from these solutions, the shearing forces orient these bundles in the direction of the applied force, and the resulting fibers are composed of highly oriented, fully extended chains, which can also crystallize easily. Thus, a product with a high modulus is obtained.

Fibers must be prepared from solutions of PPD-T in sulfuric acid, and conventional spinning techniques have proved ineffective. Two innovations have helped to solve this problem. First, it was found that, on heating with H_2SO_4, PPD-T formed a stable complex that melts at around 343 K and has a (1:10) PPD-T:H_2SO_4 composition. By using the complex, much higher concentrations could be used in the spinning process than were previously possible. Then, a new method called *dry-jet-wet-spinning* was developed, in which an air gap is left between the spinneret opening and the cold water quench bath. This design allows the chains time to orient in the solution after emerging from the spinneret hole and before final quenching into the fiber form. Spinning PPD-T in this way produces fibers with remarkable properties.

This material, developed by DuPont, has the trade name Kevlar and when formed as described earlier, gives a Kevlar 29 grade. An improved version, Kevlar 49, can be produced by hot-drawing the fiber in an inert atmosphere at temperatures above 520 K, and a third grade, Kevlar "Hp," has been produced with intermediate properties.

Comparison of the stress–strain curves for these aramids and for other fiber-forming materials is shown in Figure 15.12(a) and Figure 15.12(b); it demonstrates the superior properties of PPD-T. It is believed that the reason for the high strength (≈ 2.6 GPa) and modulus (60 to 120 GPa) lies in the three-dimensional order coming from both longitudinal and radial orientations. The PPD-T fiber can then be pictured as in Figure 15.13, with ordered sheets of hydrogen-bonded, extended chains radiating from the fiber core.

The aramids are very resistant to heat and only begin to decompose and char at temperatures in excess of 670 K. Kevlar is often combined with carbon fibers and embedded in epoxy resins to form hybrid composites that have the ability to withstand catastrophic impact and so find use in aircraft body and wing structures. Other applications include rope and cable manufacture, where the high strength per unit weight is a distinct advantage. Compared with steel, Kevlar has a much higher breaking strength, but is six times lighter, thus providing a tremendous saving in weight. This makes for ease of handling, and it can be used as cables for mooring lines, off-shore drilling platforms, parachute lines, fishing lines, mountaineering ropes, and pulley ropes. Other uses include cord for reinforcing car tires, and in protective clothing and body armor.

15.12 POLYETHYLENE

The influence of chain branching on the properties of a polymer is usefully illustrated with reference to polyethylene. Chemically, this is one of the simplest of the synthetic polymers with the repeating unit $-(CH_2-CH_2)_n-$, but it can be prepared in several different ways, and these determine the extent of chain branching in the product.

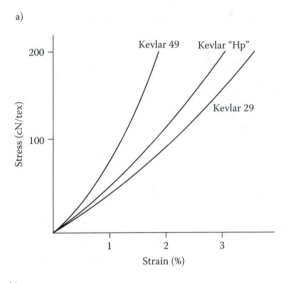

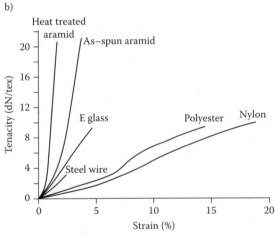

FIGURE 15.12 (a) Comparison of the stress–strain behavior of the three grades of Kevlar. (b) Comparison of the tenacity of Kevlar fibers with glass, steel, polyester, and nylon as a function of percentage strain. (From Tanner, D. et al., The Kevlar story, *Angew. Chem. Int. Ed. Engl. Adv. Mater.*, 28, 649, 1989. With permission from Verlag Chemie.)

Polyethylene now tends to be marketed in three general grades: high-density polyethylene (HDPE), linear low-density polyethylene (LLDPE or 1-LDPE), and low-density polyethylene (LDPE). The essential structural differences for each are shown in Table 15.7.

HDPE, prepared using organometallic catalysts, is a structurally regular chain material with very few small branch points (less than 7 per 1000 carbon atoms). Because of this regularity, the polymer chains can pack efficiently, resulting in a highly crystalline material with a correspondingly high density. The polymer is used to manufacture bottles, crates, and pipes.

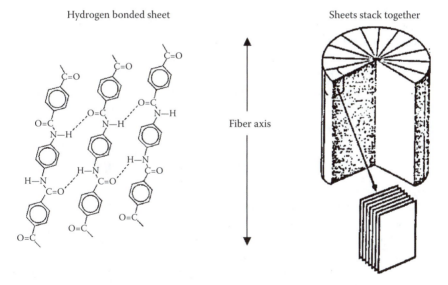

HDPE

LLDPE

LDPE

Hydrogen bonded sheet

Fiber axis

Sheets stack together

FIGURE 15.13 A representation of the arrangement of sheets of poly(*p*-phenylene tereph-thalamide) to form the fiber structure. (Adapted from Tanner, D. et al., The Kevlar story, *Angew. Chem. Int. Ed. Engl. Adv. Mater.,* 28, 649, 1989. With permission from Verlag Chemie.)

LDPE, prepared by a high-pressure, radical-initiated polymerization process, is a highly branched polymer with approximately 60 branch points per 1000 carbon atoms. It has a much lower crystalline content and density, and has good film-forming

TABLE 15.7
Comparison of Various Polyethylene Grades

Property	LDPE	LLDPE	HDPE
Melting point (K)	383	393–403	>403
Density (g/cm³)	0.92	0.92–0.94	0.94–0.97
Film tensile strength (MPa)	24	37	43

properties, so that its largest application is as film for packaging and cable coatings. It has, however, greater permeability to gases (CO_2, O_2, N_2, etc.) than HDPE.

The property gap that exists between HDPE and LDPE has been filled by LLDPE. This polymer can be prepared by solution- or gas-phase polymerization, and is actually a copolymer of ethylene with 8 to 10% of an α-olefin, such as but-1-ene, pent-1-ene, hex-1-ene, or oct-1-ene. This produces a chain with a controlled number of short-chain branches and densities intermediate between HDPE and LDPE, thereby allowing it to be prepared in various grades by controlling the type of the comonomer. Thus, the use of oct-1-ene gives a lower-density product than that obtained when but-1-ene is incorporated in the chain because the longer (hexyl) branch in the former pushes the chains further apart than the ethyl branch of the latter, hence lowering the packaging efficiency of the chains.

LLDPE is now beginning to compete with LDPE in film blowing and casting applications because of its superior resistance to puncture by hard particles. It also has better qualities of toughness and lower brittle temperatures than LDPE, and is now used to replace blends of HDPE and LDPE.

Table 15.7 shows a number of important property comparisons and indicates how the branching plays a significant role in determining the properties and, hence, the end uses of the three polymers.

15.13 ELASTOMERS AND CROSS-LINKED NETWORKS

Rubberlike elasticity and its associated properties have already been discussed in some detail (see Chapter 14), and only a brief summary of the relevant features will be given.

The fundamental requirements of any potential elastomer are that the polymer be amorphous with a low cohesive energy and that it be used at temperatures above its glass transition. The polymer in the elastic region is characterized by a low modulus (about 10^5 N m^{-2}) and, for useful elastomers, by large reversible extensions. This reversibility of the slippage of flow units requires a chain in which there is a high localized mobility of segments but a low overall movement of chains relative to one another. The first requirement is satisfied by flexible chains with a low cohesive energy, which are not inclined to crystallize (although the development of some crystalline order on stretching is advantageous). The second requirement, prevention of chain slippage, is overcome by cross-linking the chains to form a three-dimensional network.

15.13.1 Cross-Linking

Cross-linking provides anchoring points for the chains, and these points restrain excessive movement and maintain the position of the chain in the network. This is not confined to elastomers, however, and the improved material qualities that result are also found in the cross-linked phenol-formaldehyde, melamine, and epoxy resins.

When a sample is cross-linked, (1) the dimensional stability is improved, (2) the creep rate is lowered, (3) the resistance to solvent increases, and (4) because T_g is raised, it becomes less prone to heat distortion. All these effects tend to intensify as the cross-link density is increased and can be controlled by adjusting the number of cross-links in a sample.

15.13.2 Creep in Cross-Linked Polymers

The creep response depends mainly on the temperature and the cross-link density. At temperatures below T_g, cross-linking has little effect on the properties of the material, but above T_g, the secondary creep, arising from irreversible viscous flow, is reduced or eliminated by cross-linking.

Creep is a function of the elastic modulus, the mechanical damping, and the difference between ambient temperature and T_g. The thermosetting resins usually have a high modulus, low damping characteristics, and T_g well above ambient; consequently, the creep rate is low and they have good dimensional stability.

The effect of increasing the cross-link density on these parameters is illustrated for a phenol-formaldehyde resin in Figure 15.14. Above T_g, the modulus is a function of the extent of cross-linking; the damping peaks shift to higher temperatures as T_g increases and eventually become difficult to detect. This shows the extent to which cross-link density can affect the physical behavior.

15.13.3 Additives

Many elastomers are subject to oxidative degradation and can be protected to some extent by the addition of antioxidants, such as amines and hydroquinones.

The abrasion resistance can also be improved by adding a filler to reinforce the elastomer, and carbon black is widely used for this purpose. Fillers (e.g., glass fiber, mica, sawdust, etc.) are also used in thermosetting resins as reinforcement.

15.14 PLASTICS

So far, attention has been focused predominantly on fiber and elastomer requirements because it is considerably more difficult to be specific about the qualities desired in a plastic material when the range of applications covered is much more extensive. The general principles relating to the control of T_m, T_g, modulus, etc. can all be applied to the formation of a specific type of plastic, and we shall simply try to illustrate briefly the diversity of problems encountered in the field of plastic utilization.

The conflict between low creep and high impact strength mentioned earlier is not confined to fibers, but is also a problem encountered in plastic selection. It is an important point to consider for the engineering requirements of the material when

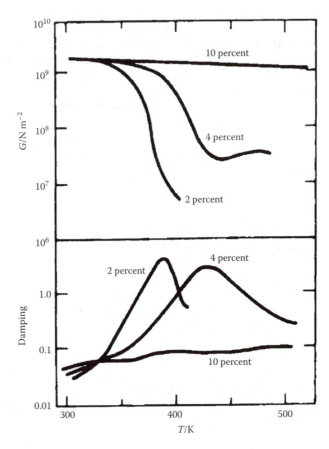

FIGURE 15.14 Influence of cross-linking on the dynamic mechanical response of phenol-formaldehyde resin. Concentrations of the cross-linking agent hexamethylenetetramine are shown alongside the appropriate curve. (After Nielsen, E., *Mechanical Properties of Polymers and Composites,* Marcel Dekker, 1974.)

the ability to absorb energy is desirable but is at odds with the equally desirable qualities of high rigidity and low creep. The problem to be faced, then, is how to make a brittle, glassy polymer tougher, i.e., how to limit the modulus or tensile strength. In general, an increase in crystallinity (and consequently, the modulus) tends to make a plastic more brittle. Crystallinity can be controlled by copolymerization or branching, and the brittleness can be tempered using one or other of these modifications. Alternatively, an elastomeric component can be introduced, which will improve the impact strength by reducing the rigidity and yield stress. This has been used in high-impact polystyrene (HIPS) or acrylonitrile-butadiene-styrene (ABS) copolymers, in which the elastomeric component is above its T_g under prevailing environmental conditions and acts as a second phase. This leads to an increased damping efficiency, which is manifest in the appearance of a second low-temperature damping maximum in the damping curve. This is seen in Figure 15.15 for high-impact polystyrene-butadiene copolymer (SBR rubber), whose T_g is 213 K.

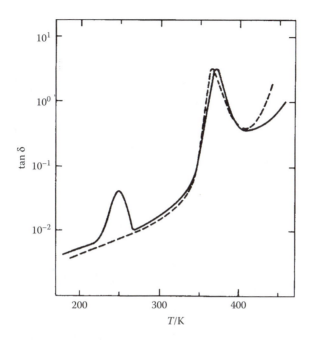

FIGURE 15.15 Damping curves for polystyrene (- - -) and high-impact polystyrene (——). The latter has an internal friction peak below ambient. The damping tan δ is plotted against *T*.

The phenomenon is similar to the toughening effect in semicrystalline polymers caused by the strengthening of the amorphous regions with crystalline cross-links; but in the latter case, the two-phase aspect arises from the existence of crystalline and amorphous regions.

Although orientation is most important in fiber formation, it can also improve the response of a brittle polymer and increase its ductility. This is particularly true in film preparation or molding, where viscous flow is inclined to introduce a certain degree of chain alignment at some stage in the process.

Interchain interactions also affect performance, and poly(oxymethylene) has a higher modulus in the glassy state than polyethylene, presumably because of the polar attractions between the chains.

When faced with the problem of selecting a plastic for a given purpose, a design engineer must be concerned with the properties of the material, the ease of processing or fabrication, the behavior under the environmental conditions to which the product will be subjected (i.e., the thermal range) and, of course, the economic factors. Each problem has to be treated as a specific case, and familiarity with structure–property relations aids the selection. The illustrations are limited to two widely differing aspects.

15.14.1 PLASTIC SELECTION FOR BOTTLE CRATE MANUFACTURE

The difficulties encountered when choosing a suitable plastic for a particular use arise mainly because each case is associated with a unique combination of properties. A good example, concerning bottle crate manufacture, has been cited by Willbourn (1969).

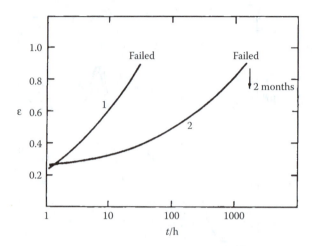

FIGURE 15.16 Comparison of loading tests using 1000-kg load on crates made from (a) high-density polyethylene and (b) poly(ethylene *b*-propylene) copolymer. The percentage compressive strain ε, is plotted against time *t*. (After Willbourn, A.H., *Plastics and Polymers*, 417, 1969. With permission.)

High-density polyethylene was chosen for the manufacture of beer crates in West Germany because it was the cheapest plastic available that was sufficiently tough and rigid for the purpose. It was also found to have a satisfactory creep response and good impact resistance, down to 253 K, which is adequate for continental winter temperatures. This plastic and the crate design were suitable for the use pattern in West Germany, where crates were usually piled 12 high.

When these crates were used in the U.K., where the practice is to stockpile 20 to 36 high for much longer periods, a rapid rate of crate failure was experienced. The change in conditions necessitated a new choice of plastic. This had to have better creep properties and a higher rigidity, but did not have to retain those good qualities at temperatures below 263 K because of the milder U.K. winters. Poly(vinyl chloride) was considered, but it is too difficult to mold; polystyrene and polypropylene have good creep characteristics, but these deteriorate at lower temperatures. The problem was solved by using poly(propylene-*b*-ethylene) copolymers, which have good toughness and mechanical response in the required temperature range. This is shown in Figure 15.16, where the high-density polyethylene failed under a load of 1000 kg in 29 h, but the copolymer lasted for 2 months. In this instance, both environmental conditions and industrial practice were important factors.

15.14.2 MEDICAL APPLICATIONS

The use of polymeric materials in the medical field is growing and gives rise to problems peculiar to the mode of application. Prosthesis is one of the major medical interests, and certain plastic replacement parts are now commonly used. High-density polyethylene is a successful replacement part for damaged hip joints and is employed as the socket, which accommodates a steel ball cemented to the femur using poly(methyl methacrylate). Artificial corneas can be prepared from poly(methyl methacrylate), and sections of artery are replaced by woven nylon or terylene tubes. Heart valves have been made from polycarbonates; and even artificial hearts, made from silicone rubber, have met with limited success. Plastic replacements for nose and ear cartilage, body absorbing sutures, and the use of polymeric membranes for dialysis in artificial kidney machines are only a few examples in a steadily growing list of uses.

The selection of suitable polymers for medical use focuses attention on the inertness of the polymer, its mechanical properties, and the extent of its biostability. It is useless implanting a polymer in the body that will be rejected or will degrade to produce toxic materials. The sample should also be pure and free of plasticizer, which might leach out and cause harmful side effects. The polymer has to be resistant to mechanical degradation and particularly abrasion, in case the abraded particles act as irritants. These conditions tend to limit the choice.

The use of polymers as adhesives is of particular interest. One reason for using the α-cyano acrylate esters as tissue adhesives has been the observation that there is a progressive, nontoxic absorption of the substances by the body. Interest in the use of polymers as reagents, which actively take part in the body functions, is now being evaluated.

External to the body, "hydrogels" are used as contact lenses, and their use may be extended to implantation. They are composed of cross-linked networks of hydroxyl methacrylate copolymers that swell when in contact with water.

Films and membranes are also used. A patient can be encased in a poly(vinyl chloride) tent while germ-free air is pumped through the canopy. It has also been suggested that films which allow the selective passage of oxygen in one direction could be used as oxygen tents. These would be similar to the silicone membranes that allow predominant passage of oxygen from water to air, and have been used to make a cage capable of supporting nonaquatic life underwater, in an air atmosphere extracted from the water. This expanding field will no doubt demand new polymers with specific applications.

15.15 HIGH-TEMPERATURE SPECIALITY POLYMERS

Although many of the more common polymers are remarkably resistant to chemical attack and stable when subjected to mechanical deformation, few can withstand the destructive effects of intense heat. To overcome this low thermal stability, chains incorporating (1) thermally unreactive aromatic rings, (2) resonance-stabilized systems, (3) cross-linked "ladder" structures, and (4) protective side groups have been synthesized. Many of these new structures have met with considerable success, and several of the aramids described in Section 15.11 are capable of maintaining 50% of ambient tensile strength at temperatures in excess of 550 K. They are also difficult to ignite and can be used to make excellent fire-resistant clothing. Particularly useful in this respect are the poly(benzimidazole)s, which are characterized by the presence of the unit in the chain. They have been known since 1961, when they were first synthesized by Marvel and Vogel, but one member in particular has achieved commercial prominence. This is the material prepared by a two-step reaction, involving first a melt polycondensation reaction between tetraaminobiphenyl and diphenyl isophthalate to give a prepolymer foam, which is crushed and heated under N_2 at 530 to 700 K to close the ring and generate the poly(benzimidazole) structure (PBI).

PBI

The polymer is soluble and can be dry-spun from dimethyl acetamide/LiCl solutions as a fiber that has exceptional heat resistance. It does not melt or ignite in air at temperatures up to 830 K, and above this temperature, while it degrades, produces virtually no smoke and undergoes carbonization. Short-term heating up to 670 K does not alter the tensile strength, and the mechanical properties are retained down to at least 150 K.

The PBI fiber has good moisture absorption characteristics (15% at ambient temperature and 65% relative humidity), making it better than cotton and comfortable to wear. When mixed with aramid fibers, it makes superior heat-protective clothing. PBI can also be molded to produce castings with high performance qualities over the temperature range 110 to 700 K.

Other examples of semiladder-like polymers include the poly(benzoxazole)s, poly(phenylene-1,3,4-oxadiazole)s, poly(quinoxaline)s, and poly(s-triazine)s. The high-temperature performance of these and other structures discussed are shown in Table 15.8.

Greater stability can be achieved if the vulnerable single bonds, which are susceptible to degradation-causing chain scission, can be eliminated. This can be done by preparing a ladder structure, a typical example being poly(imidazopyrolone). It is easily seen that single-bond scissions at points A along the chain do not lead to complete chain scission. This can only be brought about by two single bonds being broken as at points B, i.e., two bonds on opposite sides of the chain and between the same two "rungs" of the ladder. As this process has a low probability, many of these polymers can resist temperatures over 850 K, and so can effectively compete with some metals.

During the 1970s a number of noncommodity, specialty polymers were developed that display superior properties of heat resistance, impact resistance, high

TABLE 15.8
Structures of Polymers with Exceptional High-Temperature Performance

Polymer	Upper Service Temperature [K]
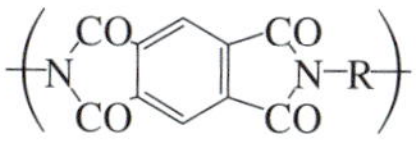 polyimide	570–620
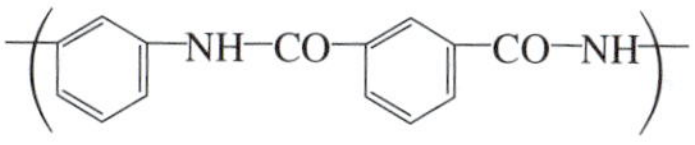 aromatic polyamide (aramid)	470–520
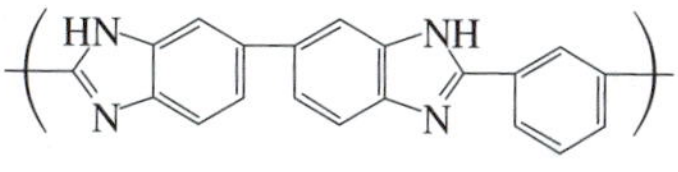 polybenzimidazole	520–570
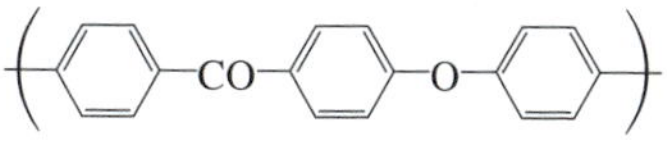 polyetheretherketone	510–530
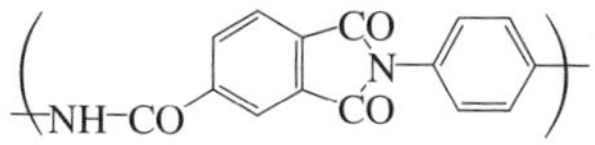 polyamide-imide	490–510
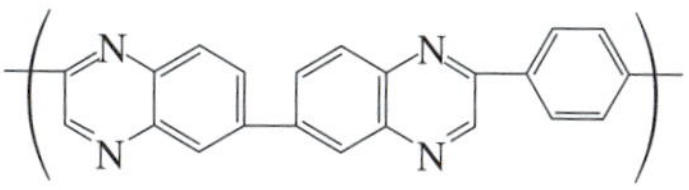 polyquinoxaline	670–720
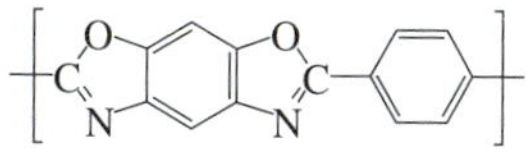 poly (p-phenylene-benzobisoxazole)	600–800
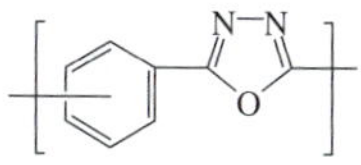 poly(oxadiazole)s	480–600

tensile strength, and stiffness. These are used either as single materials or to form the reinforcing component of a composite, blend, or alloy. Several of the more common groups, which have been classified as engineering plastics, are listed in Table 15.9, together with some of their applications.

Acetal is actually polyoxymethylene $+O-CH_2+_n$, a highly crystalline ($T_m \approx 450$ K) polymer prepared from formaldehyde. It can be stabilized by end-capping the polymer using acetic anhydride to acetylate the terminal hydroxyl groups and prevent polymer degradation by chain "unzipping." The material has good abrasion resistance, a reasonably high heat distortion temperature (383 K), and is not attacked by polar solvents. (In this context, the heat distortion or deflection test defines the temperature at which a standard size sample of polymer [5 × 1/2 × 1/8 in.]

TABLE 15.9
Engineering Resins Generally Used in Conditions of High Heat, Impact, or Moisture

Resin	Typical Applications
Acetal	Sinks, faucets, electrical switches, gears, aerosol bottles, meat hooks, lawn sprinklers, ballcocks, shaver cartridges, zippers, telephone push buttons
Polycarbonate	Helmets, power tool housings, battery cases, safety glass, automobile lenses, 5-gal bottles
Polyphenylene-sulfide	Electrical connectors, coil forms, lamp housings
Polysulfone	Electrical connectors, meter housings, coffee makers, camera bodies, automobile switch and relay bases, light-fixture sockets, fuel cell components, battery cases, medical supplies
Modified poly-phenylene oxide	Automobile dashboards, pumps, showerheads, plated automobile grilles and trim, appliance housings, wiring splice devices, protective shields
Polyimide	Radomes, printed circuit boards, turbine blades
Polyamide-imide	Valves, gears, pumps, high-temperature magnet wire enamels

distorts under a flexural load of 66 or 264 psi placed at its center. This is often found to be 10 to 20 K lower than T_g for an amorphous polymer, but may be much closer to T_m in crystalline polymers). Polyacetal is often used as a composite with a glass filler that raises the heat distortion temperature to 423 K.

In the polycarbonates, the most widely used member of the group is poly(bisphenol A carbonate).

This polymer can be prepared by the interfacial polycondensation of bisphenol A alkali salt dissolved in the water phase and phosgene ($COCl_2$) dissolved in methylene chloride. It can be used either as the pure polymer or in blends, particularly with acrylonitrile-butadiene-styrene (ABS) copolymers. The bisphenol A structure appears in other combinations, e.g., in a polysulfone copolymer (see Table 15.10) and in aromatic polyesters with phthalic acid moieties

or in poly(ether-imides). Poly(ester-carbonate)s, with the general structure

can also be prepared where the group X is an alkylene, ether, sulfide, or sulfone group.

TABLE 15.10
Commercially Important Aromatic Poly(sulfone)s

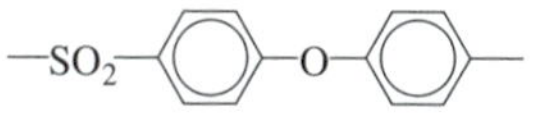 $-SO_2-\bigcirc-O-\bigcirc-$	Polyethersulfone
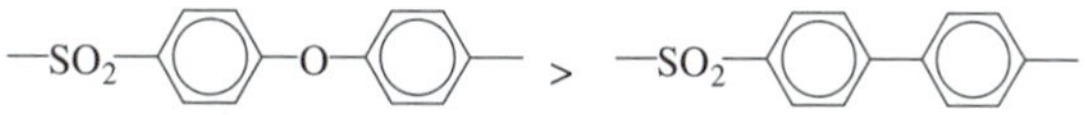 $-SO_2-\bigcirc-O-\bigcirc-$ > $-SO_2-\bigcirc-\bigcirc-$	Polyethersulfone
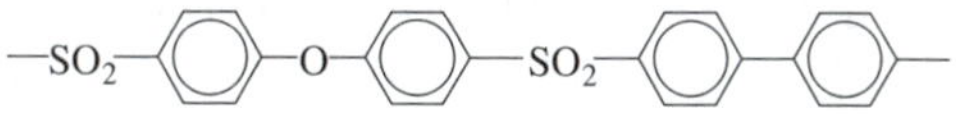 $-SO_2-\bigcirc-O-\bigcirc-SO_2-\bigcirc-\bigcirc-$	Polyarylsulfone
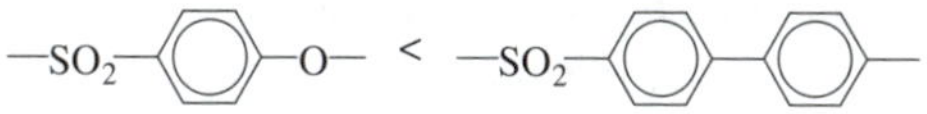 $-SO_2-\bigcirc-O-$ < $-SO_2-\bigcirc-\bigcirc-$	Polyarylsulfone
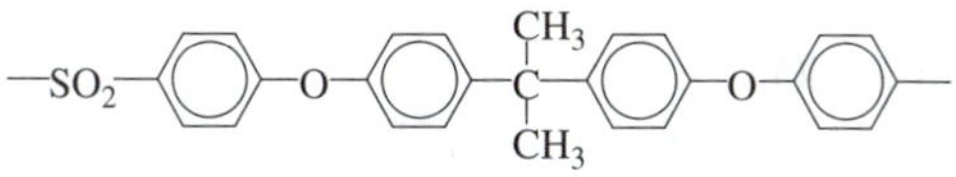 $-SO_2-\bigcirc-O-\bigcirc-\overset{CH_3}{\underset{CH_3}{C}}-\bigcirc-O-\bigcirc-$	Polysulfone

---

The polysulfones form another large group, and some of the commercially important structures are shown in Table 15.10. Poly(phenylene sulfone) tends to be too intractable for easy processing, and the copolymer structures are more useful. These are usually amorphous materials with high T_g values, typically in the range of 465 to 560 K. They are thermally stable, show good mechanical properties — particularly creep resistance — and are resistant to attack by dilute acids and alkalis. They can, however, dissolve in polar solvents, and solvent attack may also cause environmental stress cracking.

The polyarylene sulfones can be synthesized using an electrophilic substitution reaction

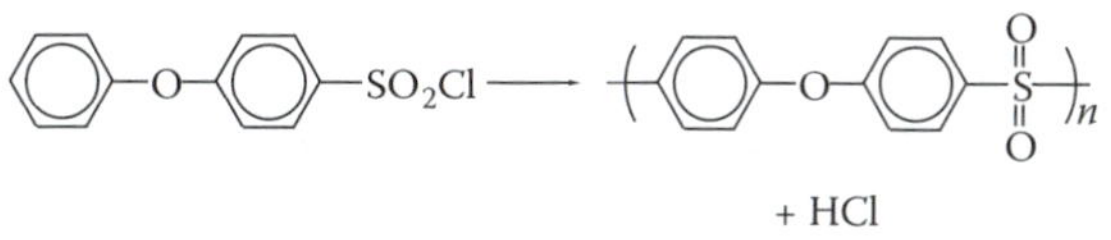

$$\bigcirc-O-\bigcirc-SO_2Cl \longrightarrow \left(\bigcirc-O-\bigcirc-\overset{O}{\underset{O}{\overset{\|}{S}}}\right)_n$$

$$+ HCl$$

which gives a *p*-substituted product. Although this method uses rather costly starting materials, other routes tend to give a mixture of *o*- and *p*-substitution. As the brittleness of the product increases with greater degrees of *o*-substitution, more economical variations of the first method have been developed.

Another sulfur-containing material, poly(phenylene sulfide)

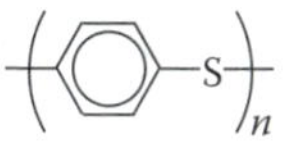

$$\left(\bigcirc-S\right)_n$$

is highly crystalline ($T_m = 563$ K), with a $T_g \approx 470$ K. It has good thermo-oxidative stability, is resistant to solvents, and when prepared as a composite with glassfiber has a heat distortion temperature of 520 K.

Poly(oxa-2,6-dimethyl-1,4-phenylene) (PPO)

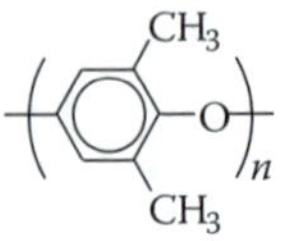

is prepared by the oxidative coupling of 2,6-dimethyl phenol, but it can also be made as a copolymer with styrene grafted on at the synthesis stage. PPO itself does not normally crystallize to any great extent from the melt, but it has a good heat deflection temperature (>370 K), possesses good self-lubricating qualities, and is an excellent electrical insulation material. It is widely used as a blend with polystyrene or a polyamide to form materials in the NORYL series.

Polyimides form the most important group of thermally stable polymers and are characterized by the presence of the

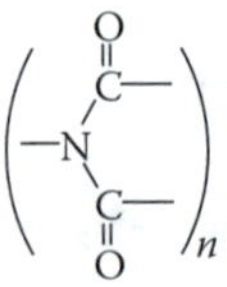

group in the structure. Several different synthetic routes have been developed, and a large variety of different materials have been prepared. The standard approach involves the polycondensation of pyromellitic dianhydride with a diamine such as 4,4′-diamino diphenyl ether, and in this case a polyimide marketed under the name KAPTON is produced.

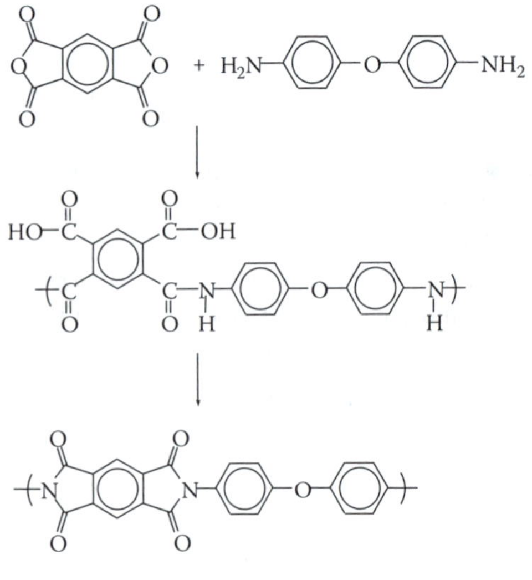

This synthesis involves a two-step reaction, the first of which yields an intermediate poly(amic acid) that is soluble in polar solvents. In the second, heating to temperatures of about 570 K effects ring closure to form the insoluble, intractable polyimide. A more direct synthetic method, avoiding the ring closure step, makes use of diisocyanates. KAPTON has an extremely high heat distortion temperature of 630 K and shows exceptional thermo-oxidative resistance.

NASA has developed a polyimide, LARC-TPI.

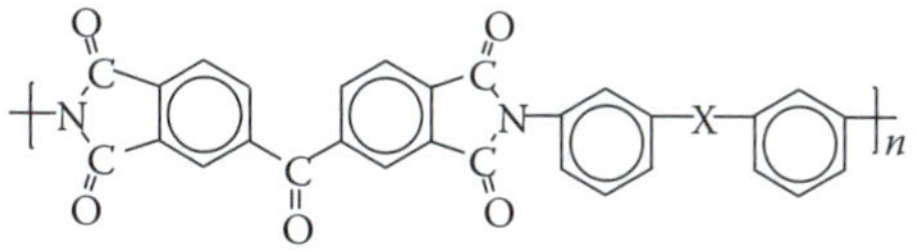

When X = (C=O), this is a semicrystalline material with T_m = 623 K and T_g = 519 K. On heating above T_m, the crystallinity is destroyed, and a totally amorphous polymer is formed. The film tensile strength and modulus at 298 K are 135.8 MPa and 3.72 GPa, respectively. When the group X is $+SO_2+$, a completely amorphous material is obtained, with a higher T_g of 546 K, film tensile strength of 62.7 MPa, and modulus of 4.96 GPa at 298 K.

These high T_g and T_m values can make the polymers difficult to handle, and improved processability can be obtained by introducing flexible groups in the chain. If the group X is an ether oxygen, T_g is lowered and the crystallinity reduced. Similarly, incorporation of $+O+CH_2CH_2+_2O+$ in place of X leads to a material that is more easily processed, is amorphous with a T_g = 428 K and at 298 K has a film tensile strength and modulus of 86.2 MPa and 2.7 GPa, respectively.

Copolymerization can also make the polyimides more tractable. Thus, poly-amide-imides such as KERMEL can be produced.

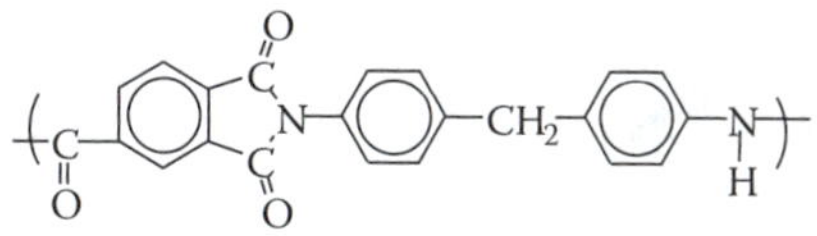

Another example of this group, marketed as TORLON, has the structure

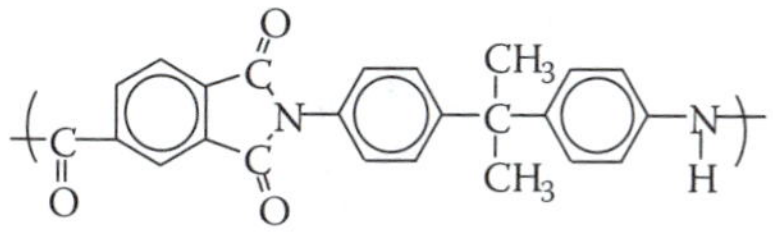

with a film (298 K) tensile strength of 186 MPa, modulus of 4.6 GPa, and a heat-distortion temperature of 555 K.

Polyether-imides with good melt-processing properties are also available, but tend to have inferior and high-temperature mechanical properties when compared with the preceding examples. Typical of this group is the following:

This is known as ULTEM and has lower values of heat distortion temperature (473 K), tensile strength (9 MPa), and modulus (3.2 GPa) at 298 K.

The aromatic polyketones form an important group of high-performance plastics as they exhibit extremely good high-temperature resistance and are melt-processable. A number of examples are shown in Table 15.11, and of these the best known is poly(ether ether ketone), or PEEK, structure (III). This is a crystalline polymer ($T_m = 607$ K and $T_g = 416$ K), with a heat distortion temperature of 433 K. The latter can be increased to 588 K for PEEK-glass fiber composites. The material is insensitive to hydrolysis and is particularly stable for long periods in hot water. This contrasts with the polyimides, which are sensitive to hydrolytic degradation. PEEK has found very special uses in the nuclear industry as a coating for wires and cables and as blow-molded containers. Other applications include composites with carbon fibers for use in the aerospace industry.

15.16 CARBON FIBERS

Although originally studied for its high-temperature qualities, the carbon fiber is, at present, used to advantage mainly in low-temperature situations. The fibers are prepared by converting oriented acrylic fibers into aligned graphite crystal fibers in a two-stage process. In the first stage, the acrylic fiber is oxidized, under tension to prevent disorientation of the chains, by heating in a current of air at 490 K for several hours. This is thought to lead to cyclization and the formation of a ladder polymer.

idealized structure

The second stage involves heating the fibers for a further period at 1770 K to eliminate all elements other than carbon. This *carbonization* is believed to involve cross-linking of the chains to form the hexagonal graphite structure, and this final heat treatment can affect the mechanical properties to a marked extent, as shown in Figure 15.17. The major application so far is in composite structures where they act as extremely effective reinforcing fibers. These reinforced plastic composites find uses in the aircraft industry, in the small-boat trade, and as ablative composites.

15.17 CONCLUDING REMARKS

The systematic study of structure–property relations provides an understanding of many of the fundamentals of the subject and can lead to quick dividends, as shown by the following example of the advanced art of fiber engineering.

TABLE 15.11
Typical Aromatic Poly(ether ketone)s

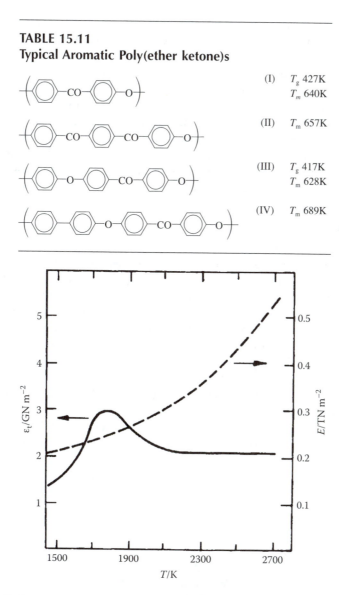

(I)	T_g 427K
	T_m 640K
(II)	T_m 657K
(III)	T_g 417K
	T_m 628K
(IV)	T_m 689K

FIGURE 15.17 Mechanical properties, the tensile strength ε_t, and Young's modulus E of carbon fibers as a function of the graphitizing temperature. (After Bailey, J.E. and Clarke, A.J., *Chem. Britain*, 6, 484, 1970.)

A sheep grows a wool fiber that possesses a corkscrew crimp in the dry state and affords the animal a bulky insulation blanket. When it rains, the fiber becomes wet and loses the crimp; the wool strands then bed down to form a close-packed, rain-tight covering capable of reducing the loss of body heat in the damp conditions. This *evolutionary* wool fiber, whose properties are derived from its bicomponent nature, was simulated in the laboratory by preparing a two-component acrylic fiber, in which each component had a different hydrophilicity. This feat was achieved by fiber scientists in only a few months by making use of their knowledge of structure and behavior.

Of course, nature was first and, at best, only reasonable facsimiles of some natural products can be synthesized in the laboratory. The scientist is not yet able to match the sophistication of many naturally occurring macromolecules, which are no longer simply materials, but working functional units. The complexity of the interrelation between structure and function in many proteins and nucleoproteins is the ultimate in molecular design, and although it will be some time before we can hope to reach this level in synthesis, progress in understanding the relations in simpler systems is a step in the right direction.

PROBLEMS

1. Based on their work on fibers, Carothers and Hill concluded that the lower limit of molecular weight for formation of polyester fibers is 12,000, and perhaps lower for polyamides, but higher molecular weight is recommended for polyethylene fibers (Hill and Walker, 1948). Justify this conclusion.

2. Values of the glass transition for a series of nylon polymers are given in Table 15.12. Using data in Table 15.3, establish whether a correlation exists between T_m and T_g. Is this consistent with expectations from Figure 15.2?

3. For the polyamides listed in Table 15.1, construct a plot of melting temperature vs. m and discuss the observed trend based on the chemical structure and arrangement of the chains.

4. Consider the following polymer structures and, for any given pair of n and m values, explain which one you expect to have the higher melting point.

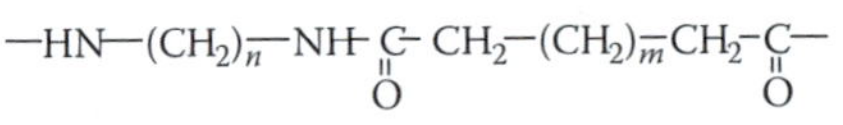

$$-HN-(CH_2)_n-NH-\overset{O}{\underset{\|}{C}}-CH_2-(CH_2)_m-CH_2-\overset{O}{\underset{\|}{C}}- \qquad -HN-(CH_2)_n-NH-\overset{O}{\underset{\|}{C}}-O-(CH_2)_m-O-\overset{O}{\underset{\|}{C}}-$$

Polyamide 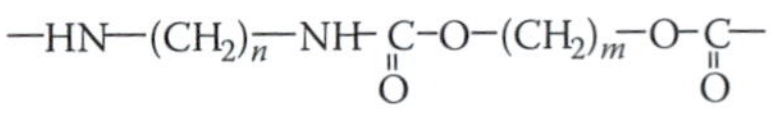   Polyurethane

---

TABLE 15.12
Glass Transition Temperature T_g of Linear Aliphatic Polyamides

Monadic Nylon	T_g (K)
6	327
7	325
8	323
9	319
10	315
11	315
12	313

---

5. Place the following polyurethane structures in order of decreasing melting temperature and explain your choice.

Repeat Unit
—HN—(CH$_2$)$_6$—NH—C—O—(CH$_2$)$_3$—⟨benzene⟩—(CH$_2$)$_3$—O—C— with C=O (O below)
—HN—(CH$_2$)$_8$—NH—C—O—(CH$_2$)$_{10}$—O—C— with C=O groups (O below)
—HN—(CH$_2$)$_8$—NH—C—O—(CH$_2$)$_2$—⟨benzene⟩—(CH$_2$)$_2$—O—C— with C=O (O below)

6. Consider a random copolymer of vinylidene fluoride and chlorotrifluoro-ethylene. By considering the glass transition of poly(vinylidene fluoride) as 235 K and the glass transition of poly(chlorotrifluoroethylene) as 319 K, estimate the T_g of the random copolymer containing the two monomers in the same weight percentage.

7. In Figure 15.18, plots of glass transition temperature vs. copolymer composition are given for (a) vinyl chloride/methyl acrylate (VC/MA) and (b) vinylidene chloride/methyl acrylate (VDC/MA). The behavior of copolymers with random and regularly alternating structures is compared.

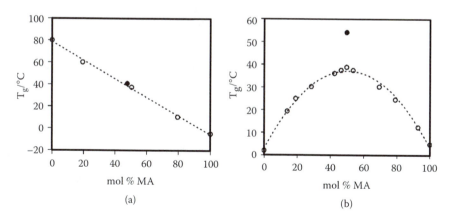

FIGURE 15.18 Glass transition temperature monomer composition behavior of (a) VC/MA and (b) VDC/MA copolymers: (●) regularly alternating and (○) random. (Data from Hirroka, M. and Kato, T., *J. Polym. Sci., Polym. Lett. Ed.*, 9, 199, 1971. Adapted from Tonelli, A.E., *Macromolecules*, 8, 544, 1975.)

With reference to the variations predicted by the Fox equation, discuss the experimentally observed T_g/composition trend for these copolymers. Suggest possible reasons for the difference of the T_g composition curves.

8. Statistical copolymers of p-nitrostyrene (p-NS) and styrene (S) were synthesized by free-radical bulk polymerization and copolymer compositions measured by elemental analysis. Glass transitions from differential scanning calorimetry at 20 K/min are:

Weight Fraction of p-NS	T_g(K)
0.000	374
0.069	376
0.076	379
0.172	379
0.286	389
0.317	390
0.378	390
0.473	403
0.580	408
0.698	415
1.000	461

Comment on the variation of the glass transition with copolymer composition, and establish whether the T_g vs. composition curve follows the predicted trend from the Flory–Fox equation.

9. Addition of a low-molecular-weight compound as diluent lowers the polymer–glass transition. By considering that the free volumes of the polymer and diluent are additive, Kelley and Bueche (1961) derived the following relationship between T_g and the volume fractions of polymer, v_p, and diluent v_d.

$$T_g = \frac{v_p T_{gp} (\alpha_l - \alpha_g) + v_d T_{gd} \alpha_d}{v_p (\alpha_l - \alpha_g) + v_d \alpha_d}$$

where $(\alpha_l - \alpha_g)$ is set equal to 4.8×10^{-4} K^{-1}. For the system poly(methyl methacrylate) $(T_{gp} = 104.7°C)$/diethyl phthalate:

v_p	T_g(°C)
0.500	−4.7
0.580	2.6
0.670	8.9
0.750	38.4
0.830	58.4
1.000	104.7

Confirm that the above relationship holds, and find the values of α_d and T_{gd}.

10. Estimate the decrease in glass transition temperature resulting from addition of 20 wt% di-n-butylphthalate (DBP) $(T_g = 178$ K$)$ to poly(vinyl chloride) $(T_g = 355$ K$)$. Compare, on a single graph, the expected DSC and modulus vs. temperature traces for the pure and plasticized polymers.

11. Draw the repeat unit structures of the homopolymers and copolymers prepared from monomers (i) and (ii).

(i)

$$HO-\overset{O}{\underset{\|}{C}}-(CH_2)_4-\overset{O}{\underset{\|}{C}}-OH \qquad H_2N-(CH_2)_6-NH_2 \qquad HO-\overset{O}{\underset{\|}{C}}-\langle\bigcirc\rangle-\overset{O}{\underset{\|}{C}}-OH$$

adipic acid   hexamethylene diamine   terephthalic acid

(ii)

$$HO-\overset{O}{\underset{\|}{C}}-(CH_2)_8-\overset{O}{\underset{\|}{C}}-OH \qquad H_2N-(CH_2)_6-NH_2 \qquad HO-\overset{O}{\underset{\|}{C}}-\langle\bigcirc\rangle-\overset{O}{\underset{\|}{C}}-OH$$

sebacic acid   hexamethylene diamine   terephthalic acid

The melting temperature vs. composition data for the polyhexamethylene sebacamide/terephthalamide copolymers reported in the following text show the expected minimum, but this is not the case for the polyhexamethylene adipamide/terephthalamide copolymers (Edgar and Hill, 1951). Construct a plot of melting temperature T_m vs. copolymer composition. With reference to the polymer chemical structures, explain the different trends observed for the sebacamide and adipamide data.

Polyhexamethylene Adipamide/Terephthalamide Copolymers		Polyhexamethylene Sebacamide/Terephthalamide Copolymers	
Mole Fraction of Terephthalic Component	$T_m(°C)$	Mole Fraction of Terephthalic Component	$T_m(°C)$
0	264	0	228
0.094	267	0.06	225
0.128	272	0.107	217
0.283	278	0.170	206
0.383	285	0.219	208
0.482	296	0.272	225
		0.326	239
		0.377	261
		0.432	275
		0.479	301

12. KRATON® polymers are high-performance elastomers used to improve the properties of a wide range of end products. KRATON G polymers are triblock copolymers, with a random ethylene/propylene copolymer forming the saturated midblock and styrene forming the two other blocks (styrene-ethylene/propylene-styrene, SEPS).

    a. Given that the glass transition of polyethylene is 180 K and the glass transition of polypropylene is 250 K, calculate T_g for the midblock random copolymer containing the two monomers in the same weight percentage.

    b. The glass transition of the polystyrene block is 373 K. Sketch the trace that would be recorded using differential scanning calorimetry for KRATON G when the volume fraction of styrene is 70%.

    c. Predict the modulus vs. temperature behavior of this block copolymer.

13. The introduction of 6-hydroxy-2-naphthoic acid (HNA) to poly(hydroquinone terephthalate) leads to changes in the melting temperature, as shown in Figure 15.19. Explain these results.

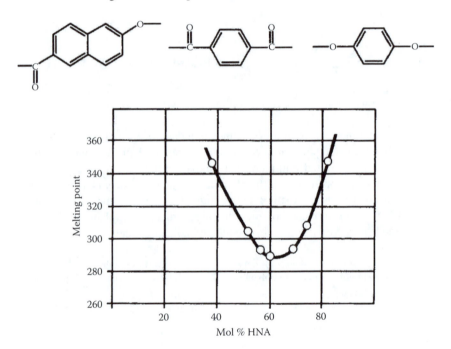

FIGURE 15.19 Melting point of copolyesters of 6-hydroxy-2-naphthoic acid (HNA) with terephthalic acid (TA) and hydroquinone. (Reproduced with permission from *Hoechst High Chem. Magazine.*)

14. Anionic polymerization of butadiene, followed by hydrogenation, provides a route to compositionally homogeneous materials similar in structure to ethylene/1-butene copolymers. The copolymer composition can be varied by addition of polar modifiers to the polymerization mixture, thus generating materials with differing mole fractions of 1,2 addition, x_{12}.

Thermal properties (final melting temperature T_m and secondary transition temperature T_x), density at 23°C, crystalline volume fraction, α_v, and tensile modulus at 23°C, E_s, of hydrogenated polybutadiene samples were reported by Krigas et al. (1985):

x_{12}	$T_x(°C)$	$T_m(°C)$	$\rho\,(23°C)(g\,cm^{-3})$	α_v	$E_s(GPa)$
0		135	0.946	0.615	0.5
0.08	–30	106.0	0.910	0.443	0.077
0.18	–40	91.0	0.895	0.311	0.030
0.23	–45	75.0	0.884	0.219	0.014
0.30	–51	72.0	0.877	0.153	0.007
0.42	–61	28.0	0.862	0.025	—
0.51	–62	–3.0	0.860	—	—
0.57	–62	—	0.861	—	—
0.86	–42	—	0.865	—	—
0.99	–27	—	0.872	—	—

Discuss the observed changes on the physical parameters with chemical structure and, therefore, the degree of branching.

15. One of the characteristics of block copolymers is that they tend to phase separately. The triblock styrene-*block*-butadiene-*block*-styrene has been extensively investigated. Consider the following experimental evidence and qualitatively predict the relative stress–strain behavior at the various compositions.

a. At 13% polystyrene, there is no phase separation between the blocks, and the material is very soft.

b. At 28% polystyrene, the material behaves as a thermoplastic elastomer; it elongates to 600% and then stiffens considerably.

c. At 80% polystyrene, the behavior is similar to that of polystyrene.

REFERENCES

Bailey, J.E. and Clarke, A.J., *Chem. Britain*, 6, 484, 1970.

Drumm, M.F., Dodge, C.W.H., and Nielsen, L.E., *Ind. Eng. Chem.*, 48, 76, 1956.

Edgar, O.B. and Hill, R., *J. Polym. Sci.*, 8, 1, 1951.

Gaudiana, R.A., Minns, R.A., Sinta, R., Weeks, N., and Rogers, H.G., Amorphous rigid rod polymers, *Prog. Polym. Sci.*, 14, 47, 1989.

Hegenrother, P.M., *Polym. J.*, 19, 73, 1987.

Hill, R., *Fibres from Synthetic Polymers*, Elsevier, 1953.

Hill, R. and Walker, E.E., *J. Polym. Sci.*, 3, 609, 1948.

Hirroka, M. and Kato, T., *J. Polym. Sci., Polym. Lett. Ed.*, 9, 199, 1971.

Kelley, F.N. and Bueche, F., *J. Polym. Sci.*, 50, 549, 1961.

Krigas et al., *J. Polym. Sci., Polym. Phys. Ed,.* 23, 509, 1985.

Lopez, L.C. and Wilkes, G.L., Poly(phenylene sulphide), *Rev. Macromol. Chem. Phys.*, C29, 83, 1989.

Nielsen, E., *Mechanical Properties of Polymers and Composites*, Marcel Dekker, 1974.

Richards, R.B., *J. Appl. Chem.*, 1, 370, 1951.

Tanner, D., Fitzgerald, J.A., and Phillips, B.R., The Kevlar story, *Angew. Chem. Int. Ed. Engl. Adv. Mater.*, 28, 649, 1989.

Tonelli, A.E., *Macromolecules*, 8, 544, 1975.

Willbourn, A.H., *Plastics and Polymers*, 417, 1969.

BIBLIOGRAPHY

Allen, G. and Bevington, J.C., Eds., *Comprehensive Polymer Science,* Vol. 2, 7, Pergamon Press, 1989.

Bawn, C.E.H., Structure and performances, *Plastics and Polymers*, 373, 1969.

Bloch, B. and Hastings, G.W., *Plastics in Surgery,* Thomas, 1967.

Bruce-Black, W., Structure-property relationships in high temperature fibres, *Trans. NY Acad. Sci.*, 32, 765, 1970.

Critchley, J.P., Knight, G.J., and Wright, W.W., *Heat Resistant Polymers*, Plenum Press, 1983.

Dyson, R.W., Ed., *Speciality Polymers*, Blackie and Son, 1987.

Dyson, R.W., Ed., *Engineering Polymers*, Blackie and Son, 1989.

Elias, H.G. and Vohwinkel, F., *New Commercial Polymers 2*, Gordon and Breach Science, 1986.

Folkes, M.J., Ed., *Processing, Structure and Properties of Block Copolymers*, Elsevier Applied Science, 1985.

Goodman, I., *Synthetic Fibre Forming Polymers*, Royal Institute of Chemistry, 1967.

Hearle, J.W.S. and Peters, R.H., *Fibre Structure*, Butterworths, 1963.

Lewin, M. and Preston, J., *High Technology Fibers.* Part A (1985) and Part B (1989), Marcel Dekker.

Mascia, L., *The Role of Additives in Plastics*, Edward Arnold, 1974.

McIntyre, J.E., *The Chemistry of Fibres*, Edward Arnold, 1971.

Mittal, K.L., Ed., *Polyimides: Synthesis, Characterization and Applications*, Plenum Press, 1984, Vol. 1, 2.

Moncrieff, R.W., *Man-Made Fibres*, John Wiley and Sons, 1963.

Seymour, R.B. and Carraher, C.E., *Structure-Property Relationships in Polymers*, Plenum Press, 1984.

Seymour, R.B. and Kirshenbaum, G.S., Eds., *Performance Polymers: Their Origin and Development*, Elsevier, 1986.

Tobolsky, A.V. and Mark, H., *Polymer Science and Materials*, Wiley-Interscience, 1971, chap. 14, 15.

Wilson, D., Ed., *Polyimides,* Blackie and Son, 1989.

16 Polymers for the Electronics Industry

16.1 INTRODUCTION

The application of specialty polymers in electronics and photonics is extensive, both in a "passive" role — when they act as insulators, encapsulating agents, adhesives, and materials for integrated circuit (IC) fabrication — and in an "active" way as electronic and photonic conductors, or as active material in nonlinear optics.

Some of these uses are rather mundane, in a chemical sense, but are necessary as an integral part of the whole process, e.g., in the packaging and protection of fragile ICs to avoid damage or to prevent the detrimental effects of humidity and corrosion. Epoxy novolacs or silicone-epoxy thermosetting resins are ideal for these purposes and are used to encase the IC that may already have been treated with a barrier coating of room-temperature vulcanizing (RTV) silicone rubber to prevent moisture absorption. Other applications are more interesting to the chemist as they require thought and ingenuity in the molecular design of the polymer.

The properties of polymers that make them an essential part of microelectronics engineering will be discussed under two main headings: polymer resists and conducting polymers. These will best illustrate why progress in this area could not have been made without exploiting the unique features of polymeric materials. This will be followed by a brief discussion of some photonic applications.

16.2 POLYMER RESISTS FOR IC FABRICATION

ICs are arguably among the most important products of the modern electronics industry. They are built up from various arrangements of transistors, diodes, capacitors, and resistors that are individually constructed on a flat silicon or gallium arsenide substrate by selective diffusion of small amounts of materials into particular regions of the semiconductor substrate, and by metallization of the paths linking the active circuit elements. The patterns defining these regions and the linking pathways must first be drawn by a lithographic process on a layer of resist material, and then transferred onto the substrate by an etching process. In this context, the *lithographic process* is the art of making precise designs on thin films of resist material by exposing them to a suitable form of patterned radiation, e.g., ultraviolet, electron beam, x-ray, or ion beam, with the formation of a latent image on the resist that can subsequently be developed by treatment with solvents or plasma. The *resist* is a material, usually polymeric, that is sensitive to, and whose properties (either chemical or physical) are changed by exposure to the electromagnetic radiation. It must also be resistant (hence, the name), after development, to the etching process and protect the areas it still covers, while allowing the exposed regions of the substrate to be attacked. In this way, a pattern is transferred into the substrate and the remaining resist material is removed.

16.3 THE LITHOGRAPHIC PROCESS

Several steps are involved in the lithographic process, and these are shown schematically in Figure 16.1. If the substrate chip is silicon, it is first oxidized to produce a thin surface layer of SiO_2 (step 1). A solution of the polymer resist is then spun evenly onto this surface and baked to remove the solvent and form a thin film of the resist (step 2), approximately 0.5 to 2 μm thick. The next stage (step 3) is the exposure of the resist to electromagnetic radiation either through a patterned mask or by direct "writing" if the radiation source is an electron beam. Depending on the radiation and the nature of the polymer resist, the exposed regions are either rendered soluble if the polymer is degraded — this is called a *positive-acting resist* — or insoluble if the polymer is cross-linked — this is called a *negative-acting resist*. A positive or negative pattern can then be developed by treatment with solvents that dissolve the exposed regions in the positive resist (step 4a), or the unexposed regions in the case of the negative-acting resist (step 4b), so producing the template for the

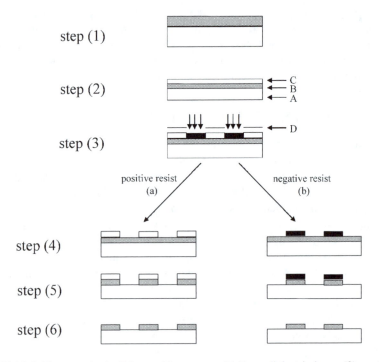

FIGURE 16.1 The steps in the lithographic process. (1) Form dielectric layer, (2) coat with polymer resist, (3) expose to electromagnetic radiation, (4) develop pattern, (5) etch, (6) strip resist. A: The substrate (silicon, etc.), B: thin dielectric layer (e.g., SiO_2), C: polymer resist layer, and D: mask.

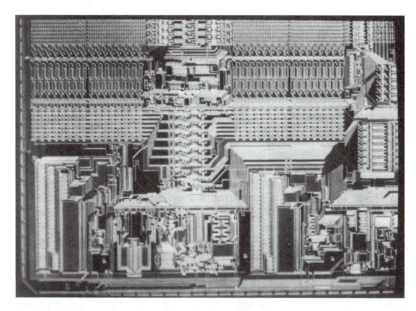

FIGURE 16.2 An example of a very large-scale integrated (VLSI) circuit chip manufactured by IBM. (Reproduced with permission from *Hoechst High Chem. Magazine*, 1989).

pattern that is to be etched onto the substrate. Etching (step 5) can be achieved by treatment with buffered HF or by using dry plasma etching methods. In either case, the polymer must protect the regions of the chip it still covers while allowing the exposed areas of the substrate to be attacked. Once the pattern is transferred in this way, the remaining resist is stripped off and discarded (step 6).

As designers move toward greater miniaturization and increasing device complexity, there is a need for smaller feature sizes. Thus, the minimum feature sizes on metal-oxide semiconductor (MOS) random access memory (RAM) devices may be ~3.5 μm if a very large-scale integrated (VLSI) circuit 256K-RAM device is to be made. For such a chip, which is only a $\frac{1}{2}$ in. square of silicon, to be constructed, 256,000 places where an electrical charge can be located must be defined together with the connecting circuitry. This puts a severe strain on the ability to define these features accurately, which is a function of both the wavelength of the radiation and, more importantly, the response of the resist to the radiation. Now the ingenuity of the polymer chemist is called upon to design suitable resist materials. The manufacture of VLSI devices requires the use of short-wavelength radiation, but many circuits are still made by photolithography involving irradiation with UV light. An example of a VLSI device is shown in Figure 16.2.

16.4 POLYMER RESISTS

The response of a polymer to radiation and the resolution obtained are not easy to predict, but there are certain criteria that the candidate resist should meet. The most important of these are:

1. Adequate sensitivity to the radiation used
2. The ability to adhere to the substrate and be easily removed after etching
3. Possession of a high T_g, particularly if it is a positive-working resist, to prevent flow and distortion of developed patterns
4. Resistance to etching reagents

Resists can be judged using performance criteria such as (1) sensitivity and (2) resolution (contrast).

16.4.1 SENSITIVITY

This can be defined as the amount of incident energy, i.e., the flux per unit area measured in C/cm^2, required to effect enough chemical change in the resist to ensure that after development, the desired relief image is obtained. This can be measured by plotting the log of the radiation dose, D, against the normalized film thickness after development. The response curves for both positive- and negative-working resists are showing in Figure 16.3.

Sensitivity increases as the dose required to produce the image decreases; i.e., the lower the dose required to produce the desired image, the greater the sensitivity. The sought-after range of sensitivities is approximately 0.01 to 1.0 $\mu C/cm^2$.

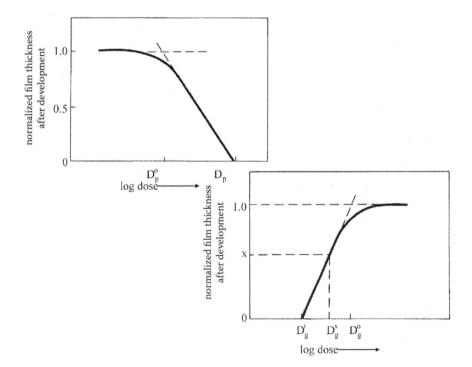

FIGURE 16.3 Normalized film thickness plotted against log (dose) to produce typical sensitivity curves for (a) a positive- and (b) a negative-acting polymer resist.

16.4.2 RESOLUTION

Resolution (γ), or contrast, has different definitions for positive and negative resists. For positive resists, γ_p is a function of both the rate of degradation and the rate of change of solubility of the resist on exposure, whereas for a negative resist, γ_n is a function of the rate of gel formation. Numerical values are obtained from the slope of the linear portion of the response curve and are given by:

$$\gamma_p = \log \left[\frac{D_p}{D_p^o} \right]^{-1} \tag{16.1}$$

$$\gamma_n = \log \left[\frac{D_g^o}{D_g^i} \right]^{-1} \tag{16.2}$$

where D_g^i is the onset of gel formation, D_g^o is the dose required to produce 100% initial film thickness, and D_p is the dose required to effect complete solubility in the exposed regions while leaving the unexposed region insoluble.

We can now examine some of the systems that have proved to be successful in this application.

16.5 PHOTOLITHOGRAPHY

Many polymers are altered on exposure to ultraviolet radiation, and this has led to the development of photolithographic techniques using conventional UV radiation from a mercury vapor lamp with an emission spectrum of $\lambda = 430$ nm, 405 nm, and 365 nm. When resolution of features of less than ~2 μm is required, then deep UV sources with $\lambda \approx 150$ to 250 nm may be used, provided resists that absorb in this wavelength region can be selected.

16.5.1 POSITIVE PHOTORESISTS

For pattern development, a positive-working resist normally depends on an increase in the solubility of the resist in the exposed region relative to the unexposed areas. As the wavelengths used in the near-UV range are not sufficiently energetic to induce bond scission, the solubility must be altered by some other means.

A photoresist widely used in the electronics industry is the two-component system comprising a short chain Novolac resin, which acts as the film-forming agent, mixed with 20 to 50 wt% of a naphthoquinone diazide photosensitive compound. This sensitizer is insoluble in basic solutions and is a sufficiently large molecule to act as a "dissolution inhibitor," preventing the acidic Novolac film from being dissolved by aqueous alkaline solutions in which the resin is normally soluble. Exposure to UV radiation converts the diazide into indene carboxylic acid that is now soluble in base solution, thereby rendering the whole of the exposed region soluble, whereas the unexposed regions remain resistant and insoluble. This reaction involves elimination of N_2 followed by a Wolff rearrangement, after which the presence of small amounts of water in the resist completes the conversion of the ketene to the acid.

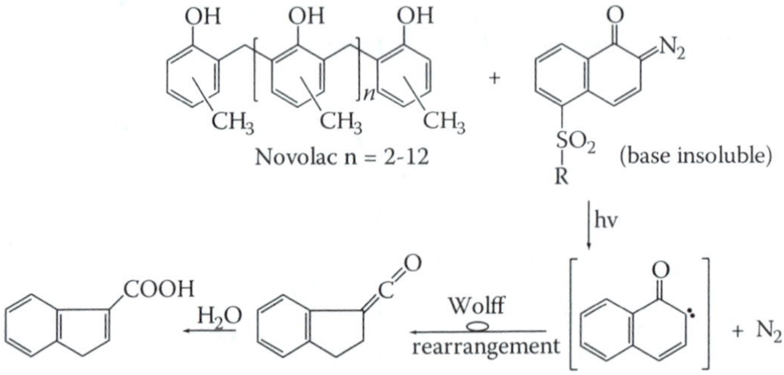

Novolacs tend to absorb too strongly in the deep UV region (~250 nm), but other systems have been developed specifically for this purpose. A base-soluble, poly(methyl methacrylate-*stat*-methacrylic acid) copolymer that is transparent in the deep UV can be mixed with a photosensitive, base-insoluble, dissolution inhibitor *o*-nitrobenzyl-cholate(I). Exposure to deep UV photolyzes the ester to produce cholic acid and *o*-nitrosobenzyl alcohol as shown:

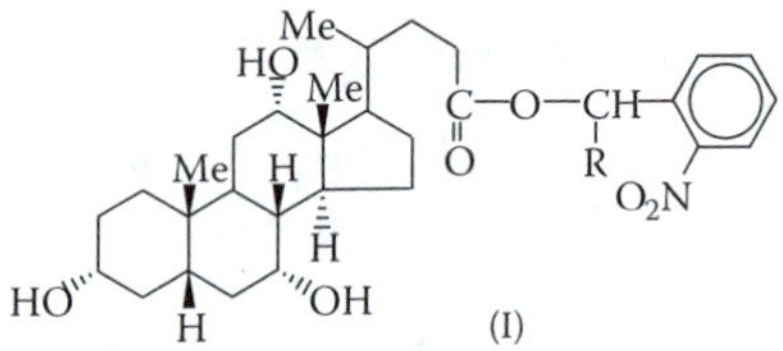

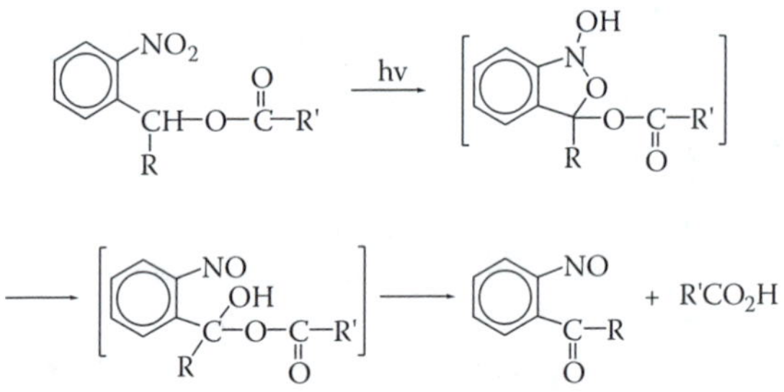

This system functions in the same way as the Novolac resist and gives a positive tone pattern.

16.5.2 NEGATIVE PHOTORESISTS

This type of resist has been the mainstay of the microelectronics industry when resolution down to 2 μm is adequate but is not so useful for finer work. The most commonly used systems are prepared from mixtures of cyclized polyisoprene mixed with a suitable photosensitive compound such as an aromatic diazide. This acid-catalyzed cyclization of polyisoprene produces a complex mixture of structures, but leads to a higher T_g material, with improved film-forming properties.

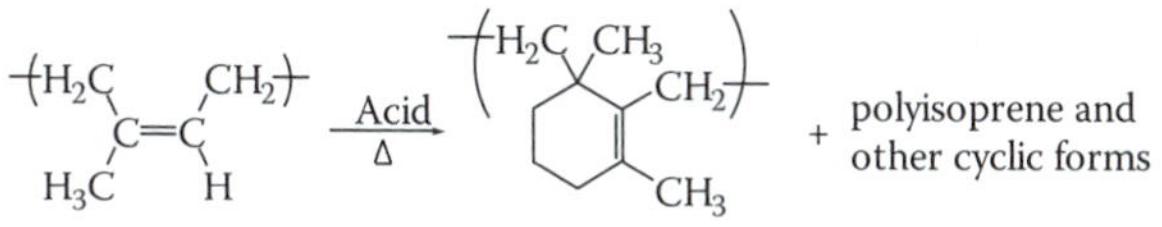

The bisazide forms a bisnitrene on irradiation that then cross-links the polyisoprene by reacting with the double bonds or the allylic hydrogens in the exposed regions to produce an insoluble matrix, e.g.,

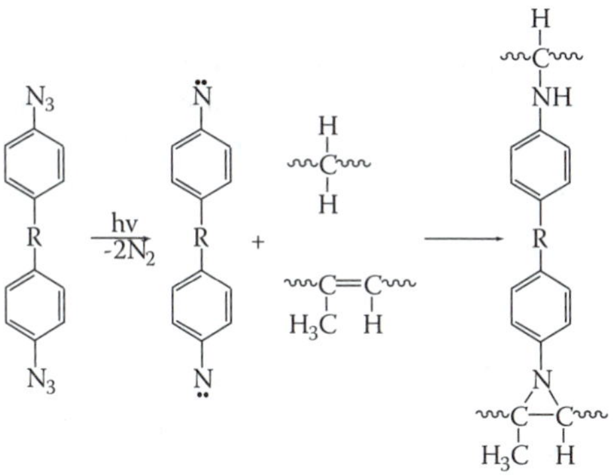

A second system makes use of the potential for poly(vinyl cinnamate) to form cross-links upon radiation, e.g.,

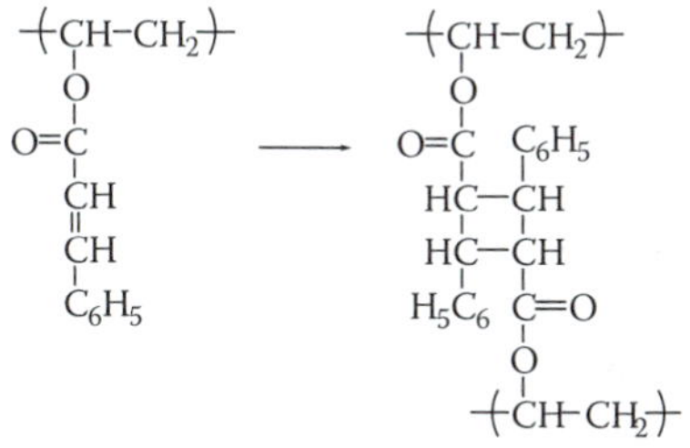

This photodimerization can be sensitized by Michler's ketone, 4,4'-bis(dimethyl-amino)benzophenone.

Greater resolution can be obtained using deep UV radiation, but now other resists must be used as the conventional ones are optically opaque in this region. Poly(methyl methacrylate) can be used as a positive resist, as can several of its derivatives. In each case, the carbonyl groups absorb at 215 nm, and this leads to chain scission and degradation.

$$+CH_2-\underset{\underset{OCH_3}{\overset{|}{C=O}}}{\overset{\overset{CH_3}{|}}{C}}-CH_2-\underset{\underset{OCH_3}{\overset{|}{C=O}}}{\overset{\overset{CH_3}{|}}{C}}\!\!+_n \xrightarrow{\;h\nu\;} +CH_2-\underset{\underset{OCH_3}{\overset{|}{C=O}}}{\overset{\overset{CH_3}{|}}{C}}\!-CH_2\!-\underset{\underset{OCH_3}{\overset{|}{C=O}}}{\overset{\overset{CH_3}{|}}{C}}\!\!+_n \longrightarrow$$

$$+CH_2-\underset{\underset{}{\overset{|}{CH_3}}}{\overset{\overset{}{|}}{C}}=CH_2 \;+\cdot\underset{\underset{OCH_3}{\overset{|}{C=O}}}{\overset{\overset{CH_3}{|}}{C}}\!\!+ \;\;+\;\; CO,\, CO_2,\, CH_3^{\cdot},\, CH_3O^{\cdot}$$

A negative tone resist can be obtained by using an image reversal technique with the Novolac-naphthoquinone diazide system. This procedure makes use of the normal steps for creating a positive resist, but on treating with a base and baking at temperatures > 350 K, a base-catalyzed decarboxylation of the indene carboxylic acid occurs, forming an indene derivative that is a photo-insensitive dissolution inhibitor. The complete resist is now subjected to a flood exposure of all areas by the UV source, and this converts the naphthoquinone diazide in the previously unexposed regions into the acid form. This renders these regions soluble, and development produces the negative tone pattern. These events are summarized in Figure 16.4.

Other positive resists sensitive to these wavelengths include the alkyl and aryl sulfones.

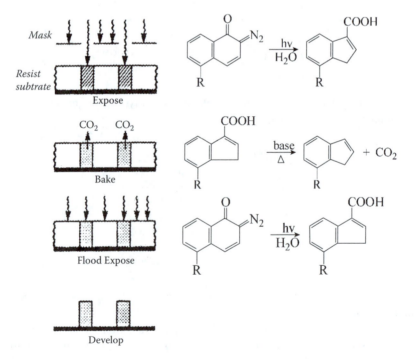

FIGURE 16.4 Reverse imaging using a Novolac-naphthoquinone diazide resist mixture exposed to UV radiation. (From Bowden, M.J. and Turner, S.R., Eds., *Electronic and Photonic Applications of Polymers,* American Chemical Society, 1988. With permission.).

16.6 ELECTRON BEAM SENSITIVE RESISTS

The inherent limitations of photolithography, caused by diffraction problems when resolutions of less than 1 μm are required, is overcome, at least in part, by turning to electron beam and x-ray lithography with much shorter wavelengths on the order of 0.5 to 5 nm. As the photon energy of an electron beam is high enough to break virtually all the bonds likely to be found in a polymer resist, the reactions involved are much less selective than those encountered in some of the photoresists. Thus, both degradation and cross-linking may take place in the same polymer on exposure to an electron beam, and the behavior of the resist as a positive- or negative-working system will depend on which of these processes dominates. This may be a function of the exposure time and intensity of the radiation, such that a positive-acting resist may begin to cross-link and transform into a negative-acting resist on prolonged exposure.

16.6.1 POSITIVE RESISTS

Most polymers that are positive resists tend to depolymerize via a monomer-unzipping action when degraded, and poly(methyl methacrylate) (PMMA) is typical of this type. Unfortunately, the sensitivity of PMMA to electron beam radiation is low, and in an attempt to improve this feature, PMMA derivatives have been prepared by replacing the α-methyl group with more polar electron-withdrawing substituents, e.g., Cl, CN, and CF_3, to assist electron capture (Figure 16.5). Modification of the

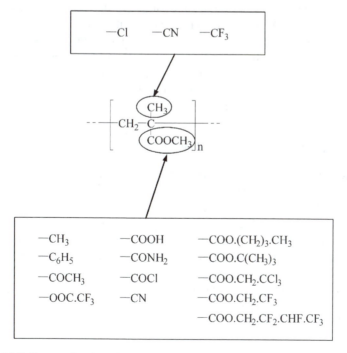

FIGURE 16.5 Groups that have been used to increase the sensitivity of positive-working electron resists.

ester group has also been a strategy, but in each case the presence of the quaternary carbon atom is perhaps the single most important feature for the resist to be a positive-working system because of its susceptibility to chain scission. Poly(alkene sulfone)s are also very sensitive positive resists, prepared by an alternating copolymerization reaction between sulfur dioxide and an appropriate alkene.

$$\left(CH_2-CH-\overset{\overset{O}{\|}}{\underset{\underset{\|}{O}}{S}}\right)\ \xrightarrow{\varepsilon\text{-beam}}\ H_2C=CH\ +\ SO_2$$

$$\underset{\underset{CH_3}{|}}{(CH_2)_n}\qquad\qquad\underset{\underset{CH_3}{|}}{(CH_2)_n}$$

Exposure to an electron beam source cleaves the polymer chains at the weak C–S bond with liberation of SO_2, and in certain cases such as poly(2-methyl pentene sulfone), there is almost complete vaporization of the exposed regions when a 20-kV electron beam source is used. The major limitation of this group of resists is their poor resistance to dry etching.

16.6.2 Negative Resists

In general, negative-acting resists tend to give poorer resolution but are faster and tougher than positive-working resists. This difference in speed arises from the fact that although only a few cross-links will make a polymer insoluble, a positive resist may require extensive fragmentation before it can be developed successfully. Good negative-working resists should have cross-linking sites such as double bonds, epoxy groups, and possibly phenyl rings to delocalize and absorb the energy of the electron beam, thereby protecting the chain from scission. Some examples of useful systems are shown in Figure 16.6. Halogenated aromatic polymers form another group of interest (Table 16.1), where the introduction of a halogen atom increases the sensitivity of structures based on polystyrene.

16.7 X-RAY AND ION SENSITIVE RESISTS

One drawback in electron beam lithography stems from the fact that much of the interaction between the polymer and the electron beam occurs as a result of low-energy secondary electrons being produced in the film. These are scattered beyond the definition of the exposing beam and can produce unwanted reactions in regions adjacent to the primary exposure. This is known as the *Proximity Effect* and can cause undercutting and overlapping of closely spaced features. In x-ray and ion beam lithography, the secondary electrons produced have lower energies and shorter path lengths and, consequently, the proximity effect is less pronounced. These techniques may find greater use in the future. The electron micrograph of an exposed and developed x-ray resist is shown in Figure 16.7.

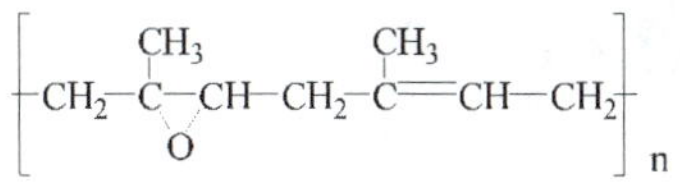

Epoxidized polyisoprene (0.05 μC/cm^2)

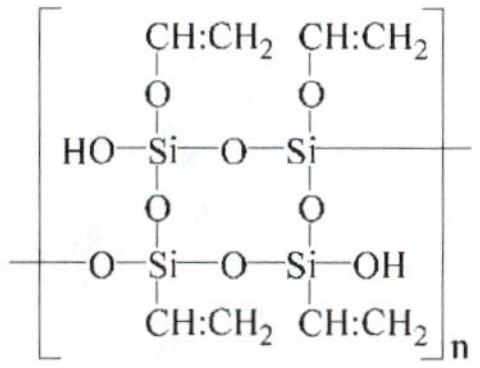

polvinylcyclosiloxane (4 μC/cm^2)

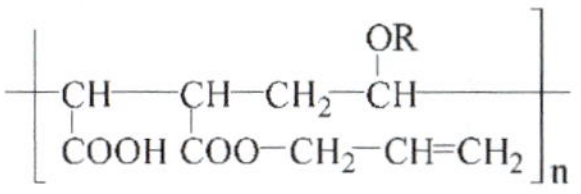

Allyl half-ester of maleic anhydride/
vinyl ether copolymer (0.1 μC/cm^2)

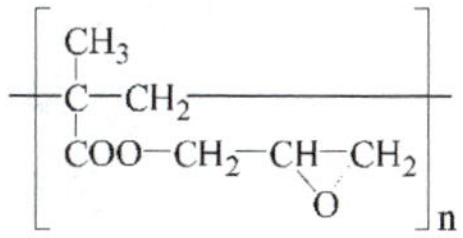

Poly-glycidyl methacrylate (1μC/cm^2)

FIGURE 16.6 Some examples of sensitive negative-working polymers resists. (From Clark, M.G., Materials for optical storage, *Chem. Ind.*, 258, 1985. With permission.)

TABLE 16.1
Sensitivity of Some Halogenated Aromatic Polymers

Polymer	Sensitivity
Poly(styrene-*stat*-4-chlorostyrene)	6 μC/cm^2
Poly(styrene-*stat*-4-chloromethylstyrene)	1 μC/cm^2
Poly(3-bromo-9-vinylcarbazole)	2 μC/cm^2

16.8 ELECTROACTIVE POLYMERS

Inherently, organic polymers with all-carbon backbones are insulators and can be used as encapsulating materials when a medium of high resistivity is required, such as in coatings for cables and electrical wiring. It has been found that the resistivity can be decreased if a composite of the polymer with carbon black or finely divided metal is fabricated, but the conduction in these cases takes place via the filler and not through the polymer, which merely acts as a supporting matrix. Incorporation of a filler can also reduce the mechanical strength of the polymer.

In 1977, the first major breakthrough was achieved when it was discovered that polyacetylene, which is a very poor conductor in the pure state, could be turned into a highly conductive polymer by conversion to the salt on reacting it with I_2. The result was a dramatic increase of over 10^{10} in conductivity. As conduction appears

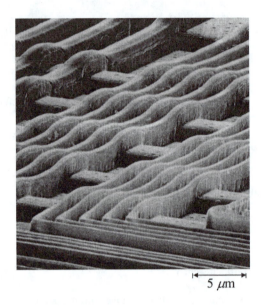

5 μm

FIGURE 16.7 Pattern cut in a polymer resist using x-ray lithography, showing the steep edges and resolution capable with this technique. (From *Hoechst High Chem. Magazine*, 1989. With permission.)

to be due to movement of electrons through the polymer, this discovery has added an exciting new dimension to the rapidly expanding area of synthetic metals. Other polymers that display similar characteristics are usually polyconjugated structures that are insulators in the pure state but when treated with an oxidizing or a reducing agent can be converted into polymer salts with electrical conductivities comparable to metals. Some idea of the possible range of conductivities (σ) is given in Figure 16.8, where σ varies from 10^{-18} S cm^{-1} for a good polymeric insulator (e.g., poly tetrafluoroethylene) up to $\sigma \sim 10^6$ S cm^{-1} for a metallic conductor (e.g., copper).

16.9 CONDUCTION MECHANISMS

Electrical conductivity is a function of the number of charge carriers of species "i" (n_i), the charge on each carrier (ε_i), and carrier mobilities (μ_i), described by the relation $\sigma = \Sigma\mu_i.n_i.\varepsilon_i$, where the unit of conductivity is S cm^{-1}. Conduction in solids is usually explained in terms of band theory, which postulates that when atoms or molecules are aggregated in the solid state, the outer atomic orbitals containing the valence electrons are split into bonding and antibonding orbitals, and mix to form two series of closely spaced energy levels. These are usually called the *valence band* and the *conduction band,* respectively. If the valence band is only partly filled by the available electrons, or if the two bands overlap so that no energy gap exists between them, then application of a potential will raise some of the electrons into empty levels, where they will be free to move throughout the solid thereby producing a current. This is the description of a conductor. If, on the other hand, the valence band is full and is separated from the empty conduction band by an energy gap,

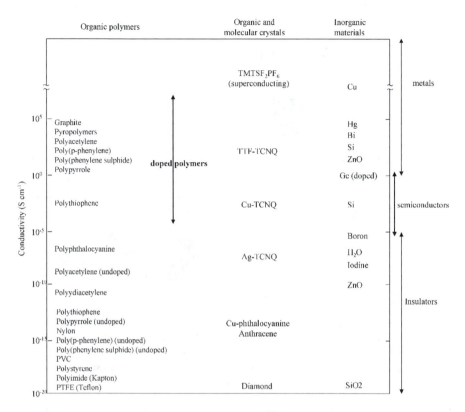

FIGURE 16.8 Conductivity ranges for polymers (doped and undoped), inorganic materials, and molecular crystals.

then there can be no net flow of electrons under the influence of an external field unless electrons are elevated into the empty band, and this will require a considerable expenditure of energy. Such materials are either semiconductors or insulators, depending on how large the energy gap may be, and the majority of polymers are insulators. The band model then assumes that the electrons are delocalized and can extend over the lattice (see Figure 16.9).

When we consider electronic conduction in polymers, it will become clear that band theory is not totally suitable because the atoms are covalently bonded to one another, forming polymeric chains that experience weak intermolecular interactions. Thus, macroscopic conduction will require electron movement, not only along chains but also from one chain to another.

16.10 PREPARATION OF CONDUCTIVE POLYMERS

Polymers have the electronic profiles of either insulators or semiconductors; thus, the band gap in a fully saturated chain such as polyethylene is 5 eV and decreases to about 1.5 eV in the conjugated system polyacetylene. The respective intrinsic conductivities are $\sim 10^{-17}$ S cm^{-1} and $\sim 10^{-8}$ S cm^{-1}, both of which are very low.

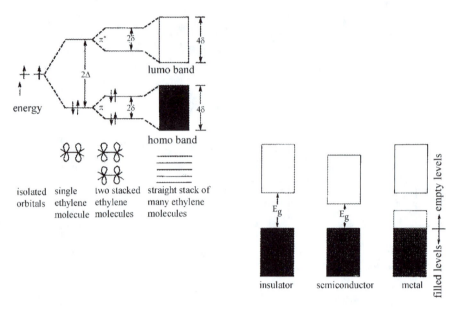

FIGURE 16.9 Schematic diagram showing the principles of band theory as described in the text. The dark regions represent the bands filled with electrons, and the light regions represent the bands that are available for conduction. The energy gap between filled and empty states is E_g. (Adapted from Cowan, D.O. and Wiygul, F.M., The organic solid state, *Chem. Eng. News*, 28, 1986. With permission from the American Chemical Society.)

Conducting polymers can be prepared either by oxidizing or reducing the polymer using a suitable reagent (see Table 16.2). The band theory model would explain the increased conductivity as either removal of electrons from the valence band by the oxidizing agent, leaving it with a positive charge, or donation of an electron to the empty conduction band by a reducing agent. These processes are called *p-type doping* and *n-type doping,* respectively. This explanation is an oversimplification, as conductivity in polymers is associated with charge carriers that do not have free spins rather than the expected unpaired electrons detected in metals, so a modified model must be developed. This will be explained when the individual conducting polymers are described.

Whereas the addition of a donor or acceptor molecule to the polymer is called *doping,* the reaction that takes place is actually a redox reaction and is unlike the doping of Si or Ge in semiconductor technology, where there is substitution of an atom in the lattice. The terminology in common use will be retained here, but it should be remembered that the doping of conductive polymers involves the formation of a polymer salt, and this can be effective either by immersing the polymer in a solution of the reagent or by electrochemical methods.

The reactions can be represented in the generalized case for oxidation by

$$\underset{\text{Red}}{\overset{\text{Ox/A}^-}{P_n \;\rightleftharpoons\; [P_n^+ A^-]}} \; \underset{\text{Red}}{\overset{\text{Ox/A}^-}{\rightleftharpoons\; [P^{2+} 2A^-]}}$$

TABLE 16.2
Structures and Conductivity of Doped Conjugated Polymers

Polymer	Structure	Typical Methods of Doping	Typical Conductivity $(S\ cm)^{-1}$
Polyacetylene	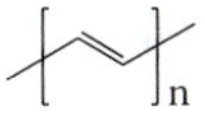	Electrochemical, chemical (AsF$_5$, I$_2$, Li, K)	500–1.5×10^5
Polyphenylene	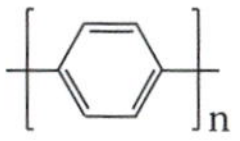	Chemical (AsF$_5$, Li, K)	500
Poly(phenylene sulfide)	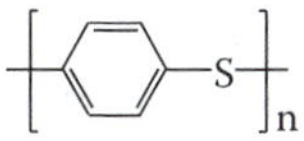	Chemical (AsF$_5$)	1
Polypyrrole	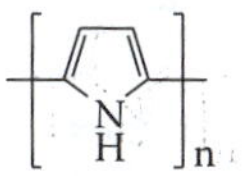	Electrochemical	600
Polythiophene	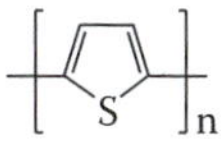	Electrochemical	100
Poly(phenyl-quinoline)	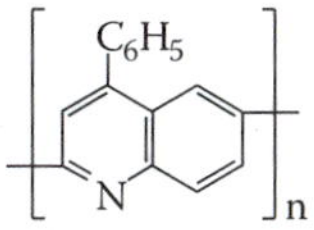 C$_6$H$_5$	Electrochemical, chemical (sodium naphthalide)	50

Source: Adapted from Cowan, D.O. and Wiygul, F.M., The organic solid state, *Chem. Eng. News*, 28, 1986. With permission of the American Chemical Society.

where P_n represents a section of the polymer chain. The first step is the formation of a cation (or anion) radical called a *soliton* or a *polaron* — the distinction will be explained later. This step may then be followed by a second electron transfer with the formation of a dictation (or dianion) known as a *bipolaron*. Alternatively, after the first redox reaction, charge transfer complexes may form between charged and neutral segments of the polymer when possible:

$$[P_n^{\bullet +} A^-] + P_m \rightarrow [(P_n P_m)^{\bullet +} A^-]$$

These general principles can best be illustrated by examining specific examples, in particular polyacetylene, which has been studied intensively.

16.11 POLYACETYLENE

Polyacetylene can be made by a number of synthetic routes, but Ziegler catalysts figured prominently in the early work. Indeed, the synthesis of the material that subsequently opened the way to the discovery of conducting polyacetylene was a fortuitous accident in which acetylene gas was passed through a heptane solution of the Ziegler catalyst $Ti(OC_4H_9)_4/Al(C_2H_5)_3$ that was vastly in excess of the amounts normally used. The polyacetylene that formed at the gas–liquid interface was a lustrous flexible polycrystalline film, rather than the powder usually obtained. This has become known as "Shirakawa" polyacetylene and has a predominantly cis conformation when formed at temperatures of 195 K. On raising the temperature of the film, isomerization to the more stable trans form takes place. The polymer is infusible, insoluble, usually contaminated by catalyst residues, and tends to become brittle and dull when exposed to air owing to slow oxidation. These features make it difficult to process or handle, and attempts have been made to either improve the polymer or make derivatives or precursors that are soluble in organic solvents.

Many of these problems have been solved by Feast (1985), who developed a very elegant synthetic method, now commonly known as the *Durham route*. This is a two-stage process in which soluble precursor polymers are prepared by a metathesis ring-opening polymerization reaction, and these are subsequently heated to produce poly-acetylene by a thermal elimination reaction. An example of the method is given below:

A refinement of the process involves photochemical conversion of I(a) into III,

which on polymerization produces a precursor that is stable at room temperature and can be converted to *trans*-polyacetylene on heating at 330 to 340 K. The advantages of the Durham route are: (a) contaminating catalyst residues can be removed because the precursor polymers are soluble and can be purified by disso-lution and precipitation and (b) the precursors can be drawn and oriented or cast as films prior to conversion to the all-trans form of polyacetylene. This allows some degree of control over the morphology of the final product, which in the pristine state appears to be fibrous and disordered. As conductivity can be maximized by alignment of the polymer chains, stretching the film will assist this process, which can best be accomplished using the prepolymer.

We can now examine the mechanism of conduction in polyacetylenes in relation to the structure and doping procedures. In a polyconjugated system, the π-orbitals are assumed to overlap, and form a valence and a conduction band as predicted by band theory. If all the bond lengths were equal, i.e., delocalization led to each bond having equal partial double bond character, then the bands would overlap and the polymer would behave like a quasi-one-dimensional metal having good conductive properties. Experimental evidence does not substantiate this, and reference to the physics of a monatomic one-dimensional metal, with a half-filled conduction band, has shown that this is an unstable system and will undergo lattice distortion by alternative compression and extension of the chain. This leads to alternating atom pairs with long and short interatomic distances found along the chain. The effect is embodied in the Peierls theorem, which states that a one-dimensional metal will be unstable and that an energy gap will form at the Fermi level because of this lattice distortion, so that the material becomes an insulator or a semiconductor. This break in the continuity of the energy bands is caused by the use of elastic energy during lattice distortion, which is compensated by a lowering of the electronic energy and formation of a band gap (see Figure 16.10). The analogy with polyacetylene then becomes obvious, and it is found that single- and double-bond alternation persists in the chain, leading to an energy gap between the valence and conduction bands. The trans structure of polyacetylene is also unique, as it has a twofold degenerate ground state in which sections A and B are mirror images

and the single and double bonds can be interchanged without changing the energy. Thus, if the cis structure begins to isomerize to the trans geometry from different

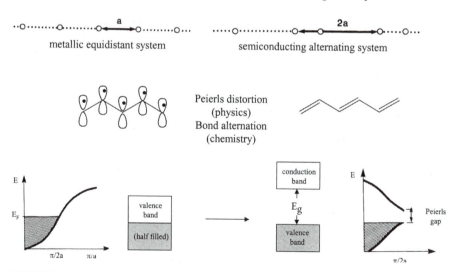

FIGURE 16.10 Schematic diagram illustrating the Peierls distortion that leads to formation of an energy gap and production of a semiconductor rather than a conductor. (Adapted from Bowden, M.J. and Turner, S.R., Eds., *Electronic and Photonic Applications of Polymers*, American Chemical Society, 1988. With permission.)

locations in a single chain, an A sequence may form and eventually meet a B sequence, as shown, but in doing so, a free radical is produced. This is a relatively stable entity, and the resulting defect in the chain is called a *neutral soliton*, which corresponds, in simple terms, to a break in the pattern of bond alternation, i.e., it separates the degenerate ground state structures. The electron has an unpaired spin and is located in a nonbonding state in the energy gap, midway between the two bands. It is the presence of these neutral solitons that gives *trans*-polyacetylene the characteristics of a semiconductor with an intrinsic conductivity of about 10^{-7} to 10^{-8} S cm^{-1}.

The conductivity can be magnified by doping. Exposure of the film to dry ammonia gas leads to a dramatic increase in σ to $\sim 10^3$ S cm^{-1}. Controlled addition of an acceptor or p-doping agent, such as AsF_5, Br_2, I_2, or $HClO_4$, removes an electron and creates a positive soliton (or a neutral one if the electron removed is not the free electron). In chemical terms, this is the same as forming a carbenium ion that is stabilized by having the charge spread over several monomer units. Similarly, a negative soliton can be formed by treating the polymer with a donor or n-doping agent that adds an electron to the midgap energy level. This can be done by dipping the film in the tetrahydrofuran (THF) solution of an alkali metal naphthalide, or by an electrochemical method.

Neutral soliton — Free radical

Positive soliton — Carbocation (carbenium-ion)

Negative soliton — Carbanion

At high doping levels, the soliton regions tend to overlap and create new midgap energy bands that may merge with the valence and conduction bands, allowing freedom for extensive electron flow. Thus, in polyacetylene, the charged solitons are responsible for making the polymer a conductor.

16.12 POLY(*p*-PHENYLENE)

The poly(*p*-phenylene) structure has all the characteristics required of a potential polymer conductor, but it has proved difficult to synthesize high-molecular-weight material. One possible method is polycondensation,

$$ n\ Mg + n\ Br{-}\langle\bigcirc\rangle{-}Br \longrightarrow \left[\langle\bigcirc\rangle\right]_n + n\ MgBr_2 $$

but this only yields oligomeric material, and even this is insoluble. A novel route developed by workers at ICI has solved these problems by again making use of a tractable intermediate polymer. Radical polymerization of 5,6-dihydroxycyclohexa-1,3-diene IV leads to a soluble precursor polymer that can be processed prior to the final thermal conversion into poly(p-phenylene).

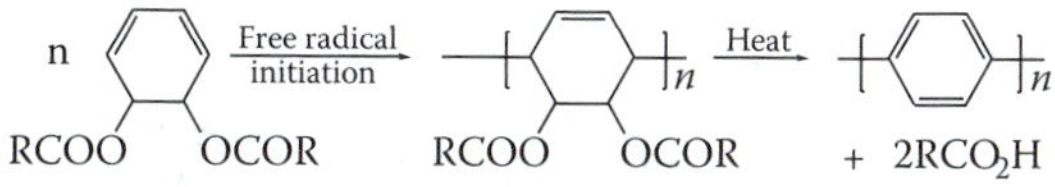

$$n \quad \underset{RCOO \quad OCOR}{\bigcirc} \xrightarrow[\text{initiation}]{\text{Free radical}} \underset{RCOO \quad OCOR}{-[\bigcirc]_n} \xrightarrow{\text{Heat}} -[\bigcirc]_n + 2RCO_2H$$

(IV)   Precursor polymer   Poly(p-phenylene)

The material is an insulator in the pure state but can be both n- and p-doped, using methods similar to those used for polyacetylene. However, as poly(p-phenylene) has a higher ionization potential, it is more stable to oxidation and requires strong p-dopants. It responds well to AsF_5, with which it can achieve conductivity levels of about 10^2 S cm^{-1}. In contrast, Br_2 and I_2 are ineffective. An interesting variation in the preparation of conducting poly(p-phenylene) involves a one-step reaction, in which crystalline p-phenylene oligomers (e.g., p-terphenylene) are exposed to AsF_5 vapor. These assume a metallic blue luster and can then be polymerized to yield a highly conducting polymer. Poly(p-phenylene) is very stable and can withstand temperatures up to 720 K in air without degrading.

Examination of the structure shows that the soliton defect cannot be supported in poly(p-phenylene), as there is no degenerate ground state. Instead, the two nearly equivalent structures are the benzenoid and quininoid forms, which have different energies.

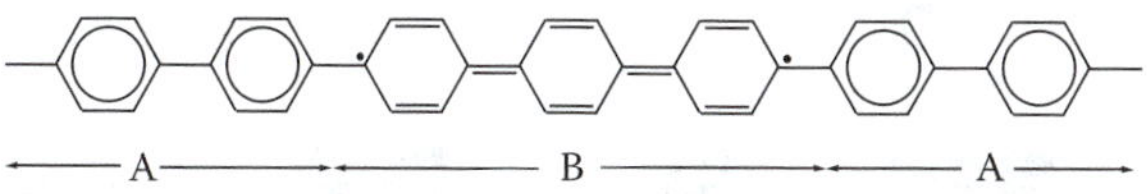

The benzenoid sections A have a lower energy than the quininoid section B, which must be limited by the benzenoid structures, and so the band gap of 3.5 eV is higher than that in polyacetylene.

In the band theory model, it is assumed that conduction occurs because the mean free path of a charge carrier extends over a large number of lattice sites, and that the residence time on any one site is small compared with the time it would take for a carrier to become localized. If, however, a carrier is trapped, it tends to polarize the local environment, which relaxes into a new equilibrium position. This deformed section of the lattice and the charge carrier then form a species called a *polaron*. Unlike the soliton, the polaron cannot move without first overcoming an energy barrier, so movement is by a hopping motion. In poly(p-phenylene), the solitons are trapped by the changes in polymer structure because of the differences in energy, and so a polaron is created, which is an isolated charge carrier. A pair of these charges is called a *bipolaron* and, on doping, the chemical equivalents are the radical ion and di-ion, respectively. In poly(p-phenylene) and most other poly-conjugated conducting polymers, the conduction occurs via the polaron or bipolaron (see Figure 16.11).

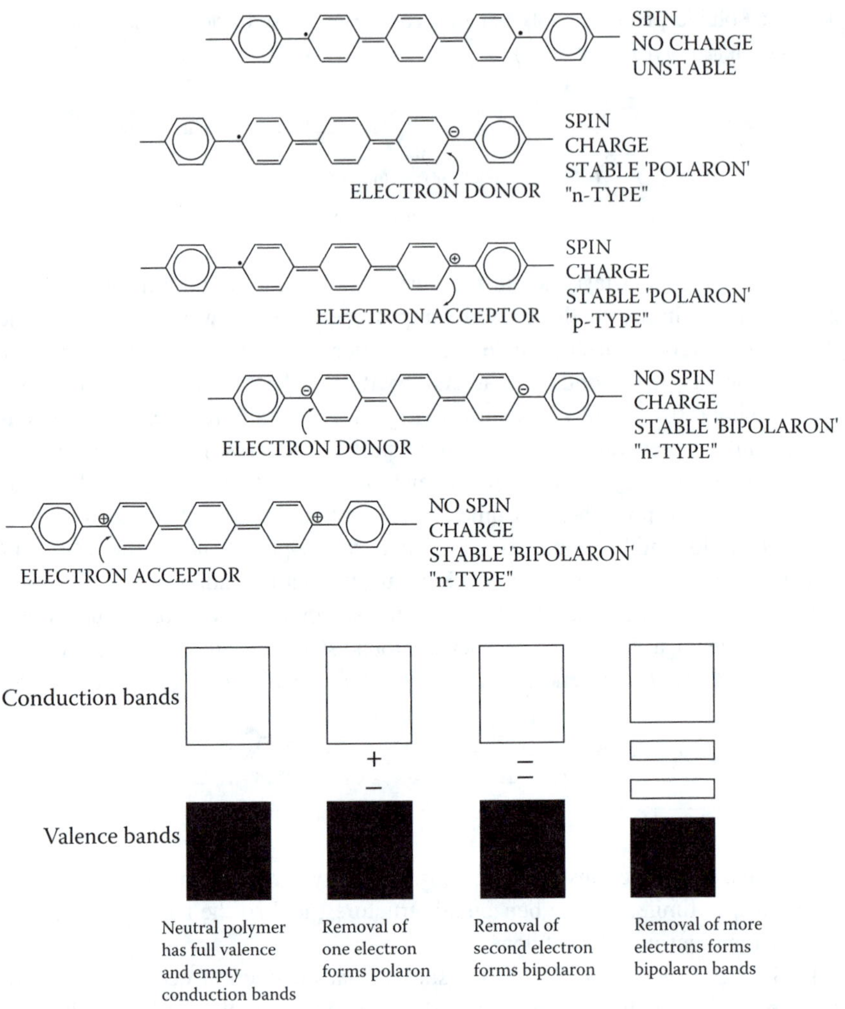

SPIN
NO CHARGE
UNSTABLE

SPIN
CHARGE
STABLE 'POLARON'
"n-TYPE"

ELECTRON DONOR

SPIN
CHARGE
STABLE 'POLARON'
"p-TYPE"

ELECTRON ACCEPTOR

NO SPIN
CHARGE
STABLE 'BIPOLARON'
"n-TYPE"

ELECTRON DONOR

NO SPIN
CHARGE
STABLE 'BIPOLARON'
"n-TYPE"

ELECTRON ACCEPTOR

Conduction bands

Valence bands

| Neutral polymer has full valence and empty conduction bands | Removal of one electron forms polaron | Removal of second electron forms bipolaron | Removal of more electrons forms bipolaron bands |

FIGURE 16.11 Illustration of polaron and bipolaron structures in poly(*p*-phenylene) and the proposed band structure for the oxidized (p-type) polymer. (Partially adapted from Cowan, D.O. and Wiygul, F.M., The organic solid state, *Chem. Eng. News*, 28, 1986. With permission from the American Chemical Society.)

16.13 POLYHETEROCYCLIC SYSTEMS

Several useful polymeric structures based on the repeat unit

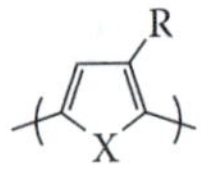

have been studied, where R = H, alkyl, etc., and X = NH, S.

16.13.1 POLYPYRROLE

The polymerization of pyrrole can be carried out electrochemically by anodic oxidation, during which process simultaneous doping occurs. Typically, electrolysis of a solution of pyrrole (0.06 M) and $(Et)_4N^+BF_4^-$ (0.1 M) in acetonitrile containing 1% water leads to deposition of an insoluble blue-black film of the polymer at the anode. The film contains BF_4^-, has a conductivity of about 10^2 S cm^{-1}, and the composition shown below:

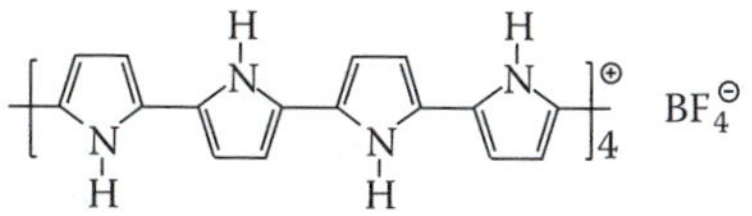

The material is amorphous but is insoluble in organic solvents. It has good stability in air with the conductive properties retained up to about 570 K and no significant loss in conductivity over prolonged periods. Copper–bronze films of poly(pyrrole perchlorate), with conductivities of ~40 S cm^{-1}, have been prepared electrochemically under dry-box and oxygen-free conditions, using $AgClO_4$ in acetonitrile. The stoichiometry of the films was

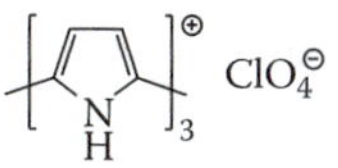

Yellow-green films of neutral polypyrrole can be prepared by the electrochemical reduction of the perchlorate films. This is an insulator with σ ~10^{-10} S cm^{-1}.

The neutral polymer can be reoxidized by exposure either to air (when the films turn black in about 15 min) or to Br_2, I_2, and $FeCl_3$ vapor. Immersion of the polymer in metal salt solutions of Ag^+, Cu^{2+}, or Fe^{3+} also makes the polymer conductive.

The N-substituted derivatives of pyrrole produce much poorer conductors, and bulky substituents tend to produce powders rather than films.

16.13.2 SULFUR COMPOUNDS

Polyheterocyclics containing sulfur are also of interest. Poly(2,5-thienylene) can be prepared as a light green powder by electrochemical oxidative coupling. In the neutral state, this compound has σ ~10^{-11} S cm^{-1}, but on exposure to I_2, the conductivity is raised to σ ~10^{-1} S cm^{-1}. Of particular interest in this group is poly(isothianaphthalene)

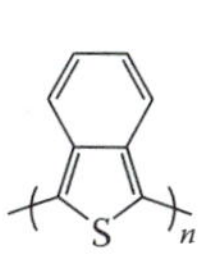

which has an unusual combination of properties. It is a good conductor when doped with a very narrow band gap (1 eV), but it also forms a transparent film.

16.14 POLYANILINE

Reaction of aniline with ammonium persulfate in aqueous HCl produces polyaniline as a dark blue powder with a conductivity of 5 S cm^{-1}. The structure of the conducting form of the polymer is believed to be the di-iminium salt.

Electrochemical oxidation of aniline in aqueous HBF$_4$ produces a clear, dark green cohesive film on a platinum foil anode. Reduction with methanolic alkali solution produces the neutral polymer, which is an insulator with σ ~10^{-11} S cm^{-1}.

16.15 POLY(PHENYLENE SULFIDE)

Whereas some conducting polymers can be processed by synthesis via a soluble precursor, poly(phenylene sulfide) is soluble or can be melt-processed. Doping with AsF$_5$ (which can be accelerated by the presence of AsF$_3$) forms a conductive polymer salt with σ ~1 S cm^{-1}, but also tends to make the film brittle.

16.16 POLY(1,6-HEPTADIYNE)

Green-gold lustrous films of poly(1,6-heptadiyne)

can be prepared by the cyclopolymerization of 1,6-heptadiyne using a Ziegler catalyst. The polymer is amorphous and has the structural repeat unit shown, but doping only achieves σ ~10^{-1} S cm^{-1}, and the films are quite unstable.

16.17 APPLICATIONS

Doped polyacetylene can act as an electrode and can be used in a rechargeable battery. Metal electrodes are continually subjected to dissolution and redeposition during the charge–discharge cycles, resulting in mechanical wear. This makes the use of polymer electrodes attractive because the ions can enter or leave without significant disturbance of the polymer structure. Although polyaniline electrodes have also been developed, polypyrrole-salt films are the most promising for practical application. This is due to both their stability and their self-supporting characteristics, which are particularly useful in the design of flat, space-saving cells. The use of these films also extends to flexible conductor tracks for contact bridges in switches and to electrochromic displays for optical memories.

The feasibility of all-plastic batteries has been demonstrated by construction of such a battery from two layers of polyacetylene salt, sandwiching a film of polycarbonate impregnated with $LiClO_4$. All-plastic batteries represent a combination of electrical conductivity with the lightweight corrosion-resistant properties of many plastics — an attractive prospect.

As well as battery application, conducting polymers may also find uses as electromagnetic shielding because they tend to absorb low-frequency radiation, or as parts of solar cells and semiconductors. Their use as heating elements in thin-wall coverings and in wire and cable applications is also being investigated.

16.18 PHOTONIC APPLICATIONS

Devices designed to transmit information by means of photons now incorporate polymeric materials with the appropriate structures. Passive applications include the coating of optical fibers by UV curable epoxy acrylates, thermal curing silicones, or heat-shrinkable polyethylene, to protect them from mechanical wear. Polymers have also been used in the manufacture of wave guides. Suitable polymers can now be synthesized having active nonlinear optical properties, which depend on the electronic excitation of the π-electron system.

16.19 LIGHT-EMITTING POLYMERS

In 1990, Friend and co-workers in Cambridge demonstrated that some conjugated, semiconducting polymers, in addition to carrying an electrical current, can be stimulated to emit light. A thin film of poly(phenylene vinylene) was sandwiched between two electrodes, one with a high work function and the other with a low work function, and a voltage applied. Light emission was observed in the yellow-green region of the visible spectrum, which led to the conclusion that suitable polymers exhibiting electroluminescence could be utilized in the fabrication of light-emitting displays such as polymer light-emitting diodes (PLEDs).

Poly(phenylene vinylene) (PPV), is an excellent material for this purpose; it has good electroluminescence qualities and it is mechanically stable, but it is also insoluble and intractable, making film formation difficult. This has been overcome by developing synthetic routes that produce a soluble precursor which can form films easily prior to *in situ* conversion to PPV. One such method is

Subsequent work has developed several derivatives of PPV that are soluble in organic solvents and can be readily cast as thin films. These are produced by substituting in the various positions R_1 to R_5; thus,

$R_1=R_2=$phenyl, poly(diphenyl phenylene vinylene) DP-PPV

$R_1=R_4=OC_6H_{13}$; $R_5=CN$; CN-PPV

$R_2=-O-CH_2-CH(C_2H_5)C_4H_9$; $R_3=OCH_3$: MEH−PPV

These and other derivatives have different band gaps, and as the band gap controls the wavelength of the emitted light, they can cover a wide range of the visible electromagnetic spectrum and even into the infrared region.

Other groups of polymers now used include polythiophene and several of its derivatives, and polyfluorenene, which is an efficient emitter of blue light. Tunable band gapwidths, and hence tunable wavelengths of electroluminescence, have been achieved using bends of ring-substituted polythiophenes.

16.19.1 APPLICATIONS

These materials can be used in the manufacture of flat panel and flexible displays, photovoltaic cells, field-effect transistors, Schottky diodes, and polymer light-emitting diodes.

To make a PLED, the following requirements should be met:

a. The conjugated polymer should be in the form of an amorphous, thin film (<100 nm) sandwiched between two metallic electrodes to form the anode and the cathode.
b. As it is preferable to have an asymmetric injection of electrons, a low-work-function cathode is used (calcium, aluminum), and as at least one transparent electrode is essential, indium tin-oxide (ITO) is used as a high-work-function anode.
c. Application of a voltage injects electrons from the cathode into one side of the polymer film, and positive "holes" from the anode into the other side.
d. The polymer acts as a recombination center and some, but not all, of the electrons drop into the holes and combine to form a high-energy state that decays with the release of excess energy as light.
e. The wavelength of this electroluminescence depends on the band gap in the polymer.

Although a single polymer layer is sufficient, improved efficiencies can be obtained using double layers of polymer. Thus, a layer of CN-PPV can be used on top of a PPV layer. As the CN group is an electron-withdrawing group, this enhances the electron injection and the CN-PPV acts as the electron transport layer, next to the cathode. It also has a lower band gap than PPV, so the recombination with the positive holes takes place in the PPV layer, which is next to the anode. The light emitted then passes through the transparent anode and can be seen as part of a display.

These devices have quantum efficiencies (i.e., number of photons emitted per number of electrons injected) of 1 to 4%, fast switching and emission times, are lightweight, and can be activated by voltages as low as 3 V.

16.20 NONLINEAR OPTICS

The application of polymers as active components can be achieved when they exhibit nonlinear optical properties and can participate as an integral part of a device. Nonlinear optical properties in a polymer depend on the electronic excitation of a π-electron system that results in the alteration of the phase, frequency, or amplitude of the incident radiation to give a new electromagnetic radiation field. Thus, if a local electric field E is applied to a molecule, the induced polarization P (a scalar quantity) is expressed as

$$P = \alpha E + \beta E^2 + \gamma E^3 + \ldots \tag{16.3}$$

where the tensor quantities are, α, the linear polarizability, and β and γ, the second- and third-order nonlinear electronic susceptibilities. The latter nonlinear quantities are small but can be detected when intense laser sources are used. A comparable expression for the macroscopic nonlinear effects for an assembly of molecules can be written as

$$P = \chi_{1J}^{(1)} E_J + \chi_{IJK}^{(2)} E_J E_K + \chi_{IJKL}^{(3)} E_J E_K E_L + \ldots \tag{16.4}$$

where χ now refers to the properties of the ensemble, and the terms E_J, etc. are components of the electric field strength. Here, $\chi_{1J}^{(1)}$ is related to the refractive index of the medium in linear optics, and the terms $\chi_{IJK}^{(2)} E_K$ and $\chi_{IJKL}^{(3)} E_K E_L$ have the same dimensions. Thus, materials with nonzero values of $\chi^{(2)}$ or $\chi^{(3)}$ undergo a change in refractive index when placed in an electric or optical field.

Application of an electric field polarizes the molecules in the medium, and these act as scattering centers for the radiation. If the medium has an asymmetric response to the applied field, then effects such as second harmonic generation, in which the frequency of the incident radiation is doubled, or a linear electro-optic (Pockels) effect may be observed. A paramagnetic amplification may also occur where irradiation by two incident fields of frequency v_a and v_b results in mixing, with amplification of the weaker wave. These effects arise when the material is $\chi^{(2)}$ active, and for a molecule to exhibit these properties, it should (1) have an extended conjugated π-electron system, (2) possess an electron-donating and an electron-accepting group to promote intramolecular charge transfer, and (3) crystallize in a noncentrosymmetric fashion to ensure polarization of the molecule in the crystal. A schematic representation would be

where there is a change in the dipole moment between the ground and the excited states. The acceptor A and the donor D sites then provide a push-pull action to distort the electron density and polarize the molecule. Molecules that can form polar crystals with high values of $\chi^{(2)}$ have the following structures:

Thus, for a polymer to exhibit second-order, nonlinear optical effects, it must satisfy these conditions or, alternatively, have the asymmetry induced by electric field poling.

In the former case, polydiacetylenes offer an attractive possibility; they can be prepared by solid-state polymerization of the crystalline monomers to form polymer crystals, and some derivatives have been found to have quite large $\chi^{(2)}$ values. Provided the packing distances of the monomers in the crystal are suitable, polymerization can be initiated thermally, or by UV or γ radiation, to form polydiacetylenes by 1,4 addition.

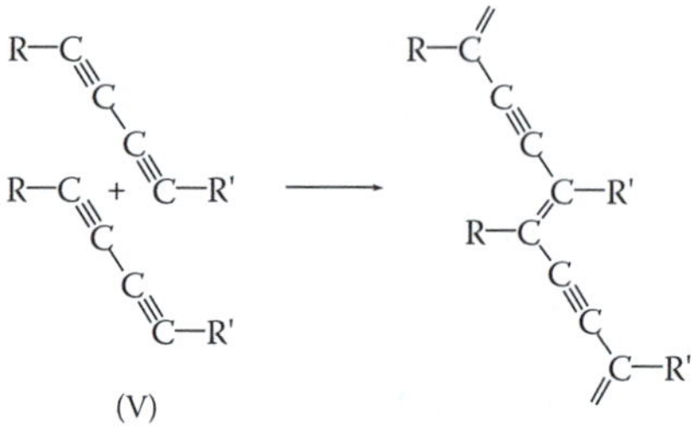

(V)

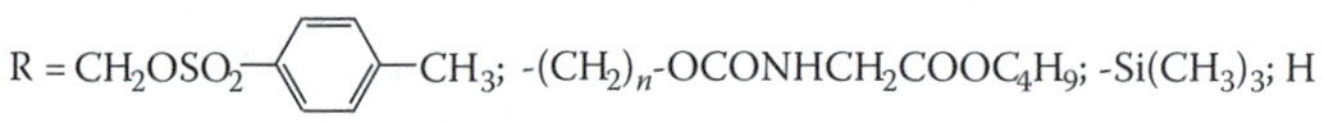

$$R = CH_2OSO_2-\!\!\!\bigcirc\!\!\!-CH_3;\ -(CH_2)_n-OCONHCH_2COOC_4H_9;\ -Si(CH_3)_3;\ H$$

If a monomer such as V is used, where $R \neq R'$, then the polymer crystals obtained may have second-order, nonlinear optical properties.

The other route that can be used to prepare polymer films with large $\chi^{(2)}$ is by doping them with molecules possessing large β values. The molecules are aligned in the film by applying an external DC field at temperatures above the T_g of the polymer, and then quenching this into the glassy state while maintaining the field. These molecularly doped, poled polymers can produce $\chi^{(2)}$ values between 10^{-6} and 10^{-8} esu, which is comparable to $LiNbO_3$ or GaAs crystals. Examples of this approach, which have been reported, are the doping of PMMA with the azo dye Disperse Red I (VI)

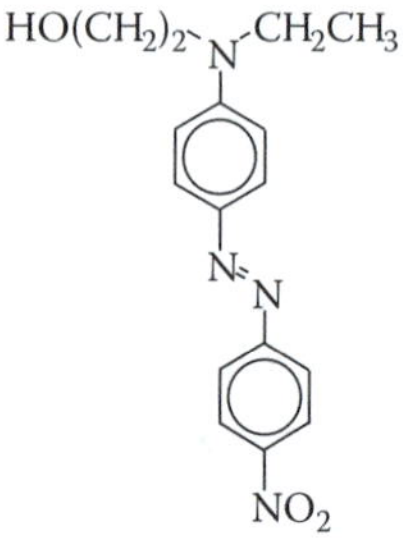

(VI) 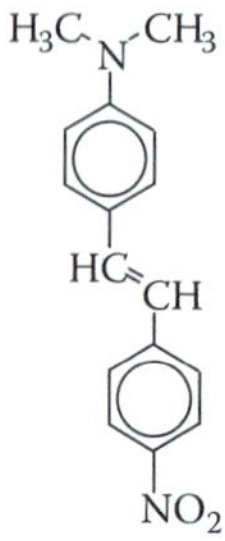   (VII)

and the dispersion of 4,4'-N,N-dimethyl amino-nitrostilbene (VII) in a liquid crystalline copolymer VIII followed by poling in the nematic liquid crystalline phase prior to quenching.

$$R^1 = -(CH_2)_6-O-\langle\bigcirc\rangle-\overset{O}{\underset{\|}{C}}-O-\langle\bigcirc\rangle-CN$$

$$R^2 = -(CH_2)_6-O-\langle\bigcirc\rangle-\overset{O}{\underset{\|}{C}}-O-\langle\bigcirc\rangle-OCH_3$$

(VIII)

In the latter system, there appears to be competition between alignment and thermal motion, so the best results were obtained when the poling was carried out close to the $T_g \sim 298$ K rather than at higher temperatures; the nematic to isotropic transition was $T_i = 373$ K. More recent attempts have been made to improve the poled systems by incorporating the nonlinear, optically active molecules into the polymer chain structure and comb-branch liquid crystalline polymers with R' and the new group

$$R^3 = -(CH_2)_6-O-\langle\bigcirc\rangle-X-\langle\bigcirc\rangle-NO_2$$

$$X = -\overset{O}{\underset{\|}{C}}-O- \quad \text{or} \quad -CH=CH-$$

This provides an all-polymeric material for subsequent poling.

Whereas some of the polydiacetylenes show second-order, nonlinear optical properties, those with R = R', such as poly[bis(p-toluene sulfonate)diacetylene], have a centrosymmetric crystal habit, and $\chi^{(2)}$ is zero. They do have finite $\chi^{(1)}$ and $\chi^{(3)}$ values, in common with materials such as polyacetylene, polypyrrole, and other conjugated polymers. Important nonlinear optical properties occur when a large $\chi^{(3)}$ term is observed, which includes the quadratic electro-optic (Kerr) effect, frequency tripling, optical phase conjugation, and optical bistability resulting from changes in the refractive index of the medium. These may prove useful in the development of photonic switches. The values of $\chi^{(3)}$ are found to improve when there is good alignment of the polymer chains in the crystal lattice, and this is enhanced in the direction of chain orientation. It should be noted that there are no symmetry constraints on a molecule, which can exhibit $\chi^{(1)}$ or $\chi^{(3)}$ properties.

16.21 LANGMUIR–BLODGETT FILMS

The preparation of ordered thin films can be of considerable interest in the construction of electronic devices and in the study of model membrane systems. One method that has gained in popularity is the Langmuir–Blodgett (LB) technique, in which molecules with a hydrophilic head and a hydrophobic tail can form a monolayer at an air–water interface, and then be transferred onto a solid surface. This latter process can be achieved either by dipping a glass slide (or some other substrate) vertically into the trough containing the monolayer on the surface, as shown schematically in Figure 16.12 or, alternatively, by using a rotating substrate to give a horizontal transfer method. Either way can be used to build up monolayers after one dipping,

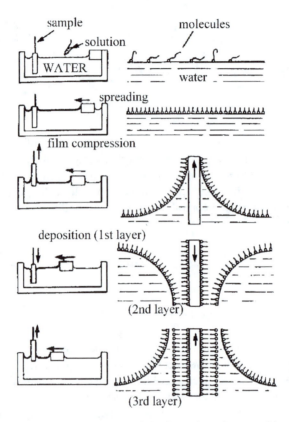

FIGURE 16.12 Schematic representation of the formation of mono-, bi-, and tri-layers of molecules from a Langmuir–Blodgett trough. (From Barraud, A., *Non-Linear Optical Properties of Organic Molecules and Crystals,* Chemla, D.S. and Zyss, J., Eds., Vol. 1, No. 4, Academic Press, 1987, p. 359. With permission from Academic Press.)

or multilayers by repeated passage of the substrate into the film. Polymeric films can be prepared by selecting a molecule with a polymerizable unit (double or triple bonds) that can be converted, after film formation, into a polymeric structure by thermal treatment or exposure of the film to UV or γ radiation. Several possible monomer structures are shown schematically in Figure 16.13.

Polydiacetylenes with appropriate amphiphilic structures can be used to fabricate thin films in this way, e.g., heptadeca-4,6-diyne-1-ol and the corresponding acid. It is often found that the stability of the monolayers is a function of the structure of the molecules. It has also been found advantageous to use the neutralized form of acid derivatives, and diacetylene monocarboxylic acids generally form much more stable films if the cadmium salt is used initially, where the Cd^{2+} ion is then present in the aqueous phase as a counterion.

Films prepared by the LB method have found application in nonlinear optics, but they have also found applications in nanolithography. The miniaturization of ICs requires high resolution, and electron beams have been used, as explained in Section

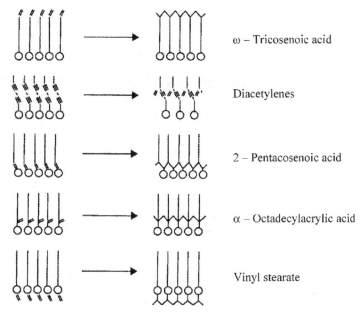

FIGURE 16.13 Schematic of possible monomer structures that can be polymerized in thin films formed using the Langmuir–Blodgett method.

16.6, for this purpose. Some limitations have been experienced, arising from the unwanted exposure obtained from the scattering of secondary electrons, which reduces the pattern definition on the resist. This can be improved by using much thinner resist films, and hence shorter exposure times are required. Conventional spin-coating techniques do not always guarantee that the resist film will be free of unacceptable defects such as pinholes that can spoil the subsequent pattern, and thin LB films can be prepared that are superior in this respect. Improved resolution has been obtained from resists prepared using polymerized ultrathin (45 nm) LB films of ω-tricosenoic acid, $CH_2 = CH(CH_2)_{20}COOH$, and α-octadecyl acrylic acid.

The technique, in general, shows great promise in the area of molecular electronics, where precise control of the molecular structure is paramount.

16.22 OPTICAL INFORMATION STORAGE

Polymeric materials can be used for optical information storage, and some are ideally suited for the manufacture of optical video or digital audio disks. The information is normally transferred to the polymer using a monochromatic laser by one of four possible methods:

1. ablative — "hole burning"
2. bubble formation
3. texture change
4. bilayer alloying

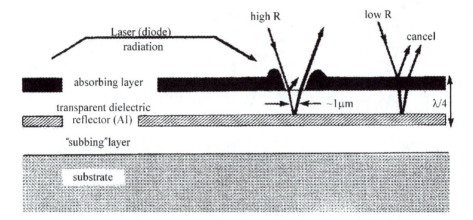

FIGURE 16.14 Trilayer structure of a typical ablative-mode optical disk. (From Clark, M.G., Materials for optical storage, *Chem. Ind.*, 258, 1985. With permission.)

Technique (1) is commonly employed and involves the creation, on the polymer surface, of a series of small pits that have different lengths and frequency of spacing. The information can be retrieved by measuring the intensity and modulation of light reflected from the pattern of pits on the disk surface.

The disks themselves must be fabricated from materials that have the following characteristics:

1. Dimensional stability
2. Isotropic expansion
3. Optical clarity
4. Low birefringence

In addition, the surface should be free of contaminating particles and occlusions that would interfere with the information retrieval process. A typical ablative-mode optical disk has the structure shown in Figure 16.14. The substrate is an optically transparent material such as polycarbonate, PMMA, poly(ethylene terephthalate), or poly(vinyl chloride) topped by a "subbing" layer to provide an optically smooth surface for the recording layer. A metal reflector (typically aluminum) is then incorporated next to a transparent dielectric medium such as poly(α-methyl styrene) and, finally, the absorbing layer where the information pits are created is added. The latter can be a metal–polymer composite (silver particles in a gel) or a dye molecule dispersed in a polymer matrix such as squaryllium dyes, which act as infrared absorbers for GaAs lasers, typically

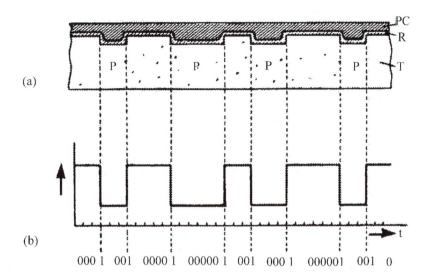

FIGURE 16.15 (a) Typical cross-sectional structure of an optical disk and (b) the intensity profile of the read-out system in binary code as a function of time (t). Here, T is the transparent polymer support, R is the reflective metal layer, PC is the protective coating, and P represents a depression created for information storage. (Reproduced with permission from Hüthig and Wepf Verlag.)

The absorbing layer is then protected by a transparent overcoating of cross-linked poly(dimethyl siloxane). This will produce a "direct read after write," or DRAW, disk that is nonerasable. A compact disk profile made in a similar way, and where the information can be read out in digital form, is shown in Figure 16.15.

Polymers can also be used to manufacture lenses and screens for projection television systems. These are most conveniently made from PMMA, or combinations of glass and PMMA, to counteract the high thermal expansion of the polymer. The use of ultraviolet curable coatings for lens replication and protective layers is widespread, and these systems are based on diacrylate or dimethacrylate monomers mixed with photoinitiators such as

$$H_2C=C-\underset{\underset{O}{\|}}{C}-O-R_2-O-\underset{\underset{O}{\|}}{C}-\overset{R_1}{\underset{}{C}}=CH_2$$

acetophenone or benzilketals, where $R_1 = CH_3$, H and R_2 can often be

$$-(OCH_2CH_2)_n-O-\underset{\underset{CH_3}{\overset{CH_3}{|}}}{\underset{}{C}}-O-(CH_2CH_2-O)_n-$$

which tends to have low shrinkage after cure while retaining good optical and thermal properties.

16.23 THERMORECORDING ON LIQUID CRYSTALLINE POLYMERS

The principle of using side chain liquid crystalline polymers as optical storage systems has been established. This has been demonstrated using a polymer film prepared from a side-chain polymer showing nematic liquid crystalline characteristics with the structure

$$\text{w(CH}_2-\text{CH)}_n\text{w} \\ \text{COO}-(\text{CH}_2)_5-\text{O}-\langle\bigcirc\rangle-\langle\bigcirc\rangle-\text{CN}$$

The mesogenic side groups are first oriented by application of an electric field to the polymer above the glass transition temperature, such that homeotropic alignment is obtained. On cooling below the T_g, the alignment is locked into the glassy phase, and a transparent film that will remain stable on removal of the electric field is produced. If this film is now exposed to a laser beam, localized heating occurs at the point where the beam impinges on the film and the material passes into the isotropic melt state. This results in the local loss of the homeotropic orientation and, on cooling, an unoriented region with a polydomain texture, which scatters light and produces a nontransparent spot, forms in the film. Information can then be "written" onto the film and can subsequently be erased by raising the temperature of the whole film to regain the isotropic, disordered, melt state. The system is illustrated in Figure 16.16.

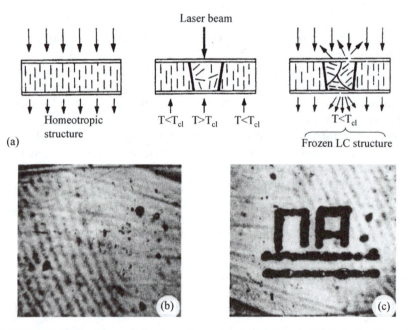

FIGURE 16.16 Thermal recording using a homeotropically aligned side-chain liquid crystal formed as a glassy film with the liquid crystalline state frozen into the glass. A laser beam is used to address the film (b) by producing local heating and disorder, which is subsequently frozen in by cooling below T_g(c). (From Platé, N.A. and Shibaev, V.P., *Comb-Shaped Polymers and Liquid Crystals*, Plenum Press, 1987. With permission of Plenum Publishers and the authors.)

Polymers are superior to low-molar-mass liquid crystalline molecules in this respect as they can retain the orientation longer when cooled into the glassy state before the electric field is switched off, whereas the low-molar-mass materials lose orientation rapidly.

REFERENCES

Ballard, D.G.H., Courtis, A., Shirley, I.M., and Taylor, S.C., A biotech route to poly(phenylene), *J. Chem. Soc. Chem. Commun.*, 954, 1983.

Barraud, A., *Non-Linear Optical Properties of Organic Molecules and Crystals,* Chemla, D.S. and Zyss, J., Eds., Vol. 1, No. 4, Academic Press, 1987, p. 359.

Bowden, M.J. and Turner, S.R., Eds., *Electronic and Photonic Applications of Polymers,* American Chemical Society, 1988.

Clark, M.G., Materials for optical storage, *Chem. Ind.*, 258, 1985.

Cowan, D.O. and Wiygul, F.M., The organic solid state, *Chem. Eng. News*, 28, 1986.

Etemad, S., Heegor, A.J., and MacDiarmid, A.G., Polyacetylene, *Ann Rev. Phys. Chem.*, 33, 443, 1982.

Feast, W.J., Synthesis and properties of some conjugated potentially conductive polymers, *Chem. Ind.*, 263, 1985.

Garito, A.F. and Wong, K.Y., Non-linear optical processes in organic and polymer structures, *Polym. J.*, 19, 51, 1987.

Gossink, R.G., Polymers for audio and video equipment, *Angew. Makromol. Chem.*, 145/146, 365, 1986.

Kraft, A., Grimsdale, A.C., and Holmes, A.B., Electroluminescent conjugated polymers-seeing polymers in a new light, *Angew. Chem. Int. Ed.*, 37, 402, 1998.

Platé, N.A. and Shibaev, V.P., *Comb-Shaped Polymers and Liquid Crystals*, Plenum Press, 1987.

Potember, R.S., Hoffman, R.C., Hu, H.S., Cocchiaro, J.E., Viands, C.A., and Poehler, T.O., Electronic devices from conducting organics and polymers, *Polym. J.*, 19, 147, 1987.

Roberts, E.D., Resists used in lithography, *Chem. Inc.*, 251, 1985.

Roberts, G.G., *Adv. Phys.* 34, 475, 1985.

Williams, D.J., Organic polymers and non-polymeric materials with large optical nonlinearities, *Angew. Chem. Int. Ed. Engl.*, 23, 690, 1984.

BIBLIOGRAPHY

Barraud, A., *Non-Linear Optical Properties of Organic Molecules and Crystals,* Chemla, D.S. and Zyss, J., Eds., Vol. 1, No. 4, Academic Press, 1987, p. 359.

Bowden, M.J. and Turner, S.R., Eds., *Electronic and Photonic Applications of Polymers,* American Chemical Society, 1988.

Davidson, T., Ed., *Polymers in Electronics*, ACS Symposium, Series 242, American Chemical Society, 1983.

Goosey, M.T., Ed., *Plastics for Electronics*, Elsevier Applied Science, 1985.

Kuzmany, H., Mehring, M., and Roth, S., *Electronic Properties of Polymers and Related Compounds*, Springer-Verlag, 1985.

Platé, N.A. and Shibaev, V.P., *Comb-Shaped Polymers and Liquid Crystals*, Plenum Press, 1987.

Seanor, D.A., Ed., *Electrical Properties of Polymers,* Academic Press, 1982.

Skotheim, T.A., Ed., *Handbook of Conducting Polymers*, Vol. I, II, Marcel Dekker, 1986.

Thompson, L.F., Willson, C.G., and Bowden, M.J., Eds., *Introduction to Microlithography*, ACS Symposium, Series 219, American Chemical Society, 1983.

Thompson, L.F., Willson, C.G., and Frechet, J.M.T., Eds., *Materials for Microlithography*, ACS Symposium, Series 266, American Chemical Society, 1984.

Index